Catharanthus in Drug Discovery

Sunil Kumar • Bikarma Singh

Catharanthus in Drug Discovery

From Molecules to Medicine

Springer

Sunil Kumar
Department of Chemistry
Sant Kavi Baba Baijnath Post
Graduate College
Harakh, Uttar Pradesh, India

Bikarma Singh
Botanical Garden, Plant Conservation
and Agrotechnology Division
CSIR-National Botanical Research Institute
Lucknow, Uttar Pradesh, India

ISBN 978-981-95-5047-0 ISBN 978-981-95-5048-7 (eBook)
https://doi.org/10.1007/978-981-95-5048-7

This Springer imprint is published by the registered company Springer Nature Singapore Pte Ltd.
The registered company address is: 152 Beach Road, #21-01/04 Gateway East, Singapore 189721, Singapore

Preface

Botanical drugs, derived from naturally occurring plant substances, are rich sources of bioactive compounds with significant therapeutic potential. These natural compounds have long played a critical role in promoting human health and treating diseases. As we advance into the twenty-first century, the importance of plant-based medicine is being rediscovered and re-emphasized within scientific and medical communities. Historically, plants formed the cornerstone of traditional medicine systems around the world and laid the foundation for many modern pharmaceuticals. Despite this deep-rooted connection, the pharmaceutical industry in recent decades has shifted its focus predominantly toward synthetic drug development. While this approach has yielded several important therapeutic agents, it has also led to a slowdown in novel drug discoveries, particularly in the face of rising drug resistance and complex chronic illnesses. This has reignited global interest in botanical therapeutics, not only as alternatives to synthetic drugs but also as complementary treatments. Now, botanicals are increasingly available in the form of dietary supplements, herbal remedies, drugs, or even as ingredients in cosmetics and functional foods.

In this context, the book *Catharanthus in Drug Discovery: From Molecules to Medicine* presents a comprehensive exploration of *Catharanthus*, commonly known as "Sadabahar" or Madagascar periwinkle. This plant is renowned for its rich phytochemical profile, particularly for producing alkaloids like vincristine and vinblastine, which are used in the treatment of various cancers. The book delves into the botany, phytochemistry, pharmacological properties, clinical applications, and future potential of this remarkable plant. By highlighting the intricate interplay between biodiversity and human health, the book aims to inspire both the scientific community and industry stakeholders to embrace a more holistic and sustainable approach to drug discovery. In response to the healthcare needs of growing populations—especially in developing regions—there is increasing collaboration between biologists, genetic engineers, chemists, and pharmacologists. Researchers are developing transgenic plants to enhance the yield and efficacy of medicinal compounds, while also exploring underutilized herbs for their pharmacological benefits. Pharmaceutical companies are actively investigating plant sources from biodiverse environments such as the Himalayan highlands, where unique climatic conditions foster the growth of rare and potent medicinal plants. These efforts have the potential to yield breakthrough treatments for chronic diseases, including cardiovascular disorders, asthma, inflammatory conditions, and pain management.

This book brings together the insights of leading scientists to offer a cohesive and accessible resource that bridges academic research with practical applications in pharmaceutical development. With 14 updated and extensively revised chapters, it explores modern approaches to drug discovery and ethnobotany, emphasizing the therapeutic potential of plant-based compounds. The content is designed to inspire innovation in fields such as ethnopharmacology, phytomedicine, and pharmaceutical biotechnology, making it a valuable reference for students, researchers, clinicians, and industry professionals alike. By translating complex scientific findings into actionable knowledge, the book aims to guide readers through both the historical foundations and future directions of botanical drug research. We encourage readers to actively engage with the material and contribute their feedback to further the dialogue on plant-based therapeutics. This work reflects the dedication and expertise of each contributing author, whose efforts have made this volume a rich and informative guide. We hope it serves as a useful tool for those involved in pharmaceutical R&D and inspires continued exploration in the evolving landscape of natural product drug discovery.

Harakh, Uttar Pradesh, India Sunil Kumar
Lucknow, Uttar Pradesh, India Bikarma Singh

Contents

About the Authors

Sunil Kumar is currently serving as an Assistant Professor and Head of the Department of Chemistry at Sant Kavi Baba Baijnath Government P.G. College in Harakh, Barabanki, Uttar Pradesh, India. He earned his PhD for his research on the application of mass spectrometric techniques for the qualitative and quantitative analysis of phytoconstituents, as well as the identification of chemical markers in *Phyllanthus* and *Rauwolfia* species, conducted at the sophisticated Analytical Instrument Facility of CSIR-Central Drug Research Institute in Lucknow. With 6 years of teaching and research experience, he specializes in the qualitative and quantitative analysis of phytochemicals using LC-MS/MS techniques, focusing on the isolation, quality control, and standardization of Indian medicinal plants and their herbal formulations. He has authored 6 book chapters, 2 books, and 36 papers published in reputable international journals. In addition to his research contributions, he serves as a reviewer for several international scientific journals. He is an active member of various research committees and has been invited to deliver keynote lectures at numerous conferences, seminars, and workshops. A lifetime member of the Indian Science Congress Association, he also holds the position of Assistant Zonal Officer for the Spoken Tutorial program organized by the Indian Institute of Technology Bombay.

Bikarma Singh is currently working as a Principal Scientist and Scientist In-Charge of Maharshi Kashyap Plant Propagation Campus (MKPPC) at CSIR-National Botanical Research Institute, Lucknow, one of the pioneer institute of Council of Scientific and Industrial Research (CSIR) known for largest research and development activities under Ministry of Science and Technology, Government of India. He is also serving as Associate Professor in AcSIR, New Delhi, India, and PhD Supervisor in Department of Botany, Jammu University, J&K, India. He graduated up as an Honours and Gold Medalist in Botany from North-Eastern Hill University, Shillong, in 2005, and completed doctorate degree in Botany at Gauhati University Assam and Botanical Survey of India (Shillong) Meghalaya in 2012. Prior to joining CSIR-NBRI Lucknow in 2020, he worked as Senior Scientist in CSIR-Indian Institute of Integrative Medicine, Jammu, and before that Scientist-Ecologist in WAPCOS Limited, Gurgaon, and was involved in several consultancy projects related to studies on ecology and biodiversity. He pursued R&D with passion and possesses 18 years of research experience in Plant Sciences (Botany and Natural Products), and has expertise in systematic, ethnobotany, value addition, biodiversity conservation, and EIA. He is serving as a reviewing member of several national and international scientific SCI journals, and has been a lead leader for organizing many seminars and workshops. He has authored/co-authored 11 books and 46 book chapters and has published 107 scientific research papers in different reputed journals. He has two patents on medicinal plants, and worked as PI/Co-PI/Team member for more than 50 research projects funded by governmental or non-governmental organizations. He has supervised and ignited four PhD students in Science to defend their doctorate. He is a member of several research committees and has delivered invited talk as a keynote speaker/lecturer under various themes in several conferences, seminars and workshops. He is also working for societal upliftment via implementation of CSIR Aroma mission and CSIR Floriculture mission in different states of India.

Abbreviations

13-HPL	13-Hydroperoxide lyase
13-LOX	13-Lipoxygenase
9-HPL	9-Hydroperoxide lyase
9-LOX	9-Lipoxygenase
AAE	Acetylajmalan esterase
AAT	Alcohol acyltransferase
ABC	Advanced breast cancer
ABTS	2,2′-Azino-bis(3-ethylbenzothiazoline-6-sulfonic acid)
ADH	Alcohol dehydrogenase
AI	Artificial intelligence
ALL	Acute lymphoblastic leukemia
ALP	Alanine phosphatase
AOC	Allene oxide cyclase
AOS	Allene oxide synthase
APCI	Atmospheric pressure chemical ionization
APPI	Atmospheric pressure photoionization ionization
BFC	Berlin–Frankfurt–Münster Protocol (BFM)
BHA	Butylated hydroxyanisole
C4H	Cinnamate 4-hydroxylase
CADD	Computer-aided drug design
CHI	Chalcone isomerase
CHS	Chalcone synthase
CI	Chemical ionization
COG	Children's Oncology Group
CPC	Centrifugal partition chromatography
CUPRAC	Cupric reducing antioxidant capacity
CWS	*Catharanthus* whole seed
CWS-A	*Catharanthus* whole seed–agar medium
CWS-G	*Catharanthus* whole seed–gel medium
D4H	Desacetoxyvindoline 4-hydroxylase
DAD	Diode array detector
DAT	Deacetylvindoline 4-O-acetyltransferase
DHVR	Dihydrovomilenine reductase

DLN	Dose-limiting neurotoxicity
DLT	Dose-limiting toxicity
DPPH	2,2-Diphenyl-1-picrylhydrazyl
EAE	Enzyme-assisted extraction
ED_{50}	Median effective dose
EEA	Endophytic actinomycetes
EI	Electron ionization
EMA	European Medicines Agency
EMP	Estramustine phosphate
ESI	Electrospray ionization
FDA	Food and Drug Administration
FLD	Fluorescence detector
FRAP	Ferric reducing antioxidant power
GBC	Glibenclamide
GC	Gas chromatography
GC-MS	Gas chromatography–mass spectrometry
GCP	Good Clinical Practice
GLP	Good Laboratory Practice
GLUT-2	Glucose transporter type 2
GLUT-4	Glucose transporter type 4
GLVs	Green leaf volatiles
HDI	High-dose interferon
HILIC	Hydrophilic interaction liquid chromatography
HPLC	High-performance liquid chromatography
HPLC–HRMS	High-performance liquid chromatography–high-resolution mass spectrometry
HPLC–HRMS/MS	High-performance liquid chromatography–high-resolution tandem mass spectrometry
HPLC–MS	High-performance liquid chromatography–mass spectrometry
HPLC–MS/MS	High-performance liquid chromatography–tandem mass spectrometry
HPTLC	High-performance thin layer chromatography
HTS	High-throughput screening
IC_{50}	Half-maximal inhibitory concentration
ISO	Isomerase
IUCN	International Union for Conservation of Nature
LA-NSCLC	Locally advanced non-small cell lung cancer
LBDD	Ligand-based drug design
LC_{50}	Median lethal concentration
LC_{90}	90% lethal concentration
LC–MS	Liquid chromatography–mass spectrometry
LC–MS/MS	Liquid chromatography–tandem mass spectrometry
LD_{50}	Median lethal dose

LDL	Low-density lipoprotein
LOD	Limit of detection
LOQ	Limit of quantitation
MAE	Microwave-assisted extraction
MEP	Methylerythritol phosphate (pathway)
MI	Machine learning
MIAs	Monoterpene indole alkaloids
MIC	Minimum inhibitory concentration
MS	Mass spectrometry
MTD	Maximum tolerated dose
MVA	Mevalonate (pathway)
MVP	Mitomycin, Vindesine, Cisplatin
NAMT	Norajmaline methyltransferase
NHL	Non-Hodgkin lymphoma
NMR	Nuclear magnetic resonance
NOAEL	No observed adverse effect level
NPCE	Negative-pressure cavitation extraction
NPCE–RP–HPLC	Normal phase column elution–reverse phase–high-performance liquid chromatography
OPLC	Overpressured layer chromatography
OPR	12-Oxophytodienoate reductase
ORAC	Oxygen radical absorbance capacity
PAL	Phenylalanine ammonia-lyase
PC	Pharmacokinetics
PLE	Pressurized liquid extraction
Ppm	Parts per million
PRX1	Anhydrovinblastine synthase
QSAR	Quantitative structure-activity relationship
RDA	Retro-Diels-Alder
RFS	Relapse-free survival
RP-HPLC	Reverse phase high-performance liquid chromatography
RSD	Relative standard deviation
RT	Retention time
SAM	S-adenosyl-L-methionine
SAR	Structure–activity relationship
SBDD	Structure-based drug design
SCC	Squamous cell carcinoma
SCLC	Small cell lung cancer
SFE	Supercritical fluid extraction
SGD	Strictosidine β-glucosidase
SGOT	Serum glutamate oxaloacetate transaminase
SGPT	Serum glutamate pyruvate transaminase
SIDRs	Streptozotocin-induced diabetes mellitus rats
STR	Strictosidine synthase

STZ	Streptozotocin
TC	Total cholesterol
TE	Trolox equivalent
TG	Triglycerides
TIAs	Terpene indole alkaloids
TLC	Thin layer chromatography
TPSA	Topological polar surface area
TRT	Thoracic radiotherapy
TSP–MS	Thermospray mass spectrometry
UAE	Ultrasound-assisted extraction
UDP	Uridine diphosphate
UGTs	UDP-glycosyltransferases
UHPLC	Ultra-high-performance liquid chromatography
UHPLC–ESI–MS/MS	Ultra-high-performance liquid chromatography–electrospray ionization–tandem mass spectrometry
UKALL	United Kingdom Acute Lymphoblastic Leukemia Protocol
UPLC–ESI–Q–TOF–MS/MS	Ultra-performance liquid chromatography with electrospray ionization coupled to quadrupole–time of flight tandem mass spectrometry
UPLC–MS	Ultra-performance liquid chromatography–mass spectrometry
UV	Ultraviolet
UV/Vis	Ultraviolet–visible
VAMP	Vinblastine, Dactinomycin, Methotrexate, and Prednisone Protocol
VC	Vinorelbine–cisplatin
VCR	Vincristine
VESPER	Vinflunine in Muscle-Invasive Bladder Cancer Trial
VIP	Vasoactive intestinal peptide
VLB	Vinblastine
VLDL	Very low-density lipoprotein
VOCs	Volatile organic compounds
VR	Vomilenine reductase

List of Figures

List of Tables

Current Perspectives of Drug Discovery: Aspects and Applications

1

Abstract

Drug discovery is a dynamic and vital process aimed at developing new therapeutic compounds to address unmet medical needs. This chapter provides an overview of the evolving landscape, tracing the journey from traditional medicinal practices to modern scientific approaches. Key methods include ethnopharmacology, natural product screening, serendipitous discoveries, and the synthesis of compound derivatives for pharmacological testing. With advances in technology, the field now relies on high-throughput screening (HTS), along with omics-based approaches such as genomics, proteomics, and metabolomics. In silico methods, including computer-aided drug design and the integration of artificial intelligence (AI) and machine learning (ML), are revolutionizing how potential drug candidates are identified and optimized. Additionally, chromatographic techniques play a crucial role in the rapid and efficient screening of bioactive metabolites. Together, these tools and strategies represent a comprehensive, multidisciplinary framework that continues to drive innovation in modern drug discovery.

Keywords

Medicine · Drug · Discovery · Learning

1.1 Introduction

A drug is any substance or chemical that influences the physiological functions of the human body. The process of drug discovery and development involves identifying and developing compounds with therapeutic potential. In modern medicine, this process focuses on the systematic identification, characterization, and optimization of bioactive or druggable molecules, often through advanced scientific techniques

S. Kumar, B. Singh, *Catharanthus in Drug Discovery*,
https://doi.org/10.1007/978-981-95-5048-7_1

and screening methods. In contrast, traditional systems of medicine—such as Ayurveda, Siddha, Unani, Traditional Chinese Medicine (TCM), and various indigenous practices—emphasize the discovery of remedies through the use of natural resources. These systems rely on centuries-old knowledge and involve identifying, preparing, and combining natural substances like herbs, minerals, and animal products for therapeutic use. While modern and traditional approaches differ in methodology, both aim to discover safe and effective treatments for human ailments, highlighting the importance of integrating ancient wisdom with contemporary science in the ongoing pursuit of new drugs (Patwardhan and Mashelkar 2009; Xu 2006).

In recent times, drug discovery and development has become a highly complex, interdisciplinary, and systematic process aimed at developing and optimizing safe and effective therapeutic candidates through scientific research, technological advancement, and adherence to regulatory standards. Herbal drugs or medicinal plants consist of a complex matrix of both known and unknown phytochemicals, rather than a single active compound. While some phytochemicals are present in high concentrations, many exist in trace amounts. The pharmacological effects of these herbal drugs are typically attributed to one or more active phytochemicals—sometimes known, sometimes not. Both macro- and microconstituents may contribute to therapeutic efficacy, making the comprehensive analysis of multiple phytochemicals essential for quality control. Herbal drugs can be used individually or in combinations, with each preparation containing numerous bioactive compounds. Occasionally, a single compound accounts for the full therapeutic effect. Therefore, the identification, quantification, and isolation of active phytochemicals using modern analytical instrumentation has become increasingly important (Chihomvu et al. 2024).

Traditional extraction methods such as maceration, percolation, Soxhlet extraction, decoction, infusion, microwave-assisted extraction, enzyme-assisted extraction, and supercritical fluid extraction are often labor-intensive, time-consuming, and come with inherent limitations. In recent years, advanced analytical techniques have been increasingly employed for phytochemical investigations of medicinal plants. These modern approaches surpass traditional methods in efficiency and are well suited for isolating and characterizing a wide range of secondary metabolites. Despite these advancements, only about 10% of phytochemicals from medicinal plants have been explored using these techniques. Rapid screening and detailed phytochemical analysis now play a critical role in the drug discovery and development process from plant-based sources (Egbuna et al. 2019).

Spectroscopic/spectrometric and chromatographic techniques are currently very useful for the identification and assessment of components present in complex matrices. Separation techniques such as thin-layer chromatography (TLC), high-performance thin-layer chromatography (HPTLC), high-performance liquid chromatography (HPLC), and gas chromatography (GC), combined with powerful detection techniques such as mass spectrometry (MS), are established methods for rapid identification and characterization. The holistic method needs a very different approach than regular analysis because it involves creating a profile that includes many different elements instead of just focusing on a few specific molecules, aiming to find new biomarkers and treatment targets. Liquid chromatography–mass spectrometry (LC–MS) is a sophisticated and extensively utilized method for metabolite profiling and fingerprinting. The elemental composition determination of a phytochemical is a crucial tool for identification and characterization. The advanced LC–MS instruments can achieve precise mass measurements, defined as being within 5 ppm of the theoretical mass. LC and GC are frequently favored for an in-depth investigation of phytochemicals inside complex matrices. Excellent-throughput analysis and downsizing using gas chromatography–mass spectrometry (GC–MS) or LC–MS, which offer excellent resolution in a brief analysis duration at minimal operational expenses, are preferred (Ahmad Dar et al. 2020).

The introduction of sub-2-μm columns in HPLC and the emergence of ultrahigh-performance liquid chromatography (UHPLC) have marked significant advancements in the profiling of crude extracts, enabling faster and more efficient separations. However, these rapid separations challenge liquid chromatography detectors, which must respond swiftly and accurately. Various ionization techniques have been developed to meet these demands, including atmospheric pressure chemical ionization (APCI), atmospheric pressure photoionization (APPI), electrospray ionization (ESI), and laser-based methods such as atmospheric pressure laser-induced acoustic desorption chemical ionization (AP/LIAD-CI). Among these, ESI remains the most widely employed ionization method for analyzing a broad range of phytochemicals. The advent of ambient ionization techniques has further revolutionized MS by allowing direct sample analysis in open environments outside the instrument. Concurrently, advances in MS have improved resolution (up to 24 million) and mass accuracy (sub-ppb levels), with routine instruments achieving >40,000 resolution and < 1 ppm accuracy. The rise of "-omics" sciences—such as metabolomics and genomics—adds further complexity, necessitating analytical methods that can comprehensively detect and quantify metabolites in complex matrices. Therefore, complementary use of both preparative and non-preparative methods is essential, as no single technique can meet all requirements for sensitivity, resolution, and throughput (Zhang et al. 2018a, b, c, d, e, f; Chen et al. 2023).

1.2 Bridge Between Traditional and Modern Approach for Drug Discovery

Traditional approaches to drug discovery have historically relied heavily on natural sources such as medicinal plants, microorganisms, animals, and marine organisms. These resources formed the bedrock of early pharmacological practices across diverse ancient civilizations. Various traditional systems of medicine—Ayurveda and Siddha (India), TCM, Unani (Greco-Arabic), Egyptian medicine, Native American practices, African traditional medicine, Greek (Hippocratic) medicine, Roman medicine, Islamic medicine, Kampo (Traditional Japanese Medicine), European herbalism, and Shamanic traditions—have all contributed significantly to the foundational knowledge of drug discovery. These healing traditions identified

Table 1.1 Drugs identified through serendipity and traditional knowledge systems

S. no.	Source	Compound names	Common/botanical names	Action
1.	Plant-based	Quinine	*Cinchona* trees	Malaria
		Morphine	*Papaver somniferum*	Analgesic, painkiller
		Artemisinin	Artemisia annua	Fevers
		Ephedrine	*Ephedra* species	Respiratory conditions
		Atropine	*Atropa bella-donna*	Pupil dilation
		Aspirin	Willow bark	Pain and inflammation
		Reserpine	*Rauwolfia serpentina*	Hypertension
		Digitalis	*Digitalis purpurea*	Heart failure treatment
2.	Microbial	Penicillin	*Penicillium notatum*	Bacterial infections
		Streptomycin	*Streptomyces* bacteria	Tuberculosis
		Vancomycin	*Amycolatopsis orientalis*	Antibacterial
3.	Marine	Ziconotide	Venom of the cone snail	Severe pain management

and used numerous plant-derived substances to manage and treat a wide array of health conditions (Table 1.1).

For instance, quinine extracted from the bark of *Cinchona pubescens* was used to treat malaria, while morphine, derived from *Papaver somniferum* (opium poppy), served as a powerful analgesic. Notably, non-plant natural sources have also played vital roles in drug development. Penicillin, derived from the fungus *Penicillium notatum*, revolutionized the treatment of bacterial infections. Likewise, ziconotide, isolated from cone snail venom, is employed in pain management, and captopril, modeled after snake venom peptides, is used to treat hypertension. Several landmark drugs were discovered through keen observation, serendipity, or ethnomedical insight. Artemisinin, derived from *Artemisia annua*, is another potent antimalarial drug with roots in TCM. Similarly, aspirin, initially sourced from *Salix babylonica* (willow tree), has long been used for treating fevers, pain, and inflammation. Some discoveries occurred entirely by accident—penicillin was famously identified due to mold contamination, while warfarin was developed after studying cattle poisoning cases caused by spoiled sweet clover. The concept of zoopharmacognosy—animals' self-medication practices—has also inspired drug discovery. For example, chimpanzees chew the leaves of the *Aspilia* species to expel intestinal parasites. Many drugs were identified from nonmedical contexts: nitroglycerin was adapted for angina relief, chlorpromazine for psychiatric disorders, and botulinum toxin, initially studied in food poisoning, found applications in cosmetic treatments. Traditional home remedies like ephedrine (from *Ephedra*) and reserpine (from *Rauwolfia serpentina*) evolved into respiratory and antihypertensive drugs, respectively. Some well-recognized potential drugs are presented in Figs. 1.1, 1.2 and 1.3). Additionally, microbial interactions yielded streptomycin and vancomycin, key antibiotics for infectious diseases. Collectively, these traditional approaches reflect the deep and multifaceted relationship between nature, cultural practices, and empirical observation, laying the groundwork for modern drug discovery.

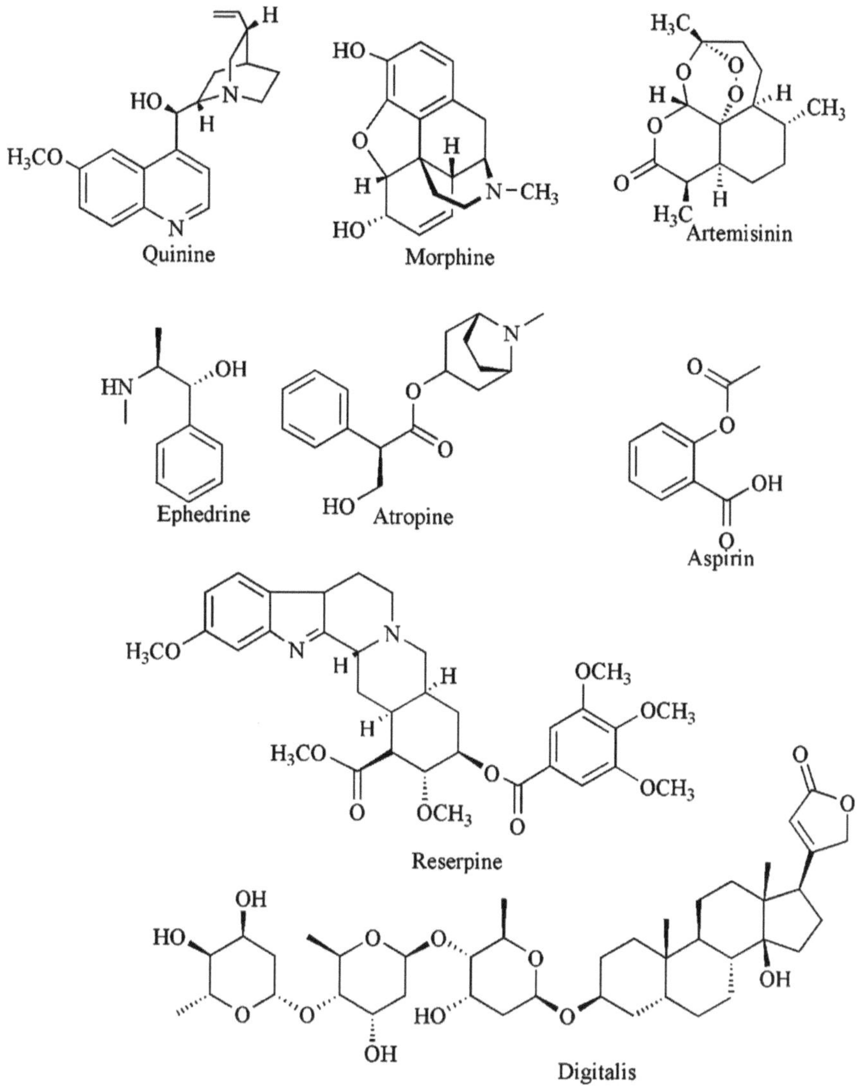

Fig. 1.1 Structure of some important plant-based drugs discovered using the traditional system of medicine

1.3 Modern Approaches for Drug Discovery

The modern approach to drug discovery and development, as illustrated in Figs. 1.4, 1.5, and 1.6, represents a paradigm shift from traditional empirical methods to more precise, technology-driven strategies. Over the last few decades, significant advancements in science and technology have revolutionized the way new therapeutic agents are identified, optimized, and validated. Traditional trial-and-error

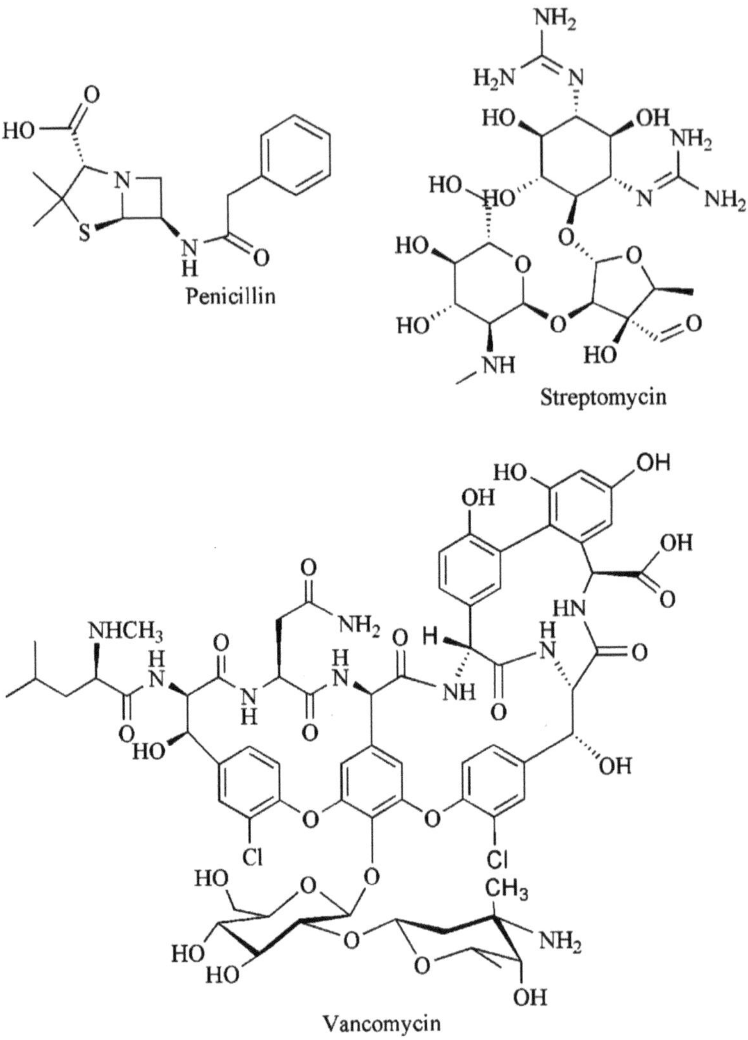

Fig. 1.2 Structure of penicillin, streptomycin, and vancomycin drugs discovered from microbial community

approaches, which relied heavily on natural products and empirical observations, have been largely replaced or supplemented by rational drug design and data-driven methodologies.

Contemporary drug discovery now extensively incorporates computational tools such as AI and ML, which enable rapid prediction of drug–target interactions, optimization of lead compounds, and minimization of off-target effects. Techniques like computer-aided drug design (CADD), structure-based drug design (SBDD), and ligand-based drug design (LBDD) play a crucial role in identifying and refining potential drug molecules based on the three-dimensional structure of biological

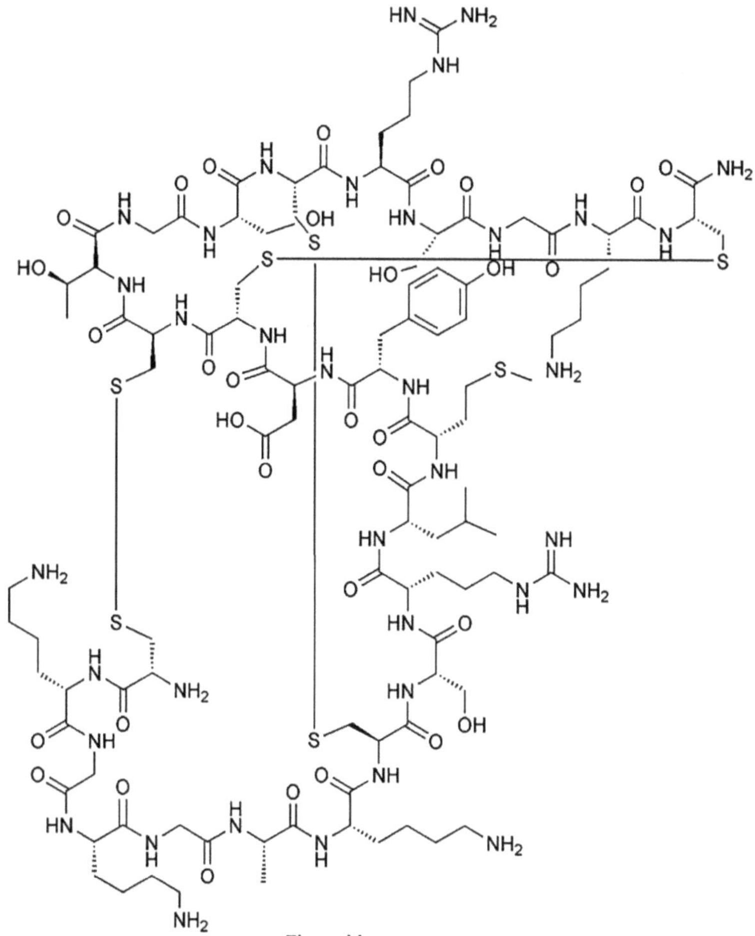

Fig. 1.3 Structure of ziconotide drugs discovered from a marine source

targets or known ligands. In addition, molecular docking, molecular dynamics simulations, and emerging tools like quantum computing are being employed to predict binding affinities and simulate the dynamic behavior of drug–receptor interactions at the atomic level.

HTS further accelerates drug discovery by enabling the rapid evaluation of thousands of compounds for biological activity in a short period. When integrated with omics-based approaches—including genomics, proteomics, metabolomics, and transcriptomics—HTS helps identify novel drug targets and understand disease pathways more comprehensively.

Analytical techniques such as HPLC coupled with MS are critical in identifying and quantifying bioactive compounds, especially from complex mixtures like medicinal plant extracts. High-performance liquid chromatography–mass

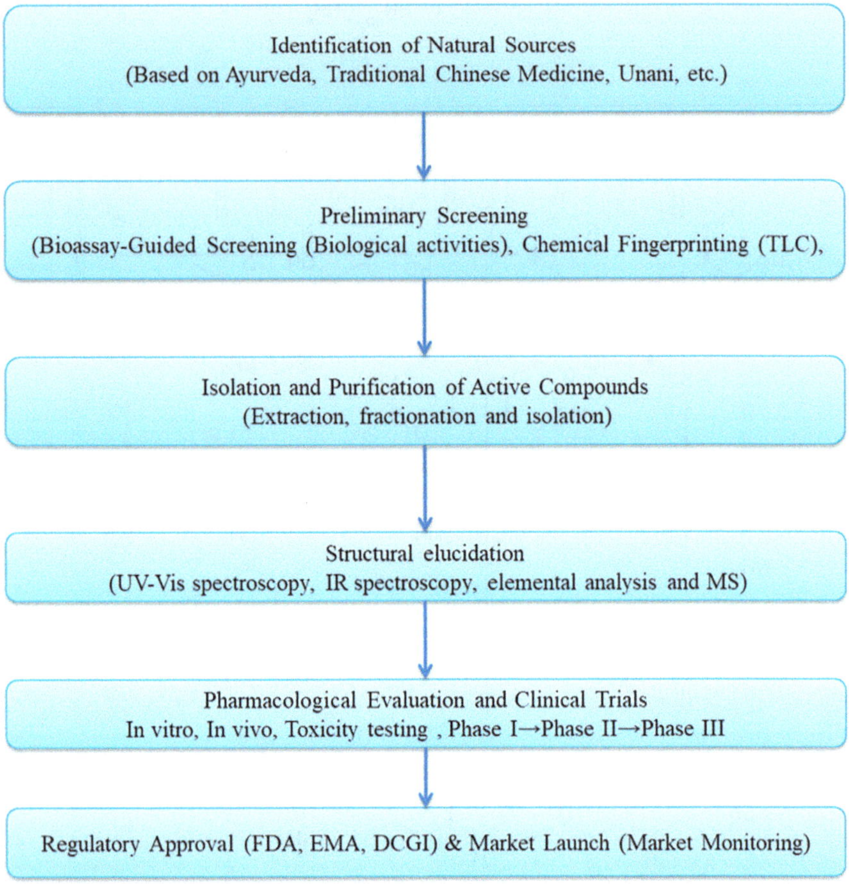

Fig. 1.4 Schematic representation of drug identification protocols from the traditional system of medicine

spectrometry (HPLC–MS) offers high accuracy, sensitivity, and specificity, combining the powerful separation capability of HPLC with the selective detection and structural elucidation potential of MS. This integration of computational, analytical, and high-throughput technologies marks a new era in drug discovery—one that is faster, more targeted, and increasingly personalized.

1.4 Current Drug Development Challenges

Despite substantial advancements in science and technology, drug discovery and development remain highly complex, expensive, and time-intensive processes. One of the major challenges lies in the high attrition rate across different stages of drug development—from initial compound screening to preclinical studies and clinical

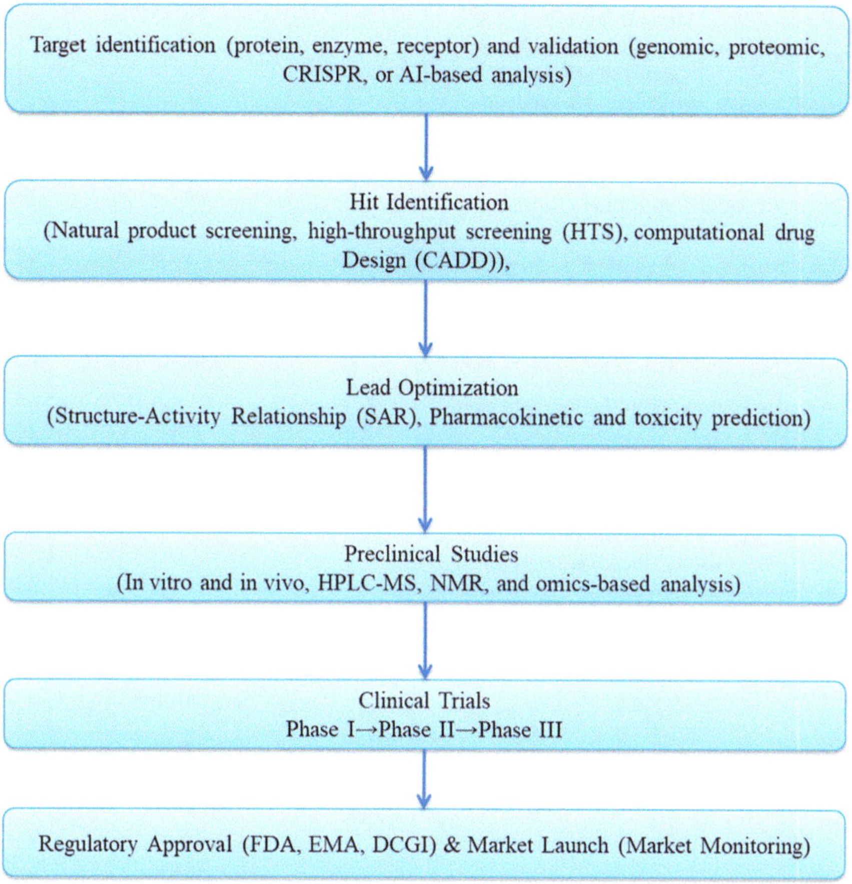

Fig. 1.5 Schematic representation of drug identification protocols using proteins and genomics

trials. Phytochemical investigations, especially those involving medicinal plants, are particularly intricate due to the complex nature of plant extracts and the difficulty in isolating and characterizing individual bioactive compounds. Moreover, ensuring the reproducibility and standardization of plant-based formulations adds another layer of complexity. Clinical trials, a critical step in validating drug efficacy and safety in humans, are not only resource-intensive but also face ethical, regulatory, and logistical hurdles. They require large financial investments and long durations, often spanning several years, with no guarantee of success. Many promising candidates fail in late-stage trials due to unforeseen toxicity, inadequate efficacy, or poor pharmacokinetics. In addition, limited funding, especially for research on natural products and rare diseases, further impedes progress. Regulatory approvals and intellectual property issues also complicate the translation of laboratory findings into marketable therapeutics. Thus, despite modern innovations, drug discovery

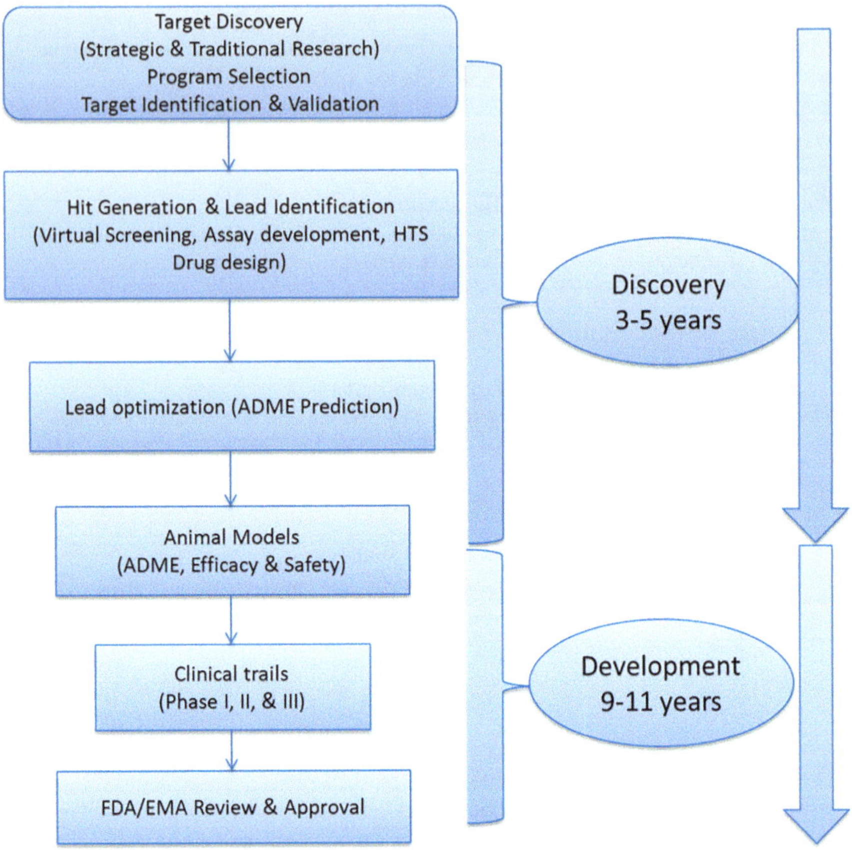

Fig. 1.6 Schematic representation of drug discovery timelines

continues to face multifaceted challenges that demand strategic planning, interdisciplinary collaboration, and sustained investment to overcome.

1.4.1 Complexity and Synergetic Effect of Phytochemicals

Medicinal plants are rich sources of structurally diverse phytochemicals, including alkaloids, flavonoids, terpenoids, glycosides, tannins, and phenolics, among others. While many compounds may be present within a single plant extract, only a few are typically responsible for the observed therapeutic effects. In some cases, a single bioactive compound exerts the primary pharmacological action. However, in many instances, multiple compounds work together to enhance the overall efficacy of the preparation through what is known as the synergistic effect. This phenomenon occurs when the combined action of two or more phytochemicals produces a therapeutic outcome greater than the sum of their individual effects. The complexity of these chemical interactions poses significant challenges for drug discovery, as isolating and identifying synergistic relationships requires advanced analytical techniques and bioassays. Moreover, maintaining the integrity and balance of active constituents during extraction, formulation, and standardization is critical. Understanding the synergistic behavior of phytochemicals is essential for developing effective, safe, and consistent plant-based therapeutics.

1.4.2 Lack of Standardization, Quality Control, and Validation

The content of bioactive compounds is affected by species, geographical, seasonal, and soil conditions, climate change, gender, harvesting time, and plant parts. The drying, extraction, and storage methods and conditions also play an important role in therapeutic effect and can significantly change the concentration of bioactive compounds. Therefore, proper standardization and quality control are essential to ensure consistent potency, safety, and efficacy. The knowledge of traditional medicines is often established on anecdotal evidence, and cultural and historical reliance rather than on scientific validation. Many herbal medicines have not undergone extensive and comparative preclinical and clinical trials with standard compounds to confirm their safety and efficacy, due to limited funding for research.

1.4.3 Extraction, Isolation, and Purification

Large-scale extraction or isolation of bioactive compounds from medicinal plants is very difficult because of the low or trace amount of compounds synthesized in the plants. During extraction and isolation process, some compounds become degraded due to heat, light, and oxidation; therefore, it is very difficult to isolate chemically and thermally unstable compounds. The potency and purity of compounds can also be affected by heavy metals, pesticides, and microbial contaminants. The advanced isolation and purification requires techniques for isolation and extraction process of phytochemicals from plants and microbes.

1.4.4 Low Bioavailability and Pharmacokinetics Issues

One of the major limitations in the development of plant-based drugs is their low bioavailability and complex pharmacokinetic behavior. The bioavailability of a compound refers to the proportion that enters the bloodstream and reaches the site of action in an active form. A key factor influencing bioavailability is the solubility of the compound. Poorly soluble phytochemicals often face challenges in dissolution and absorption within the gastrointestinal tract, resulting in reduced therapeutic efficacy. Conversely, certain compounds that dissolve rapidly may be quickly metabolized and eliminated, minimizing their time of action and reducing clinical effectiveness. These pharmacokinetic limitations significantly impact the development and use of natural compounds as reliable therapeutic agents.

To overcome these challenges, researchers are increasingly turning to advanced drug delivery systems that enhance solubility, stability, and sustained release. Technologies such as nanoformulations, liposomes, and phytosomes offer promising solutions. These delivery systems can encapsulate phytochemicals, protect them from degradation, and promote controlled absorption and targeted delivery in the body. By improving the bioavailability and pharmacokinetic profile of plant-based compounds, these innovative strategies can enhance their clinical potential, paving the way for more effective and standardized herbal medicines in modern therapeutic applications.

1.4.5 Ethical and Legal Challenges and Regulatory Hurdles

The development of plant-based pharmaceuticals is often hindered by a range of ethical, legal, and regulatory challenges. One of the most pressing issues is the management of intellectual property rights (IPR) and the unethical exploitation of indigenous knowledge, commonly referred to as biopiracy. Many local and indigenous communities possess centuries-old traditional knowledge of medicinal plants and their uses. However, due to a lack of access, awareness, or resources, these communities rarely patent their knowledge. In contrast, pharmaceutical companies often capitalize on this traditional wisdom—documented or orally transmitted—by developing drugs based on such information without offering due recognition, benefit-sharing, or compensation to the original knowledge holders. This exploitation raises significant ethical concerns and contributes to distrust between scientific institutions and indigenous communities.

Another major obstacle is the lack of a unified global regulatory framework for plant-based drugs. Unlike synthetic drugs, which are rigorously regulated by bodies such as the US Food and Drug Administration (FDA) and the European Medicines Agency (EMA), herbal medicines often fall under fragmented national guidelines. Different countries classify, regulate, and evaluate herbal products based on their own criteria, resulting in inconsistencies in quality, efficacy, and safety standards. This regulatory diversity creates barriers for international research collaboration and market approval. Therefore, there is a growing need to establish a globally harmonized regulatory authority or framework for evaluating and approving plant-based pharmaceuticals, ensuring scientific credibility, public safety, and ethical integrity across nations.

1.4.6 Environmental and Sustainability Development

Deforestation, habitat destruction, and overexploitation of medicinal plants lead to the loss of biodiversity and environmental degradation. There is a need to cultivate the plants using controlled agricultural practices instead of collecting them from the wild for sustainable cultivation.

1.4.7 Cost and Time

The process of discovery and development is very expensive and takes a long time.

References

Ahmad Dar A, Sangwan PL, Kumar A (2020) Chromatography: an important tool for drug discovery. J Sep Sci 43(1):105–119

Chen CY, Li YH, Li Z, Lee MR (2023) Characterization of effective phytochemicals in traditional Chinese medicine by mass spectrometry. Mass Spectrom Rev 42(5):1808–1827

Chihomvu P, Ganesan A, Gibbons S, Woollard K, Hayes MA (2024) Phytochemicals in drug discovery—a confluence of tradition and innovation. Int J Mol Sci 25(16):8792

Egbuna C, Kumar S, Ifemeje JC, Ezzat SM, Kaliyaperumal S (eds) (2019) Phytochemicals as lead compounds for new drug discovery. Elsevier, Amsterdam

Patwardhan B, Mashelkar RA (2009) Traditional medicine-inspired approaches to drug discovery: can Ayurveda show the way forward? Drug Discov Today 14(15–16):804–811

Xu X (2006) New concepts and approaches for drug discovery based on traditional Chinese medicine. Drug Discov Today Technol 3(3):247–253

Zhang A, Sun H, Wang X (2018a) Mass spectrometry-driven drug discovery for development of herbal medicine. Mass Spectrom Rev 37(3):307–320

Zhang QW, Lin LG, Ye WC (2018b) Techniques for extraction and isolation of natural products: a comprehensive review. Chin Med 13(1):20

Zhang D, Li X, Hu Y, Jiang H, Wu Y, Ding Y, Yu K, He H, Xu J, Sun L, Qian F (2018c) Tabersonine attenuates lipopolysaccharide-induced acute lung injury via suppressing TRAF6 ubiquitination. Biochem Pharmacol 154:183–192

Zhang F, Yan C, Wei C, Yao Y, Ma X, Gong Z, Liu S, Zang D, Chen J, Shi FD, Hao J (2018d) Vinpocetine inhibits NF-κB-dependent inflammation in acute ischemic stroke patients. Transl Stroke Res 9:174–184

Zhang Y, Zhao Y, Guo J, Cui H, Liu S (2018e) Anticancer activity of safranal against colon carcinoma is due to induction of apoptosis and G2/M cell cycle arrest mediated by suppression of mTOR/PI3K/Akt pathway. J BU ON 23(3):574–578

Zhang YW, Kong XY, Wang JH, Du GH (2018f) Vinblastine and vincristine. In: Natural small molecule drugs from plants. Springer, Singapore, pp 551–557

Botanical Aspects of Genus *Catharanthus* G. Don: Species Diversity and Distribution

2

Abstract

The genus *Catharanthus* G. Don, belonging to the family Apocynaceae, comprises a group of small, herbaceous flowering plants known for both their ornamental value and potent medicinal properties, particularly their anticancer activity. The genus consists of nine recognized species, namely, *Catharanthus roseus, C. coriaceus, C. lanceus, C. longifolius, C. makayensis, C. ovalis, C. pusillus, C. scitulus,* and *C. trichophyllus*, most of these species are endemic to the island of Madagascar, which is considered as the center of origin and diversity for *Catharanthus*. From this native region, several species, especially *C. roseus* (Madagascar periwinkle), have been introduced and cultivated in tropical and subtropical regions worldwide due to their therapeutic significance and adaptability. All species in the genus exhibit relatively small, colourful flowers and are adapted to a wide range of ecological conditions. The present chapter focuses on providing a comprehensive account of the species diversity of medicinal species, geographical distribution, morphological traits, and taxonomic classification of the genus *Catharanthus*. Additionally, it explores the ecological adaptations that have allowed these species to thrive in various environments. These botanical insights are crucial for understanding the genetic diversity and conservation strategies of the genus, as well as for supporting its continued use in pharmaceutical and horticultural applications.

Keywords

Genus diversity · Endemic · Madagascar · Apocynaceae

S. Kumar, B. Singh, *Catharanthus in Drug Discovery*,
https://doi.org/10.1007/978-981-95-5048-7_2

2.1 General

The botanical study of plants encompasses their classification, structure, physiology, ecology, distribution, and economic importance. In this context, the genus *Catharanthus* G. Don, belonging to the family Apocynaceae, represents a group of evergreen shrubs and small, herbaceous perennials. These species are notable for both their ornamental appeal and significant medicinal value, particularly in the pharmaceutical industry for producing anticancer compounds. The genus *Catharanthus* comprises nine recognized species, which include *C. roseus* (L.) G. Don, *C. coriaceus* Markgr., *C. lanceus* (Bojer ex A.DC.) Pichon, *C. longifolius* (Pichon) Pichon, *C. makayensis* Allorge et al., *C. ovalis* Markgr., *C. pusillus, C. scitulus* (Pichon) Pichon, and *C. trichophyllus* (Bak.) Pichon. With the exception of *C. pusillus*, which is native to India and Sri Lanka, all species are endemic to Madagascar, a biodiversity hotspot located off the eastern coast of Africa (Jacobs et al. 2004; Shokeen et al. 2007; Das and Sharangi 2017; Allorge et al. 2015). Details of the species and their distribution are given in Table 2.1.

The genus is closely related to the genera *Vinca* and *Amsonia* and derives its name *Catharanthus*, meaning "pure flower." These species have adapted to a variety of ecological niches and are typically found in dry forests, coastal zones, and rocky terrains. Their global distribution today spans across Asia (India, China, Thailand, Indonesia, the Philippines, Sri Lanka), Africa (Kenya, Nigeria, South Africa), Australia, Fiji, New Zealand, the Americas (Florida, Texas, Brazil, Mexico, Argentina), and parts of Europe, especially Mediterranean regions. Despite this wide distribution, several species face threats of extinction due to habitat destruction, deforestation, and climate change (Fig. 2.1).

Plants of the genus *Catharanthus* typically grow between 50 and 100 cm in height and display either an erect or spreading growth habit. A distinctive feature of this genus, shared with others in the Apocynaceae family, is the presence of milky latex in the stems. These plants are often cultivated for their attractive flowers, which bloom in various colors throughout the year, enhancing their value as ornamentals. Among all species, *C. roseus* is the most extensively studied and widely utilized, both horticulturally and medicinally. Native to Madagascar but now cultivated globally, *C. roseus* has a long history of use in traditional medicine systems. Over the last five decades, scientific interest in this species has surged due to its rich profile of over 150 terpenoid indole alkaloids (TIAs). Many of these compounds possess potent pharmacological activities. Notable examples include vincristine and vinblastine, two bisindole alkaloids with powerful anticancer properties, and ajmalicine, an alkaloid used as an antihypertensive agent. These compounds have played a transformative role in cancer chemotherapy and cardiovascular medicine. *Catharanthus* species exhibit wide ecological adaptability, unique morphological features, and a rich profile of secondary metabolites. Conserving these species and further exploring their phytochemical potential is essential for the future.

Table 2.1 Nativity, ecology, and distribution range of genus *Catharanthus* across the globe. (https://wfoplantlist.org)

Sr. no.	Species	Native range	Ecology and habitat	Global distribution	Common names
1.	*Catharanthus roseus* (L.) G.Don (=*Vinca rosea* L.)	East and South Madagascar	Subshrub or shrub and grows primarily in the seasonally dry tropical biome	Bangladesh	Nayantara
				Brazil	Boa-noite, Congorca
				Cook Islands	Tiare-tupapaku-kimo
				Dominica	Caca poule
				French Guiana	Pervenchede de Madagascar
				Guatemala	Chatilla
				Guyana	Periwinkle
				India	Ainskati, Billaganneru, Nayantara, Nityakalyani, Periwinkle, Sadabahar, Sadaphul, Ushamanjairi
				Jamaica	Periwinkle
				Japan	Nichinich-so, Nichinichi-so
				Kenya	Maua
				Madagascar	Madagascan periwinkle
				Mexico	Ninfa
				Pakistan	Sada-bahar
				Peru	Chavelita
				Philippines	Atay-biya, Chichirica, Kantotan, Tsitsirika
				Rodrigues Islands	Saponaire
				Sri Lanka	Minimal, Patti-poo
				Thailand	Phaeng phoi farang, Phang-puai-fa-rang
				United States	Periwinkle
				Venda	Liluvha
				Vietnam	Dua can
				West Indies	Brown man's fancy, white tulip

(continued)

Table 2.1 (continued)

Sr. no.	Species	Native range	Ecology and habitat	Global distribution	Common names
2.	*C. coriaceus* Markgr.	Central Madagascar	Perennial or subshrub and grows primarily in the desert or dry shrubland biome	Mirinarivo, Ambalavao Itremo (Madagascar)	–
3.	*C. lanceus* (Bojer ex A.DC.) Pichon (=*Vinca lancea* Bojer ex A.DC.)	Central Madagascar	Perennial undershrub and grows primarily in the desert or dry shrubland biome	Central plateau around Antananarivo and south wards to Fianarantsoa, Madagascar	–
4.	*C. longifolius* (Pichon) Pichon (=*Lochnera longifolia* Pichon)	Central and South Madagascar	Subshrub or shrub and grows primarily in the desert or dry shrubland biome	Andringitra to the Mandrare Basin in the Southeast Madagascar	–
5.	*C. pusillus* (Murray) G.Don (=*Vinca pusilla* Murray)	Indian subcontinent	Annual herbs and grows primarily in the seasonally dry tropical biome	India and Sri Lanka	Sadabahar
6.	*C. trichophyllus* (Baker) Pichon (=*Vinca trichophylla* Baker)	E. Madagascar	Subshrub and grows primarily in the desert or dry shrubland biome	Northwestern part of Madagascar; Canary Islands	–
7.	*C. makayensis*	Madagascar	Grows as annual herbs, primarily in the seasonally dry tropical biome	Allorge, Phillipson & Razakamal	–
8.	*C. ovalis* Markgr.	Central Madagascar	Perennial shrubs, and grows primarily in the desert or dry shrubland biome	Central plateau and extending to the Isalo massif in the southwest	–

(continued)

Table 2.1 (continued)

Sr. no.	Species	Native range	Ecology and habitat	Global distribution	Common names
9.	*Catharanthus scitulus* (Pichon) Pichon (=*Lochnera scitula* Pichon)	South and Southeast Madagascar	Annual herbs, and grows primarily in the desert or dry shrubland biome	Makay massif	–

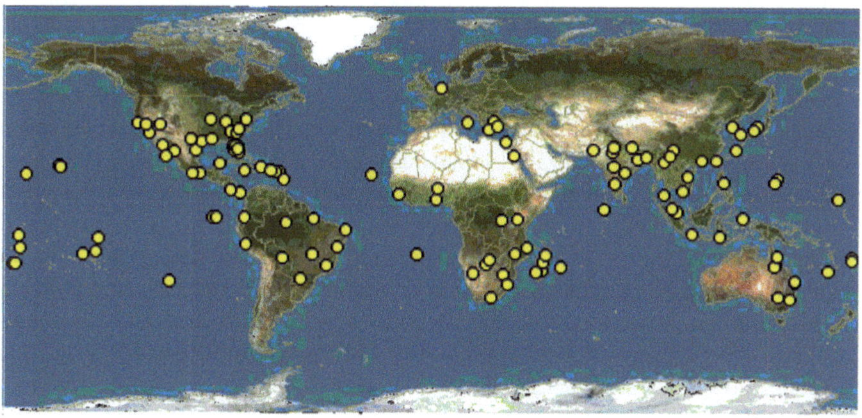

Fig. 2.1 Global distribution of genus *Catharanthus* G. Don

2.2 Botanical Characteristics and Geographical Distribution of Genus *Catharanthus*

All species of the genus *Catharanthus* are believed to have originated from Madagascar, an island located off the southeastern coast of Africa. Due to varying geographical conditions, climatic influences, and evolutionary pressures, these species have developed notable morphological diversity, such as differences in leaf shape, flower color, and plant stature. Over time, several species—especially *C. roseus*—have been introduced and naturalized in various tropical and subtropical regions across the globe. Today, the genus is widely distributed in parts of Asia, Africa, the Americas, Australia, and Europe, where these plants thrive in warm climates and are cultivated for ornamental and medicinal purposes.

2.2.1 *Catharanthus roseus* (L.) G. Don

C. roseus, previously known as *Vinca rosea*, is a perennial herbaceous plant widely cultivated in tropical and subtropical regions across the globe. Commonly referred to as the Madagascar periwinkle, this species is native to Madagascar but has been naturalized in a wide array of countries, including India, Sri Lanka, China, Australia, Brazil, Mexico, Kenya, South Africa, the Philippines, Pakistan, Thailand, Vietnam, the United States, and various regions in Europe, the West Indies, and Oceania. In India and Sri Lanka, it grows abundantly in gardens and the wild, while in China, it is widely cultivated for both ornamental and medicinal purposes (Zenk et al. 1977; Das and Sharangi 2017; Kotakadi et al. 2013; Pereira et al. 2009).

C. roseus typically grows to a height of 8–24 in and spreads about 1 ft., although some trailing varieties can extend up to 2 ft. in length. The plant is much-branched and develops a slightly woody base as it matures. Its leaves are oblong, thick, and leathery, arranged either alternately or in opposite pairs. They range in size from 2.5 to 9.0 cm in length and are attached by a short petiole measuring about 1–1.8 cm. The foliage is typically glabrous (hairless), with a silky green appearance and a prominent pale midrib. The flowers are single, never double, and bloom in a variety of hues including pink, white, and pale pink with a characteristic dark violet center (Svoboda et al. 1959; El-Ghazaly 1990; Allorge et al. 2015). Two widely recognized cultivars are "Rosea," which bears pink flowers, and "Alba," which produces white flowers. These vibrant flowers contribute to the plant's popularity as an ornamental species. Although *C. roseus* is highly valued in traditional and modern medicine for its alkaloid content, the plant is known to be slightly toxic, especially if ingested in large quantities. Despite its toxicity, it continues to be cultivated globally for its aesthetic appeal, pharmacological properties, and its role in landscape design and floral displays in warm climates.

2.2.2 *Catharanthus coriaceus* Markgr

C. coriaceus is a rare and endemic species of Madagascar, first described by Markgraf in 1970. Despite its formal description in the twentieth century, herbarium specimens of the plant were collected as early as 1946, with further samples gathered during subsequent decades—1955, 1956, 1963, 1964, 1969, 1975, 1976, 1980, and 1984. These collections were primarily concentrated in regions around Ambatofinandrahana-Itremo, Arivonimamo, and Ambalavao, areas known for their rich biodiversity and unique floral assemblages. In 1978, *C. coriaceus* was included in the International Union for Conservation of Nature (IUCN) Plant Red Data Book as an endangered species. According to the IUCN, it exists as a "very small population of very limited distribution" in Madagascar and is "acutely threatened by bushfires and land clearing activities," which are persistent threats to many native Malagasy species. Despite this early recognition of its endangered status, no formal conservation measures were known to have been implemented at the time, and there was no information about planned conservation strategies to ensure the species' survival. Several studies attempted to estimate the distribution of *C. coriaceus*, their reported areas did not consistently align with actual collection sites. Notably, the Itremo region, which has yielded the most specimens, was overlooked in their assessments, highlighting inconsistencies and gaps in early distribution data.

Morphologically, *C. coriaceus* is characterized by leaves that are typically twice as long as the internodes and rounded at the base. Its flowers occur in pairs, a feature that helps in its identification among other species within the genus. Given its critically limited range and ongoing threats, *C. coriaceus* urgently requires targeted conservation planning, habitat protection, and further taxonomic and ecological research to ensure its survival (Hammer 1994).

2.2.3 *Catharanthus lanceus* (Bojer ex A.DC.) Pichon

Catharanthus lanceus is a perennial herbaceous plant belonging to the Apocynaceae family, cultivated and found growing in several countries, particularly in tropical and subtropical regions. Native to Madagascar, this species is valued for both its ornamental appeal and ecological adaptability. The plant can attain a height of up to 1 m (approximately 3 ft) and is commonly known by its vernacular name, "lance-leaf periwinkle." One of the key distinguishing features of *C. lanceus* is its fragrant flowers, which exhibit a unique color transformation. The flowers typically have cream-colored corolla lobes that gradually change to pink, reddish-violet, or pink-magenta as they mature, making the plant visually striking and attractive for ornamental cultivation.

C. lanceus thrives in a variety of natural habitats, demonstrating a high level of ecological tolerance. It is often found growing in volcanic or burned soils, along hillside rocks, and in open woodlands, particularly on ridges where sunlight is abundant. Its natural distribution spans elevations ranging from 750 to 2000 m (approximately 2500–6600 ft), indicating its adaptability to both mid- and high-altitude environments. Due to its ability to colonize disturbed and nutrient-poor soils, *C. lanceus* plays a potential role in ecological restoration and landscaping in degraded environments. Although less studied than *C. roseus*, this species may also possess important phytochemical properties that warrant further research, especially given its resilience and distinct morphological traits. Conservation awareness is essential due to habitat loss in its native range (Farnsworth et al. 1967).

2.2.4 *Catharanthus longifolius* (Pichon) Pichon

C. longifolius is a perennial herbaceous plant belonging to the family Apocynaceae. While most species of the genus *Catharanthus* are endemic to Madagascar, *C. longifolius* is endemic to India, making it a botanically unique and regionally significant species. The plant is characterized by its more or less woody stems, which range in height from approximately 40 to 150 cm, giving it a robust, upright growth habit. *C. longifolius* typically grows in tropical and subtropical climates and is adapted to a variety of ecological conditions. Although not as extensively cultivated as *C. roseus*, it is both harvested from the wild and cultivated locally for its medicinal value. The plant has been traditionally used in folk medicine for treating various ailments, although scientific studies on its phytochemistry and pharmacological potential remain limited. Its long, narrow leaves and growth form also contribute to its ornamental appeal in home gardens and local landscapes. Given its restricted distribution and dependence on natural habitats, *C. longifolius* may face threats from habitat degradation, overharvesting, and land-use changes. Thus, further botanical research, phytochemical exploration, and conservation measures are essential to ensure its sustainable use and long-term preservation in its native range (Yasin et al. 2019).

2.2.5 *Catharanthus pusillus* (Baker) Pichon

C. pusillus is a perennial herbaceous plant belonging to the Apocynaceae family and is widely known by various vernacular names across India and other regions of the world. Unlike other *Catharanthus* species that thrive in tropical zones, *C. pusillus* is well adapted to mild climates, where it prefers partial shade and moist, well-drained soils. It is typically found growing in woodlands, grassy bluffs, and disturbed habitats such as cemeteries, roadside areas, and open fields. The plant blooms regularly from April to May and may flower sporadically from May to September, depending on environmental conditions (Navitha et al. 2012). The plant can grow up to 1 m tall and produces a characteristic white latex when injured. It features a carrot-shaped taproot extending up to 70 cm, and its stems and branches are reddish and almost quadrangular in shape. The internodes are significantly longer than the leaves, giving the plant a loose, spreading structure. The leaves are simple, opposite, and entire, typically with 1–3 stipules at each side of the leaf base. The petioles are short, around 1–3 mm, and support oblong to narrowly ovate blades, measuring 1–4.5 cm in length and 3–13 mm in width. Leaf texture ranges from herbaceous to thinly leathery, and both surfaces are shiny and glabrous. The flowers, which are bisexual and fragrant, are borne in the leaf axils and are 5-merous, showing regular floral symmetry. The pedicels measure 5–32 mm, and the sepals, partially fused at the base, are 5–10 mm long. The corolla tube, 15–22 mm long, is cylindrical with a constricted throat, lined with a ring of hairs just below the stamen insertion. The petal lobes are ovate to obovate, exhibiting shades of pink, reddish, magenta, violet, or cream with a pinkish base. Stamens are inserted just below the throat, with very short filaments. The ovary is superior, composed of two narrowly oblong carpels. The style is slender (10–16 mm), topped by a cylindrical pistil head. The fruit consists of striate, glabrous follicles, 1.5–5 cm long, that are dehiscent and green. The seeds are black, oblong, and grooved on one side, with 10–20 seeds per follicle (Zárate et al. 2001).

2.2.6 *Catharanthus trichophyllus* (Bak.) Pichon

C. trichophyllus is a perennial herb about 3.3 ft. (1 m) tall. It secretes white latex. The stems are about 1 m tall, and the branches are square, winged, and reddish or purplish in color.. The leaves are oppositely arranged with hairy, pointed oval blades up to 3.4 in (8.5 cm) long. They are each accompanied by several stipules. Flowers occur singly or sometimes in pairs within the leaf axils. The calyx is approximately 0.4 in (1 cm) long with narrow lobes 5 in long. The corolla has a tubular throat about 0.8 in (2 cm) long, opening into five lobes, each about 0.7 in long. The flowers are generally white and sometimes pink, red, or purple, with yellowish centers and unpleasant smell. The fruit consists of a pair of greenish or purplish follicles about 2.8 in (7 cm) long, each fruit containing 10–20 seeds (Mukhopadhyay and Cordell 1981a, b; Davioud et al. 1989).

2.2.7 *Catharanthus makayensis* Allorge, Phillipson and Razakamal

Recently, a new species called *C. makayensis* was reported. It differs from all other species in the genus due to its hexagonal cross section of stems and the arrangement of flowering and fruiting characters. Its 5–6 mm long filiform sepals are clearly much narrower and longer than those of all other reported species, with the exception of *C. coriaceus*, which is highly distinct with its pendulous fruits, as opposed to the erect fruits of *C. makayensis*. *C. makayensis* and *C. roseus* are the only species to have a magenta, rather than white to pale yellow eye. The corolla lobes and flower buds of *C. makayensis* are pure white or tinted pale pink, whereas *C. roseus* typically has pink lobes. The relationships of the new species with the eight other species of *Catharanthus* are discussed, and an identification key to the Malagasy species of the genus is provided (Allorge et al. 2015).

2.2.8 *Catharanthus ovalis* Markgr

C. ovalis is a perennial herbaceous plant that typically grows up to 40 cm in height. Endemic to Madagascar, this species is often found in its natural habitat, where it is harvested from the wild for local medicinal applications. Although less widely studied than some other *Catharanthus* species, *C. ovalis* holds traditional value in herbal medicine practices. Its sustainable use depends on careful management due to collection from wild. The plant's modest size and perennial nature allow it to persist in its native environment, contributing to the biodiversity of the region (Shahriar et al. 2024).

2.2.9 *Catharanthus scitulus* (Pichon) Pichon

C. scitulus is a lesser-known flowering plant species belonging to the genus *Catharanthus* within the family Apocynaceae. It is endemic to Madagascar, meaning it is naturally found only in this specific geographic region. The species is adapted to the island's unique ecological conditions and contributes to the botanical richness of Madagascar's flora. Like other members of the genus, *C. scitulus* is likely to exhibit ornamental and potential medicinal properties, although it remains relatively underexplored. Due to habitat loss and environmental pressures, its conservation status may be of concern, highlighting the need for further study and preservation efforts.

References

Allorge L, Phillipson PB, Razakamalala R (2015) Catharanthus makayensis L. Allorge, Phillipson & Razakamal. (Apocynaceae), a new species from Madagascar. Candollea 70(1):61–66

Das S, Sharangi AB (2017) Madagascar periwinkle (Catharanthus roseus L.): diverse medicinal and therapeutic benefits to humankind. J Pharmacogn Phytochem 6(5):1695–1701

Davioud E, Kan C, Hamon J, Tempé J, Husson HP (1989) Production of indole alkaloids by in vitro root cultures from Catharanthus trichophyllus. Phytochemistry 28(10):2675–2680

El-Ghazaly G (1990) Development of pollen grains of Catharanthus roseus (Apocynaceae). Rev Palaeobot Palynol 64(1–4):165–174

Farnsworth NR, Blomster RN, Buckley JP (1967) Catharanthus alkaloids XIII: antineoplastic and hypotensive activity of alkaloid fractions and certain alkaloids from Catharanthus lanceus. J Pharm Sci 56(1):23–27

Hammer ML (1994) The status and distribution of Catharanthus coriaceus Markgraf (Apocynaceae). Biodivers Conserv 3(6):501–511

Jacobs DI, Snoeijer W, Hallard D, Verpoorte R (2004) The Catharanthus alkaloids: pharmacognosy and biotechnology. Curr Med Chem 11(5):607–628

Kotakadi VS, Rao YS, Gaddam SA, Prasad TNVKV, Reddy AV, Gopal DS (2013) Simple and rapid biosynthesis of stable silver nanoparticles using dried leaves of Catharanthus roseus Linn. G. Donn and its antimicrobial activity. Colloids Surf B: Biointerfaces 105:194–198

Mukhopadhyay S, Cordell GA (1981a) Catharanthus alkaloids. XXXV. Isolation of leurosidine N′b-oxide from Catharanthus roseus. J Nat Prod 44(5):611–613

Mukhopadhyay S, Cordell GA (1981b) Catharanthus alkaloids. XXXVI. Isolation of vincaleukoblastine (VLB) and periformyline from Catharanthus trichophyllus and pericyclivine from Catharanthus roseus. J Nat Prod 44(3):335–339

Navitha A, Helen Sheeba DA, Ramesh C, Sartaj Banu M (2012) Hypoglycemic and anti-diabetic activity of ethanolic extract of Catharanthus pusillus (Murray) G. Don. IOSR J Pharm 2(4):17–21

Pereira DM, Ferreres F, Oliveira J, Valentão P, Andrade PB, Sottomayor M (2009) Targeted metabolite analysis of Catharanthus roseus and its biological potential. Food Chem Toxicol 47(6):1349–1354

Shahriar S, Shermin SA, Hasnat H, Hossain F, Han A, Geng P, Alam S, Mamun AA (2024) Chemico-pharmacological evaluation of the methanolic leaf extract of Catharanthus ovalis: GC–MS/MS, in vivo, in vitro, and in silico approaches. Front Pharmacol 15:e1347069

Shokeen B, Sethy NK, Kumar S, Bhatia S (2007) Isolation and characterization of microsatellite markers for analysis of molecular variation in the medicinal plant Madagascar periwinkle (Catharanthus roseus (L.) G. Don). Plant Sci 172(3):441–451

Svoboda GH, Neuss N, Gorman M (1959) Alkaloids of Vinca rosea Linn. (Catharanthus roseus G. Don.). V. Preparation and characterization of alkaloids. J Am Pharm Assoc 48(11):659–666

Yasin NA, Khan WU, Ahmad SR, Ahmad A, Akram W, Ijaz M (2019) Role of Acinetobacter sp. cs9 in improving growth and phytoremediation potential of catharanthus longifolius under cadmium stress. Pol J Environ Stud 28(1):435–443

Zárate R, Dirks C, van der Heijden R, Verpoorte R (2001) Terpenoid indole alkaloid profile changes in Catharanthus pusillus during development. Plant Sci 160(5):971–977

Zenk MH, El-Shagi H, Arens H, Stöckigt J, Weiler EW, Deus B (1977) Formation of the indole alkaloids serpentine and ajmalicine in cell suspension cultures of Catharanthus roseus. In: Plant tissue culture and its biotechnological application. Springer, Berlin, Heidelberg, pp 27–43

Associated Traditional Knowledge and Applications of Various *Catharanthus* Species

3

Abstract

Various *Catharanthus* species have long been valued in traditional medicinal systems across various cultures, notably in African Traditional Medicine and Indian Ayurveda. These plants are widely utilized for their therapeutic properties, with preparations made from different parts—such as leaves, roots, and the whole plant—being used in diverse forms including infusions, pastes, and extracts. Traditional healers employs *Catharanthus* plant-based remedies to address a wide range of health conditions. These include infectious diseases, immune system disorders, hormonal imbalances, metabolic dysfunctions, cardiovascular and neurological issues, respiratory ailments, and reproductive health concerns. The breadth of applications highlights the plant's significance in holistic and integrative medicine practices. Among the different species, *C. roseus* and *C. pusilla* hold particular importance due to their potent medicinal value in the treatment of cancer and diabetes. The plant contains several bioactive alkaloids, such as vincristine and vinblastine, which are well known for their anticancer properties and are used in modern chemotherapy. Additionally, traditional formulations have shown benefits in managing blood sugar levels, making the species a valuable resource for diabetes care. The widespread use and enduring relevance of *Catharanthus* species in traditional medicine underscore their importance not only in cultural heritage but also in modern pharmacological research. These plants serve as a bridge between traditional knowledge systems and contemporary therapeutic development, warranting further scientific exploration and validation. In this chapter, associated traditional knowledge and applications of various *Catharanthus* species are presented.

Keywords

Indigenous knowledge · Beliefs · Utilization

S. Kumar, B. Singh, *Catharanthus in Drug Discovery*,
https://doi.org/10.1007/978-981-95-5048-7_3

3.1 Introduction

Traditional knowledge of medicinal plants is accumulated based on either herbal medicine systems (Ayurveda (India), TCM (China), Native American Medicine, and African Traditional Medicine) or practices of indigenous wisdom and skills of local communities, used for healing and health purposes. According to Ayurvedic and Siddha medicine (India), various preparations of plant parts of the species of *Catharanthus* are used for the treatment of diabetes, skin disorders, fever, insect bites, and high blood pressure. Similarly, TCM uses their cooling properties to treat infections and inflammation. Local communities of different countries also treat malaria, dysentery, indigestion, diarrhea, and toothache. Its plant parts (whole plant, leaf, stem, flowers, and root) have been used to treat a wide variety of diseases since ancient times. An overview of different plant parts, preparation, and their traditional uses of *Catharanthus* is summarized and presented in Fig. 3.1 and Table 3.1. Europeans used the plant for minor illnesses like headaches and diabetes. In Ayurveda, stems, leaves, and roots of the plants have been used to treat diabetes, hypertension, asthma, gastrointestinal problems, and problems in female sexuality medicines. However, smoking dried leaves of the plant may cause side effects like incoordination, prickling of the skin, and hallucinations, and excessive inhalation

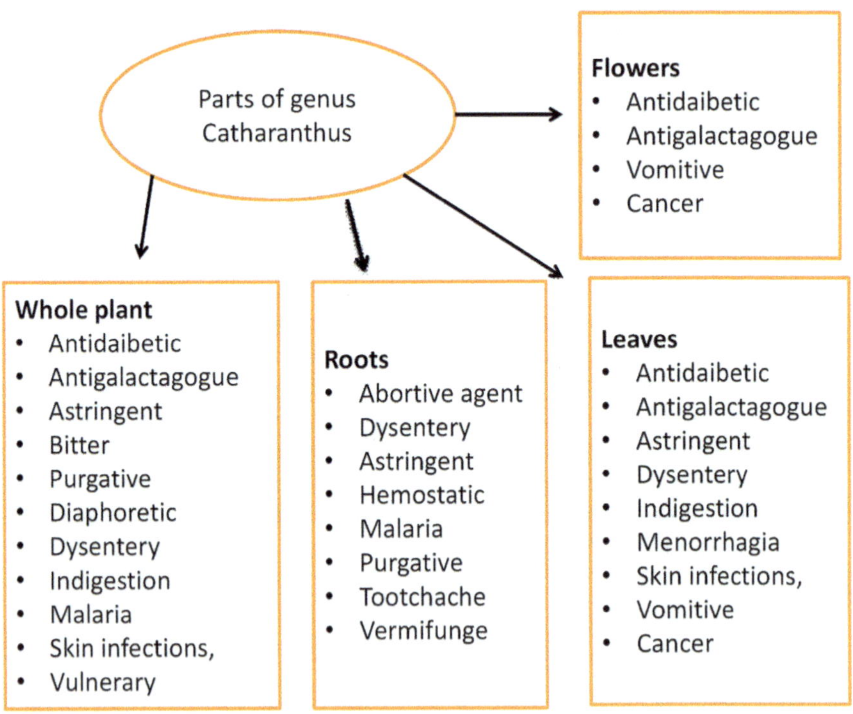

Fig. 3.1 Schematic representation of the uses of *Catharanthus* in disease treatment

Table 3.1 Traditional knowledge associated with different *Catharanthus* species in treatment of human disease

Sr no.	Species	Plant parts (preparation used)	Ethnomedicinal uses	Country	References
1.	*C. roseus/C. pusillus*	Whole plant	Muscle relieve pain, depression of the central nervous system, and wasps stings	India	Ross (2003), Das and Sharangi (2017), Kumar et al. (2022), Senbagalakshmi et al. (2017)
2.	*C. roseus*	Whole plant	Nose bleed, bleeding gums, mouth ulcers, and sore throats	India	Senbagalakshmi et al. (2017)
3.	*C. roseus/C. pusillus*	The hot water extract of dried entire plant is taken orally	Cancer	India	Senbagalakshmi et al. (2017), Aslam et al. (2010a)
4.	*C. roseus*	Hot water extract of dried leaves is taken orally	Hodgkin's disease	India	Aslam et al. (2010a, b)
5.	*C. roseus*	The root extract is taken orally	Menorrhagia	India	Senbagalakshmi et al. (2017), Aslam et al. (2010a, b)
6.	*C. roseus/C. lanceus*	Root	Tonic and stomachic	India	Senbagalakshmi et al. (2017), Aslam et al. (2010a, b)
7.	*C. roseus/C. pusillus/C. lanceus*	The infusion of the leaves	Menorrhagia, toothache, memory loss, and blood circulation	India	Senbagalakshmi et al. (2017), Aslam et al. (2010a, b)
8.	*C. roseus*	The plant was boiled to make a poultice	Stop bleeding	Hawaii	Aslam et al. (2010a, b), Das and Sharangi (2017), Kumar et al. (2022)
9.	*C. roseus*	Hot water extract of dried root	Emmenagogue	Philippines	Aslam et al. (2010a, b), Das and Sharangi (2017), Kumar et al. (2022)

(continued)

Table 3.1 (continued)

Sr no.	Species	Plant parts (preparation used)	Ethnomedicinal uses	Country	References
10.	*C. roseus/C. pusillus*	Hot water extract of leaves	Diabetes mellitus, amenorrhea	Philippines	Aslam et al. (2010a, b), Das and Sharangi (2017), Kumar et al. (2022)
11.	*C. roseus*	Hot water extract of root	Abortion, emmenagogue, dysmenorrheal with scanty flow abortion	Philippines	Aslam et al. (2010a, b), Das and Sharangi (2017), Kumar et al. (2022)
12.	*C. roseus*	Hot water extract of leafy stems	Diabetes	West Indies	Aslam et al. (2010a, b), Das and Sharangi (2017), Kumar et al. (2022)
13.	*C. roseus/C. pusillus*	Leaves	Regulating blood sugar, reducing toothaches, and high blood pressure, soothing insect bites, improving memory, and circulation, healing wounds	Sri Lanka	Senbagalakshmi et al. (2017)
14.	*C. roseus*	Whole plant	Diabetes, cancer		Senbagalakshmi et al. (2017)
15.	*C. roseus/C. lanceus*	Decoction of dried root	Fevers and malaria	Brazil	Aslam et al. (2010a, b), Senbagalakshmi et al. (2017)
16.	*C. roseus/C. lanceus/C. pusillus*	Hot water extract of root bark, dried entire plant	Diabetes mellitus	Brazil	Aslam et al. (2010a, b), Senbagalakshmi et al. (2017)
17.	*C. roseus*	Hot water extract of dried entire plant	Diabetes	Thailand	Aslam et al. (2010a, b), Senbagalakshmi et al. (2017)
18.	*C. roseus*	An extract of the flower	Eyewash for the eyes of infants	Rodriguez Islands	Aslam et al. (2010a, b), Senbagalakshmi et al. (2017)
19.	*C. roseus/C. pusillus*	Decoction of dried leaves	Diabetes, hypertension, and cancer	Cook Islands	Aslam et al. (2010a, b), Senbagalakshmi et al. (2017)

(continued)

Table 3.1 (continued)

Sr no.	Species	Plant parts (preparation used)	Ethnomedicinal uses	Country	References
20.	*C. roseus*	Medicinally it is used to treat	Kidney problems, diabetes, and arthritis	Marshall Islands	Aslam et al. (2010a, b), Senbagalakshmi et al. (2017)
21.	*C. roseus/C. pusillus/C. coriaceus*		Cure stomach ulcer, sedative, and anticancer	Japan	Aslam et al. (2010a, b), Senbagalakshmi et al. (2017)
22.	*C. roseus*	Root infusion	Diabetes vincaria	Mexico	Aslam et al. (2010a, b), Senbagalakshmi et al. (2017)
23.	*C. roseus/C. pusillus*	Infusion of whole extract of the aerial parts	Orally for diabetes	Kenya	Aslam et al. (2010a, b), Senbagalakshmi et al. (2017)
24.	*C. roseus*	Decoction of dried roots	Stomach problem	Kenya	Aslam et al. (2010a, b), Senbagalakshmi et al. (2017)
25.	*C. roseus*	Hot water extract of dried leaves	Diabetes	Kenya	Aslam et al. (2010a, b), Senbagalakshmi et al. (2017)
26.	*C. roseus*	Hot water extract of dried entire plant	Antigalactagogue	England	Aslam et al. (2010a, b), Senbagalakshmi et al. (2017)
27.	*C. roseus*	Hot water extract of the leaf	Diabetes	England	Aslam et al. (2010a, b), Senbagalakshmi et al. (2017)
28.	*C. roseus/C. pusillus*	Take the plant	For high blood pressure	Dutch	Aslam et al. (2010a, b), Senbagalakshmi et al. (2017)
29.	*C. roseus*	Hot water extract of aerial parts	Menstrual regulator	Vietnam	Aslam et al. (2010a, b), Senbagalakshmi et al. (2017)
30.	*C. roseus*	Hot water extract of entire plant	Antigalactogogue	Vietnam	Aslam et al. (2010a, b), Senbagalakshmi et al. (2017)

(continued)

Table 3.1 (continued)

Sr no.	Species	Plant parts (preparation used)	Ethnomedicinal uses	Country	References
31.	*C. roseus*	The bitter and astringent leaves	Used as vomitive	France	Aslam et al. (2010a, b), Senbagalakshmi et al. (2017)
32.	*C. roseus/C. pusillus*	Roots	Purgative, vermifuge,depurative, hemostatic, and toothache remedy	France	Aslam et al. (2010a, b), Senbagalakshmi et al. (2017)
33.	*C. roseus*	Hot water extract of entire plant	Antigalactagogue	France	Aslam et al. (2010a, b), Senbagalakshmi et al. (2017)
34.	*C. roseus*	Hot water extract of dried leaves	Diabetes	Guatemala	Aslam et al. (2010a, b), Senbagalakshmi et al. (2017)
35.	*C. roseus*	Hot water extract of the leaf	Combat primary inertia in childbirth; diabetes	Dominica	Aslam et al. (2010a, b), Senbagalakshmi et al. (2017)
36.	*C. roseus*	Whole plant	Used for diabetes	Cuba	Aslam et al. (2010a, b), Das and Sharangi (2017), Kumar et al. (2022)
37.	*C. roseus/C. pusillus/C. coriaceus*	Hot water extract of dried entire plant	Cancer, heart disease, and leishmaniosis	Peru	Aslam et al. (2010a, b), Senbagalakshmi et al. (2017)
38.	*C. roseus*	Take the plant for	High blood pressure	Venda	Aslam et al. (2010a, b), Senbagalakshmi et al. (2017)
39.	*C. roseus*	Water extract of dried root	Venereal disease	Venda	Aslam et al. (2010a, b), Senbagalakshmi et al. (2017)
40.	*C. roseus*	Hot water extract of dried leaves	Diabetes	Jamaica	Aslam et al. (2010a, b), Senbagalakshmi et al. (2017)
41.	*C. roseus/C. pusillus*	The flower extract	Eyewash for the eyes of infants	Jamaica	Aslam et al. (2010a, b), Senbagalakshmi et al. (2017)

(continued)

Table 3.1 (continued)

Sr no.	Species	Plant parts (preparation used)	Ethnomedicinal uses	Country	References
42.	*C. roseus/C. pusillus*	Gargling with an infusion of the plant	Relieve pain from a sore throat, laryngitis, and chest complaints and ease Lung congestion and inflamation	United States	Aslam et al. (2010a, b), Das and Sharangi (2017), Senbagalakshmi et al. (2017),
43.	*C. roseus*	An extract from the flowers	Eye irritation and infection	Caribbean	Aslam et al. (2010a, b), Senbagalakshmi et al. (2017)
44.	*C. roseus*	Whole plant	Diabetes and sedative	Caribbean	Aslam et al. (2010a, b), Senbagalakshmi et al. (2017)
45.	*C. roseus/C. pusillus*	Decoction of dried leaves	Diabetes mellitus	Europe	Aslam et al. (2010a, b), Senbagalakshmi et al. (2017)
46.	*C. roseus*	Hot water extract of dried ovules	Diabetes	Pakistan	Aslam et al. (2010a, b), Senbagalakshmi et al. (2017)
47.	*C. roseus*	Use the stalks and leaves	Dysmenorrhea	Indonesia	Aslam et al. (2010a, b), Senbagalakshmi et al. (2017)
48.	*C. roseus*	Hot water extract of dried leaves	Menorrhagia and diabetes	Australia	Aslam et al. (2010a, b), Senbagalakshmi et al. (2017)
49.	*C. roseus/C. pusillus*	Hot water extract of root bark	Febrifuge	Australia	Aslam et al. (2010a, b), Senbagalakshmi et al. (2017)
50.	*C. roseus*	Decoction of whole plant	Diabetes	Myanmar	Aslam et al. (2010a, b), Senbagalakshmi et al. (2017)
51.	*C. roseus*	Expressed juice together with the same amount of honey	Anemia	Myanmar	Aslam et al. (2010a, b), Senbagalakshmi et al. (2017)

(continued)

Table 3.1 (continued)

Sr no.	Species	Plant parts (preparation used)	Ethnomedicinal uses	Country	References
52.	*C. roseus*	Crush fresh leaves	Ulcers, inflammation	Myanmar	Aslam et al. (2010a, b), Senbagalakshmi et al. (2017)
53.	*C. roseus/C. coriaceus*		Traditionally used for diabetes	Spain	Aslam et al. (2010a, b), Senbagalakshmi et al. (2017)
54.	*C. roseus*	Hot water extract of the aerial parts	Menstrual regulator, malaria	China	Aslam et al. (2010a, b), Kumar et al. (2022)
55.	*C. roseus*		Astringent, diuretic, and cough remedy	China	Aslam et al. (2010a, b), Kumar et al. (2022),
56.	*C. roseus*	Hot water extract of leaves	Diabetes and rheumatism	Mozam	Aslam et al. (2010a, b), Senbagalakshmi et al. (2017)
57.	*C. roseus*	Hot water extract of root	Hypertensive and febrifuge	Mozam	Aslam et al. (2010a, b), Senbagalakshmi et al. (2017)
58.	*C. roseus/C. lanceus*	Hot water extract of dried leaves	Menorrhagia and diabetes	Africa	Kumar et al. (2022)
59.	*C. roseus*	Plant decoction	Diabetes	Malaysia	Das and Sharangi (2017)
60.	*C. roseus*	Decoction of dried entire plant	Diabetes mellitus and liver diseases	Taiwan	
61.	*C. roseus/C. lanceus*	The bitter and astringent leaves and roots	Vomitive, purgative, vermifuge, depurative, hemostatic, and toothache remedies	Madagascar	Aslam et al. (2010a, b), Das and Sharangi (2017)
62.	*C. roseus*	The juice of leaves	Indigestion and dyspepsia	Mauritius	Aslam et al. (2010a, b), Das and Sharangi (2017)
63.	*C. roseus*	Leaves	Menorrhagia and rheumatism	Africa	Das and Sharangi (2017)
64.	*C. roseus*	Decoction of flower	Asthma, tuberculosis, and flatulence	Bahamas	Das and Sharangi (2017), Kumar et al. (2022)

(continued)

Table 3.1 (continued)

Sr no.	Species	Plant parts (preparation used)	Ethnomedicinal uses	Country	References
65.	*C. lanceus/C. pusillus*		Toothache, fever, and skin diseases, and also as a diuretic	India	Chandran and Saj (2015)

may cause kidney and nervous system problems. Conventionally, in India, seven flowers/leaves are used at a time for treatment of different diseases, whereas in the Cook Islands, boiled leaves with water and, in the West Indies, roots of plants infused in whiskey are used to control diabetes (Nammi et al. 2003; Das and Sharangi 2017; Chaturvedi et al. 2022; Kumar et al. 2022; Nisat et al. 2024).

3.2 Various *Catharanthus* Species and Their Medicinal Uses

3.2.1 *Catharanthus roseus*

C. roseus, commonly known as Madagascar periwinkle, is a highly valued plant recognized for its ornamental appeal and extensive traditional medicinal uses. Across different cultures and regions, various parts of the plant—including leaves, roots, flowers, and latex—have long been used in folk medicine to treat numerous ailments. Traditional healers frequently employ *C. roseus* to manage conditions such as diabetes, high blood pressure, and gastrointestinal disorders like diarrhea (Govindasamy and Srinivasan 2012; Rasineni et al. 2010). The leaves are particularly significant in herbal treatments and are used to address a wide range of health issues including heavy menstrual bleeding, painful periods, joint pain, upset stomach, indigestion, diabetes, hypertension, cancer, and skin conditions. Additionally, they are employed to treat infectious diseases like dengue fever, malaria, and skin infections resulting from insect bites. Leaf extracts are also used for minor ailments such as toothaches, sore throats, eye irritation, and breathing difficulties, and they are known to possess antiviral, calming, and anti-inflammatory properties. Roots of *C. roseus* are traditionally used for managing hypertension, general weakness, and mental restlessness. They are believed to have sedative, tranquilizing, and tonic effects. Moreover, the plant is used as an emetic (to induce vomiting) and the roots serve as purgative, depurative, vermifuge, and hemostatic agents, as well as being applied for toothache relief.

In the Philippines, decoctions of leaves are used to manage diabetes, and immature leaves help relieve stomach cramps. Root decoctions are employed to treat intestinal parasitism. In Mauritius, leaf infusions are used to relieve indigestion and dyspepsia. In Indian states like Orissa and Assam, leaf juice is applied to wasp stings, while roots and leaves are also used for their anticancer properties. Globally, *C. roseus* finds widespread application. In Indo-China, floral extracts are used to treat dysmenorrhea; in Cuba and Jamaica, as eyewash for infants; and in the

Bahamas, for treating asthma. Bermuda uses it for hypertension, while in Suriname and across the Caribbean, it helps with eye infections, malaria, menstrual pains, and excessive sweating. In Uganda, leaf distillations are administered for digestive ulcers, whereas in India, powdered leaves are consumed with milk for treating abscesses. In Togo, a root decoction is used to treat dysmenorrhea.

European traditions have long valued the plant for treating diabetes and hyperglycemia. In China, it is widely used for managing blood cancers, hypertension, lymphoma, and other ailments. In Central and South America, it is a common household remedy for colds, inflammation, and sore throats. Modern studies continue to support and expand the therapeutic potential of *C. roseus* (Ross 2003; Das and Sharangi 2017; Kumar et al. 2022; Senbagalakshmi et al. 2017).

3.2.2 *Catharanthus pusillus*

Catharanthus pusillus is widely recognized in traditional medicine for its broad therapeutic applications. Various parts of the plant—including roots, leaves, and latex—are utilized to manage a wide range of ailments. It is traditionally used to treat skin and liver diseases, leprosy, dysentery, ulcers, tumors, earaches, intestinal worms, and childhood meningitis. Additionally, it is employed for conditions such as heart disease, rheumatism, infantile malnutrition, hernia, dyspepsia, and asthma. In certain practices, individuals suffering from hysteria consume approximately 20 g of fruit powder daily for 3 months as part of a therapeutic regimen. A boiled decoction of the dried plant in oil is also commonly used for treating various ailments, either applied topically or ingested depending on the condition being addressed. These diverse uses highlight the plant's prominence in ethnomedicine and its potential for future pharmacological exploration (El-Sayed and Cordell 1981).

3.2.3 *Catharanthus coriaceus*

The infusion or decoction obtained from leaves are used to manage problems related excessive bile like bilious fevers and diarrhea. The bitter blossoms are taken by individuals with diabetes, who masticate them to suppress their hunger (https://tropical.theferns.info).

3.2.4 *Catharanthus lanceus*

C. lanceus is traditionally used for its diverse medicinal properties. A decoction of the root is taken orally to relieve toothache and reduce fever. Additionally, it is employed as a purgative to cleanse the digestive system, as an oxytocic to stimulate uterine contractions, and as a vermifuge to expel intestinal worms. These applications reflect the plant's significance in traditional medicine systems (Source: tropical.theferns.info).

3.2.5 *Catharanthus longifolius*

C. longifolius is traditionally valued for its medicinal properties, particularly for its anthelmintic (worm-expelling) use. In local healing practices, preparations made from the aerial parts of the plant—typically in the form of extracts—are administered to eliminate intestinal worms. This remedy is commonly employed in regions where access to conventional deworming medications is limited. The plant's efficacy in treating parasitic infections highlights its importance in ethnobotanical medicine and supports its continued use in traditional healthcare systems. Further research may help validate and expand its potential applications in modern phytotherapy (Source: tropical.theferns.info).

3.2.6 *Catharanthus trichophyllus*

C. trichophyllus holds significant traditional medicinal value, particularly in Madagascar, parts of Africa, and India. In these regions, various parts of the plant—especially the leaves and aerial parts—are commonly prepared as decoctions or infusions and used in the management of blood sugar levels and treatment of malaria. These traditional remedies are integral to local health practices and are often employed in areas with limited access to modern healthcare. In addition to its internal applications, the plant is also valued for its topical use. Crushed leaves or paste derived from them are applied directly to the skin for treating various skin ailments. This external application is known to help alleviate symptoms of skin diseases, as documented by Davioud et al. (1989) and Sevestre-Rigouzzo et al. (1992). The wide range of therapeutic applications of *C. trichophyllus* underscores its importance in ethnomedicine and supports continued interest in its pharmacological potential (https://tropical.theferns.info).

3.2.7 *Catharanthus makayensis*

C. makayensis, an endemic species native to Madagascar, holds potential medicinal value within traditional healing practices. Though less studied than other *Catharanthus* species, it is believed to possess bioactive compounds with therapeutic properties. Local healers have used extracts from its leaves and stems to treat common ailments such as inflammation, skin irritations, and digestive disturbances. Preliminary phytochemical analyses suggest the presence of alkaloids, indicating possible pharmacological relevance similar to *C. roseus*. However, comprehensive scientific validation is still lacking. Continued ethnobotanical and pharmacological research may uncover valuable compounds for use in modern therapeutic applications (https://tropical.theferns.info).

3.2.8 *Catharanthus ovalis*

C. ovalis is a lesser-known species within the *Catharanthus* genus, traditionally recognized for its unique and varied medicinal properties. In some cultural practices, extracts of this plant were historically consumed as a trial poison—likely to assess their potency or as part of ritualistic practices (Stöckigt and Soll 1980). Despite this, the plant has demonstrated potential therapeutic uses. Ethnobotanical reports indicate that extracts from the aerial parts of *C. ovalis* are effective in expelling intestinal worms, highlighting its anthelmintic properties (Shahriar et al. 2024). This suggests that the plant may contain bioactive compounds capable of disrupting parasitic life cycles. Although scientific research on *C. ovalis* remains limited compared to its relative *C. roseus*, its traditional applications point to possible pharmacological value. Further studies are needed to isolate and identify its active constituents and evaluate their efficacy and safety for use in modern medicine (https://tropical.theferns.info).

3.2.9 *Catharanthus scitulus*

C. scitulus, a lesser-known species within the *Catharanthus* genus, possesses medicinal potential similar to its relative *C. roseus*. Traditionally, it has been used in folk remedies for its antimicrobial, anti-inflammatory, and wound-healing properties. Preparations made from its leaves and stems are applied to treat skin infections, minor cuts, and inflammation. Some ethnobotanical reports suggest its use in managing blood sugar levels and improving digestion. While not as extensively studied as *C. roseus*, *C. scitulus* holds promise for its bioactive compounds, making it a potential candidate for further pharmacological and therapeutic research (https://tropical.theferns.info).

References

Aslam J, Khan SH, Siddiqui ZH, Fatima Z, Maqsood M, Bhat MA, Nasim SA, Ilah A, Ahmad IZ, Khan SA, Mujib A (2010a) Catharanthus roseus (L.) G. Don. An important drug: its applications and production. Pharm Glob 4(12):1–16

Aslam J, Mujib A, Fatima Z, Sharma MP (2010b) Variations in vinblastine production at different stages of somatic embryogenesis, embryo, and field-grown plantlets of Catharanthus roseus (L.) G. Don, as revealed by HPLC. In Vitro Cell Dev Biol – Plant 46(4):348–353

Chandran A, Saj OP (2015) Phytochemical studies on the medicinal herb Catharanthus pusillus. World J Pharm Res 5(1):152–171

Chaturvedi V, Goyal S, Mukim M, Meghani M, Patwekar F, Patwekar M, Khan SK, Sharma GN (2022) A comprehensive review on Catharanthus roseus L. (G.) Don: clinical pharmacology, ethnopharmacology and phytochemistry. J Pharmacol Res Dev 4(2):17–36

Das S, Sharangi AB (2017) Madagascar periwinkle (Catharanthus roseus L.): diverse medicinal and therapeutic benefits to humankind. J Pharmacogn Phytochem 6(5):1695–1701

Davioud E, Kan C, Hamon J, Tempé J, Husson HP (1989) Production of indole alkaloids by in vitro root cultures from Catharanthus trichophyllus. Phytochemistry 28(10):2675–2680

El-Sayed A, Cordell GA (1981) Catharanthus alkaloids. XXXIV. Catharanthamine, a new antitumor bisindole alkaloid from Catharanthus roseus. J Nat Prod 44(3):289–293

Govindasamy C, Srinivasan R (2012) In vitro antibacterial activity and phytochemical analysis of Catharanthus roseus (Linn.) G. Don. Asian Pac J Trop Biomed 2(1):155–158

Kumar S, Singh B, Singh R (2022) Catharanthus roseus (L.) G. Don: a review of its ethnobotany, phytochemistry, ethnopharmacology and toxicities. J Ethnopharmacol 284:e114647

Nammi S, Boini MK, Lodagala SD, Behara RBS (2003) The juice of fresh leaves of Catharanthus roseus Linn. reduces blood glucose in normal and alloxan diabetic rabbits. BMC Complement Altern Med 3(1):1–4

Nisat UT, Jahan N, Afrin SR, Akter B, Nahar A, Islam MR, Hossain MK (2024) Phytochemical, ethnopharmacological, and medicinal importance of Catharanthus roseus (Apocyanaceae): a mini-review. South Asian Res J Nat Prod 7:87–101

Rasineni K, Bellamkonda R, Singareddy SR, Desireddy S (2010) Antihyperglycemic activity of Catharanthus roseus leaf powder in streptozotocin-induced diabetic rats. Pharm Res 2(3):195–201

Ross IA (2003) Catharanthus roseus. In: Medicinal plants of the world. Humana Press, Totowa, pp 175–195

Senbagalakshmi P, Rao MV, Kumar TS (2017) In vitro studies, biosynthesis of secondary metabolites and pharmacological utility of Catharanthus roseus (L.) G. Don.: a review. In: Catharanthus roseus. Springer, Cham, pp 153–199

Sevestre-Rigouzzo M, Nef-Campa C, Ghesquière A, Chrestin H (1992) Genetic diversity and alkaloid production in Catharanthus roseus, C. trichophyllus and their hybrids. Euphytica 66(1):151–159

Shahriar S, Shermin SA, Hasnat H, Hossain F, Han A, Geng P, Alam S, Mamun AA (2024) Chemico-pharmacological evaluation of the methanolic leaf extract of Catharanthus ovalis: GC–MS/MS, in vivo, in vitro, and in silico approaches. Front Pharmacol 15:e1347069

Stöckigt J, Soll HJ (1980) Indole alkaloids from cell suspension cultures of Catharanthus roseus and C. ovalis. Planta Med 40(9):22–30

Separation and Isolation Techniques: From Traditional to Modern Methods

4

Abstract

Separation and isolation techniques are fundamental to the drug discovery programs pertaining to the identification and development of bioactive compounds from natural and synthetic sources. Traditionally, these methods relied heavily on simple processes such as solvent extraction, distillation, precipitation, and filtration, which were often labor-intensive and limited in selectivity and efficiency. Over time, these classical approaches laid the groundwork for more refined and sophisticated techniques. Modern advancements have introduced a wide array of precise and efficient methodologies, including chromatography (e.g., TLC, HPTLC, HPLC, GC), electrophoresis, membrane filtration, solid-phase extraction, and supercritical fluid extraction. These techniques enable the separation of complex mixtures into pure components based on differences in physical and chemical properties such as polarity, size, charge, and solubility. The integration of modern analytical tools like MS and NMR (Nuclear Magnetic Resonance) spectroscopy has further enhanced the capability to isolate, identify, and characterize compounds with high accuracy and sensitivity. This progression from traditional to modern methods has significantly improved the scope and efficiency of compound isolation, supporting advancements in pharmaceuticals, biotechnology, environmental science, and natural product research. The evolution of separation techniques reflects the growing demand for accuracy, speed, and scalability in both academic and industrial settings, ultimately contributing to innovation in drug discovery and material science.

Keywords

Separation techniques · HPTLC · HPLC · GC · Soxhlet extraction · Drug discovery

S. Kumar, B. Singh, *Catharanthus in Drug Discovery*,
https://doi.org/10.1007/978-981-95-5048-7_4

4.1 Introduction

Drug discovery has evolved into a complex, interdisciplinary, and highly structured process that integrates scientific innovation, technological advancements, and strict regulatory compliance to develop and optimize safe and effective therapeutic candidates. Herbal drugs, derived from plants, are inherently complex matrices composed of hundreds of known and unknown phytochemicals, rather than a single active component. While some phytochemicals are present in relatively high concentrations, many others occur in trace amounts. The compound responsible for the therapeutic effect may be either known or unidentified, and in many cases, the pharmacological activity arises from the synergistic action of multiple macro- and microlevel phytochemicals. Hence, analyzing multiple constituents is considered more rational and effective for quality control. Herbal formulations may consist of single herbs or mixtures, each contributing a multitude of chemical constituents, making their analysis particularly challenging. It is no longer sufficient to focus solely on a single active compound when assessing therapeutic efficacy. Therefore, the identification, quantification, and isolation of various classes of bioactive phytochemicals have become essential. This demands the application of modern analytical instrumentation techniques, which offer high precision, sensitivity, and efficiency, thereby enabling comprehensive profiling and standardization of herbal drugs for both research and clinical use (Korfmacher 2005; Locatelli et al. 2012; Sinha et al. 2023).

Modern analytical techniques have significantly advanced phytochemical investigations, surpassing traditional methods in isolating and characterizing a wide range of secondary metabolites present in plant extracts. Despite this progress, only about 10% of higher plant species have been chemically characterized using these advanced approaches. Rapid screening, identification, and quantification of bioactive phytochemicals from medicinal plants or herbal formulations are now essential steps in the efficient isolation and analysis of compounds from crude extracts. A critical step in this process is *dereplication*, which involves identifying known compounds in crude extracts before isolation, thereby avoiding redundant, time-consuming procedures and excessive solvent use. Recent technological developments in purification, isolation, and structural elucidation have enabled the formulation of effective strategies for the quality control and standardization of herbal products. These strategies aim to maintain the chemical consistency and homogeneity of plant extracts, ensuring reliability and efficacy in derived products or formulations. A wide array of analytical methods, including chromatography, spectroscopy, and mass spectrometry, is now employed for phytochemical analysis, pharmacological research, and product standardization. Metabolomics has emerged as a powerful tool in natural product research, emphasizing the comprehensive analysis of the metabolome—the complete set of low- and intermediate-molecular-weight compounds in a biological sample. Within metabolomics, two major approaches are employed: *metabolic profiling*, which targets the identification and quantification of specific groups of metabolites, and *metabolic fingerprinting*, which classifies samples based on their biological origin or relevance. These approaches have broad applications in the quality control of herbal drugs, disease research, and toxicity evaluation, offering deep insights into the complex nature of plant-based therapeutics (Najmi et al. 2022; Nasim et al. 2022).

Spectroscopic, spectrometric, and chromatographic techniques are currently very useful and fundamental steps for the identification and assessment of components present in a complex matrix. Phytochemical analysis is usually performed using typical techniques such as TLC, HPTLC, HPLC, GC, and, more recently, MS and NMR. Recently, significant improvements have been made in the separation and resolution of analytical techniques. High-resolution instruments are now becoming available and welcomed by a broader group of researchers. The holistic approach needs a more diverse procedure than traditional analysis, as instead of focusing on a few selected and targeted compounds, it is imperative to generate a profile of a large number of constituents, aiming at the identification of novel biomarkers and drug targets. LC-MS is an advanced and most popular technique for metabolite profiling and fingerprinting. The elemental composition of a compound or phytochemical is one of the most valuable tools of information for identification and characterization purposes. Accurate mass measurements, which means, by definition, that the measured mass should be within 5 ppm of the theoretical mass, can be obtained from the latest generation. Chromatographic procedures are often preferred for a detailed analysis of the metabolome from a complex matrix. This technique has been automated and coupled with online separation, along with detection such as GC–MS or HPLC–MS. Advances in hyphenated GC–MS or HPLC–MS can improve the quality control of natural products through qualitative and quantitative analysis. High-throughput analysis and miniaturization by GC–MS or HPLC–MS that provide high resolution in a short analysis time with low operational costs are desirable. The introduction of sub-2-μm columns for HPLC and the development of UHPLC proved important steps forward for crude extract profiling. The speed of separation provided by these new separation methods challenges the liquid chromatography detectors that have to provide faster responses (Salim et al. 2008; Javad and Naz 2018; Ahmad Dar et al. 2020).

LC-MS has witnessed remarkable advancements over the past decade, becoming one of the most powerful and versatile tools in analytical chemistry. Numerous ionization techniques have contributed to this progress, including APCI, APPI, laser desorption ionization, and atmospheric pressure laser-induced acoustic desorption chemical ionization (AP/LIAD-CI). Among these, ESI stands out as the most widely used method for ionizing a broad spectrum of phytochemicals due to its efficiency and versatility. A significant leap in LC-MS development has come from the introduction of ambient ionization techniques, which allow direct analysis of samples in open environments outside the mass spectrometer, thus minimizing sample preparation. Additionally, substantial improvements in mass resolution, accuracy, and separation capabilities—particularly through the integration of UHPLC—have been achieved. Current instruments can reach up to 24 million resolutions, with mass accuracies in the subparts per billion (ppb) range, while user-friendly systems commonly deliver resolutions above 40,000 and mass accuracies below 1 ppm. HPLC–MS systems have become increasingly accessible due to simplified operation, enhanced instrument robustness, and reduced maintenance requirements, making them ideal for routine use across research laboratories. Their sensitivity, reliability, and cost-effectiveness have made MS a standard tool in natural product research. Moreover, the rapid expansion of "-omics" fields—such as metabolomics and genomics—has placed new demands on analytical techniques, requiring the ability to detect and quantify vast numbers of metabolites within complex matrices. To meet these demands, both preparative and nonpreparative approaches must be employed complementarily, as no single method alone can meet the full spectrum of sensitivity, specificity, and throughput required in modern analytical science (Tolonen et al. 2009; Yang et al. 2009a, b; Zhang et al. 2018).

4.2 Extraction, Isolation, and Characterization of Phytochemicals

Extraction, isolation, and characterization of bioactive compounds are essential steps following the collection and appropriate drying of plant material, forming the foundation of natural product research. These processes help in identifying therapeutic agents present in medicinal plants. Broadly, the techniques used can be categorized into two types: traditional and modern (or advanced) methods. Traditional methods include maceration, decoction, infusion, and Soxhlet extraction, which have been used for centuries in herbal medicine. Modern techniques, on the other hand, involve advanced tools like ultrasound-assisted extraction (UAE), microwave-assisted extraction (MAE), supercritical fluid extraction (SFE), pressurized liquid extraction (PLE), and enzyme-assisted extraction (EAE) and chromatographic and spectroscopic methods for precise isolation and structural characterization.

4.2.1 Traditional Methods for Extraction, Isolation, and Characterization

Since ancient times, traditional methods for the extraction, isolation, and characterization of phytochemicals have served as the foundation of pharmacognosy and phytochemistry. These techniques have played a crucial role in guiding early drug discovery and remain valuable tools today. Even with the advent of advanced analytical technologies such as HPLC and HPLC–MS, traditional healers and researchers continue to rely on classical methods for identifying plant-based compounds. Though relatively simple, these approaches laid the groundwork for modern natural product chemistry and are still useful in resource-limited settings or for preliminary screening. Isolation refers to the process of separating individual compounds or groups of compounds from a complex plant matrix. Conventional methods rely on chemical and physical properties such as polarity, solubility, volatility, and reactivity for separation. Characterization was traditionally performed using organoleptic, chemical, and physical techniques to determine a compound's class, structure, and purity. However, these methods are often time-consuming, labor-intensive, and lack specificity and sensitivity, making it difficult to resolve the complex matrices and structural intricacies found in plant extracts (Harborne 1998; Raaman 2006).

4.2.1.1 Maceration

Maceration is a simple and cost-effective method of extraction commonly used in phytochemical studies. It involves soaking dried and powdered plant material in a suitable solvent—such as water, ethanol, or methanol—at room temperature for several hours or days. Occasional shaking or stirring in a conical flask helps improve the extraction yield. Although widely used due to its ease and minimal equipment requirements, maceration is time-consuming and generally less efficient, especially for extracting thermo-labile or poorly soluble compounds. It is most suitable for initial, small-scale extractions in traditional or resource-limited settings.

4.2.1.2 Percolation

Percolation is a continuous solid–liquid extraction process. It is based on the principle of mass transfer. In this process, phytochemicals dissolve in a suitable solvent according to the polarity of targeted compounds and move under gravitational force. The powdered material is moistened with a small quantity of solvents and packed evenly into a cylindrical percolator (glass or stainless steel). Solvents such as ethanol, methanol, water, or hydroalcoholic mixtures are commonly used as the mobile phase and poured on top and allowed to pass through the bottom with phytochemicals. The solvent used for extraction is known as a menstruum. Percolation extraction is comparatively efficient, ensuring maximum yield with minimal thermal degradation of thermolabile phytochemicals. It requires careful setup and is time-consuming, and not very suitable for very fine powders. It is most commonly used for the preparation of tinctures and herbal extracts for pharmaceutical and cosmetic industries and standardization.

4.2.1.3 Infusion and Decoction

Infusion and decoction are traditional extraction methods used to obtain active phytochemicals from plant materials. In infusion, soft plant parts such as leaves, flowers, and stems are soaked in hot or cold water to extract polar compounds. Hot infusion is more commonly used than cold, as it is quicker and more efficient. Infusion is particularly suitable for volatile phytochemicals and requires minimal equipment. In contrast, decoction involves boiling harder plant parts like roots, bark, seeds, or woody stems in water to extract compounds such as alkaloids, glycosides, and tannins. This method is more effective for dense, woody materials but less suitable for heat-sensitive compounds, as prolonged boiling may degrade them. Decoction is also more time-consuming than infusion. Both techniques are widely used in traditional medicine and serve as foundational methods for preliminary phytochemical extraction, especially in settings with limited access to modern equipment.

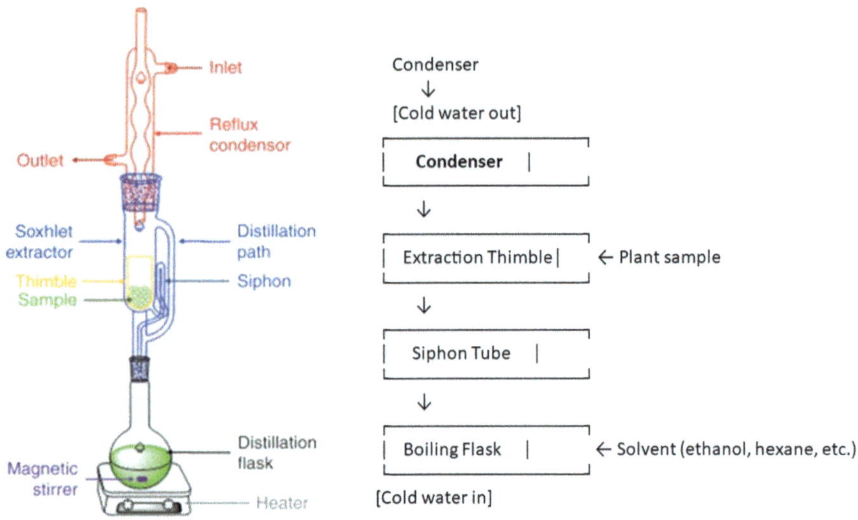

Fig. 4.1 Protocols for the extraction of phytochemicals using Soxhlet extraction techniques

4.2.1.4 Soxhlet Extraction

Soxhlet extraction is a semicontinuous, efficient, and exhaustive technique used for the extraction of phytochemicals from plant materials. It operates on the principle of repeated refluxing and siphoning, allowing the sample to be continuously washed with fresh portions of solvent such as ethanol, hexane, or chloroform. Compared to maceration, it requires less solvent and provides better extraction efficiency.

In this method, the solvent is heated to evaporate, condensed in the condenser, and drips onto the plant sample. Once the extraction chamber fills to a certain level, the solvent carrying the dissolved compounds siphons back into the boiling flask, and the cycle repeats. Soxhlet extraction is suitable for isolating various classes of phytochemicals, including alkaloids, flavonoids, essential oils, and lipids. However, it is time-consuming, consumes large volumes of solvent, and is not ideal for heat-sensitive compounds due to prolonged exposure to elevated temperatures (Fig. 4.1).

4.2.1.5 Thin-Layer Chromatography (TLC)

TLC is a simple, fast, and inexpensive analytical technique used to separate and identify phytochemicals in a mixture of compounds. It is commonly used in natural product chemistry, organic synthesis, herbal drug analysis, pharmaceutical chemistry, food, cosmetics, and forensics. Adsorption chromatography is the foundation of TLC. A stationary phase—typically a thin coating of alumina or silica gel on a plate made of glass, plastic, or aluminum—is subjected to a sample mixture. By means of capillary action, a mobile phase (solvent) ascends the plate and transports the components of the sample at varying rates according to their solubility in the mobile phase and adsorption to the stationary phase. It has low resolution and sensitivity compared to HPTLC or HPLC with manual application, which may reduce reproducibility.

4.2.2 Modern Separation Techniques for Characterization and Isolation

Advanced separation and characterization techniques for purification, qualitative, and quantitative analysis of phytochemicals from complex natural matrices have become indispensable tools in natural product research. These advanced methods offer low to high sensitivity, high selectivity, and throughput, enabling the identification and quantitation of trace quantities of phytochemicals from complex plant extracts. They have significantly improved the workflows for isolation, characterization, and quantitation in terms of accuracy, repeatability, and efficiency (van Beek 1999; Waksmundzka-Hajnos and Sherma 2010; Nasiri et al. 2021; Panda et al. 2022).

4.2.2.1 High-Performance Thin-Layer Chromatography (HPTLC)

HPTLC is a relatively sophisticated and advanced method for separation, identification, and quantitation of phytochemicals from medicinal plants, which offer high-resolution separation, accuracy, reproducibility, and automation. It is widely used for both qualitative and quantitative analysis in the food, pharmaceutical, and herbal industries. The principle of HPTLC is almost the same as that of TLC, while the process of HPTLC is more controlled and precise due to automated instrumentation. The separation of phytochemicals is based on differential affinity toward a stationary phase and a mobile phase. HPTLC plates, sample applicator, chromatography chamber, densitometer scanner, and data software are the main components of the instrument.

4.2.2.2 High-Performance Liquid Chromatography (HPLC)

HPLC is an advanced type of liquid chromatography used to separate, identify, quantify, and purify phytochemicals in a mixture of compounds with high resolution, sensitivity, reproducibility, and precision. It is extensively used in the pharmaceutical industry, biotechnology, environmental chemistry, food industries, and natural products research. HPLC is partition chromatography. The separation of compounds carried out through interactions with the stationary (column packing material) and the mobile phase (solvent system) under high pressure. The interaction with the stationary and the mobile phase depends on polarity, size, charge, and affinity. Compound has more interaction with the stationary phase elute relatively slow. Solvent reservoirs (mobile phase), pump, injector, column, detector, and data system are the main components of an HPLC System. Solvent reservoirs hold solvents such as water, methanol, and acetonitrile allowed to isocratic (constant composition) or gradient (changing composition over time) system. Pump maintains and ensures a steady and consistent flow rate and allows mobile phase high pressure (typically 500–6000 psi). There are mixing of a fixed volume of sample and mobile phase in the injector at constant flow. The column is an important part of HPLC and is packed with a stationary phase material such as silica. The stationary phase material decides the type of column, such as reverse-phase, normal-phase, ion-exchange, and size exclusion. Separated compounds allow to detect and quantifies the chemical constituents present in the sample. Ultraviolet–visible (UV/Vis) (most widely used), fluorescence, refractive index, and mass spectrometer (MS) are the detectors. Mobile phase composition, stationary phase, flow rate, temperature, and sample size are common factors affecting separation and resolution.

4.2.2.3 High-Performance Liquid Chromatography–Mass Spectrometry (HPLC–MS)

HPLC–MS is a powerful hyphenated analytical technique that combines the separation efficiency of HPLC with the precise detection capabilities of MS. This technique allows for the identification and quantification of both polar and nonpolar compounds, including those with high molecular weights, based on their mass-to-charge ratio (m/z). The integration of HPLC with MS offers high sensitivity, selectivity, and the ability to generate structural information about target molecules. Soft ionization methods such as ESI and APCI are most commonly used for the analysis of phytochemicals, as they preserve molecular structure during ionization.

HPLC–MS, coupled with tandem mass spectrometry (MS/MS), is particularly valuable for structural elucidation of complex and diverse phytochemicals. It enables both qualitative and quantitative analysis, as well as dereplication—helping to quickly identify known compounds and avoid redundant isolation. This technique is widely used to analyze secondary metabolites such as alkaloids, flavonoids, phenolic acids, terpenoids, saponins, and glycosides, even when present at trace levels. HPLC–MS plays a vital role across various fields, including natural product research, pharmacognosy, quality control, drug development, and metabolomics, as summarized in Table 4.1.

4.2.2.4 Gas Chromatography–Mass Spectrometry (GC–MS)

Like HPLC–MS, the GC–MS is a powerful hyphenated analytical technique that integrates the separation efficiency of GC with the sensitive detection capabilities of MS. It is widely used for the analysis of volatile and semivolatile phytochemicals, offering excellent sensitivity, selectivity, and reproducibility. In GC–MS, vaporized phytochemicals are injected into a capillary column along with an inert carrier gas such as helium or nitrogen. Compounds are separated based on their volatility and interaction with the column's stationary phase. The main components of a GC–MS system include the injector, GC column, oven, carrier gas, ion source, mass analyzer, detector, and data system. The two most commonly used ionization methods are electron ionization (EI) and chemical ionization (CI). GC–MS is especially valued for its high speed, reliability, and ability to provide detailed structural information, making it an essential tool for phytochemical profiling and quality control of herbal drugs (Table 4.2).

Table 4.1 Applications of HPLC-MS methods in different areas

S. no.	Field	Subfield	Detection, identification, and determination
1.	Pharmaceutical industry	Drug discovery and development	Lead compounds
		Pharmacokinetics and pharmacodynamics (PK/PD)	Absorption, distribution, metabolism, and excretion (ADME)
		Metabolism	Parent drugs and their metabolites
		Adulteration/impurity profiling	Trace-level impurities in drug substances and products
2.	Natural products research	Phytochemical profiling	Secondary metabolites
		Dereplication	Known compounds
		Standardization	Quality control
3.	Proteomics and peptidomics		Proteins and peptides
4.	Clinical and biomedical applications	Therapeutic drug monitoring (TDM)	Drug concentrations
		Biomarker discovery	Metabolic or protein-based biomarkers
		Toxicology	Poisons, drugs of abuse, and toxic metabolites
		Vitamins and steroid hormone analysis	Low-abundance bioactive molecules in blood or serum
5.	Food and nutrition sciences	Food safety and nutritional quality	Food additives, preservatives, colorants, and contaminants
6.	Environmental analysis		Pesticides, herbicides, pharmaceutical residues, and industrial pollutants
7.	Metabolomics		Understand metabolic pathways, stress responses, and disease states.
8.	Forensic science		Illicit drugs, toxic substances, and explosives
9.	Chemical and material science		Polymers, resins, and organic molecules
10.	Regulatory testing	Quality control	Batch-to-batch consistency

Table 4.2 Applications of GC-MS across diverse scientific and industrial domains

Field	Applications
Phytochemistry	Essential oil profiling, volatile metabolite analysis
Pharmaceuticals	Residual solvent analysis, drug metabolism studies
Clinical diagnostics	Biomarkers, steroid analysis, breath testing
Environmental science	Pesticides, POPs, VOCs in air/water/soil
Forensic science	Drugs, explosives, toxins, arson investigation
Food industry	Flavors, additives, contamination, adulteration detection
Materials/chemical industry	Solvent quality, polymer breakdown products

References

Ahmad Dar A, Sangwan PL, Kumar A (2020) Chromatography: an important tool for drug discovery. J Sep Sci 43(1):105–119

Harborne AJ (1998) Phytochemical methods a guide to modern techniques of plant analysis. Springer Science & Business Media, Berlin

Javad S, Naz S (2018) Chromatographical techniques in phytochemical research. In: Phytochemistry. Apple Academic Press, Palm Bay, pp 287–306

Korfmacher WA (2005) Foundation review: principles and applications of LC-MS in new drug discovery. Drug Discov Today 10(20):1357–1367

Locatelli M, Governatori L, Carlucci G, Genovese S, Mollica A, Epifano F (2012) Recent application of analytical methods to phase I and phase II drugs development: a review. Biomed Chromatogr 26(3):283–300

Najmi A, Javed SA, Al Bratty M, Alhazmi HA (2022) Modern approaches in the discovery and development of plant-based natural products and their analogues as potential therapeutic agents. Molecules 27(2):349

Nasim N, Sandeep IS, Mohanty S (2022) Plant-derived natural products for drug discovery: current approaches and prospects. Nucleus 65(3):399–411

Nasiri A, Jahani R, Mokhtari S, Yazdanpanah H, Daraei B, Faizi M, Kobarfard F (2021) Overview, consequences, and strategies for overcoming matrix effects in LC-MS analysis: a critical review. Analyst 146(20):6049–6063

Panda D, Dash BP, Manickam S, Boczkaj G (2022) Recent advancements in LC-MS based analysis of biotoxins: present and future challenges. Mass Spectrom Rev 41(5):766–803

Raaman N (2006) Phytochemical techniques. New India Publishing, New Delhi

Salim AA, Chin YW, Kinghorn AD (2008) Drug discovery from plants. In: Bioactive molecules and medicinal plants. Springer Berlin Heidelberg, Berlin, Heidelberg, pp 1–24

Sinha D, Odoh UE, Ganguly S, Muhammad M, Chatterjee M, Chikeokwu I, Egbuna C (2023) Phytochemistry, history, and progress in drug discovery. In: Phytochemistry, computational tools and databases in drug discovery. Elsevier, Amsterdam, pp 1–26

Tolonen A, Turpeinen M, Pelkonen O (2009) Liquid chromatography–mass spectrometry in in vitro drug metabolite screening. Drug Discov Today 14(3–4):120–133 van Beek TA (1999) Modern methods of secondary product isolation and analysis. Chemicals from plants. Perspectives on plant secondary products. Imperial College Press and World Scientific Publishing, Singapore, pp 91–186

Waksmundzka-Hajnos M, Sherma J (eds) (2010) High performance liquid chromatography in phytochemical analysis. CRC Press, Boca Raton

Yang D, Pornpattananangkul D, Nakatsuji T, Chan M, Carson D, Huang CM, Zhang L (2009a) The antimicrobial activity of liposomal lauric acids against Propionibacterium acnes. Biomaterials 30(30):6035–6040

Yang M, Sun J, Lu Z, Chen G, Guan S, Liu X, Jiang B, Ye M, Guo DA (2009b) Phytochemical analysis of traditional Chinese medicine using liquid chromatography coupled with mass spectrometry. J Chromatogr A 1216(11):2045–2062

Zhang A, Sun H, Wang X (2018) Mass spectrometry-driven drug discovery for development of herbal medicine. Mass Spectrom Rev 37(3):307–320

Monoterpene Indole Alkaloids: Identification, Characterization, and Isolation Aspects

5

Abstract

Monoterpene indole alkaloids (MIAs), a prominent subgroup of terpene indole alkaloids (TIAs), exhibit remarkable structural diversity and significant pharmacological potential. Found predominantly in plants from the *Apocynaceae*, *Loganiaceae*, and *Rubiaceae* families, MIAs are biosynthesized through the condensation of tryptamine and secologanin, catalyzed by strictosidine synthase. This reaction leads to the formation of strictosidine, a central intermediate that diversifies into a wide range of MIAs. This communication provides a comprehensive overview of the analytical methodologies used for the identification, characterization, and isolation of MIAs. Modern analytical techniques such as HPTLC, HPLC, UHPLC, GC, and their coupling with mass spectrometry (HPLC–MS and GC–MS) are widely utilized. These methods offer high sensitivity, specificity, and resolution, enabling detailed profiling of complex plant extracts. Such approaches have proven effective in both qualitative and quantitative assessments of MIAs, aiding in the discovery of novel compounds and their therapeutic applications. We emphasize their role in isolating key MIAs such as vincristine, vinblastine, ajmalicine, and serpentine from *Catharanthus* and related species, thereby contributing to advancements in natural product chemistry and drug development.

Keywords

Monoterpene indole alkaloids · *Catharanthus* · Analytical techniques

S. Kumar, B. Singh, *Catharanthus in Drug Discovery*,
https://doi.org/10.1007/978-981-95-5048-7_5

5.1 Introduction

Monoterpene indole alkaloids (MIAs) or Terpene indole alkaloids (TIAs) represent a highly diverse and significant subclass of indole alkaloids, encompassing over 2000 known compounds. These specialized secondary metabolites are primarily isolated from plant families such as *Apocynaceae*, *Loganiaceae*, and *Rubiaceae*. Biosynthetically, TIAs originate from the condensation of two precursor molecules—tryptamine and secologanin—catalyzed by the key enzyme strictosidine synthase, which plays a central role in the formation of the intermediate strictosidine (O'Connor and Maresh 2006; Shigemori et al. 2003; DeLuca and Kurz 1988; Düsman et al. 2004). This intermediate then undergoes a complex series of transformations leading to a broad spectrum of bioactive alkaloids. Numerous analytical methodologies have been developed and refined to facilitate the identification, characterization, and quantification of TIAs in plant extracts. Common techniques include HPTLC, HPLC, UHPLC, GC, and their coupling with mass spectrometry (HPLC–MS and GC–MS). These tools enable precise separation and structural elucidation of TIAs, even in complex mixtures. In our previous investigation, we successfully identified and characterized a total of 72 TIAs from crude plant extracts using advanced HPLC–MS/MS (high-performance liquid chromatography–tandem mass spectrometry) analysis. Among the *Catharanthus* species, only a limited number of TIAs—approximately 130—have been reported to date. Notable compounds include vincristine, vinblastine, vindesine, vinorelbine, catharanthine, vindoline, ajmalicine (raubasine), serpentine, tabersonine, lochnericine, 19-hydroxy-tabersonine, 16-methoxy-2,3-dihydro-3-hydroxytabersonine, ajmaline, tetrahydroalstonine, and horhammericine. This chapter provides a comprehensive overview of the analytical strategies employed for the study of TIAs. We summarize various approaches, including HPTLC, HPLC, UHPLC, GC, HPLC–MS, and GC–MS, focusing on their applications in the qualitative and quantitative analysis of TIAs across multiple plant sources.

5.2 Qualitative Analysis of Monoterpene Indole Alkaloids

A synthesis of expertise in chromatography, mass spectrometry, and phytochemistry is essential for the analysis of HPLC–MS/MS spectra. Fundamental protocols include the examination of chromatographic separation, identification of parent ions (precursor ions such as $[M + H]^+$ or $[M–H]^-$), analysis of fragmentation patterns (MS/MS), determination of accurate mass and isotopic distributions, comparison with libraries and databases, validation against standards (when accessible), utilization of diagnostic ions or neutral loss scans, and consideration of source-specific phytochemistry. Metabolite assignments were performed by analyzing the retention time, MS, and MS/MS data, which encompassed accurate mass, isotopic distribution, and fragmentation patterns in positive ion modes, of the identified chemicals as recorded in the literature. Two principal categories of notable fragment ions were produced throughout the fragmentation processes. The biosynthetic routes demonstrate that monoterpene indole alkaloids comprise two constituents: indole and terpene. Two types of fragment ions related to the indole and terpene parts are identified for TIAs. Spectroscopic and spectrometric parameters are presented in Table 5.1.

5.2.1 Identification and Characterization of Monoterpene Indole Alkaloids by Mass Spectrometry

There are many methods that have been reported for the identification and characterization of reported on the basis of MS and MS/MS (Zhou et al. 2005; Chen et al. 2013; Zhang et al. 2014; Chen et al. 2017; Kumar et al. 2018a; Jeong and Lim 2018). Ajmalicine, catharanthine, tetra-hydroalstonine, akuammigine, catharanthine, 3-isoajmalicine, pleiocarpamine, akuammiline, tabersonine, 20-hydroxytabersonine, lochnericine, horhammericine, serpentine, minovincinine, vindolinine, 20-epivindolinine, vindolinine-*N* t-oxide, and 20-epivindolinine-Nb-oxidewere were detected in positive ionization mode as a $[M + H]^+$ in extracts and *C. roseus* tissue cultures using LC-MS (Fernández-Pérez et al. 2013; Auriola et al. 1990; Schiel et al. 1987; O'Connor 2012; Kumar et al. 2018a; Shokeen et al. 2007; Kutney et al. 1980; Deus-Neumann and Zenk 1984; Hanafy et al. 2016). Vindoline, catharanthine, serpentine, and ajmalicine were identified and characterized in *C. roseus* cell line C20hi using ultra-performance liquid chromatography–mass spectrometry (UPLC-MS) by comparisons of the MS/MS spectra and the retention times with those of authentic standards. The $[M + H]^+$ ions of vindoline, serpentine, catharanthine, and ajmalicine were reported at *m/z* 457, 349, 337, and 353 (He et al. 2011a, b; Chen et al. 2011, 2013; Zhang et al. 2017a, b; Cao et al. 2005). Lochnericine, 21-hydroxycyclolochnerine, and 19(S)-epimisiline were isolated from hairy root cultures and root leaf and stems of *C. roseus* and identified and characterized by mass spectrometry. The molecular ion peak for 19(S)-epimisiline was observed at *m/z* 368, and its molecular formula is $C_{21}H_{24}N_2O_4$, and the fragment ions at *m/z* 214 and *m/z* 154 suggested the characteristic of aspidosperma-type alkaloids (Stöckigt et al. 2002; Akhgari et al. 2015). So electron impact mass spectrometry (EIMS) of 19(*S*)-epimisiline confirmed a vincadifformine-like skeleton (Peraza-Sánchez et al. 1998; Wesołowska et al. 2016; Gorman et al. 1959; Wang et al. 2015; Moza et al. 1964; Fernández-Pérez et al. 2013). Lochnericine showed a protonated ion at *m/z* at 353 (Carqueijeiro et al. 2018). Catharanthine, tabersonine, serpentine, and ajmalicine have showen protonated adducts at *m/z* 337, 349, and 353, respectively. Further study showed the fragment ion at *m/z* 305 from *m/z* 337 may indicate the existence of tabersonine, an isomer of catharanthine, and the fragment ion at *m/z* 321 from *m/z* 353 may indicate the existence of an isomer of ajmalicine (Auriola et al. 1990). The MS/MS analysis of catharanthine or tabersonine strongly suggested that the peak having the *m/z* 337.19 that localized in the whole stem tissue is catharanthine. MS/MS fragments corresponding to serpentine (*m/z* 349.15) and ajmalicine (*m/z* 353.18) were also detected in single-cell MS and LC–MS/MS (liquid chromatography–tandem mass spectrometry) analyses. In the *m/z*

Table 5.1 HPLC–MS and MS/MS data of different extracts of *Catharanthus* species

S. no	RT (min)	Compounds	Mol. formula	Molecular weight		Error (ppm)	Fragment ions	Distribution (like leaf, stem)	Reference
				Calculated	Observed				
1.	2.83	10-Hydroxynortetraphyllicine	$C_{19}H_{22}N_2O_2$	311.1754	311.1753	0.04	293.1633 (100), 185.0818 (7), 160.0733 (32), 146.0946 (45), 138.0914 (57), 120.0837 (46)		Kumar et al. (2018a)
2.	7.87	10-Hydroxynortetraphyllicine (isomer)	$C_{19}H_{22}N_2O_2$	311.1754	311.1754	0.14	293.1633 (100), 279.1549 (5), 262.1194 (8), 249.1374 (52), 222.1137 (11), 205.0676 (9), 194.0915 (25), 192.1156 (39), 167.0913 (13), 144.0796 (30)		Kumar et al. (2018a)
3.	4.17	3,4,5,6-Tetradehydroyohimbine	$C_{21}H_{22}N_2O_3$	351.1703	351.1704	−0.29	333.5954 (6), 321.1587 (76) 320.1614 (100), 308.1617 (47), 288.1331 (23), 260.1348 (6), 248.1382 (15)		Kumar et al. (2018a)
4.	7.36	3,4,5,6-Tetradehydroyohimbine (isomer)	$C_{21}H_{22}N_2O_3$	351.1703	351.1704	−0.07	333.1773 (5), 321.1586 (37), 289.1318 (100), 261.1349 (14), 247.1213 (49), 233.1042 (13), 207.0907 (42), 182.0817 (33), 144.0792 (5)		Kumar et al. (2018a)
5.	7.7	3,4,5,6-Tetradehydroyohimbine (isomer)	$C_{21}H_{22}N_2O_3$	351.1703	351.1705	−0.35	321.1586 (37), 289.1318 (100), 247.1218 (49), 207.0907 (42), 182.0817 (33), 144.0792 (5)		Kumar et al. (2018a)
6.	5.4	Ajmalicine	$C_{21}H_{24}N_2O_3$	353.1860	353.1863	1	–	Root, stem, and leaf	Jeong and Lim (2018)
7.	6.2	Ajmalicine	$C_{21}H_{24}N_2O_3$	353.1860	353.2	<5	144.1 (100), 160.1 (34.6), 174.1 (26.4)	Root, stem, and leaf	Zhang et al. (2014)
8.	8.3	Ajmalicine	$C_{21}H_{24}N_2O_3$	353.1860	353.2	<5	174.1 (27.5), 160.1 (38.3), 144.1 (100)	Leaf, root, and stem	Chen et al. (2013)

9.	2.98	Ajmalicine	$C_{21}H_{24}N_2O_3$	353.1860	353.1864	1.9	335.1754 (26.7), 144.0809 (100)	Stem and root	Jeong and Lim (2018)
10.	~6.4	Ajmalicine	$C_{21}H_{24}N_2O_3$	353.1860	353.2	<5	335.1 (29), 144.1 (100)	Stem	Zweil et al. (2019)
11.	~3.9	Ajmalicine	$C_{21}H_{24}N_2O_3$	353.1860	353.2	<5	335.2 (32), 144.1 (100)	Leaf and stem	Koel et al. (2020)
12.	~4.5	Ajmalicine	$C_{21}H_{24}N_2O_3$	353.1860	353.2	–	335.1 (42), 144.1 (100)	Leaf and stem	Verma et al. (2008)
13.	~6.2	Ajmalicine	$C_{21}H_{24}N_2O_3$	353.1860	353	–	335, 144 (major)	Leaf extract	Balsevich et al. (1988)
14.	~5.2	Ajmalicine	$C_{21}H_{24}N_2O_3$	353.1860	353.2	–	335.1, 144.1	Cambial meristematic cells	Verma et al. (2008)
15.	9.64	Ajmalicine (isomer)	$C_{21}H_{24}N_2O_3$	353.186	353.1864	−1.12	321.1616 (4), 222.1127 (9), 210.1093 (15), 199.1200 (3), 178.0839 (5), 144.0799 (100)		Kumar et al. (2018a)
16.	9.94	Ajmalicine†	$C_{21}H_{24}N_2O_3$	353.186	353.186	−0.19	321.1562 (5), 222.1134 (12), 210.1125 (13), 178.0837 (7), 144.0799 (100)		Kumar et al. (2018a)
17.	9.48	Ajmaline†	$C_{20}H_{26}N_2O_2$	327.2067	327.2066	0.3	309.1964 (5), 239.1554 (15), 210.1287 (30), 194.0983 (27), 182.0979 (35), 158.0974 (88), 144.0817 (100)		Kumar et al. (2018a)
18.	6.9	Alstonine	$C_{21}H_{20}N_2O_3$	348.1474	349.1547	0.2	317.1285 (24), 263.0815 (100), 235.0865 (51)	Root, stem, and leaf	Jeong and Lim (2018)
19.	7.34	Antirhine	$C_{19}H_{24}N_2O$	297.1961	297.196	0.29	154.1229 (15), 144.0798 (100), 131.9302 (8)		Kumar et al. (2018a)
20.	2.8	Aricine	$C_{22}H_{26}N_2O_4$	382.1893	383.197	1.2	369.1814 (1), 340.1949 (24), 321.1605 (49), 296.1284 (90), 160.0754 (16)	Root and stem	Jeong and Lim (2018)

(continued)

Table 5.1 (continued)

S. no	RT (min)	Compounds	Mol. formula	Molecular weight		Error (ppm)	Fragment ions	Distribution (like leaf, stem)	Reference
				Calculated	Observed				
21.	6.1	Catharanthine	$C_{21}H_{24}N_2O_2$	336.1838	337.1908	−0.8	–	Root, stem, and leaf	Jeong and Lim (2018)
22.	4.7	Catharanthine	$C_{21}H_{24}N_2O_2$	336.1838	337.2	<5	308.2 (35.4), 264.1 (26.5), 144.1 (100)	Root and stem	Zhang et al. (2014)
23.	7.6	Catharanthine	$C_{21}H_{24}N_2O_2$	336.1838	337.2	<5	308.1 (53.4), 264.1 (29.1), 144.1 (100)	Root and stem	Chen et al. (2013)
24.	~2.6	Catharanthine	$C_{21}H_{24}N_2O_2$	336.1838	337.1905	−0.9	308.2 (74), 264.1 (38), 144.1 (100)	Root and stem	Kumar et al. (2018a)
25.	1.85	Catharanthine	$C_{21}H_{24}N_2O_2$	336.1838	337.1912	2	319.1802 (28.5), 144.0812 (100), 109.0632 (41.3)	Leaf and stem	Jeong and Lim (2018)
26.	~2.2	Catharanthine	$C_{21}H_{24}N_2O_2$	336.1838	337.2	<5	308.1 (55), 144.1 (100)	C20hi cell line	He et al. (2011a, b)
27.	~3.1	Catharanthine	$C_{21}H_{24}N_2O_2$	336.1838	337.2	<5	308.1 (60), 144.1 (100)	Leaf and stem	Zweil et al. (2019)
28.	~2.1	Catharanthine	$C_{21}H_{24}N_2O_2$	336.1838	337.2	<5	308.1 (68), 144.1 (100)	Leaf extract	Koel et al. (2020)
29.	~3.9	Catharanthine	$C_{21}H_{24}N_2O_2$	336.1838	337.2	–	308.1 (64), 144.1 (100)	Leaf	Verma et al. (2008)
30.	~4.6	Catharanthine	$C_{21}H_{24}N_2O_2$	336.1838	337	–	308, 144 (major)	Leaf extract	Balsevich et al. (1988)
31.	~6.3	Catharanthine	$C_{21}H_{24}N_2O_2$	336.1838	337	–	308, 144	Leaf and stem	Auriola et al. (1991)
32.	~4.1	Catharanthine	$C_{21}H_{24}N_2O_2$	336.1838	337.2	–	308.1, 144.1	Cambial meristematic cells	Liang et al. (2018)

33.	— (DI)	Catharanthine	$C_{21}H_{24}N_2O_2$	336.1838	337	–	308, 144	Not specified	Koel et al. (2020)
34.	8.4	Catharanthine	$C_{21}H_{24}N_2O_2$	336.1838	337	–	Not specified	Leaf	Mu et al. (2012)
35.	11.02	Catharanthine†	$C_{21}H_{24}N_2O_2$	337.1911	337.191	0.24	305.1648 (5), 173.1074 (8), 165.0893 (13), 144.0800 (100),		Kumar et al. (2018a)
36.	13.27	Cathovalinine	$C_{21}H_{24}N_2O_4$	369.1809	369.1812	−0.85	351.1768 (16), 337.1546 (100), 321.1572 (13), 309.1598 (5), 292.1539 (41), 277.1539 (11), 246.1081 (6)		Kumar et al. (2018a)
37.	9.8	Deacetylcathovaline	$C_{22}H_{28}N_2O_4$	385.2122	385.2122	0.05	367.1990 (3), 335.1734 (16), 325.1894 (75), 158.0945 (100)		Kumar et al. (2018a)
38.	9.46	Deacetylvindoline	$C_{23}H_{30}N_2O_5$	415.2227	415.2234	−1.55	397.2194 (5), 365.1865 (20), 355.2010 (65), 295.1800 (5), 188.1066 (100)		Kumar et al. (2018a)
39.	11.36	Hydroxycathovaline	$C_{24}H_{30}N_2O_6$	443.2177	443.2175	0.37	365.1836 (5), 280.1347 (6), 158.0953 (100)		Kumar et al. (2018a)
40.	3.86	Hydroxytabersonine	$C_{21}H_{24}N_2O_3$	353.186	353.1862	−0.23	335.1732 (12), 321.1588 (100), 303.1522 (10), 277.1375 (11), 247.1224 (16), 228.1025 (38), 216.0966 (13),196.0757 (26), 168.0789 (60), 156.0784 (14), 144.0781 (18)		Kumar et al. (2018a)
41.	5.71	Hydroxytabersonine (isomer)	$C_{21}H_{24}N_2O_3$	353.186	353.1862	−0.65	335.1740 (9), 321.1568 (100), 303.1471 (6), 277.1342 (21), 228.1033 (35), 196.0783 (15), 168.0797 (41)		Kumar et al. (2018a)

(continued)

Table 5.1 (continued)

S. no	RT (min)	Compounds	Mol. formula	Molecular weight		Error (ppm)	Fragment ions	Distribution (like leaf, stem)	Reference
				Calculated	Observed				
42.	7.18	Hydroxytabersonine (isomer)	$C_{21}H_{24}N_2O_3$	353.186	353.1863	−0.86	335.1717 (17), 321.1630 (100), 304.1535 (13), 276.1377 (51), 248.1244 (29), 216.1012 (16), 177.0919 (29), 144.0804 (79), 177.0697 (63)		Kumar et al. (2018a)
43.	7.66	Hydroxytabersonine (isomer)	$C_{21}H_{24}N_2O_3$	353.186	353.186	−0.23	335.103 (9), 321.1499 (15), 293.1382 (14), 249.1363 (100), 234.1241 (11), 220.1001 (27), 198.1088 (22), 169.0656), 144.0776 (24)		Kumar et al. (2018a)
44.	11.13	Hydroxytabersonine (isomer)	$C_{21}H_{24}N_2O_3$	353.186	353.1861	−0.28	335.1737 (5), 224.0946 (30), 210.1101 (16), 184.0790 (13), 144.0804 (100)		Kumar et al. (2018a)
45.	11.94	Hydroxytabersonine (isomer)	$C_{21}H_{24}N_2O_3$	353.186	353.186	−0.22	336.1820 (10), 308.1635 (6), 229.1088 (14), 189.1018 (5), 165.0905 (13), 144.0801 (100)		Kumar et al. (2018a)
46.	12.31	Hydroxytabersonine (isomer)	$C_{21}H_{24}N_2O_3$	353.186	353.1862	−0.23	336.1837 (15), 308.1616 (8), 229.107 (18), 189.1011 (8), 165.0880 (16), 144.0797 (100), 135.1007 (24)		Kumar et al. (2018a)
47.	12.98	Hydroxytabersonine (isomer)	$C_{21}H_{24}N_2O_3$	353.186	353.1861	−0.89	336.1317 (8), 321.1589 (100), 144.0775 (8)		Kumar et al. (2018a)
48.	13.66	Hydroxytabersonine (isomer)	$C_{21}H_{24}N_2O_3$	353.186	353.1863	−1.5	294.1468 (8), 266.1157 (100), 160.0779 (10), 144.0814 (18), 122.0935 (24)		Kumar et al. (2018a)

49.	7.22	Lochneridine	$C_{20}H_{24}N_2O_3$	341.186	341.1861	−0.51	323.1746 (100), 279.1488 (38), 264.1312 (8), 224.1040 (14), 210.0938 (8), 174.0920 (7)		Kumar et al. (2018a)
50.	12.3	Lochneridine (isomer)	$C_{20}H_{24}N_2O_3$	341.186	341.1862	−0.78	309.1577 (100), 264.1595 (34), 235.1204 (7), 208.0982 (11)		Kumar et al. (2018a)
51.	12.45	Lochneridine (isomer)	$C_{20}H_{24}N_2O_3$	341.186	341.1858	0.48	309.1577 (100), 264.1595 (34), 235.1204 (7), 221.1057 (9), 208.0982 (11), 197.0980 (5)		Kumar et al. (2018a)
52.	2.3	Lochnerine	$C_{20}H_{24}N_2O_2$	324.1838	325.1917	2.1	307.1804 (34.2), 277.1336 (15.4), 186.0912 (27.6), 174.0910 (49.4), 146.0673 (43.1), 138.0909 (100)	Stem	Jeong and Lim (2018)
53.	11.62	Minovincinine	$C_{21}H_{26}N_2O_3$	355.2016	355.2015	−0.14	337.1922 (34), 305.1656 (41), 295.1786 (31), 277.1701 (100), 249.1376 (23), 170.0972 (7), 144.0792 (30), 136.1126 (28)		Kumar et al. (2018a)
54.	2.4	Mitraphylline	$C_{25}H_{30}N_2O_7$	368.1736	369.1814	1.4	351.1711 (3), 323.1755 (2.3), 247.1232 (2.3), 168.0805 (100)	Root and stem	Jeong and Lim (2018)
55.	4.77	Norseredamine	$C_{20}H_{24}N_2O_2$	325.1911	325.1912	0.44	307.1766 (100), 277.1329 (10), 172.0729 (5), 160.1041 (17), 152.1068 (33), 148.1096 (12), 144.0763 (5)		Kumar et al. (2018a)
56.	10.57	Norseredamine (isomer)	$C_{20}H_{24}N_2O_2$	325.1911	325.1911	0.22	293.1633 (100)		Kumar et al. (2018a)

(continued)

Table 5.1 (continued)

S. no	RT (min)	Compounds	Mol. formula	Molecular weight		Error (ppm)	Fragment ions	Distribution (like leaf, stem)	Reference
				Calculated	Observed				
57.	4.38	Perivine	$C_{20}H_{22}N_2O_3$	339.1703	339.1708	0.22	321.1570 (7), 307.1452 (19), 279.1504 (21), 262.1204 (19), 248.1071 (23), 234.1550 (19), 220.1128 (21), 184.0746 (48), 166.0844 (39), 144.0798 (100)		Kumar et al. (2018a)
58.	9.7	Perivine (isomer)	$C_{20}H_{22}N_2O_3$	339.1703	339.1704	−0.14	307.1427 (100), 262.1463 (20), 247.1193 (11)		Kumar et al. (2018a)
59.	5.94	Perivine (isomer)	$C_{20}H_{22}N_2O_3$	339.1703	339.1708	−1.56	307.1430 (100)		Kumar et al. (2018a)
60.	8.03	Perivine (isomer)	$C_{20}H_{22}N_2O_3$	339.1703	339.1705	−0.58	323.1716 (14), 305.1719 (15), 295.1773 (21), 279.1456 (37), 262.1147 (21), 238.1225 (100), 223.0940 (19), 212.0937 (56), 199.0864 (44), 186.0901 (26), 174.0942 (39)		Kumar et al. (2018a)
61.	5.8	Pleiocarpamine	$C_{20}H_{22}N_2O_2$	322.1681	323.1759	1.6	294.1494 (31), 263.1545 (51), 234.1271 (100), 204.0809 (100), 177.0959 (31), 122.0959 (30)	Root	Jeong and Lim (2018)
62.	6.51	Pleiocarpamine	$C_{20}H_{22}N_2O_2$	323.1754	323.1756	−0.74	307.1825 (8), 216.1018 (100), 156.0795 (52)		Kumar et al. (2018a)
63.	8.68	Pleiocarpamine	$C_{20}H_{22}N_2O_2$	323.1754	323.1753	0.43	291.1485 (100)		Kumar et al. (2018a)

64.	9.32	Pleiocarpamine (isomer)	$C_{20}H_{22}N_2O_2$	323.1754	323.176	−1.25	293.1644 (100)		Kumar et al. (2018a)
65.	7.3	Reserpine	$C_{33}H_{40}N_2O_9$	608.2739	609.3	<5	397.2 (71.2), 268.1 (31.2), 195.1 (100), 174.1 (42.1), 144.1 (87.3)	Root, stem, and leaf	Zhang et al. (2014)
66.	31.15	Reserpine†	$C_{33}H_{40}N_2O_9$	609.2807	609.2809	0.32	577.2563 (5), 448.1992 (26), 397.2139 (52), 365.1866 (13), 236.1291 (14), 224.1289 (6), 204.1029 (2), 195.0658 (100), 174.0921 (49)		Kumar et al. (2018a)
67.	6.3	Serpentine	$C_{21}H_{20}N_2O_3$	348.1474	349.1548	0.3	–	Root, stem, and leaf	Jeong and Lim (2018)
68.	6.9	Serpentine	$C_{21}H_{20}N_2O_3$	348.1474	349.2	<5	321.1 (19.3), 306.1 (21.1), 144.1 (100)	Leaf and stem	Zhang et al. (2014)
69.	8.9	Serpentine	$C_{21}H_{20}N_2O_3$	348.1474	349.2	<5	321.1 (30.5), 306.1 (24.3), 144.1 (100)	Leaf and stem	Chen et al. (2013)
70.	3.22	Serpentine	$C_{21}H_{20}N_2O_3$	348.1474	349.1551	2.2	321.1481 (36.4), 144.0805 (100)	Stem	Jeong and Lim (2018)
71.	~6.9	Serpentine	$C_{21}H_{20}N_2O_3$	348.1474	349.2	<5	321.1 (22), 144.1 (100)	Leaf and stem	Zweil et al. (2019)
72.	~4.3	Serpentine	$C_{21}H_{20}N_2O_3$	348.1474	349.2	<5	321.2 (28), 144.1 (100)	Leaf and stem	Koel et al. 2020
73.	~5.8	Serpentine	$C_{21}H_{20}N_2O_3$	348.1474	349.2	–	321.1 (29), 144.1 (100)	Leaf	Verma et al. (2008)
74.	~6.8	Serpentine	$C_{21}H_{20}N_2O_3$	348.1474	349	–	321, 144 (major)	Leaf extract	Balsevich et al. (1988)
75.	10.85	Serpentine (isomer)	$C_{21}H_{20}N_2O_3$	349.1547	349.1545	0.52	317.1271 (32), 289.1294 (7), 277.0949 (11), 263.0798 (100), 247.1203 (6), 235.0836 (27), 206.0818 (7)		Kumar et al. (2018a)

(continued)

Table 5.1 (continued)

S. no	RT (min)	Compounds	Mol. formula	Molecular weight		Error (ppm)	Fragment ions	Distribution (like leaf, stem)	Reference
				Calculated	Observed				
76.	6.46	Serpentine derivative	$C_{20}H_{18}N_2O_3$	335.139	335.1391	0.1	317.1325 (53), 291.1485 (18), 263.0832 (100), 235.0889 (34), 206.0895 (10), 195.0865 (6)		Kumar et al. (2018a)
77.	7.39	Serpentine derivative	$C_{20}H_{18}N_2O_3$	335.139	335.1338	0.59	317.1261 (21), 291.1461 (21), 263.0806 (100), 235.0859 (51), 219.0885 (11), 207.0900 (20)		Kumar et al. (2018a)
78.	10.13	Serpentine†	$C_{21}H_{20}N_2O_3$	349.1547	349.1547	0	317.1262 (73), 289.1342 (5), 277.0957 (5), 263.0801 (100), 247.1249 (5), 235.0848 (32), 206.0819 (7), 144.0801 (3)		Kumar et al. (2018a)
79.	1.8	Sitsirkine	$C_{21}H_{26}N_2O_3$	354.1943	355.2013	−0.9	323.1756 (100), 168.0800 (32.5), 144.0803 (45.7), 140.1068 (17.3)	Stem	Jeong and Lim (2018)
80.	5.7	Strychnine	$C_{21}H_{22}N_2O_2$	334.1681	335.1756	0.6	228.1092 (8), 196.0755 (18), 168.0804 (100), 167.0726 (24)	Stem and leaf	Jeong and Lim (2018)
81.	8.5	Tabersonine	$C_{21}H_{24}N_2O_2$	336.1837	337.1912	0.5	305.1651 (44), 277.1698 (9), 249.1387 (20), 228.1018 (25), 196.0754 (25), 168.0804 (100)	Root and stem	Jeong and Lim (2018)
82.	4.1	Tabersonine	$C_{21}H_{24}N_2O_2$	336.1838	337.1915	2.3	305.1683 (42.5), 144.0809 (100)	Leaf and root	Jeong and Lim (2018)
83.	~7.5	Tabersonine	$C_{21}H_{24}N_2O_2$	336.1838	337.2	<5	305.1 (37), 144.1 (100)	Leaf	Zweil et al. (2019)
84.	~5.0	Tabersonine	$C_{21}H_{24}N_2O_2$	336.1838	337.2	<5	305.1 (33), 144.1 (100)	Leaf extract	Koel et al. (2020)

85.	~6.7	Tabersonine	$C_{21}H_{24}N_2O_2$	336.1838	337.2	–	305.1 (31), 144.1 (100)	Leaf	Verma et al. (2008)
86.	~7.3	Tabersonine	$C_{21}H_{24}N_2O_2$	336.1838	337	–	305, 144 (major)	Leaf extract	Balsevich et al. (1988)
87.	~6.5	Tabersonine	$C_{21}H_{24}N_2O_2$	336.1838	337.2	–	305.1, 144.1	Cambial meristematic cells	Liang et al. (2018)
88.	8.3	Tetrahydroalstonine	$C_{21}H_{24}N_2O_3$	352.1787	353.1856	−1.0	293.1651 (11), 168.0802 (100), 167.0722 (15)	Root, stem, and leaf	Jeong and Lim (2018)
89.	5.52	Tetraphyllicine	$C_{20}H_{24}N_2O$	309.1961	309.1961	0.07	291.1869 (100), 276.1622 (10), 261.1434 (14), 152.1062 (66), 148.1135 (37), 144.0820 (17), 136.1125 (38), 135.1034 (71), 134.0956 (42)		Kumar et al. (2018a)
90.	2.76	Trichophylline	$C_{21}H_{22}N_2O_4$	367.1652	367.165	0.74	352.1754 (7), 293.1647 (12), 186.0695 (6), 121.0877 (100)		Kumar et al. (2018a)
91.	4.32	Trichophylline (isomer)	$C_{21}H_{22}N_2O_4$	367.1652	367.1654	−0.45	351.1712 (17), 307.1522 (11), 264.1395 (6), 160.0757 (31), 121.0869 (100)		Kumar et al. (2018a)
92.	5.93	Trichophylline (isomer)	$C_{21}H_{22}N_2O_4$	367.1652	367.1652	0.02	337.1537 (20), 306.1480 (5), 276.1207 (5), 222.1249 (10),206.1147 (11), 180.1008 (100), 156.0802 (22)		Kumar et al. (2018a)
93.	28.69	Trichophylline (isomer)	$C_{21}H_{22}N_2O_4$	367.1652	367.165	−1.5	335.1381 (100), 307.1397 (7), 228.0969 (18), 168.0782 (13)		Kumar et al. (2018a)
94.	5.06	Vellosiminol	$C_{19}H_{22}N_2O$	295.1805	295.1805	0.27	277.1708 (96), 158.0948 (39) 144.0800 (54), 138.0903 (100), 120.0806 (84)		Kumar et al. (2018a)

(continued)

Table 5.1 (continued)

S. no	RT (min)	Compounds	Mol. formula	Molecular weight		Error (ppm)	Fragment ions	Distribution (like leaf, stem)	Reference
				Calculated	Observed				
95.	9.9	Vindolidine	$C_{24}H_{30}N_2O_5$	426.2155	427.2231	0.9	367.2022 (6.4), 158.0963 (100), 143.0722 (4.7)	Stem and leaf	Jeong and Lim (2018)
96.	~8.3	Vindolidine	$C_{24}H_{30}N_2O_5$	426.2155	427.2	<5	367.2 (25), 158.1 (100), 144.1 (32)	Stem	Zweil et al. (2019)
97.	9.7	Vindoline	$C_{25}H_{32}N_2O_6$	456.226	457.2335	0.4	–	Stem and leaf	Jeong and Lim (2018)
98.	5.3	Vindoline	$C_{25}H_{32}N_2O_6$	456.226	457.2	<5	397.2 (62.4), 337.1 (25.7), 144.1 (100)	Leaf and stem	Zhang et al. (2014)
99.	9.4	Vindoline	$C_{25}H_{32}N_2O_6$	456.226	457.2	<5	397.1 (74.8), 337.2 (22.3), 144.1 (100)	Leaf and stem	Chen et al. (2013)
100.	~3.5	Vindoline	$C_{25}H_{32}N_2O_6$	456.226	457.2329	−1.1	397.2 (85), 337.2 (42), 144.1 (100)	Leaf and stem	Kumar et al. (2018a)
101.	2.25	Vindoline	$C_{25}H_{32}N_2O_6$	456.226	457.2341	1.8	397.2013 (100), 337.1715 (56.2), 144.0808 (98.3)	Leaf	Jeong and Lim (2018)
102.	~2.5	Vindoline	$C_{25}H_{32}N_2O_6$	456.226	457.2	<5	397.2 (72), 337.1 (34), 144.1 (100)	C20hi cell line	He et al. (2011a, b)
103.	~4.2	Vindoline	$C_{25}H_{32}N_2O_6$	456.226	457.2	<5	397.2 (78), 337.2 (31), 144.1 (100)	Leaf and stem	Zweil et al. (2019)
104.	~2.6	Vindoline	$C_{25}H_{32}N_2O_6$	456.226	457.2	<5	397.2 (85), 337.2 (38), 144.1 (100)	Leaf extract	Koel et al. (2020)
105.	~7.5	Vindoline	$C_{25}H_{32}N_2O_6$	456.226	–	–	–	Transformed cell cultures (hairy root and suspension lines)	O'Keefe et al. (1997)

106.	~3.2	Vindoline	$C_{25}H_{32}N_2O_6$	456.226	457.2	–	397.2 (72), 337.1 (28), 144.1 (100)	Leaf and stem	Verma et al. (2008)
107.	~5.4	Vindoline	$C_{25}H_{32}N_2O_6$	456.226	457	–	397, 144 (major)	Leaf extract	Balsevich et al. (1988)
108.	~5.1	Vindoline	$C_{25}H_{32}N_2O_6$	456.226	457	–	397, 337, 144	Leaf extract	Auriola et al. (1991)
109.	~3.5	Vindoline	$C_{25}H_{32}N_2O_6$	456.226	457.2	–	397.2 (base), 144.1	Cambial meristematic cells	Liang et al. (2018)
110.	— (DI)	Vindoline	$C_{25}H_{32}N_2O_6$	456.226	457	–	397, 337, 144	Not specified	Zhou et al. (2005)
111.	7.2	Vindoline	$C_{25}H_{32}N_2O_6$	456.226	457	–	Not specified	Leaf	Mu et al. (2012)
112.	13.91	Vindoline†	$C_{25}H_{32}N_2O_6$	457.2333	457.2332	0.24	439.2205 (5), 397.2096 (40), 379.1982 (5), 365.1840 (5), 337.1835 (6), 188.1057 (100), 162.0892 (5)		Kumar et al. (2018a)
113.	2.5	Vindolinine	$C_{21}H_{24}N_2O_2$	336.1842	337.1914	1.1	320.1654 (45), 308.1645 (2)	Root	Jeong and Lim (2018)
114.	6.28	Vindolinine	$C_{21}H_{24}N_2O_2$	337.1916	337.192	−2.82	320.1638 (100), 177.0913 (11), 144.0800 (13), 117.0691 (10)		Kumar et al. (2018a)
115.	5.44	Vindolinine	$C_{21}H_{24}N_2O_2$	337.1911	337.1911	−0.14	305.1648 (45), 180.1015 (100), 144.0806 (7)		Kumar et al. (2018a)
116.	5.63	Vindolinine (isomer) (vinpocetine derivative)	$C_{21}H_{24}N_2O_2$	337.1916	337.1916	−1.63	321.1613 (53), 320.1622 (100), 308.1633 (41), 280.1306 (13), 248.1388 (6), 177.0865 (7), 144.0792 (6)		Kumar et al. (2018a)

(continued)

Table 5.1 (continued)

S. no	RT (min)	Compounds	Mol. formula	Molecular weight		Error (ppm)	Fragment ions	Distribution (like leaf, stem)	Reference
				Calculated	Observed				
117.	6.33	Vindolinine (isomer)	$C_{21}H_{24}N_2O_2$	337.1911	337.191	0.26	320.1638 (100), 308.1653 (5), 276.1355 (5), 177.0898 (12), 144.0803 (14), 117.0693 (13)		Kumar et al. (2018a)
118.	8.61	Vindolinine (isomer)	$C_{21}H_{24}N_2O_2$	337.1916	337.1919	2.42	305.1645 (100)		Kumar et al. (2018a)
119.	9.52	Vindolinine (isomer)	$C_{21}H_{24}N_2O_2$	337.1916	337.1919	2.39	305.1633 (11), 277.1716 (8), 222.1267 (53), 180.1022 (53), 156.0796 (100), 122.0954 (35)		Kumar et al. (2018a)
120.	9.98	Vindolinine (isomer)	$C_{21}H_{24}N_2O_2$	337.1916	337.1918	2.2	307.1439 (12), 277.1665 (9), 246.1259 (12), 222.1255 (45), 206.1144 (20), 180.1007 (64), 170.0950 (19), 156.0793 (100), 144.0789 (29)		Kumar et al. (2018a)
121.	10.03	Vindolinine (isomer)	$C_{21}H_{24}N_2O_2$	337.1911	337.192	2.79	307.1467 (13), 277.1683 (5), 246.1285 (12), 222.1267 (42), 206.1147 (19), 180.1020 (67), 170.0973 (17), 156.0806 (100), 144.0806 (25)		Kumar et al. (2018a)
122.	10.76	Vindolinine (isomer)	$C_{21}H_{24}N_2O_2$	337.1916	337.1919	2.12	173.1073 (5), 165.0884 (7), 144.0794 (100)		Kumar et al. (2018a)
123.	11.66	Vindolinine (isomer)	$C_{21}H_{24}N_2O_2$	337.1916	337.192	2.95	305.1647 (35), 206.1151 (5), 180.1012 (100), 144.0786 (9)		Kumar et al. (2018a)

124.	11.74	Vindolinine (isomer)	$C_{21}H_{24}N_2O_2$	337.1911	337.1911	−0.17	305.1641 (32), 277.1678 (4), 180.1011 (100), 144.0805 (10)		Kumar et al. (2018a)
125.	3.1	Vindolinine isomer (19-S)	$C_{21}H_{24}N_2O_2$	336.1842	337.1914	1.3	320.1654 (45), 308.1645 (2)	Root	Jeong and Lim (2018)
126.	14.23	Vindorosine	$C_{24}H_{30}N_2O_5$	427.2227	427.2227	0.2	367.2006 (36), 355.1744 (5), 158.0954 (100)		Kumar et al. (2018a)
127.	5.41	Vinpocetine†	$C_{22}H_{26}N_2O_2$	351.2067	351.2068	−0.33	322.1783 (22), 323.1783 (17), 308.1657 (25), 294.1503 (100), 266.1189 (76), 253.11.02 (11), 236.1439 (20), 222.1277 (18), 208.1128 (9), 193.0941 (2)		Kumar et al. (2018a)
128.	3.6	Yohimbine	$C_{21}H_{26}N_2O_3$	354.1943	355.202	1	323.1384 (2), 224.1282 (6), 212.1279 (15), 144.0805 (100)	Root and leaf	Jeong and Lim (2018)
129.	13.56	Yohimbine (isomer)	$C_{21}H_{26}N_2O_3$	355.2016	355.2017	−0.11	323.1744 (100)		Kumar et al. (2018a)
130.	6.86	Yohimbine †	$C_{21}H_{26}N_2O_3$	355.2016	355.2018	−0.64	323.1366 (27), 251.1520 (11), 224.1301 (14), 212.1265 (23), 169.0777 (5), 144.0807 (100)		Kumar et al. (2018a)

427.22 peak, which is supposed to be demethoxyvindoline, the major MS/MS fragments had *m/z* of 158.10, 367.20, and 409.21. These peaks are similar to the vindoline MS/MS fragment pattern (*m/z* 188.10, 397.20, and 439.21), although demethoxyvindoline MS/MS fragments may deviate by *m/z* 30 for the deletion of the methoxy group and addition of a proton. At *m/z* 531.23 ion peak, detected fragment ion peaks (*m/z* 144.08, 165.05, 282.11, 320.13, 334.14, 352.16, and 514.21) were similar to the major MS/MS fragments of strictosidine that was produced by transformed yeast (Yamamoto et al. 2016). 19-S-vindolinine, vindolinine, ajmalicine and an ajmalicine isomer, tabersonine, catharanthine, serpentine, and a serpentine isomer were identified by Song et al. (1992) using HPLC–MS. Five alkaloids, including catharanthine, vindolidine, and vindoline, were identified and characterized in the commercial product. 19S-Vindolinine, vindolinine, and catharanthine showed the same *m/z* 337. They were not distinguished. Fragment ions were observed as discussed above (Zhou et al. 2005; Chen et al. 2013). Chen et al. (2013) identified vindoline, ajmalicine, and catharanthine by direct-injection ion trap-mass spectrometry. These alkaloids showed precursor ions and product ions of vindoline, ajmalicine, and catharanthine were observed at *m/z* 457 → 397, 353 → 144, and 337 → 144 by LC–IT-MS, respectively. MS provided a high selectivity and sensitivity for the determination of vindoline, ajmalicine, and catharanthine in positive mode (Chen et al. 2013; Fernández-Pérez et al. 2013). Vincarodine was identified in *C. roseus*. Vincarodine showed a number of metastable ions, one of these being for the transition *m/z* 398 → 200, indicating a very interesting loss of nine of the teniridoid-derived carbon atoms in one process. Fragmentation pathways of vincarodine are shown in Scheme 2 (Fig. 5.1).

Scheme 2 presents the formation of the base peak at *m/z* 200, resulting from the loss (as one fragment) of *m/z* 198 from the parent ion. A small fragment was observed at *m/z* 380, analyzed for $C_{22}H_{24}N_2O_4$. This characterizes a loss of water from vincarodine, possible only for the skeleton represented by structure 5. A fragment ion at *m/z* 368 ($C_{21}H_{24}N_2O_4$) is shifted by 42 mass units upon acetylation of vincarodine; i.e., a loss of 30 mass units from the parent ion still occurs. Cleavage of the tetrahydrofuran ring with loss of a hydrogen radical at C-17 and homolytic fission of the C-15-20 bond gives rise to an allylically stabilized radical at C-20. Formation of a new D ring with loss of C-15 affords *m/z* 368. Retro Diels–Alder fragmentation of the C-ring, followed by aromatization of the E-ring, leads to an observed loss of 72 mass units. Scheme 2 demonstrates the standard fragmentation pattern in this series of compounds as applied to vincarodine, resulting in the formation of *m/z* 297 and a measured loss of *m/z* 101 (Cordell and Farnsworth 1976; Cordall et al. 1974).

The formation of ions at *m/z* 200, 380, 368, and 310 are characteristic peaks of the structure of vincarodine. Scheme 2B illustrates the standard fragmentation pattern in this series of compounds as applied to vincarodine, resulting in the formation of *m/z* 297 and a measured loss of *m/z* 101 (Cordall et al. 1974; Neuss et al. 1973). Similarly, the cathaphylline showed the RDA (*retro*-Diels-Alder) fragmentation pathway, which is characteristic for helped in discussion the elucidation of the structure of cathaphylline. There were four ions that were considered important in the fragmentation patterns of the *p*-anilino acrylate alkaloids. Ions A and B arise

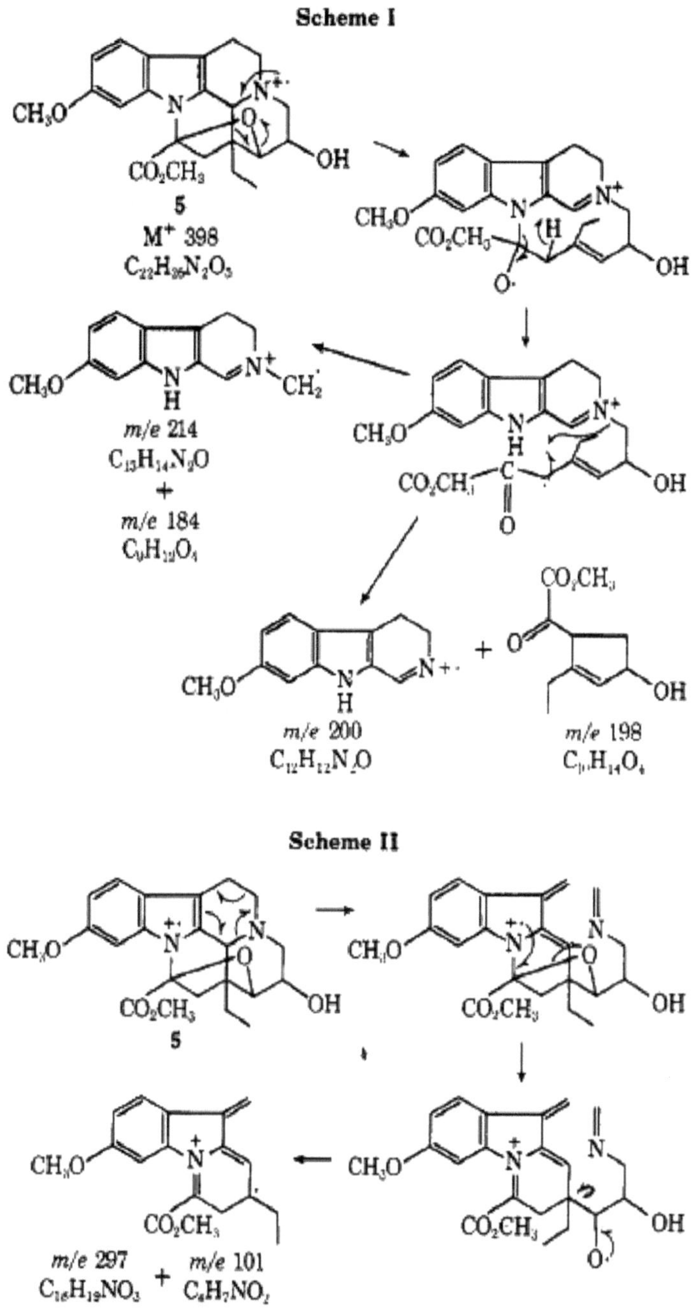

Fig. 5.1 A schematic representation showing fragmentation pathways of Vincarodine (Cordall et al. 1974)

from the RDA fragmentation of ring C and fission of the 5, 6-bond. Ions C and D arise by loss of the two-carbon side chain. A comparison of the *m/z* values was made for the compounds isolated from cathaphylline. Substitution in the indole portion leads to a change in the mass of ion A, whereas substitution in the piperidine ring leads to an increase in the mass of ions B and D. Substitution solely in the side chain affords an increase in mass for ions B and C from the "standard" values of 124 and 29 mass units (mu), respectively. Cathaphylline indicates: (a) no substitution in the indole ring, (b) no substitution in the piperidine ring, and (c) an addition of 30 mu in the two-carbon side chain. The mass spectral data for cathaphylline indicate that four oxygen atoms are present: two in the carbomethoxyl group and two in the side chain. The side chain was shown to have the molecular formula $C_2H_3O_2$ (M^+-ion D). The mass spectrum shows *m/z* 410 (M^+), in which ion B is shifted by 42 mu to *m/z* 196, indicating the presence of a hydroxyl group in the side chain. In the mass spectra of cathaphylline and acetylcathaphylline, a fragment ion was observed at *m/z* 337, corresponding to initial losses of CH_3O and $C_3H_5O_2$, respectively. Therefore, the acetylated functional group is at the terminal carbon of the side chain; ion at *m/z* 337 loses CO to give *m/z* 309 (ion D) (Cordell and Farnsworth 1976; Pluta et al. 1976; Canto-Canche and Loyola-Vargas 1999; Mukhopadhyay et al. 1983b). The mass spectrum of periformyline gave *m/z* 366 as the molecular weight compared to the calculated value of 366. The abundant fragment ions at *m/z* 194 suggested an increase of 14 mass units to the basic ion formed in the spectrum of vobasine. This could be explained by the exchange of a formyl group for the $N_{(b)}$-methyl group in vobasine to form the ion (Maloney et al. 1965; Mukhopadhyay and Cordell 1981a, b; Abraham et al. 1965). Tabersonine, serpentine, vindoline, catharanthine, and tryptamine were identified and characterized in the aerial part of *C. roseus* and *V. minor* by liquid chromatography electrospray ionization tandem mass spectrometry (LC-ESI-MS/MS). Tryptamine, serpentine, vindoline, tabersonine, and catharanthine gave fragment ions at *m/z* 161.4 → 144.2, 349.4 → 263.0, 457.6 → 188.0, 337.5 → 168.3, 337.4 → 144.2, and 355.4 → 337.2, respectively (Liu et al. 2016a, b; Fernández-Pérez et al. 2013; Song et al. 1992). Rosamine was isolated from the leaves of *C. roseus* and its structure was assigned by UV, IR (infrared), NMR, and mass spectrum. The mass spectrum of rosamine showed the molecular ion at *m/z* 352.1779 with other major fragment ions at *m/z* 335, 293, 267, 216, 184, 158, 135, and 107. The base peak was detected at *m/z* 135.1047, which is a fragment commonly found in Iboga and Aspidosperma alkaloids bearing a double bond in the piperidine ring (Ali and Bashir 1984; Bashir 1983).

5.2.2 Analysis of Monoterpene Indole Alkaloids by High-Resolution Mass Spectrometry and Liquid Chromatography–High-Resolution Tandem Mass Spectrometry (HPLC–HRMS/MS)

Various HPLC–HRMS (high-performance liquid chromatography–high-resolution mass spectrometry) and HPLC–HRMS/MS (high-performance liquid chromatography–high-resolution tandem mass spectrometry) methods reported by many

Table 5.2 Gas chromatography–mass spectrometry (GC–MS) data of TIAs d from *C. roseus* (Bhadra et al. 1993)

S. no.	Alkaloid	Molecular formula	[M⁺] m/z (intensity %)	Fragment ions (m/z, relative intensity %)	Reference
1.	Isovindolinine	$C_{21}H_{24}N_2O_2$	336 (100)	321 (3), 305 (8), 277 (5), 230 (35), 216 (30), 202 (17), 183 (13), 170 (87), 156 (36), 135 (79), 134 (88), 120 (29), 108 (14), 91 (15)	Wesołowska et al. (2016)
2.	Vindolinine	$C_{21}H_{24}N_2O_2$	336 (100)	321 (3), 305 (9), 277 (5), 230 (33), 216 (30), 202 (16), 183 (12), 170 (83), 156 (33), 135 (77), 134 (85), 120 (27), 115 (14), 107 (21), 93 (16)	Wesołowska et al. (2016)
3.	19-Epivindolinine	$C_{21}H_{24}N_2O_2$	336 (100)	321 (6), 305 (8), 277 (8), 229 (21), 202 (12), 194 (25), 180 (19), 170 (40), 156 (20), 134 (51), 122 (44), 107 (10), 91 (11), 77 (14)	Wesołowska et al. (2016)
4.	20-Epivindolinine	$C_{21}H_{24}N_2O_2$	336 (100)	321 (6), 305 (7), 277 (14), 229 (24), 202 (13), 194 (24), 180 (19), 170 (47), 156 (26), 134 (63), 122 (48), 107 (10), 91 (14), 77 (16)	Wesołowska et al. (2016)
5.	Tabersonine	$C_{21}H_{24}N_2O_2$	336 (100)	305 (9), 277 (7), 229 (22), 216 (33), 202 (6), 194 (15), 180 (14), 168 (25), 156 (26), 135 (54), 122 (31), 107 (33), 92 (5), 93 (9), 77 (12)	Wesołowska et al. (2016)
6.	Vincoline	$C_{21}H_{24}N_2O_4$	368 (2)	356 (7), 336 (100), 319 (6), 304 (16), 278 (12), 230 (18), 216 (21), 183 (22), 170 (62), 156 (36), 135 (50), 134 (64), 122 (22), 120 (24), 107 (25), 93 (13)	Wesołowska et al. (2016)
7.	Pleiocarpamine	$C_{20}H_{22}N_2O_2$	322 (96)	307 (8), 293 (4), 292 (4), 263 (100), 247 (12), 232 (34), 218 (19), 194 (14), 180 (83), 167 (15), 131 (15), 121 (5), 108 (14)	Wesołowska et al. (2016)
8.	Coronaridine	$C_{21}H_{26}N_2O_2$	338 (100)	323 (30), 309 (6), 253 (11), 214 (25), 136 (75), 135 (22), 124 (34), 122 (29), 108 (10), 96 (8), 77 (6), 41 (11)	Wesołowska et al. (2016)
9.	16-Epivindolinine-Nb-oxide	$C_{21}H_{24}N_2O_3$	352 (3)	351 (7), 338 (2), 335 (23), 334 (100), 321 (1), 319 (3), 305 (3), 293 (21), 281 (14), 275 (11), 249 (22), 227 (58), 207 (31), 194 (60), 182 (23), 167 (90), 149 (74), 127 (19), 108 (71), 92 (14)	Wesołowska et al. (2016)

(continued)

Table 5.2 (continued)

S. no.	Alkaloid	Molecular formula	[M^+] m/z (intensity %)	Fragment ions (m/z, relative intensity %)	Reference
10.	Vindolidine	$C_{24}H_{30}N_2O_5$	426 (7)	367 (20), 282 (7), 281 (20), 267 (75), 266 (5), 207 (40), 158 (100), 144 (21), 143 (27), 135 (14), 107 (15), 93 (7), 92 (11), 77 (15)	Wesołowska et al. (2016)
11.	Pericyclivine	$C_{20}H_{22}N_2O_2$	322 (100)	321 (82), 307 (29), 291 (8), 263 (31), 249 (19), 235 (11), 207 (10), 181 (12), 180 (7), 169 (69), 168 (67), 154 (13)	Wesołowska et al. (2016)
12.	Vindorosine	$C_{24}H_{30}N_2O_5$	426 (6)	366 (7), 282 (13), 279 (16), 267 (13), 266 (26), 207 (2), 158 (94), 135 (100), 121 (26), 107 (18), 93 (9), 77 (6)	Wesołowska et al. (2016)
13.	Tetrahydroalstonine	$C_{21}H_{24}N_2O_3$	352 (100)	337 (33), 293 (6), 251 (13), 223 (19), 197 (9), 184 (9), 170 (10), 169 (19), 156 (39), 129 (9), 115 (6), 77 (4), 55 (5)	Wesołowska et al. (2016)
14.	Vindoline	$C_{25}H_{32}N_2O_6$	456 (15)	397 (9), 309 (13), 296 (25), 282 (29), 188 (84), 174 (37), 161 (22), 135 (100), 121 (42), 107 (20), 93 (12)	Wesołowska et al. (2016)
15.	Ajmalicine	$C_{21}H_{24}N_2O_3$	352 (100)	337 (6), 321 (5), 293 (2), 265 (5), 225 (7), 209 (13), 184 (34), 170 (8), 169 (19), 156 (51), 143 (7), 129 (7), 115 (5), 77 (4)	Wesołowska et al. (2016)

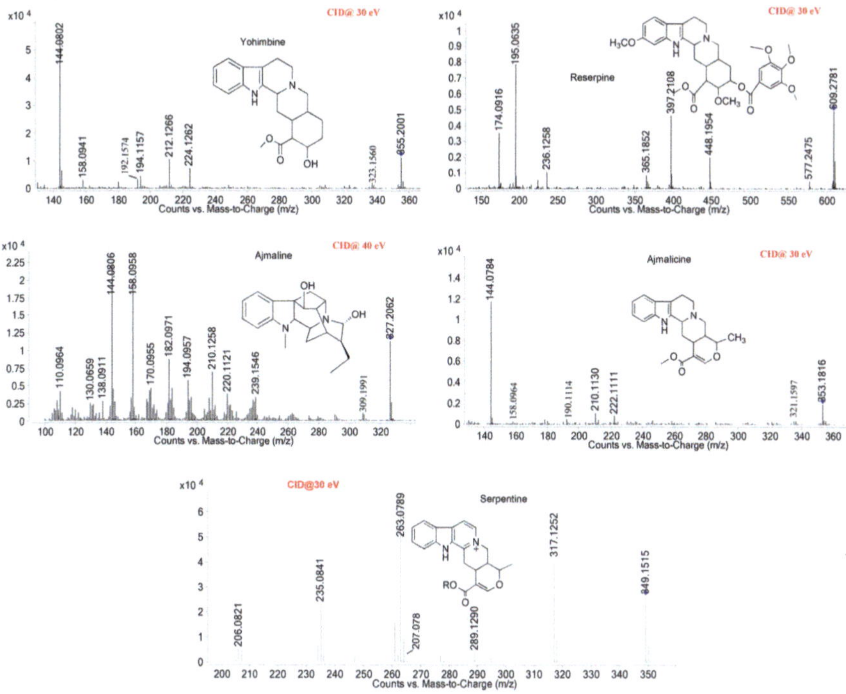

Fig. 5.2 MS/MS spectra of standard TIAs

research groups are summarized in Table 5.2 and the spectra of standard TIAs are shown in Fig. 5.2 (Balsevich et al. 1988; Auriola et al. 1991; O'Keefe et al. 1997; Verma et al. 2008; Zhou et al. 2005; Zweil et al. 2019; He et al. 2011a, b; Mu et al. 2012; Chen et al. 2013; Zhang et al. 2014; Kumar et al. 2018a; Jeong and Lim 2018; Liang et al. 2018; Koel et al. 2020). A total of 22 TIAs were separated within 16 min on a C18 column (Acquity UPLC BEH C18 column (2.1 mm × 100 mm, 1.7 μm particle size) with the mobile phase containing 0.1% formic acid +10 mM ammonium acetate in ACN/H_2O using UPLC-ESI-Q-TOF-MS/MS (ultra-performance liquid chromatography with electrospray ionization coupled to quadrupole–time of flight tandem mass spectrometry). Ajmalicine, catharanthine, serpentine, and vindoline were identified in the root, stem, and leaf by comparison of retention times, molecular ions, and fragmentation patterns with those of reference compounds. Additional TIAs were tentatively identified by comparing chromatograms with chemical databases and literature reports (Jeong and Lim 2018).

The BPC of the root (a), stem (b), and leaf (c) hydroalcoholic extracts of *C. roseus* as determined by UPLC–ESI-Q-TOF-MS/MS. The visual appearance of chromatograms showed that the BPC of leaf extracts showed comparatively more peaks than stems and leaves (Figs. 5.3, 5.4, 5.5, and 5.6).

The chromatographic separation of TIAs was performed utilizing a Thermo Betasil C8 column (250 mm × 4.5 mm, 5 μm) at 25 °C with a mobile phase of 0.1%

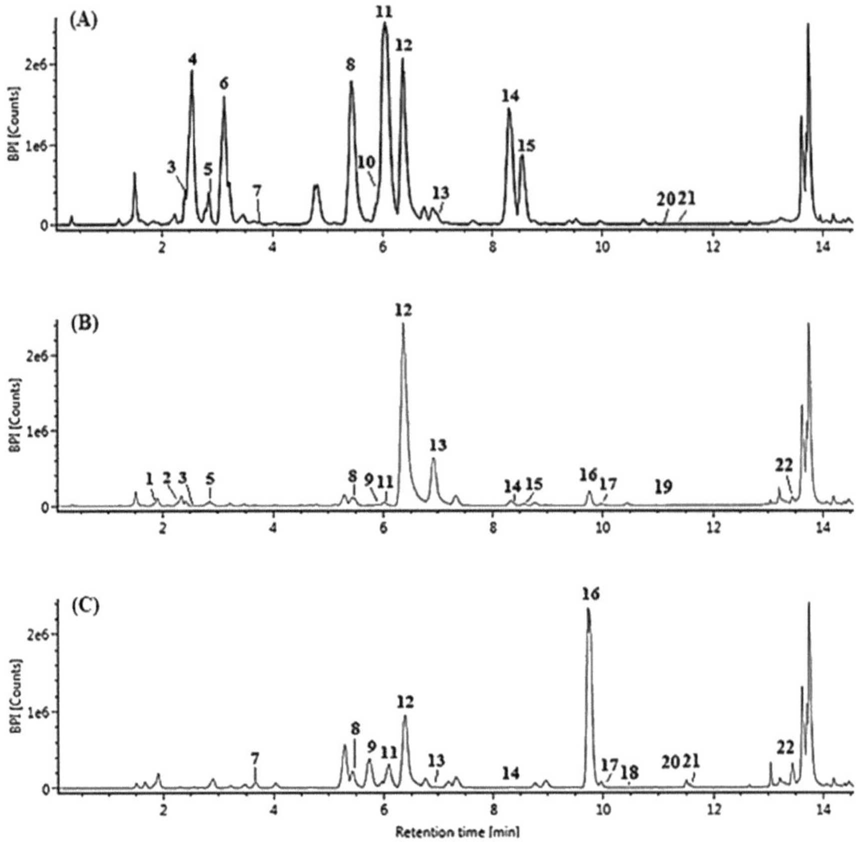

Fig. 5.3 BPC of the root (**a**), stem (**b**), and leaf (**c**) hydroalcoholic extracts of *C. roseus* using UPLC–Q-TOF (ultra-performance liquid chromatography–quadrupole time-of-flight)

formic acid in water and acetonitrile. Diagnostic fragmentation pathways for vinpocetine, vindesine, catharanthine, vinblastine, vindoline, and vincristine were determined based on their diagnostic product ions. Seventy-two TIAs were analyzed, comprising 11 TIAs distinctly recognized and characterized through comparison with established standards, while the remaining 61 TIAs were provisionally identified and characterized. The distribution of all TIAs was analyzed in ethanolic extracts from 30 samples of *C. roseus* obtained from five distinct states in India. MS/MS, in conjunction with precise mass measurements, offers an enhanced level of accuracy for the structural characterization of various chemical classes. Recently, effective methodologies have been established for the swift detection and identification of phytochemicals in crude plant extracts utilizing diagnostic fragment ions. Scheme 2 illustrates the MS/MS spectrum and fragmentation process of vindoline. We found smaller ions at *m/z* 397.2109 and 439.2223, which were created by removing CH_3COOH and H_2O from the main ion. The fragment ions at *m/z* 222.1097 and 188.1062 resulted from the elimination of $C_{11}H_{13}NO$ and $C_{11}H_{15}NO_3$ from *m/z*

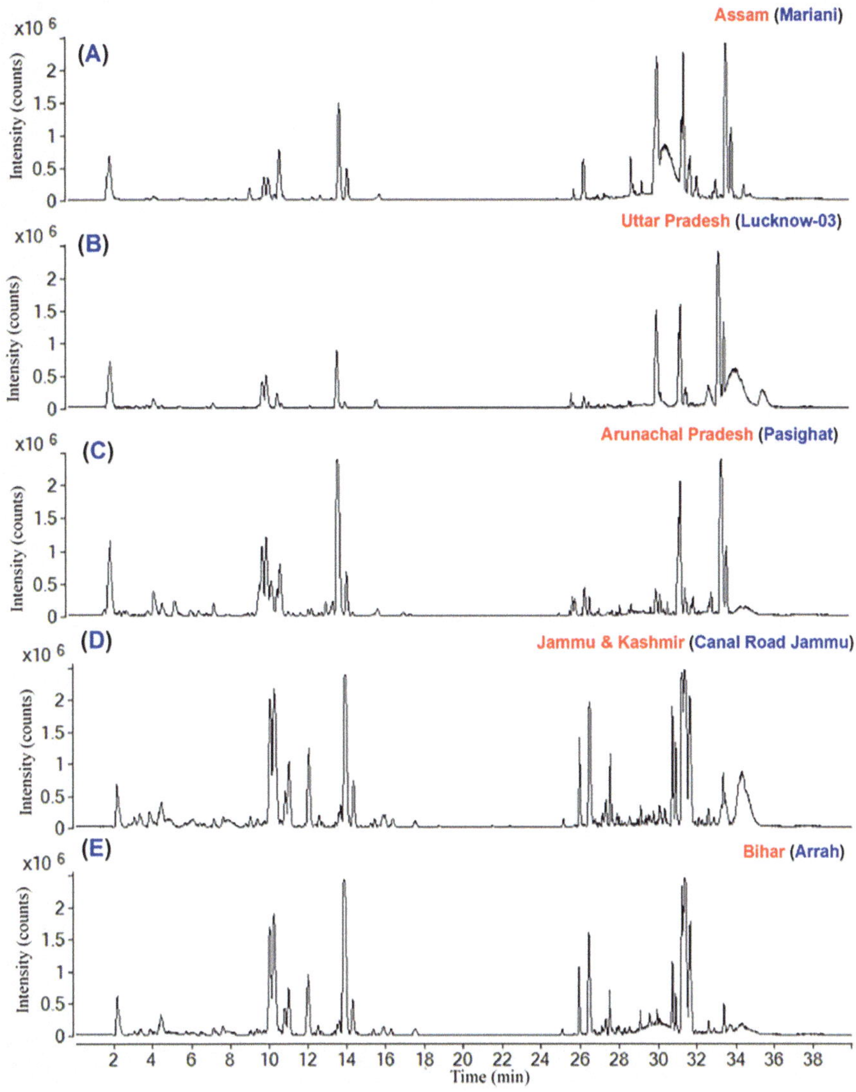

Fig. 5.4 Base peak chromatograph of ethanolic extracts of leaves collected from different geographical regions of India

397.2109 (Jeong and Lim 2018; Kumar et al. 2018a). Vindoline was found in the root and stem of *C. roseus* with a protonated ion at *m/z* 337.1914, and it had fragment ions at *m/z* 320.1654 and 308.1645 (Jeong and Lim 2018; El-Domyati et al. 2014; Zhou et al. 2005; Kumar et al. 2018a). The compounds vindoline-1, vindoline-2, vindoline-3, and vindoline-4 were found at *m/z* 369.1813, 369.1810, 369.1808, and 369.1807 as $[M + H]^+$, with times of 3.88, 8.17, 9.04, and 12.16 min, respectively, and are considered variations of vindoline. Vindoline showed fragment

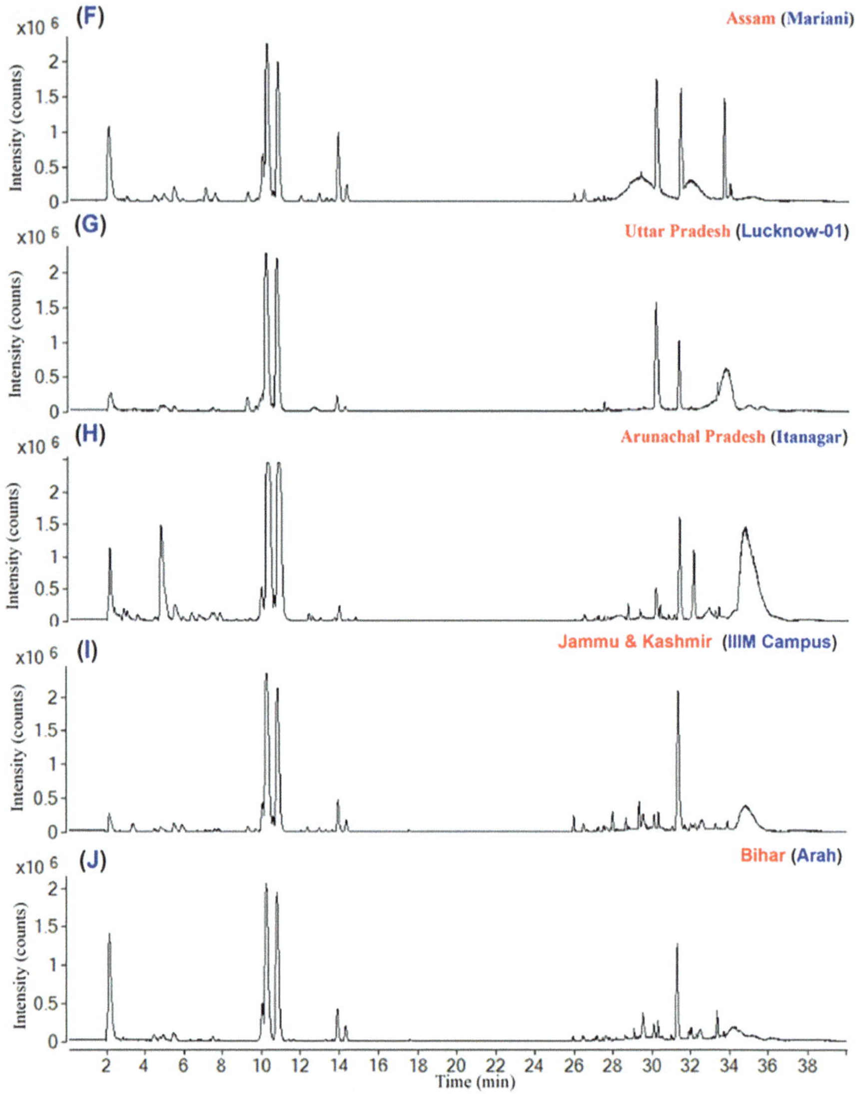

Fig. 5.5 Base peak chromatograph (BPC) of ethanolic extracts of stems collected from different geographical regions of India

ions at *m/z* 397.2096, 337.1835, and 188.1057, as shown in Scheme 5, matching the standard references (Höfer et al. 2013; Kumar et al. 2018a; Chen et al. 2018) (Fig. 5.7).

Our research group recently gathered leaves, stems, and roots of *C. roseus* from five regions in India: Assam, Uttar Pradesh, Arunachal Pradesh, Jammu & Kashmir, and Bihar. The identities and retention durations of the chosen bioactive TIAs were examined in the positive (+ESI) mode owing to the presence of basic nitrogen. The

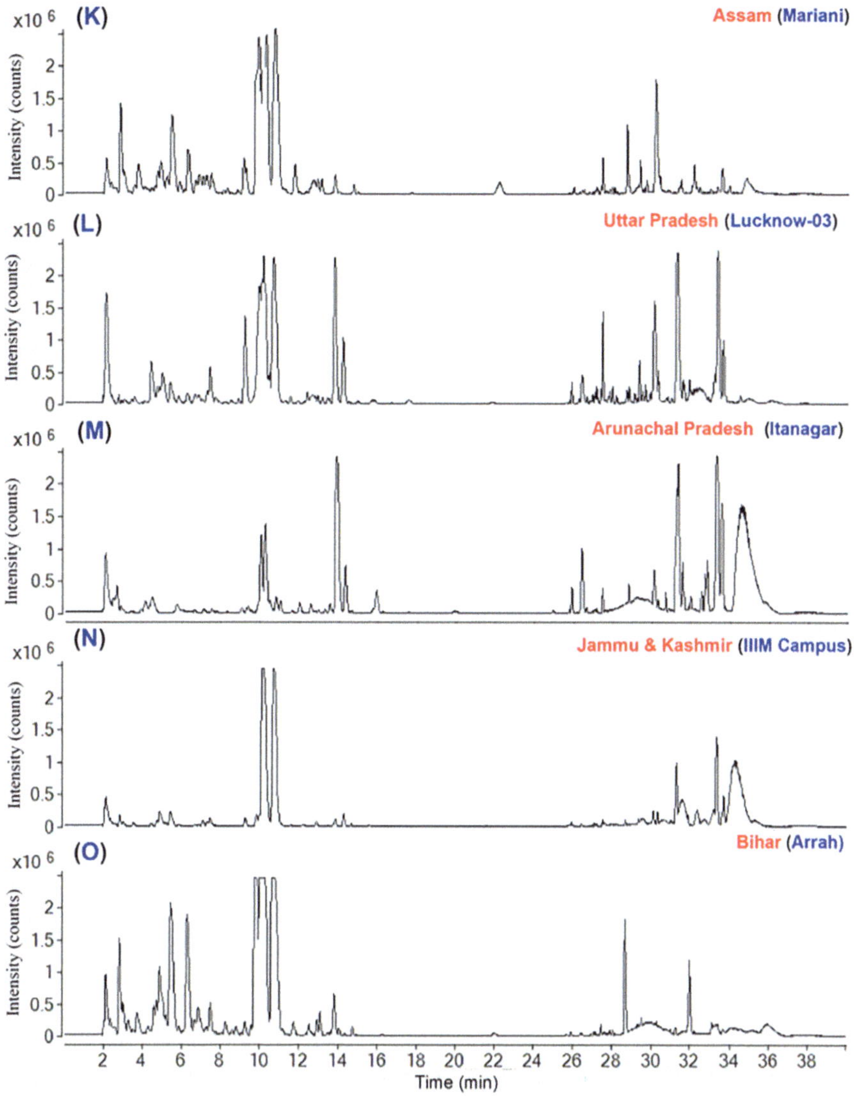

Fig. 5.6 Base peak chromatograph (BPC) of ethanolic extracts of roots collected from different geographical regions of India

comparative base peak chromatograms (BPCs) of ethanolic extracts from the leaves, stems, and roots of *C. roseus* in positive ionization mode are illustrated in Fig. 5.2. Table 5.1 presents the identities, retention times, and calculated *m/z*, observed *m/z*, molecular formulas, errors (ppm), and fragment ions for each drug. Table 5.2 presents the distribution of individual chemicals in samples collected from different geographical regions of India. Serpentine (*m/z* 349.1547) was found and described

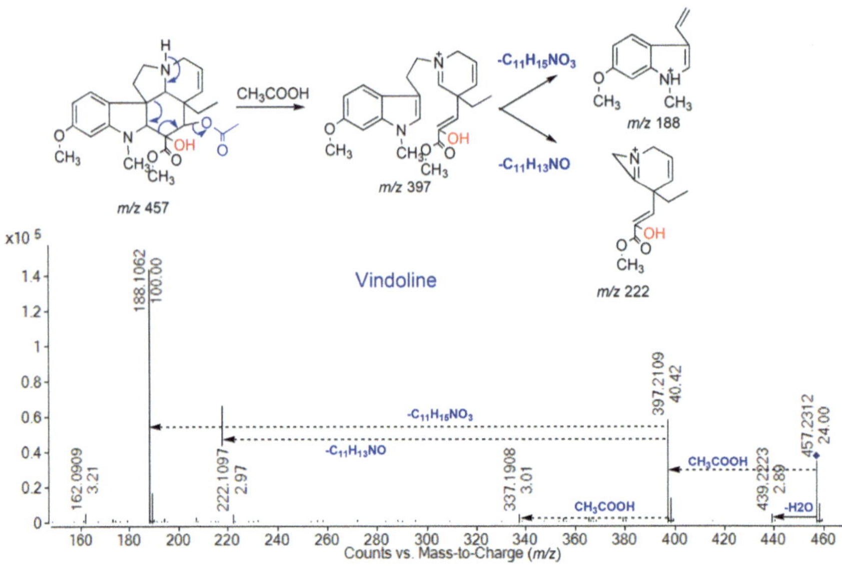

Fig. 5.7 Fragmentation pathways of vindoline and its MS/MS spectrum

by comparison with standard compounds and references from various parts of *C. roseus*. Fragment ions at *m/z* 317.1262 and 289.1342 were detected as a result of the elimination of CH_3OH and CO, respectively. Fragment ions at *m/z* 263.0801, 235.0848, and 206.0819 were generated by the elimination of $C_4H_7OCH_3$, CO, and CHO, respectively (Scheme 5.3) (Jeong and Lim 2018; Kumar et al. 2018a). Catharanthine was identified at *m/z* 337.1910 $[M + H]^+$. The MS/MS spectrum of catharanthine (Fig. 5.9) produced fragment ions at *m/z* 305.1617 and 277.1699 as a result of the loss of CH_3OH and CO. The fragment ion at *m/z* 144.0794 was identified as the base peak, resulting from the loss of $C_9H_{11}N$ from *m/z* 277.1699 (Scheme 6) (Jeong and Lim 2018; Kumar et al. 2018a) (Fig. 5.8)

A serpentine derivative was discovered at *m/z* 335.1338. The MS/MS spectra exhibited fragment ions at *m/z* 317.1261 and 291.1461, resulting from the loss of H_2O and CO_2, whereas fragment ions at *m/z* 263.0806, 235.0859, and 207.0900 were generated from the consecutive loss of CO from *m/z* 291.1461. The fragment ion at *m/z* 291.1461 produced an ion at *m/z* 219.0885 as a result of the loss of C_4H_7OH. Compound serpentine (isomer) was identified at *m/z* 349.1545 with a retention time of 10.85 min. It generated identical fragment ions to serpentine, albeit with varying relative abundances of these ions, and was discovered as an isomer of serpentine.

Ajmalicine (*m/z* 353.1860) was found, described, and validated through comparison with a standard. The MS/MS spectra of ajmalicine indicated that the fragment ion at *m/z* 321.1562 resulted from the loss of CH_3OH, whereas fragment ions at *m/z* 222.1134, 210.1125, 178.0837, and 144.0799 were generated through RDA reactions (Fig. 5.10).

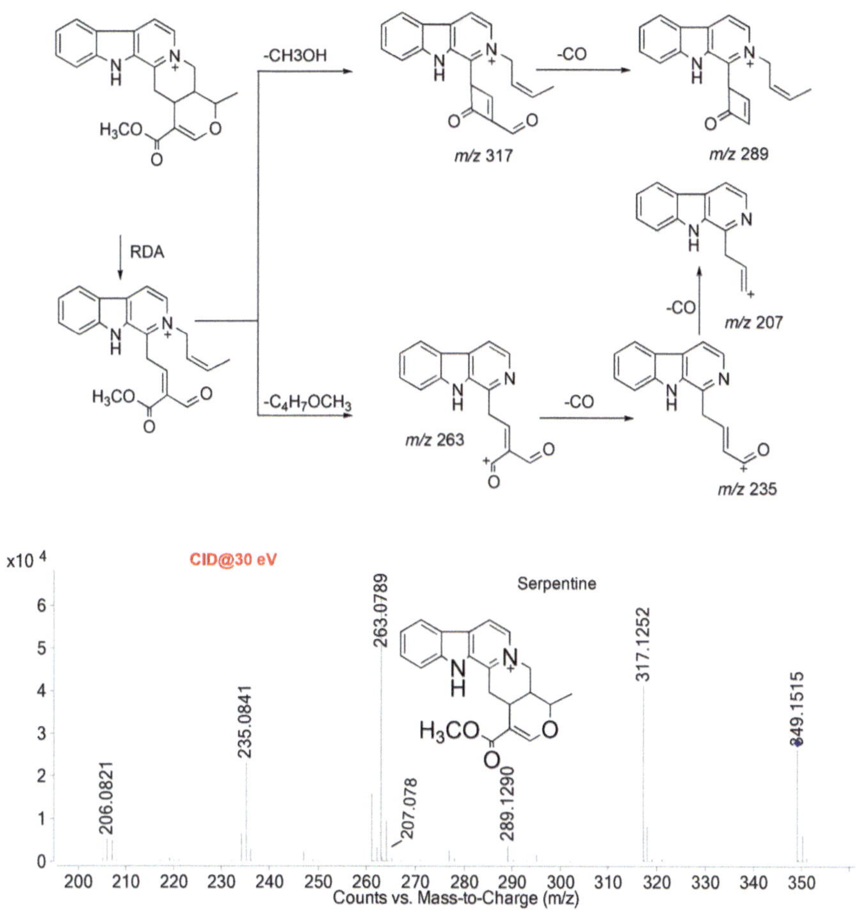

Fig. 5.8 Fragmentation pathways of serpentine and its MS/MS spectrum

Vindogentianine, vindoline, vindolidine, vindolinine, perivine, and serpentine were recovered from the leaf extract of *C. roseus*. The chemical formula of vindogentianine is $C_{35}H_{41}N_3O_8$, as determined by ESI-IT-TOF/MS, with *m/z* values of 632.3060 for $[M + H]^+$ and 676.2897 for the loss of $M + HCO_2$. Two new types of alkaloid glycosides (10-hydroxycathafoline 10-*O*-*α*-L-arabinopyranoside and 10-hydroxydeformodihydro-pseudoakuammigine 10-*O*-*α*-L-arabinopyranoside) and a new single indole alkaloid (15(S)-hydroxy-14,15-dihydrovindolinine), along with tetrahydroalstonine, ajmalicine, vindorosine, vindoline, 19(S)-vindolinine, 19(R)-vindolinine, vincapusine, yohimbine, serpentine, 18β-hydroxy-3-epi-α-yohimbine, sitsirikine, perivine, and dihydroantirhine were identified from the roots of *C. roseus*. The chemical formula of 10-hydroxycathafoline 10-*O*-α-L-arabinopyranoside is $C_{26}H_{34}N_2O_7$, as established by its HR-ESI-MS at *m/z* 487.2429 $[M + H]^+$ (Wang et al. 2012a). Aricine was detected in the roots and stems of

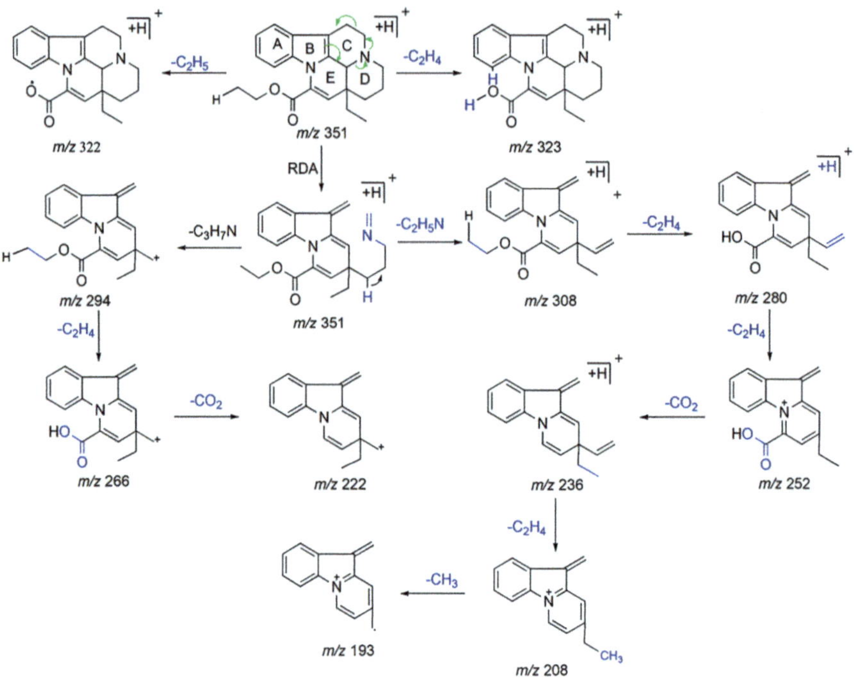

Fig. 5.9 Fragmentation pathways of catharanthine

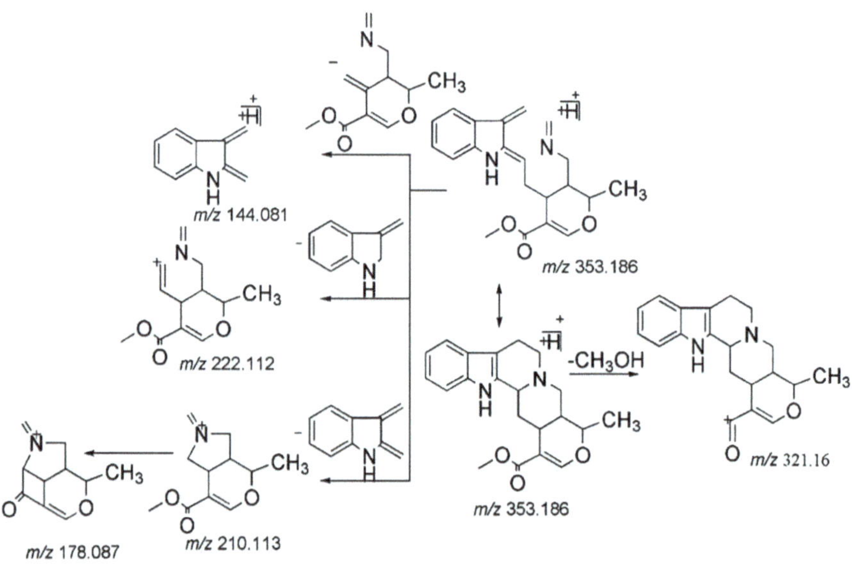

Fig. 5.10 Fragmentation pathways of ajmalicine and its MS/MS spectrum

C. roseus. The $[M + H]^+$ ion for aricine was detected at *m/z* 383.1970, with matching MS/MS fragment ions at *m/z* 369.1814, 340.1949, 321.1605, 296.1284, and 160.0754 (Jeong and Lim 2018; Strychnine exhibits a protonated molecular ion at *m/z* 335.1756 ($[M + H]^+$) along with fragment ions at 228.1092, 196.0755, 168.0804, and 167.0726 (Jeong and Lim 2018; Broecker et al. 2011a, b). Pleiocarpamine, alstonine, tetrahydroalstonine, and tabersonine were identified as protonated ions at *m/z* 323.1759, 349.1547, 353.1856, and 337.1912 in the roots, stems, and leaves of *C. roseus*, with corresponding fragment ions documented by various researchers (Jeong and Lim 2018; Sagi et al. 2016; Chen et al. 2018). The MS/MS spectra and fragmentation pathways of vinpocetine are illustrated in Scheme 5.6. The $[M + H]^+$ ion was identified at *m/z* 351.2068, and the MS/MS spectra revealed fragment ions at *m/z* 323.1783 and 322.1783, resulting from the loss of C_2H_4 and C_2H_5, respectively. The C-ring opening via RDA and subsequent elimination of C_2H_5N and C_3H_7N produced fragment ions at *m/z* 308.1657 and 294.1503, respectively. Fragment ions at *m/z* 280.1347 and 266.1189 were detected following the elimination of C_2H_4 from *m/z* 308.1657 and 294.1503, respectively. Likewise, fragment ions at 236.1439 and 222.1277 were generated by the loss of CO_2 from *m/z* 280.1347 and 266.1189, respectively. Fragment ions at *m/z* 208.1128 and 193.0941 were generated via the loss of C_2H_4 and subsequently CH_3 from *m/z* 236.1439 and 208.1128, respectively (Kumar et al. 2018a; Avula et al. 2019).

Four alkaloids, trichophylline-1, trichophylline-2 (isomer), trichophylline-3, and trichophylline-4, were identified at *m/z* 367.1650, 367.1654, 367.1650, 367.1652, and 367.1652 as $[M + H]^+$ with retention times of 2.76, 4.32, 5.93, and 28.69 min (Kumar et al. 2018a). All these alkaloids were isomeric substances that may be isomers of trichophylline. Trichophylline-1 exhibited a fragment ion at *m/z* 352.1754, resulting from the loss of CH_3 from *m/z* 367.1650. Trichophylline-2 produced fragment ions at *m/z* 351.1712 and *m/z* 307.1522, resulting from the elimination of CH_4 and CH_3COOH, respectively, from *m/z* 367.1654. Trichophylline-3 and trichophylline-4 exhibited fragment ions at *m/z* 337.1537 and 335.1381, resulting from the loss of CH_2O from *m/z* 367.1652 (Kumar et al. 2018a; Mukhopadhyay et al. 1983a). 10-hydroxynortetraphyllicine and its isomer were found at *m/z* 311.1753 and 311.1754 as $[M + H]^+$ ions, with retention times of 2.83 and 7.87 min, showing that they are isomers. Both compounds produced a fragment ion at *m/z* 293.1647 resulting from the loss of H_2O (Kumar et al. 2018a). Nine compounds, named hydroxytabersonine-1 to hydroxytabersonine-9, were found as protonated ions with *m/z* values of 353.1862, 353.1862, 353.1863, 353.1860, 353.1861, 353.1860, 353.1862, 353.1861, and 353.1863, and they were detected at retention times of 3.86, 5.71, 7.18, 7.66, 11.13, 11.94, 12.31, 12.98, and 13.66 min, showing that they are likely isomers. All the hydroxytabersonine-1 to 9 were tentatively identified as isomeric compounds to each other. Hydroxytabersonine-1, hydroxytabersonine-2, hydroxytabersonine-3, and hydroxytabersonine-4 exhibited fragment ions at *m/z* 335.1732 and 321.1588, resulting from the loss of H_2O and CH_3OH from *m/z* 353.1862, respectively. Three compounds, 3,4,5,6-tetradehydroyohimbine-1, 3,4,5,6-tetradehydroyohimbine-2, and 3,4,5,6-tetradehydroyohimbine-3, were identified at *m/z* 351.1704, 351.1704, and 351.1705, with retention times of 4.17,

7.34, and 7.7 min, respectively. Compounds 3,4,5,6-tetradehydroyohimbine-1 and 3,4,5,6-tetradehydroyohimbine-2 exhibited identical fragment ions at *m/z* 333.5954, resulting from the elimination of H_2O. The predominant fragment at *m/z* 321.1587 was detected as a result of H_2CO loss in compounds 3,4,5,6-tetradehydroyohimbine, 3,4,5,6-tetradehydroyohimbine-2, and 3,4,5,6-tetradehydroyohimbine-3 (Kumar et al. 2018a). Four isomeric molecules, perivine-1, perivine-2, perivine-3, and perivine-4, were identified with *m/z* values of 339.1708, 339.1708, 339.1705, and 339.1704, with corresponding retention times of 4.38, 5.94, 8.03, and 9.7 min, respectively. Perivine-1, perivine-2, and perivine-4 exhibited a shared ion at *m/z* 307.1452, resulting from the elimination of CH_3OH. Norseredamine, identified $[M + H]^+$ at *m/z* 325.1912, was recognized as norseredamine. Fragment ions were seen at *m/z* 307.1767, resulting from the loss of H_2O, originating from *m/z* 325.1912. Additionally, fragment ions at *m/z* 160.1041 and 144.0763 were generated by RDA cleavage and subsequent bond cleavage at various sites.

Vellosiminol (*m/z* 295.1805) was identified and classified as such. The MS/MS spectra of vellosiminol showed fragment ions at *m/z* 277.1708, which were created by losing H_2O, while RDA fragmentation and bond breaking at different points produced fragment ions at *m/z* 158.0948, 144.0800, 138.0903, and 120.0806. Compounds vindolinine-1 through vindolinine-10 were identified as isomeric compounds. They showed at *m/z* values of 337.1911, 337.1920, 337.1916, 337.1910, 337.1919, 337.1919, 337.1918, 337.1920, 337.1919, 337.192, and 337.1911, respectively, as $[M + H]^+$ ions with retention times of 5.41, 6.28, 6.33, 8.61, 9.52, 9.98, 10.03, 10.76, 11.66, and 11.74 min (Jeong and Lim 2018). The compounds vindolinine-1, vindolinine-5, and vindolinine-10 produced an identical fragment ion at *m/z* 305.1648 as a result of the loss of CH_3OH. The compounds vindolinine-1, vindolinine-5, and vindolinine-10 may be stereoisomers, as they produced identical fragment ions with varying relative abundances. In contrast, the compounds vindolinine-2, vindolinine-3, vindolinine-4, vindolinine-6, vindolinine-7, vindolinine-8, vindolinine-9, and vindolinine-10 could be structural isomers of vindolinine as they generate distinct fragment ions. The MS/MS spectra of the molecule vindolinine-2 exhibited a fragment ion at 321.1613, resulting from the loss of CH_4. The compound tetraphyllicine (*m/z* 309.1961) was recognized as tetraphyllicine. Fragment ions were seen at *m/z* 291.1869 and 276.1622 due to the loss of H_2O followed by CH_3, while retro-Diels-Alder reactions and subsequent bond breakage at other sites yielded fragment ions at *m/z* 152.1062, 148.1135, 144.0820, 136.1125, 135.1034, and 134.0956. Pleiocarpamine-1, pleiocarpamine-2, and pleiocarpamine-3 were identified at *m/z* 323.1756, 323.1753, and 323.1760, with retention times of 6.51, 8.68, and 9.32 min, respectively, as isomeric compounds. The chemical pleiocarpamine-1 exhibited a fragment ion at *m/z* 307.1825 resulting from the loss of CH_3, while pleiocarpamine-2 and pleiocarpamine-3 produced fragment ions at *m/z* 291.1485 and 293.1644, respectively, due to the loss of CH_3OH and CH_2O.

Yohimbine (*m/z* 355.2018) was identified and characterized through comparison with a standard reference. The MS/MS spectrum of yohimbine exhibited fragment ions at *m/z* 323.1366, resulting from the loss of CH_3OH, while retro-Diels–Alder (RDA) fragmentation, occurring at various bond positions, produced fragment ions

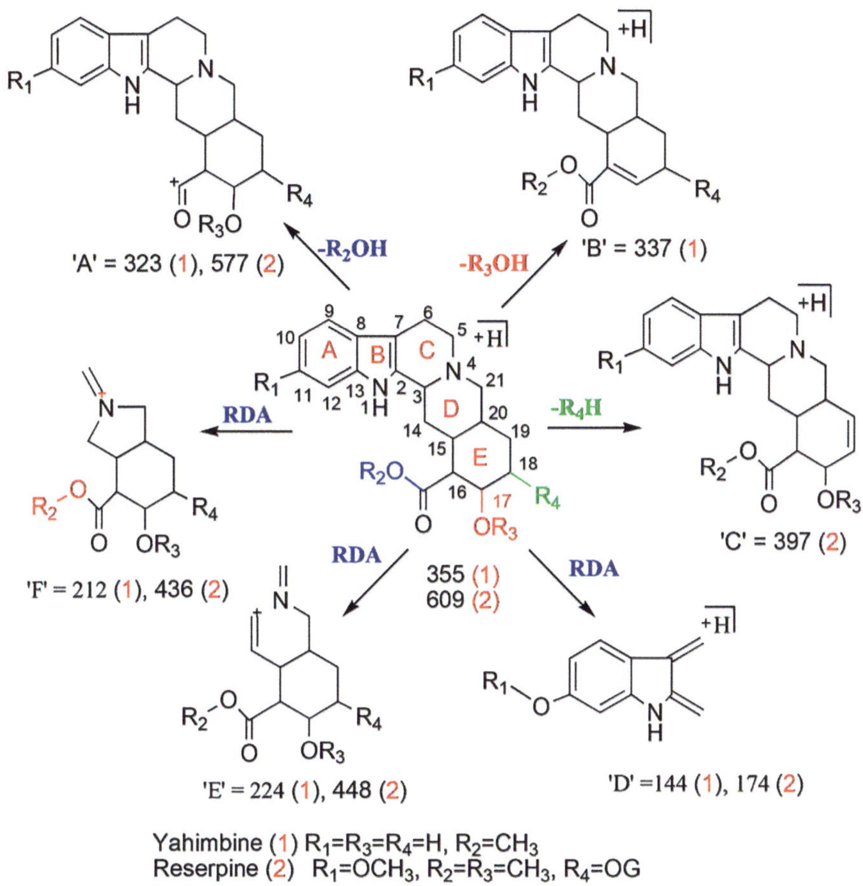

Fig. 5.11 Fragmentation pathways of yohimbine and reserpine

at *m/z* 224.1301, 212.1265, and 144.0807 (Scheme 5.6) (Jeong and Lim 2018; Lucas et al. 2015; Zhang et al. 2010; Zweigenbaum et al. 2016; Vay et al. 2019). Yohimbine-1, discovered at *m/z* 355.2017 with a retention time of 13.56 min, was recognized as an isomer of yohimbine (Fig. 5.11).

The MS/MS spectrum of the chemical cathovalinine exhibited a fragment ion at *m/z* 323.1744 resulting from the loss of CH_3OH. Hhydroxytabersonine (*m/z* 353.1863), identified as lochneridine, exhibited a fragment ion at *m/z* 323.1746 resulting from the loss of H_2O. Lochneridine-1 and lochneridine-2, observed at *m/z* 341.1860, were eluted in 12.3 and 12.45 min, respectively, and were recognized as isomers. Fragmentation was observed at *m/z* 309.1577, resulting from the loss of CH_3OH. The compound antirhine was detected as it produced fragment ions at *m/z* 154.1229 and 144.0798 as a result of RDA cleavage. Deacetylvindoline and vindorosine active compounds present *C. roseus* showed *m/z* at 415.2234 and 427.2227, respectively. Deacetylvindoline produced a fragment ion at *m/z* 397.2194 because it

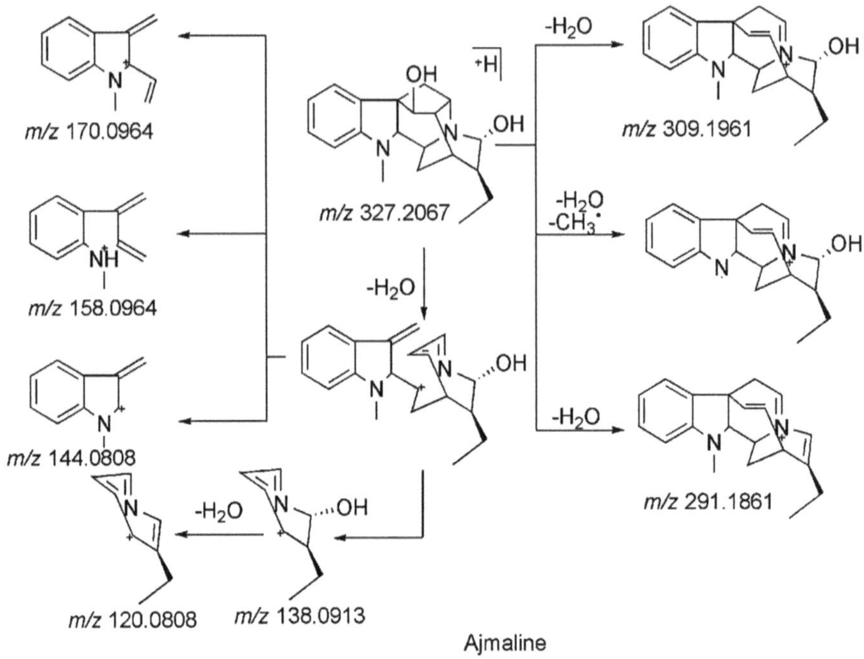

Fig. 5.12 Fragmentation pathways of ajmaline

lost a water molecule, while the fragment ion at *m/z* 367.2006 was found in vindorosine due to the loss of acetic acid from *m/z* 427.2227. Characteristic fragment ions at *m/z* 188.1066 and 158.0954 were detected in deacetylvindoline and vindorosine, respectively. Ajmaline (*m/z* 327.2066) was found and characterized using a standard reference. The fragment ion at *m/z* 309.1964 was detected as a result of the loss of H_2O. Fragment ions of RDA were detected at *m/z* 210.1287, 194.0983, 182.0979, 158.0974, and 144.0817 (Broecker et al. 2011a, b) (Fig. 5.12).

Deacetylcathovaline showed a protonated adduct at *m/z* 385.2122. The MS/MS spectrum exhibited fragment ions at *m/z* 367.1990 and 335.1734, resulting from the loss of H_2O and CH_3OH, respectively. The isomer of the chemical ajmalicine was identified at *m/z* 353.1864 with a retention period of 9.64 min, producing identical fragment ions as ajmalicine. Consequently, it was recognized as an isomer of ajmalicine. Catharanthine (*m/z* 337.1910) was identified and defined through comparison with a standard. Hydroxycathovaline (*m/z* 443.2175) was found, exhibiting a fragment ion at *m/z* 365.1836 resulting from the loss of CH_3COOH, along with another distinctive fragment ion at *m/z* 158.0953. Minovincinine showed a protonated adduct at *m/z* 355.2016. The MS/MS spectra exhibited fragment ions at *m/z* 337.1922, 305.1656, and 277.1701, resulting from the sequential loss of H_2O,

CH_3OH, and CO. Reserpine was discovered and described, exhibiting fragment ions at *m/z* 577.2563 and 397.2139 resulting from the loss of CH_3OH and $C_{10}H_{12}O_5$, while RDA fragment ions were observed at *m/z* 448.1992 and 365.1291 (Fig. 5.11). This was corroborated through comparison with the standard. 19(S)-chlorotabersonine, isomeric with 19(R)-chlorotabersonine, and 19(S)-chloro-3-oxotabersonine, isomeric with 19(R)-chloro-3-oxotabersonine, were identified as pairs of epimeric indole alkaloids containing a chlorine atom, isolated from the aerial parts of *C. roseus*. The *m/z* values observed were 371.1519 $[M + H]^+$ and 385.1306 $[M + H]^+$, accompanied by the respective molecular formula (Wang et al. 2012a, b). Sitsirkine exhibited $[M + H]^+$ at *m/z* 355.2013, with characteristic fragment ions detected at *m/z* 323.1756, 168.0800, 144.0803, and 140.1068 (Jeong and Lim 2018).

5.2.3 Analysis of Monoterpene Indole Alkaloids by Gas Chromatography and Gas Chromatography–Mass Spectrometry

GC–MS was used for the analysis of monoterpene indole alkaloids and 15 alkaloids were identified, among which vindoline, vindorosine, isovindolinine, and ajmalicine were detected in aerial parts of *C. roseus* in GC–MS analysis. Furthermore, the extraction of isovindolinine, vindolinine, and ajmalicine at 55 °C was more effective, while pericyclivine–maceration was more effective at room temperature. The isovindolinine, vindolinine, 19-epivindolinine, 20-epivindolinine, and tabersonine have the same molecular ion at *m/z* 336 due to their isomeric character. Moreover, 19-epivindolinine, 20-epivindolinine, and tabersonine showed similar fragmentation patterns and retention times due to their structure being very close to each other. Fragment ions at *m/z* 93, 107, 120, 121, 122, 134, and 135 were observed for the piperidine moiety of the molecule found in all alkaloids in mass spectra of the vindolinine class of compounds. Vindolinine and tabersonine with $[M^+]$ at *m/z* 336 are considered as catharanthine isomers. The base peak at *m/z* 336 (100%) in vincoline corresponds to the molecular ion of vindolinine (M^+, *m/z* 336). The presence of fragment ions at *m/z* 93, 107, 120, 122, and 135 shows that both compounds belong to the same group. Pleiocarpamine shows prominent signals at *m/z* 322 (M^+), *m/z* 263 (M^+-CO_2CH_3), and *m/z* 180 (quinolium ion). The mass spectrum of pericyclivine was characterized by the ions at *m/z* 322, 321, 291, 263, and 169. The spectra of ajmalicine and tetrahydroalstonine showed fragments of the β-carboline skeleton at *m/z* 184, 170, 169, and 156. Vindorosine (molecular weight, 426) showed a difference of *m/z* 30 in mass as compared to vindoline (molecular weight, 456). Moreover, vindolidine and vindorosine have the same molecular ion at *m/z* 426. The MS/MS fragmentation pathways of desacetoxyvindoline were deduced by Bhadra et al. (1993) (Wesołowska et al. 2016; Bhadra et al. 1993) (Fig. 5.13).

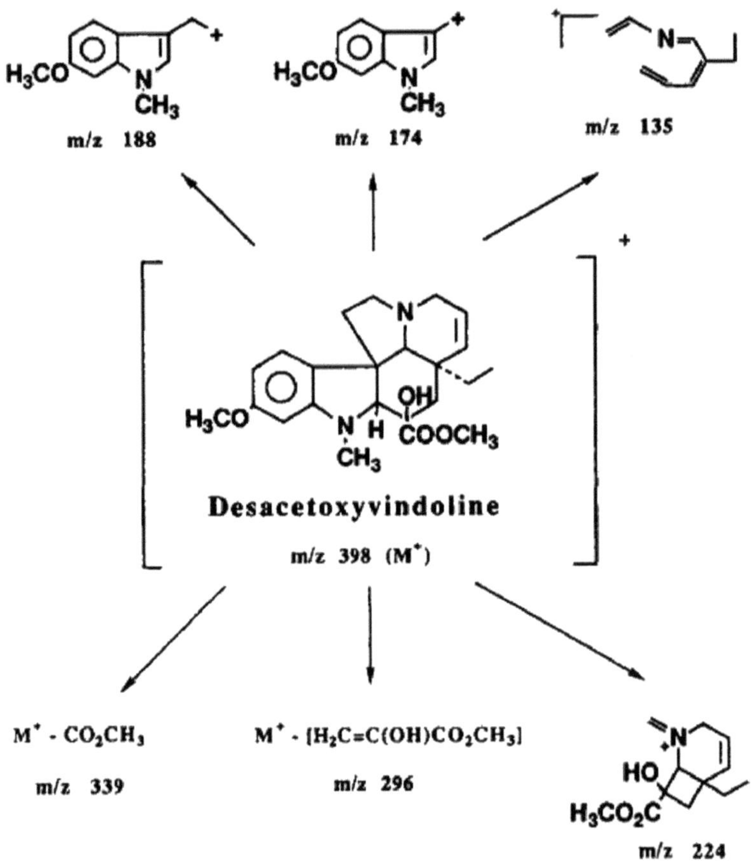

Fig. 5.13 Fragmentation pathways of desacetoxyvindoline (Bhadra et al. 1993)

5.3 Quantitative Analysis of Monoterpene Indole Alkaloids

5.3.1 Quantitative Analysis of Monoterpene Indole Alkaloids by HPTLC

In the last two decades, the HPTLC method has emerged as an important tool for the qualitative and quantitative phytochemical analysis of herbal drugs and formulations. This includes thin-layer TLC fingerprint profiles and estimation of chemical markers and biomarkers. The major advantage of HPTLC is that several samples can be analyzed simultaneously using a small quantity of mobile phase. Vincamine was separated by HPTLC using a mobile phase of a mixture of chloroform, acetone, and formic acid in a 5:1:0.5 ratio at a wavelength of 222 nm. Satisfactory separation of vincamine in the extract, with an Rf value of 0.33 ± 0.02, was reported. Regression analysis showed that the 200 and 1000 ng per spot showed a strong straight-line relationship with a correlation coefficient (R2) of 0.9957 (Viju et al. 2023). The authors claimed a good TLC method which provides a mixture of good separation of vincristine, vinblastine, vindoline, and ajmalicine using ethyl acetate:benzene:methanol:25% ammonia solution (100:5:5:3). Rf values were observed at 0.45, 0.63, 0.77, and 0.92 for vincristine, vinblastine, vindoline, and ajmalicine, respectively (Magagula et al. 2012). vinblastine and vincristine were estimated using TLC plates: silica gel 60 F254 (Merck), size 20 × 20 cm with the mobile phase ethyl acetate: benzene: ethyl alcohol: 25% ammonia solution (100:5:5:3). The wavelength scan at λ = 254, 289, 366, 400, and 500 nm were selected for both compounds with scan speed 20 mm/s. Rf values have been reported as 0.54 and 0.21 for vinblastine and vincristine, respectively in *C. roseus* (Shams et al. 2009). A comparatively simpler, rapid, and efficient validated method for the separation of a mixture of vinblastine (VLB), vincristine (VCR), leurosine, deacetoxy-VLB, N-demethyl-VLB, and deacetyl-VLB has been reported Nagy-Turák and Végh (1994) for separation on amino-bonded silica gel HPTLC plates using the mobile phase hexane:dichloromethane:acetone:2-propanol = 65:13:21:0.9 (v/v). Validation parameters showed good linearity (100–1000 ng/spot for VLB), a limit of detection (LOD) ~50 ng, precision (RSD: relative standard deviation) 2.9%. VLB content (mean) was 246.8 mg/kg (Nagy-Turák and Végh 1994).

5.3.2 Quantitative Analysis of Monoterpene Indole Alkaloids by High-Performance Liquid Chromatography

Since the last six decades, HPLC has played a progressively more important role as a separation analytical technique for phytoconstituents present in plant extracts. Its range of application is far wider, which provided the impetus that led to the rapid development of analytical methods for qualitative and quantitative. HPLC has been found to be important for the separation of alkaloids, which provide strong ultraviolet absorption, and their detection. Many different separation mechanisms have been used in the HPLC analysis of alkaloids. Liquid–liquid partition has been used effectively for the less polar alkaloids but is inappropriate for the more basic ones. Classical adsorption chromatography on silica generally creates broad asymmetrical peaks, which can be improved by the inclusion of a basic modifier, such as diethylamine, in the solvent. Far superior results are obtained if the solvent system is partially aqueous and buffered to acidic, neutral, or basic pH values. The separation mechanism is not clearly understood but can be effective. Modification of the silica with silver ions has been claimed to provide satisfactory results. Ion-exchange chromatography can give good results, but doubts have been expressed concerning the stability of ion exchangers under the conditions used to separate alkaloids. Normal-phase ion pairing and reversed-phase have been applied to alkaloid separation, but the reversed-phase equivalent proved superior to normal-phase. As stated above, indole alkaloids from extracts of *Catharanthus* were separated, identified, and detected by UV. A single wavelength, particularly 218 nm, seems to be optimal at maximizing the absorbance level for most of the alkaloids. However, 254 and 340 nm for serpentine and tabersonine-like alkaloids will allow higher precision and limits of detection, with signal-to-noise ratios 4 and 60 times higher, respectively, than at 218 nm. Therefore, a PDA detector having multiple UV channels seems preferable for quantitative studies of alkaloid mixtures. Indeed, setting the detector channel to a specific wavelength for alkaloids allows increasing the baseline by suppressing the noise from low-abundant alkaloids and the mobile phase. Moreover, as suggested by different researchers, peak identification may be more reliable using the UV–VIS spectrum obtained by PDA than just relying upon retention time (Shanks et al. 1998; Zhou et al. 2005, 2015; Zheng and Wu 2004). The LOD for vindoline and catharanthine was 10 and 0.52 μg/mL, respectively, according to the method developed by Naaranlahti et al. (1987) and the method established by Tikhomiroff and Jolicoeur (2002) (Hisiger and Jolicoeur 2007).

The corynanthe-type alkaloids, such as ajmalicine and serpentine, are fluorescent and can be detected using a fluorescence detector (Drapeau et al. 1987; Knobloch et al. 1982; Zenk et al. 1977; Monforte-González et al. 1992). Even though these alkaloids can be easily detected under fluorescence, it is more sensitive with a lower LOD at the micromolar level. The precursor's tryptophan, tryptamine, and the alkaloid ajmalicine can efficiently be detected at Ex 270/Em 370 nm, although the maximum excitation wavelength for ajmalicine is Ex 290 nm, whereas it is Ex 270 nm for both precursors. Limits of detection for tryptophan and tryptamine were lowered to 0.01 l g/mL– (Tikhomiroff and Jolicoeur 2002). Renaudin (1984) reported that the LOD can be reduced by a factor of six for ajmalicine and vindoline and even by a factor of 50 for serpentine by using fluorescence detection instead of traditional UV detection. Combining the use of fluorescence and absorbance detectors, alkaloids that co-elute but which show different physical characteristics can still be quantified by selecting appropriate wavelengths. It was possible to measure serpentine at 306 nm even when it was mixed with another unknown alkaloid (Tikhomiroff and Jolicoeur 2002). Serpentine with vinblastine has co-elution and has no effect on the serpentine response factor since no quenching occurs; they can then be distinguished from their respective fluorescence spectra (Renaudin 1984). Nevertheless, this approach requires validation for each peak with the identification of the co-eluting alkaloids and by establishing the detection conditions minimizing interference phenomena (Hisiger and Jolicoeur 2007).

TIAs are often more hydrophobic than their indole and iridoid precursors. It is advisable to establish two HPLC separation procedures, one for hydrophilic precursors and another for hydrophobic TIAs. Table 5.3 presents the various strategies that have been developed for HPLC validation. Some procedures exhibit significant similarity, differing primarily due to minor variations in equipment performance or unique laboratory research objectives. Shanks et al. (1998) revised a technique previously described by Bhadra et al. (1993) by substituting the C18 μBondapak 100 × 8 mm column (WAaters) with a C18 Bondclone, 10 μm, 300 × 3.8 mm column (Phenomenex), and employing a gradient flow mode utilizing a methanol and 5 mM $(NH_4)_2$ HPO_4 (pH 7.3) mixture (58:42, v:v). (Yang et al. 2011; Peebles et al. 2006). Altered techniques were employed for the quantification of ajmalicine, serpentine, and tabersonine; however, catharanthine could not be quantified due to interfering analytes in the UV spectrum (Verma et al. 2008; Tang et al. 2009; Thakore et al. 2015; Siddiqui et al. 2011; Benyammi et al. 2016; Zhou et al. 2015) (Table 5.3).

Table 5.3 The HPLC methods development and validation parameters

Compound	Detection	Chromatographic column	Mobile phase	Injection volume (μL)	Flow rate (mL/min)	Run time (min)	LOD/LOQ (ng/mL)	Linearity range (ng/mL)	Recovery/precision (%)	Reference
Ajmalicine	UV at 282 nm	Luna RP-C18 (2), 250 × 4.6 mm, 5 μm	Acetonitrile/1% Acetic acid	20	1.0	0–55	11.5/38.4		79%/RSD: 0.2–2.1%	Ferreres et al. (2010)
Catharanthine	UV at 282 nm	Luna RP-C18 (2), 250 × 4.6 mm, 5 μm	Acetonitrile/1% Acetic acid	20	1.0	0–55	2.4/8.0		85%/RSD: 0.2–2.1%	Ferreres et al. (2010)
Serpentine	UV at 282 nm	Luna RP-C18 (2), 250 × 4.6 mm, 5 μm	Acetonitrile/1% Acetic acid	20	1.0	0–55	3.1/10.2		56%/RSD: 0.2–2.1%	Ferreres et al. (2010)
Vindoline	UV at 297 nm	Luna C18 (250 × 4.6 mm, 5 μm)	ACN: 0.1% HCOOH (50:50)	5	1.0	8.7	2.4/8.0	20–1000	98.6/RSD: 1.62%	Hisiger and Jolicoeur (2007)
Catharanthine	UV at 297 nm	Luna C18 (250 × 4.6 mm, 5 μm)	ACN: 0.1% HCOOH (50:50)	5	1.0	6.92	2.7/9.1	20–1000	99.2/RSD: 1.43%	Hisiger and Jolicoeur (2007)
Ajmalicine	UV at 297 nm	Luna C18 (250 × 4.6 mm, 5 μm)	ACN: 0.1% HCOOH (50:50)	5	1.0	4.62	3.1/10.3	20–1000	97.5/RSD: 1.83%	Hisiger and Jolicoeur (2007)
Serpentine	UV at 297 nm	Luna C18 (250 × 4.6 mm, 5 μm)	ACN: 0.1% HCOOH (50:50)	5	1.0	5.44	3.5/11.7	20–1000	96.8/RSD: 1.94%	Hisiger and Jolicoeur (2007)
Tabersonine	UV at 297 nm	Luna C18 (250 × 4.6 mm, 5 μm)	ACN: 0.1% HCOOH (50:50)	5	1.0	7.28	2.9/9.6	20–1000	98.1/RSD: 1.71%	Hisiger and Jolicoeur (2007)

(continued)

Table 5.3 (continued)

Compound	Detection	Chromatographic column	Mobile phase	Injection volume (μL)	Flow rate (mL/min)	Run time (min)	LOD/LOQ (ng/mL)	Linearity range (ng/mL)	Recovery/precision (%)	Reference
Vindorosine	UV at 297 nm	Luna C18 (250 × 4.6 mm, 5 μm)	ACN: 0.1% HCOOH (50:50)	5	1.0	9.55	2.2/7.4	20–1000	99.0/RSD: 1.39%	Hisiger and Jolicoeur (2007)
Vinblastine	PDA at 254 nm	Microsorb-MV C18 (150 × 4.6 mm, 5 μm)	Gradient of methanol, 5 mM phosphate buffer (pH 6.6), acetonitrile	20	2.0	~6.87	23.0/69.8	32–56	Recovery: 95–103%; Intra-day RSD: 1.11–2.21%, Inter-day RSD: 1.14–2.37%	Abdul Rahim et al. (2018)
Serpentine	UV at 306 nm	Zorbax Eclipse XDB-C18 (250 × 4.6 mm)	5 mM phosphate buffer (pH 6): acetonitrile (80:20 → 20:80)	10	2.0	~20	470/–	3000–96,100	32.4 ± 4.3%	Tikhomiroff and Jolicoeur (2002)
Vincristine	UV at 220 nm	Zorbax Eclipse XDB-C18 (250 × 4.6 mm)	6 mM phosphate buffer (pH 6): Acetonitrile (80:20 → 20:80)	10	2.0	~20	420/–	2680–85,800	43.2 ± 9.9%	Tikhomiroff and Jolicoeur (2002)
Vindoline	UV at 220 nm	Zorbax Eclipse XDB-C18 (250 × 4.6 mm)	7 mM phosphate buffer (pH 6): acetonitrile (80:20 → 20:80)	10	2.0	~20	520/–	3050–97,600	39.2 ± 3.2%	Tikhomiroff and Jolicoeur (2002)
Catharanthine	UV at 220 nm	Zorbax Eclipse XDB-C18 (250 × 4.6 mm)	8 mM phosphate buffer (pH 6): acetonitrile (80:20 → 20:80)	10	2.0	~20	490/–	2840–90,800	24.5 ± 1.5%	Tikhomiroff and Jolicoeur (2002)

Vinblastine	UV at 220 nm	Zorbax Eclipse XDB-C18 (250 × 4.6 mm)	9 mM phosphate buffer (pH 6): acetonitrile (80:20 → 20:80)	10	2.0	~20	610/–	3700–118,000	88.2 ± 16.2%	Tikhomiroff and Jolicoeur (2002)
Tabersonine	UV at 330 nm	Zorbax Eclipse XDB-C18 (250 × 4.6 mm)	10 mM phosphate buffer (pH 6): acetonitrile (80:20 → 20:80)	10	2.0	~20	93/–	2720–87,100	34.5 ± 1.2%	Tikhomiroff and Jolicoeur (2002)
Tryptophan	Fluorescence (270/370 nm)	Zorbax Eclipse XDB-C18 (250 × 4.6 mm)	100 mM phosphate buffer (pH 2): acetonitrile (85:15)	10	1.8	~13	43/–	1770–56,600	54.8 ± 8.4%	Tikhomiroff and Jolicoeur (2002)
Tryptamine	Fluorescence (270/370 nm)	Zorbax Eclipse XDB-C18 (250 × 4.6 mm)	101 mM phosphate buffer (pH 2): acetonitrile (85:15)	10	1.8	~13	34/–	1520–48,500	85.4 ± 23.4%	Tikhomiroff and Jolicoeur (2002)
Secologanine	UV at 238 nm	Zorbax Eclipse XDB-C18 (250 × 4.6 mm)	102 mM phosphate buffer (pH 2): acetonitrile (85:15)	10	1.8	~13	750/–	2730–87,400	110.8 ± 5.3%	Tikhomiroff and Jolicoeur (2002)
Ajmalicine	Fluorescence (270/370 nm)	Zorbax Eclipse XDB-C18 (250 × 4.6 mm)	103 mM phosphate buffer (pH 2): acetonitrile (85:15)	10	1.8	~13	970/–	3630–116,000	98.3 ± 7.81%	Tikhomiroff and Jolicoeur (2002)
Vincristine	PDA @ 220 nm	Symmetry® C18 (150 × 4 mm, 5 µm)	Solvent A: MeOH:ACN:Am. acetate:TEA (13:32:55:0.2); solvent B: MeOH:ACN:Am. acetate:TEA (19:46:35:0.2); System 2: Am. acetate 0.005 M, pH 7.0	10	0.4 → 1.2	~33	1.0 µg (1000 ng/mL)	50–200 µg/mL		Uniyal et al. (2001)

(continued)

Table 5.3 (continued)

Compound	Detection	Chromatographic column	Mobile phase	Injection volume (µL)	Flow rate (mL/min)	Run time (min)	LOD/LOQ (ng/mL)	Linearity range (ng/mL)	Recovery/precision (%)	Reference
			Solvent A: MeOH:ACN:Am. acetate:TEA (13:32:55:0.2); solvent B: MeOH:ACN:Am. acetate:TEA (19:46:35:0.2); system 2: Am. acetate 0.005 M, pH 7.1							
			Solvent A: MeOH:ACN:Am. acetate:TEA (13:32:55:0.2); solvent B: MeOH:ACN:Am. acetate:TEA (19:46:35:0.2); system 2: Am. acetate 0.005 M, pH 7.2							

Catharanthine	PDA @ 220 nm	Symmetry® C18 (150 × 4 mm, 5 µm)	Solvent A: MeOH:ACN:Am. acetate:TEA (13:32:55:0.2); solvent B: MeOH:ACN:Am. acetate:TEA (19:46:35:0.2); system 2: Am. acetate 0.005 M, pH 7.3	10 µL	0.4 → 1.2	~33	0.5 µg (500 ng/mL)	50–200 µg/mL		Uniyal et al. (2001)
Vinblastine	PDA @ 220 nm	Symmetry® C18 (150 × 4 mm, 5 µm)	Solvent A: MeOH:ACN:Am. acetate:TEA (13:32:55:0.2); solvent B: MeOH:ACN:Am. acetate:TEA (19:46:35:0.2); system 2: Am. acetate 0.005 M, pH 7.4	10	0.4 → 1.2	~33	0.5 µg (500 ng/mL)	50–200 µg/mL		Uniyal et al. (2001)
Vindoline	PDA @ 220 nm	Symmetry® C18 (150 × 4 mm, 5 µm)	Solvent A: MeOH:ACN:Am. acetate:TEA (13:32:55:0.2); solvent B: MeOH:ACN:Am. acetate:TEA (19:46:35:0.2); system 2: Am. acetate 0.005 M, pH 7.5	10	0.4 → 1.2	~33	0.5 µg (500 ng/mL)	50–200 µg/mL		Uniyal et al. (2001)

(continued)

Table 5.3 (continued)

Compound	Detection	Chromatographic column	Mobile phase	Injection volume (μL)	Flow rate (mL/min)	Run time (min)	LOD/LOQ (ng/mL)	Linearity range (ng/mL)	Recovery/precision (%)	Reference
Vincristine	UV at 254 nm	Chromolith RP-18e (100 × 4.6 mm i.d.)	Acetonitrile: 0.1 M phosphate buffer (pH 3.5) with 0.5% glacial acetic acid (21:79, v/v)		1.2	16.57	8/25 μg/mL	0.25–25 μg/mL	96%/CV: 2.50%	Gupta et al. (2005)
Vinblastine	UV at 254 nm	Chromolith RP-18e (100 × 4.6 mm i.d.)	Acetonitrile: 0.1 M phosphate buffer (pH 3.5) with 0.5% glacial acetic acid (21:79, v/v)		1.2	26.93	14/46 μg/mL	0.25–25 μg/mL	97%/CV: 1.82%	Gupta et al. (2005)
Catharanthine	UV at 254 nm	Chromolith RP-18e (100 × 4.6 mm i.d.)	Acetonitrile: 0.1 M phosphate buffer (pH 3.5) with 0.5% glacial acetic acid (21:79, v/v)		1.2	6.81	18/56 μg/mL	0.25–25 μg/mL	98%/CV: 1.33%	Gupta et al. (2005)
Vindoline	UV at 254 nm	Chromolith RP-18e (100 × 4.6 mm i.d.)	Acetonitrile: 0.1 M phosphate buffer (pH 3.5) with 0.5% glacial acetic acid (21:79, v/v)		1.2	13.22	10/32 μg/mL	0.25–25 μg/mL	98%/CV: 1.13%	Gupta et al. (2005)
Vinblastine	UV at 254 nm	C18 (250 × 4.6 mm, 5 μm)	Acetonitrile:Water (30:70, v/v)	20	1.0	10	LOD: 2.0, LOQ: 5.0	50–1000	97.2–98.1%; RSD < 2.0%	Verma et al. (2008)

Vincristine	UV at 254 nm	C18 (250 × 4.6 mm, 5 μm)	Acetonitrile:Water (30:70, v/v)	20	1.0	10	LOD: 2.0, LOQ: 5.0	50–1000	96.8–98.3%; RSD < 2.0%	Verma et al. (2008)
Vincristine	DAD (λ = 297 nm)	C18, 250 × 4.6 mm, 5 μm	Acetonitrile/0.2% formic acid in water (37:63, v/v)	10	1.0	7.5	C. roseus: 29.5	100–800	C. roseus: 97.1–101.9% Serum: 98.8–103.0%	Zhang et al. (2014)
							Serum: 12.4		C. roseus: 97.1–101.9% Serum: 98.8–103.0%	
Vinblastine (VLB)	DAD (λ = 297 nm)	C18, 250 × 4.6 mm, 5 μm	Acetonitrile/0.2% formic acid in water (37:63, v/v)	10	1.0	7.5	*C. roseus*: 32.1.	100–800	C. roseus: 97.1–101.9% Serum: 98.8–103.0%	Zhang et al. (2014)
							Serum: 18.5		C. roseus: 97.1–101.9%serum: 98.8–103.0%	
Vindoline (VDL)	DAD (λ = 297 nm)	C18, 250 × 4.6 mm, 5 μm	Acetonitrile/0.2% formic acid in water (37:63, v/v)	10	1.0	7.5	*C. roseus*: 49.3	100–800	C. roseus: 97.1–101.9% Serum: 98.8–103.0%	Zhang et al. (2014)
							Serum: 27.2/		C. roseus: 97.1–101.9% Serum: 98.8–103.0%	

(continued)

Table 5.3 (continued)

Compound	Detection	Chromatographic column	Mobile phase	Injection volume (μL)	Flow rate (mL/min)	Run time (min)	LOD/LOQ (ng/mL)	Linearity range (ng/mL)	Recovery/precision (%)	Reference
Catharanthine (CAT)	DAD (λ = 297 nm)	C18, 250 × 4.6 mm, 5 μm	Acetonitrile/0.2% formic acid in water (37:63, v/v)	10	1.0	7.5	*C. roseus*: 33.7.	100–800	*C. roseus*: 97.1–101.9% Serum: 98.8–103.0%	Zhang et al. (2014)
							Serum: 17.8		C. roseus: 97.1–101.9% Serum: 98.8–103.0%	
Yohimbine (YHB)	DAD (λ = 297 nm)	C18, 250 × 4.6 mm, 5 μm	Acetonitrile/0.2% formic acid in water (37:63, v/v)	10	1.0	7.5	*C. roseus*: 30.2	100–800	C. roseus: 97.1–101.9%serum: 98.8–103.0%	
							Serum: 13.7		*C. roseus*: 97.1–101.9% Serum: 98.8–103.0%	Zhang et al. (2014)

The quantification of ajmalicine, serpentine, tabersonine, hörhammericine, and lochnericine was effectively achieved using isocratic elution at a flow rate of 1 ml min^{-1} utilizing a Bondclone column (Phenomenex, USA), C18 (300 × 3.8 mm, 10 μm). The metabolism of alkaloids in damaged sprouts was documented by Fernández-Pérez et al. (2013), utilizing the methodology established by Aerts et al. (1994) in their investigation of the impact of methyl jasmonate vapor on *C. roseus*. Tikhomiroff and Jolicoeur (2002) devised a method for screening secondary metabolites in hairy roots. Recent analyses of alkaloid accumulation in several plant cultivars and their suspension cultures employed several modified methodologies (Dutta et al. 2005). Successful separation of a mixture of ajmalicine, serpentine, and catharanthine tabersonine, vindoline, vinblastine, and vincristine can be accomplished with a saline buffer with potassium phosphate or ammonium acetate as the aqueous phase and methanol or acetonitrile as the organic phase (Bhadra et al. 1993; Tikhomiroff and Jolicoeur 2002). These approaches may reference the previously published research of Renaudin (1984) regarding the separation of ajmalicine, catharanthine, serpentine, vindoline, and vinblastine. Additional research indicates the application of ion-pairing additives, including triethylamine (TEA) or trifluoroacetic acid (TFA), are frequently employed to obscure nonderivatized silanol groups (Aerts et al. 1994; St-Pierre and De Luca 1995; Lee and Shuler 2000; Zhao et al. 2001; Uniyal et al. 2001; El-Sayed et al. 2004).

TIAs typically reveal that a C_{18} column (ideally after processing several thousand samples) exhibits diminished retention of alkaloids with low pKa (e.g., ajmalicine) while demonstrating enhanced retention of alkaloids with high pKa (such as tryptamine and serpentine) compared to a new column (Renaudin 1984). This behavior is attributed to the deterioration of the hydrophobic stationary phase caused by the uncapping of silanol groups after the breaking of the C18 alkyl chains. Amine modifier agents in the mobile phase influence the interaction between the stationary phase and basic analytes (e.g., alkaloids), thereby altering the elution sequence. Competitive amines can improve peak shape at neutral pH and diminish peak tailing (Nawrocki 1997; McCalley 2002). The pH of the mobile phase is a critical parameter, with an acidic pH (pH 3–4) facilitating improved peak shape for basic alkaloids. At low pH, the majority of silanol groups remain undissociated, hence constraining ion-exchange interactions, even in the presence of protonated alkaloids (McCalley 2002). To our knowledge, alkaline pH (pH 10) has been examined solely by several researchers, including Barthe et al. (2002), Rischer et al. (2006), Shams et al. (2009), Ataei-Azimi et al. (2008), and Liu et al. (2011). Commercially available columns exist for usage at severe alkaline pH; however, it is important to note that older generation columns (e.g., C8, phenyl, and C18) are unsuitable for high pH applications due to potential degradation of the stationary phase. In all instances, whether using acidic or basic mobile phases, methodologies should utilize a pH significantly different from the pKa of the target compounds, as minor fluctuations in mobile phase preparation can alter the protonation state of the alkaloids, thereby affecting the elution profile and the consistency of retention times (Uniyal et al. 2001; McCalley 2002; Shams et al. 2009; Ataei-Azimi et al. 2008; Liu et al. 2011). Barthe et al. (2002) discovered that the elution order of vindoline and catharanthine was inverted by altering the mobile phase pH from 5 to 10. A prudent amalgamation of buffer and mobile phase pH can be advantageous when formulating a separation procedure targeting certain peaks. The mobile phase pH of 4.7, as opposed to 3.9, facilitated the separation of strictosidine and nonpolar alkaloids from an extract of Tabernaemontana divaricata (Dagnino et al. 1995, 1996). Isocratic elution yielded adequate separation of alkaloids. Five alkaloids were isolated in a single run for *C. roseus* metabolites (Bhadra et al. 1993; Rijhwani and Shanks 1998). Moreover, a markedly enhanced separation of the alkaloids was attained using gradient elution, facilitating effective separation without prolonging HPLC run time (Naaranlahti et al. 1987; Tikhomiroff and Jolicoeur 2002). Examination of a crude extract from *C. roseus* hairy roots utilizing the gradient technique established by Tikhomiroff and Jolicoeur (2002) over a duration of 23 min. This review illustrates the efficacy of a gradient method for the separation of intricate combinations of *C. roseus* alkaloids, enabling the simultaneous isolation of up to 12 recognized compounds in a single run. A method utilizing negative-pressure cavitation extraction (NPCE) followed by reverse-phase RP-HPLC was devised and optimized for the extraction and measurement of vindoline and catharanthine from *C. roseus* leaves. About 60-mesh particles were utilized, with 80% ethanol, a negative pressure of −0.075 MPa, a solid-to-liquid ratio of 1:20, an extraction duration of 30 min, and three extraction cycles. Under these optimized circumstances, the extraction yields of vindoline and catharanthine are 0.5783 mg/g DW and 0.2843 mg/g DW, respectively, along with 0.018 mg/g DW and 0.126 mg/g DW. The extraction yields are comparable to those obtained using the established ultrasonic extraction method and exceed the yields from maceration extraction and heated reflux extraction. The method asserts that normal phase column elution—reverse phase—high-performance liquid chromatography (NPCE-RP-HPLC) serves as an exceptional option for the extraction and measurement of vinca alkaloids in both pilot and industrial-scale applications (Mu et al. 2012).

Kohl et al. (1982) established HPLC methods for the determination of alkaloids from *C. roseus* tissue cultures. Ajmalicine and catharanthine were identified as principal alkaloids, while tetrahydroalstonine, akuammigine, 3-isoajmalicine, pleiocarpamine, akuammiline, tabersonine, 20-hydroxytabersonine, lochnericine, horhammericine, minovincinine, vindolinine, 20-epivindolinine, and 20-epivindolinine-Nb-oxide may be classified as minor or trace alkaloids in *C. roseus* tissue cultures (Fernández-Pérez et al. 2013). The RP-HPLC validation method was established by Mu et al., utilizing a HiQ sil C18W column (4.6 mm × 250 mm) with isocratic elution. The mobile phase consisted of solution A (methanol/acetonitrile in a 4:1 ratio) and solution B (water/diethylamine in a 986:14 ratio, with pH adjusted to 7.2 using H_3PO_4), at a flow rate of 2.0 mL min^{-1} over a duration of 25 min. The validation method utilized for the extraction of vindoline and catharanthine from *C. roseus* leaves was compared with other established extraction techniques. Under optimum conditions, the extraction yields of vindoline and catharanthine were 0.5783 mg/g DW and 0.2843 mg/g DW, respectively (Mu et al. 2012). The cosolvent marginally enhanced the extraction yields of indole alkaloids, including vindoline. The crude alkaloid extract was fractionated using a μBondapak C18 column (30 cm × 3.9 mm). The run was conducted utilizing a methanol/acetonitrile/5 mM diammonium hydrogen phosphate (pH 7.3–7.5) mixture at varying ratios, with a flow rate ranging from 0.5 to 21 mL/min. The exceptional maximum concentration of vindoline, 58 wt%, was achieved at the minimum temperature of 35 °C and the maximum pressure of 300 bar (Song et al. 1992; Shams et al. 2009; Ataei-Azimi et al. 2008; Liu et al. 2011). Digvijay et al. (2004) performed investigations on the concurrent measurement of terpenoid indole alkaloids and their precursors in the leaves and roots of various genotypes utilizing reversed-phase liquid chromatography. The concentrations of serpentine varied from 0.0034% to 0.6490%, catharanthine from 0.003% to 0.0992%, and ajmalicine from 0.0085% to 0.2910% (dry weight), while tabersonine was not detected. The presentation of the data as percentages complicates direct comparison significantly.

5.3.3 Quantitative Analysis of Monoterpene Indole Alkaloids by High-Performance Liquid Chromatography Mass Spectrometry

Various modifications in HPLC methods, including mobile phase, flow rate, injection volume, pH, and column selection, have been elucidated to enhance sensitivity, selectivity, reproducibility, and peak shape. However, alterations in detection have further augmented the efficacy of HPLC methods as a formidable analytical technique (Nakagawa et al. 1987; Theodoridis et al. 2012; Mathias et al. 2017). The HPLC procedures should be streamlined by optimizing many characteristics, including reduced run time, enhanced sensitivity, selectivity, reproducibility, peak shape, and the ability to integrate more advanced detectors. The determination of molecular weight for each peak is advantageous for analyzing plant or cell extracts, as opposed to UV and fluorescence detection (Gómez-Ariza et al. 2005; Nobilis et al. 2003; Kaufmann et al. 2010). The utilization of thermospray mass spectrometry (TSP-MS), electrospray ionization mass spectrometry (ESI-MS), or matrix-assisted laser desorption with time-of-flight (MALDI-TOF) significantly improves the detection of low-abundance alkaloids in complex matrices, such as plant extracts, with greater precision than UV and fluorescence detection (Zschocke et al. 1998; Gómez-Ariza et al. 2005; Chee et al. 1996; Chandra et al. 2001; Papagiannopoulos et al. 2004). Amine modifiers can contaminate the mobile phase as masking agents; hence, their usage is not advised for ESI-MS detection. The modifiers saturate the detector, resulting in an observable ion-suppression phenomenon, which significantly diminishes the instrument's sensitivity (Antignac et al. 2005; Jessome and Volmer 2006; Colizza et al. 2016; Gustavsson et al. 2001). The choice of columns incorporating a carbamate internal group in the bonded phase has enhanced the separation. Peak tailing was found to be diminished for basic compounds in comparison to conventional C18 alkyl packing (Uniyal et al. 2001; O'Gara et al. 1999; Verpoorte and Niessen 1994). To the best of your knowledge, Auriolat and his team have previously described a quantitative approach for the analysis of indole alkaloids from complicated mixtures by combining HPLC–MS with thermospray. The detection limit was 100 pg per injection, and the response exhibited linearity up to 200 ng for catharanthine, tabersonine, serpentine, and ajmalicine (Auriola et al. 1990; Stöckigt et al. 2002). Semiquantitative quantification of TIAs based on molecular weight (m/z) in *Catharanthus* has been conducted using flow injection electrospray ionization mass spectrometry (Favretto et al. 2001; Zhou et al. 2005). Nevertheless, the majority of prevalent HPLC techniques utilize nonvolatile buffers, such as phosphate buffer, which are incompatible with MS detection (Law and Temesi 2000). Various HPLC–MS methods have been reported by many research groups and are summarized in Table 5.4.

Jeong and Lim (2018) devised a sensitive, quick, and innovative UPLC-Q-TOF-MS approach for the quantification of indole alkaloids (vincristine, vinblastine, ajmalicine, catharanthine, serpentine, and vindoline) in *C. roseus*. The alkaloids were isolated on a C_{18} column in 13 min. UPLC-Q-TOF-MS demonstrated a LOD ranging from 1 to 10 ng/mL and a limit of quantification from 3 to 30 ng/mL. The validated method was utilized for the quantification of extracts from the root, stem, and leaves of *C. roseus*. The concentrations of bisindole alkaloids (vincristine and vinblastine) were elevated in the aerial portions, but the concentrations of mono-indole alkaloids (ajmalicine, catharanthine, serpentine, and vindoline) were elevated in the roots (Jeong and Lim 2018).

Chen et al. (2013) demonstrated comparative chromatographic separations and detection of five alkaloids utilizing a DL-C18 column. MS exhibited superior selectivity and sensitivity for the measurement of five alkaloids in positive mode compared to ultraviolet detection. The separation, identification, and quantification of the five components in *C. roseus* were thoroughly achieved using HPLC with UV and MS detection. The Student's *t*-test at a 95% confidence level was employed to assess these approaches, revealing no significant difference between them. LC–MS demonstrated superior sensitivity, precision, and increased sample loading capacity. Nonetheless, LC–UV (liquid chromatography–ultraviolet detection) frequently exhibited superior precision (Chen et al. 2013). Consequently, the established LC–UV and LC–MS methodologies are appropriate for quantifying five active alkaloids in *C. roseus*. Vindoline, ajmalicine, and catharanthine were quantified at 331, 92.2, and 40.60 μg/g via LC-MS, while UV detection indicated levels of 125, 76.3, and 45.2 μg/g, respectively.. To enhance the sensitivity, selectivity, and accuracy of mass spectrometry, selective ion monitoring techniques utilizing UPLC-MS have been reported. The chromatographic separations of indole alkaloids were conducted using a Waters Acquity UPLC® BEH C18 column (50 mm × 2.1 mm, i.d. 1.7 μm) at a flow rate of 0.3 ml/min, with the mobile phase comprising acetonitrile (A) and 0.1% formic acid in water (B). Detection was performed using electrospray ionization MS in positive-ion mode with selective ion monitoring. The analysis of a single sample was completed in 6 min. The detection limits were 0.46–0.70 ng/mL for cell samples and 0.10–0.16 ng/mL for medium samples. The detection linearity spanned a broad range of 1.00–6250.0 ng/mL. This method offers a more sensitive and high-throughput approach for quantifying four alkaloids in substantial sample sizes, facilitating the screening of high-production cultivars.

Table 5.4 Mass spectrometric and chromatographic and validated parameters of developed HPLC-MS methods

Compound	Precursor ion $[M + H]^+$ (m/z)	Product ion (m/z)	Chromatographic column	Mobile phase	Injection volume (μL)	Flow rate (mL/min)	Run time	Retention (min)	LOD/LOQ (ng/mL)	Linearity Range (ng/mL)	Recovery/Precision (%)	Reference
Ajmalicine	353.3	144.2	Acquity UPLC BEH C18 column (1.7 μm, 2.1 × 50 mm	Acetonitrile and 0.1% formic acid in water (gradient)	7 μL	0.4	8	~1.8	0.48	1.0–6250	88–112/RSD < 7.8	He et al. (2011a)
Ajmalicine	353.3	353.3	UPLC BEH C18 (1.7 μm, 2.1 × 50 mm)	Acetonitrile–0.1% FA water (gradient)	5 μL	0.4	~6	~1.5–2.0	0.46–0.70	1.00–6250	Recovery: ~88–112%; RSD: 1.3–7.8	He et al. (2011b)
Ajmalicine	353	144	C18 HPLC	Acetonitrile–0.1% FA water (gradient)	–	–	–	–	–	–	–	He et al. (2011a)
Ajmalicine	353	144	Acquity UPLC BEH C_{18} (1.7 μm, 2.1 × 50 mm)	ACN/0.1% FA + ammonium acetate	5	–	–	0.039–0.583	0.5–1000	99.6–104.3%	0.23–2.71%/0.40–2.90%	He et al. (2011b)
Ajmalicine	353	144	Acquity UPLC BEH C18 (2.1 × 50 mm, 1.7 μm)	0.1% FA in water (A)/ACN (B)	4 μL	0.4	15	3.12	02-Jun	10–10,000	97.2/<3.1	Jeong and Lim (2018)

(continued)

Table 5.4 (continued)

Compound	Precursor ion $[M + H]^+$ (m/z)	Product ion (m/z)	Chromatographic column	Mobile phase	Injection volume (μL)	Flow rate (mL/min)	Run time	Retention (min)	LOD/LOQ (ng/mL)	Linearity Range (ng/mL)	Recovery/Precision (%)	Reference
Ajmalicine	353	144	C18 (2.1 × 100 mm, 1.7 μm)	Water/0.1% FA (A) & ACN (B)		0.3	15	~3.4	~2/~6	5–500	97.2/<5.1	Chen et al. (2017)
Ajmalicine	353.3	144.1	BEH C18 (2.1 × 100 mm, 1.7 μm)	A: Water +0.1% FA; B: Acetonitrile	12 μL	0.2	10	4.71	0.2/0.5	0.5–100	95.5/RSD < 5.3	He et al. (2011a)
Ajmalicine	353.2	144.1	BEH C18 (50 × 2.1 mm, 1.7 μm)	0.1% FA in water:ACN (90:10, v/v)	2	0.2	10	2.79	0.550/1.666	0.1–250	97.05/RSD < 3%	Kumar et al. (2018b)
Ajmalicine	−353	144	BEH C18 (2.1 × 100 mm, 1.7 μm)	A: Water +0.1% FA; B: Acetonitrile	5 μL	0.3	~10	~5.1	0.45	1–1000	93.6/<4.8 RSD	Zhang et al. (2014)
Ajmaline	327.2	144.1	BEH C18 (50 × 2.1 mm, 1.7 μm)	0.1% FA in water: ACN (90:10, v/v)	2	0.2	10	2.43	0.039/0.118	0.1–500	98.56/RSD < 2%	Kumar et al. (2018b)

Catharanthine	337.3	144.1	Acquity UPLC BEH C18 column (1.7 µm, 2.1 × 50 mm	Acetonitrile and 0.1% formic acid in water (gradient)	6 µL	0.4	7	~2.6	0.62	1.0–6250	95–106/RSD < 6.8	He et al. (2011a)
Catharanthine	~337.3	337.3	UPLC BEH C18 (1.7 µm, 2.1 × 50 mm)	Acetonitrile–0.1% FA water (gradient)	5 µL	0.4	~6	~2.0–2.6	0.46–0.70	1.00–6250	Recovery: ~88–112%; RSD: 1.3–7.8	He et al. (2011b)
Catharanthine	337	144	Acquity UPLC BEH C18 (2.1 × 50 mm, 1.7 µm)	0.1% FA in water (A)/ACN (B)	3 µL	0.4	14	3.46	02-Jun	10–10,000	98.7/<3.0	Jeong and Lim (2018)
Catharanthine	337	144	C18 (2.1 × 100 mm, 1.7 µm)	Water/0.1% FA (A) & ACN (B)		0.3	13	~3.2	~2/~6	5–500	96.8/<4.5	Chen et al. (2017)
Catharanthine	337.3	144.1	BEH C18 (2.1 × 100 mm, 1.7 µm)	A: Water +0.1% FA; B: Acetonitrile	11 µL	0.2	9	3.62	0.1/0.3	0.3–100	96.7/RSD < 5.0	He et al. (2011a)
Catharanthine	337.2	144.1	RP-C18 column (Agilent)	ACN:Water with 0.1% formic acid	10 µL	0.3	15	6.3				Zweil et al. (2019)

(continued)

Table 5.4 (continued)

Compound	Precursor ion $[M + H]^+$ (m/z)	Product ion (m/z)	Chromatographic column	Mobile phase	Injection volume (µL)	Flow rate (mL/min)	Run time	Retention (min)	LOD/LOQ (ng/mL)	Linearity Range (ng/mL)	Recovery/Precision (%)	Reference
Catharanthine	−337	144	RP-C18 column (Agilent)	ACN:Water with 0.1% formic acid	5 µL	0.3	~10	~4.2	0.39	1–1000	97.3/<5.1 RSD	Zhang et al. (2014)
Reserpine	609.2	195.1	BEH C18 (50 × 2.1 mm, 1.7 µm)	0.1% FA in water: ACN (90:10, v/v)	2	0.2	10 min	3.38	0.340/1.032	0.5–1000	97.31/RSD < 2%	Kumar et al. (2018a, b)
Serpentine	349.3	144.1	Acquity UPLC BEH C18 column (1.7 µm, 2.1 × 50 mm	Acetonitrile and 0.1% formic acid in water (gradient)	8 µL	0.4	9	~3.0	0.46	1.0–6250	90–107/RSD < 6.5	He et al. (2011a)
Serpentine	~349.3	349.3	UPLC BEH C18 (1.7 µm, 2.1 × 50 mm)	Acetonitrile–0.1% FA water (gradient)	5 µL	0.4	~6	~2.5–3.0	0.46–0.70	1.00–6250	Recovery: ~88–112%; RSD: 1.3–7.8	He et al. (2011b)
Serpentine	349	317	Acquity UPLC BEH C_{18} (1.7 µm, 2.1 × 50 mm)	ACN/0.1% FA + ammonium acetate	5	–	–	0.039–0.583	0.5–1000	99.6–104.3%	0.23–2.71%/0.40–2.90%	

Serpentine	349	317	Acquity UPLC BEH C18 (2.1 × 50 mm, 1.7 μm)	0.1% FA in water (A)/ACN (B)	5 μL	0.4	16	3.94	01-Mar	10–10,000	96.5/<2.9	Jeong and Lim (2018)
Serpentine	349	317	C18 (2.1 × 100 mm, 1.7 μm)	Water/0.1% FA (A) & ACN (B)		0.3	14	~4.4	~1/~3	5–500	95.9/<4.8	Chen et al. (2017)
Serpentine	349.3	144.1	BEH C18 (2.1 × 100 mm, 1.7 μm)	A: Water +0.1% FA; B: Acetonitrile	13 μL	0.2	11	5.3	0.2/0.5	0.5–100	94.2/RSD < 6.0	He et al. (2011a)
Serpentine	349.2	144.1	BEH C18 (50 × 2.1 mm, 1.7 μm)	0.1% FA in water: ACN (90:10, v/v)	2	0.2	10	3.12	0.437/1.326	0.5–500	96.33/RSD < 3%	Kumar et al. (2018b)
Serpentine	–349	144	BEH C18 (2.1 × 100 mm, 1.7 μm)	Same as above	5 μL	0.3	~10	~5.8	0.42	1–1000	94.7/<4.5 RSD	Zhang et al. (2014)
Vindoline	457.2	397.2	Acquity UPLC BEH C18 column (1.7 μm, 2.1 × 50 mm	Acetonitrile and 0.1% formic acid in water (gradient)	5 μL	0.4	6	~1.3	0.7	1.0–6250	92–105/RSD < 7.2	He et al. (2011a)
Vindoline	~457.2	457.2	UPLC BEH C18 (1.7 μm, 2.1 × 50 mm)	Acetonitrile–0.1% FA water (gradient)	5 μL	0.4	~6	~1.0–1.5	0.46–0.70 (cells)	1.00–6250	Recovery: ~88–112%; RSD: 1.3–7.8	He et al. (2011b)

(continued)

Table 5.4 (continued)

Compound	Precursor ion $[M + H]^+$ (m/z)	Product ion (m/z)	Chromatographic column	Mobile phase	Injection volume (μL)	Flow rate (mL/min)	Run time	Retention (min)	LOD/LOQ (ng/mL)	Linearity Range (ng/mL)	Recovery/Precision (%)	Reference
Vindoline	457	397	Acquity UPLC BEH C_{18} (1.7 μm, 2.1 × 50 mm)	ACN/0.1% FA + ammonium acetate			–	0.039–0.583	0.5–1000	99.6–104.3%	0.23–2.71%/0.40–2.90%	Kumar et al. (2018b)
Vindoline	457	397	Acquity UPLC BEH C18 (2.1 × 50 mm, 1.7 μm)	0.1% FA in water (A)/ACN (B)	2 μL	0.4	13	4.22		10–10,000	99.8/<2.7	Jeong and Lim (2018)
Vindoline	457	397	C18 (2.1 × 100 mm, 1.7 μm)	Water/0.1% FA (A) and ACN (B)		0.3	12	~3.9	~1/~3	5–500	97.6/<5.0	Chen et al. (2017)
Vindoline	457.3	144.1	BEH C18 (2.1 × 100 mm, 1.7 μm)	A: Water +0.1% FA; B: Acetonitrile	10 μL	0.2	8	5.91	0.1/0.3	0.3–100	98.3/RSD < 4.2	He et al. (2011a)
Vindoline	457.3	144.1	BEH C18 (2.1 × 100 mm, 1.7 μm)	Water (0.1% FA)/acetonitrile (linear gradient)	10 μL	0.2	8	5.91	0.1/0.3	0.3–100	99.1/RSD < 4.0	He et al. (2011b)
Vindoline	457.3	144.1	BEH C18 (2.1 × 100 mm, 1.7 μm)	Water (0.1% FA)/acetonitrile (gradient)	10 μL	0.2	8	5.91	0.1/0.3	0.3–100	99.1/RSD < 4.0	He et al. (2012)

Vindoline	457.2	144.1	RP-C18 column (Agilent)	ACN:Water with 0.1% formic acid	10 μL	0.3	15	9.8				Zweil et al. (2019)
Vindoline	457.2	124.1	BEH C18 (50 × 2.1 mm, 1.7 μm)	0.1% FA in water:ACN (90:10, v/v)	2	0.2	10	4.29	0.254/0.771	0.5–75	96.74/RSD < 3%	Kumar et al. (2018b)
Vindoline	–457	397	Agilent ZORBAX eclipse XDB-C18	A: Water +0.1% FA; B: ACN	5 μL	0.3	~10	~6.0	0.51	1–1000	95.8/<5.3 RSD	Zhang et al. (2014)
Vindolinine	–455	397	Agilent ZORBAX eclipse XDB-C18	A: Water +0.1% FA; B: ACN	5 μL	0.3	~10	~6.3	0.5	1–1000	96.2/<5.2 RSD	Zhang et al. (2014)
Vinpocetine	458	188	Acquity UPLC BEH C_{18} (1.7 μm, 2.1 × 50 mm)	ACN/0.1% FA + ammonium acetate	5 μL	–	–	0.039–0.583	0.5–1000	99.6–104.3%	0.23–2.71%/0.40–2.90%	Zhang et al. (2014)
Vinpocetine	353.2	144.1	BEH C18 (50 × 2.1 mm, 1.7 μm)	0.1% FA in water: ACN (90:10, v/v)	2	0.2	10	4.57	0.583/1.767	1–500	95.85/RSD < 3%	Kumar et al. (2018b)
Yohimbine	811	224	Acquity UPLC BEH C_{18} (1.7 μm, 2.1 × 50 mm)	ACN/0.1% FA + ammonium acetate	5 μL	–	–	0.039–0.583	0.5–1000	99.6–104.3%	0.23–2.71%/0.40–2.90%	Zhang et al. (2014)
Yohimbine	355.2	144.1	BEH C18 (50 × 2.1 mm, 1.7 μm)	0.1% FA in water:ACN (90:10, v/v)	2	0.2	10 min	2.68	0.294/0.892	0.4–250	97.92/RSD < 2%	Kumar et al. (2018b)

Vindoline, catharanthine, serpentine, and ajmalicine were quantified in the *C. roseus* cell line C20hi using the developed and validated UPLC-MS method. The analysis of a single sample was conducted in 6 min using positive-ion mode with selective ion monitoring. In a C20hi cell sample cultivated in suspension culture under continuous light for 6 days, four alkaloids were identified at concentrations of 80.99, 238.95, 159.00, and 8.03 ng/g dry weight of cells for vindoline, catharanthine, serpentine, and ajmalicine, respectively. Serpentine was discovered at a concentration of approximately 27.50 ng/L (He et al. 2011a). Modified UPLC-MS indicated an increased number of alkaloids by elevating the flow rate and employing a different column. The quantification via the proposed UPLC-MS method indicated that light markedly enhanced vindoline production in C20hi cells, yielding around 0.49–5.51 times higher than the controls, with a maximum yield of 75.3 ng/g of dry weight (He et al. 2011b). To get enhanced selectivity and sensitivity, a multiple reaction monitoring (MRM) HPLC–MS technique was employed in positive mode. The limits of detection were reported as 0.2800 for tryptamine, serpentine, catharanthine, and vindoline; 0.0093 for tabersonine; and 0.0930 for vincamine. The lower limits of quantification were 1.38 ng/mL for tryptamine, serpentine, vindoline, and catharanthine; 0.0138 ng/mL for tabersonine; and 0.138 ng/mL for vincamine in the spiked sample. This method exhibited superior selectivity and sensitivity for the identification of vindoline, ajmalicine, and catharanthine in positive mode compared to previously published techniques. Tabersonine, serpentine, vindoline, catharanthine, and tryptamine were measured in *C. roseus* and V. minor using liquid chromatography electrospray ionization tandem mass spectrometry. The optimized technique was utilized for the quantification of active alkaloids in the foliage of *C. roseus* and V. minor. The concentrations of tryptamine, serpentine, vincamine, and tabersonine in V. minor were 4333, 146.9, 78.90, and 132.3 ng/g, respectively. The concentrations of tryptamine, serpentine, vindoline, catharanthine, and tabersonine in *C. roseus* were 3103, 316.9, 18,600, 18,080, and 777.6 ng/g, respectively. (Liu et al. 2016a, b). A rapid, sensitive, and reproducible UHPLC–ESI-MS/MS (ultra-high-performance liquid chromatography–electrospray ionization–tandem mass spectrometry) method was recently developed and validated for the simultaneous quantification of 10 anticancer TIAs in crude extracts of *C. roseus*, utilizing UHPLC coupled with electrospray ionization hybrid triple quadrupole-linear ion trap mass spectrometry (UHPLC–ESI-QqQLIT-MS/MS) in multiple reaction monitoring (MRM) mode. The LOD and limit of quantitation (LOQ) varied from 0.039 to 0.583 ng/mL and from 0.118 to 1.767 ng/mL, respectively. The validated method was employed for the concurrent quantification of terpene indole alkaloids in ethanolic extracts from 39 samples of *C. roseus* gathered from five distinct locations in India. Serpentine was identified as the most prevalent among all the analyzed substances.

The maximum concentration of serpentine was recorded in the Uttar Pradesh stem sample C-5S (50.074 mg/g), succeeded by the Bihar stem sample CR-3R (49.885 mg/g) and the Arunachal Pradesh root sample C-44R (49.005 mg/g). Serpentine content (19.701 mg/g) was identified as the most abundant compound in the aqueous extract of *C. roseus* in a previous report utilizing HPLC and

HPLC–MS/MS methods. Ajmalicine was also detected in significant concentrations in root samples C-11R (17.70 mg/g), C-44R (16.88 mg/g), and CR-3R (15.76 mg/g). Similarly, a significant concentration of vincristine was detected in CR-4R (6.52 mg/g), followed by C-41 L (0.502 mg/g) and CR-2 L (0.478 mg/g), whereas vinblastine was predominantly found in C-44R (1.514 mg/g), followed by CR-4R (1.38 mg/g) and CR-2R (1.296 mg/g). CR-2 L had a high concentration of vindoline at 19.46 mg/g, followed by CR-3 L at 18.44 mg/g. In contrast, samples C-2R (2.42 mg/g), C-5R (2.26 mg/g), and C-44R (2.16 mg/g) demonstrated comparable levels of yohimbine. Vindesine was identified in concentrations ranging from 1.544 mg/g in C-5R to 3.22 mg/g in C-2R. Minimal quantities of ajmaline (<0.14 mg/g) were identified in all samples. Vinpocetine and reserpine were identified in concentrations of less than 0.0204 and 0.0534 mg/g, respectively. *C. roseus* is a substantial source of serpentine, akin to the *Rauwolfia* species (Kumar et al. 2018b).

References

Abraham DJ, Farnsworth NR, Blomster RN, Sharkey AG Jr (1965) Structure elucidation of periformyline, a novel alkaloid from C. lanceus. Tetrahedron Lett 6(5):317–319

Aerts RJ, Gisi D, De Carolis E, De Luca V, Baumann TW (1994) Methyl jasmonate vapor increases the developmentally controlled synthesis of alkaloids in Catharanthus and Cinchona seedlings. Plant J 5(5):635–643

Akhgari A, Laakso I, Seppänen-Laakso T, Yrjönen T, Vuorela H, Oksman-Caldentey KM, Rischer H (2015) Analysis of indole alkaloids from Rhazya stricta hairy roots by ultra-performance liquid chromatography–mass spectrometry. Molecules 20(12):22621–22634

Ali I, Bashir M (1984) Isolation and structural studies on the alkaloids in flowers of Catharanthus roseus. J Nat Prod 47(3):554–555

Antignac JP, de Wasch K, Monteau F, De Brabander H, Andre F, Le Bizec B (2005) The ion suppression phenomenon in liquid chromatography–mass spectrometry and its consequences in the field of residue analysis. Anal Chim Acta 529(1–2):129–136

Ataei-Azimi A, Hashemloian BD, Ebrahimzadeh H, Majd A (2008) High in vitro production of anticancer indole alkaloids from periwinkle (Catharanthus roseus) tissue culture. Afr J Biotechnol 7(16):2834–2839

Auriola S, Naaranlahti T, Kostiainen R, Lapinjoki SP (1990) Identification of indole alkaloids of Catharanthus roseus with liquid chromatography/mass spectrometry using collision-induced dissociation. Biomed Environ Mass Spectrom 19(7):400–404

Auriola S, Naaranlahti T, Lapinjoki SP (1991) Determination of Catharanthus roseus alkaloids by high-performance liquid chromatography–isotope dilution thermospray–mass spectrometry. J Chromatogr A 554(1–2):227–231

Avula B, Bae JY, Chittiboyina AG, Wang YH, Wang M, Khan IA (2019) Liquid chromatography–quadrupole time-of-flight mass spectrometric method for targeted analysis of 111 nitrogen-based compounds in weight loss and ergogenic supplements. J Pharm Biomed Anal 174:305–323

Balsevich J, Hogge LR, Berry AJ, Games DE, Mylchreest IC (1988) Analysis of indole alkaloids from leaves of Catharanthus roseus by means of supercritical fluid chromatography/mass spectrometry. J Nat Prod 51(6):1173–1177

Barthe L, Ribet JP, Pélissou M, Degude MJ, Fahy J, Duflos A (2002) Optimization of the separation of Vinca alkaloids by nonaqueous capillary electrophoresis. J Chromatogr A 968(1–2):241–250

Bashir M (1983) Isolation of new alkaloids from Catharanthus roseus. Planta Med 49(10):124–125

Benyammi R, Paris C, Khelifi-Slaoui M, Zaoui D, Belabbassi O, Bakiri N, Meriem Aci M, Harfi B, Malik S, Makhzoum A, Desobry S (2016) Screening and kinetic studies of catharanthine and ajmalicine accumulation and their correlation with growth biomass in Catharanthus roseus hairy roots. Pharm Biol 54(10):2033–2043

Bhadra R, Vani S, Shanks JV (1993) Production of indole alkaloids by selected hairy root lines of Catharanthus roseus. Biotechnol Bioeng 41(5):581–592

Broecker S, Herre S, Wüst B, Zweigenbaum J, Pragst F (2011a) Development and practical application of a library of CID accurate mass spectra of more than 2,500 toxic compounds for systematic toxicological analysis by LC–QTOF-MS with data-dependent acquisition. Anal Bioanal Chem 400(1):101–117

Broecker S, Pragst F, Kopf G, Koch M, Wuttig S, Herre S (2011b) Semi-quantitative determination of concentrations in systematic toxicological analysis by LC-QTOF-MS. Toxichem Krimtech 78:230–239

Canto-Canche B, Loyola-Vargas VM (1999) Chemicals from roots, hairy roots, and their application. In: Chemicals via higher plant bioengineering. Springer, Boston, pp 235–275

Cao X, Tai Y, Sun C, Wang K, Pan Y (2005) Characterization of impurities in semi-synthetic vinorelbine bitartrate by HPLC-MS with mass spectrometric shift technique. J Pharm Biomed Anal 39(1–2):39–45

Carqueijeiro I, Brown S, Chung K, Dang TT, Walia M, Besseau S, de Bernonville TD, Oudin A, Lanoue A, Billet K, Munsch T (2018) Two tabersonine 6,7-epoxidases initiate lochnericine-derived alkaloid biosynthesis in Catharanthus roseus. Plant Physiol 177(4):1473–1486

Chandra A, Rana J, Li Y (2001) Separation, identification, quantification, and method validation of anthocyanins in botanical supplement raw materials by HPLC and HPLC–MS. J Agric Food Chem 49(8):3515–3521

Chee KK, Wong MK, Lee HK (1996) Optimization of microwave-assisted solvent extraction of polycyclic aromatic hydrocarbons in marine sediments using a microwave extraction system with HPLC-fluorescence detection and GC-MS. J Chromatogr A 723(2):259–271

Chen Q, Li N, Zhang W, Chen J, Chen Z (2011) Simultaneous determination of vinblastine and its monomeric precursors vindoline and catharanthine in Catharanthus roseus by capillary electrophoresis–mass spectrometry. J Sep Sci 34(20):2885–2892

Chen Q, Zhang W, Zhang Y, Chen J, Chen Z (2013) Identification and quantification of active alkaloids in Catharanthus roseus by liquid chromatography–ion trap mass spectrometry. Food Chem 139(1–4):845–852

Chen Q, Lu X, Guo X, Guo Q, Li D (2017) Metabolomics characterization of two Apocynaceae plants, Catharanthus roseus and Vinca minor, using GC-MS and LC-MS methods in combination. Molecules 22(6):997

Chen Q, Lu X, Guo X, Liu J, Liu Y, Guo Q, Tang Z (2018) The specific responses to mechanical wound in leaves and roots of Catharanthus roseus seedlings by metabolomics. J Plant Interact 13(1):450–460

Colizza K, Mahoney KE, Yevdokimov AV, Smith JL, Oxley JC (2016) Acetonitrile ion suppression in atmospheric pressure ionization mass spectrometry. J Am Soc Mass Spectrom 27(11):1796–1804

Cordall GA, Weiss SG, Farnsworth NR (1974) Structure elucidation and chemistry of Catharanthus alkaloids. XXX. Isolation and structure elucidation of vincarodine. J Org Chem 39(4):431–434

Cordell GA, Farnsworth NR (1976) Catharanthus alkaloids XXXII: isolation of alkaloids from Catharanthus trichophyllus roots and structure elucidation of cathaphylline. J Pharm Sci 65(3):366–369

Dagnino D, Schripsema J, Verpoorte R (1995) Terpenoid indole alkaloid biosynthesis and enzyme activities in two cell lines of Tabernaemontana divaricata. Phytochemistry 39(2):341–349

Dagnino D, Schripsema J, Verpoorte R (1996) Analysis of several iridoid and indole precursors of terpenoid indole alkaloids with a single HPLC run. Planta Med 62(3):278–280

DeLuca V, Kurz WG (1988) Monoterpene indole alkaloids (Catharanthus alkaloids). In: Stafford TA (ed) Phytochemicals in plant cell cultures. Academic Press, New York, pp 385–401

Deus-Neumann B, Zenk MH (1984) Instability of indole alkaloid production in Catharanthus roseus cell suspension cultures. Planta Med 50(05):427–431

Digvijay S, Shashi PR, Suchi S, Kumar RS, Raghavendra M, Sushil K (2004) Simultaneous quantification of some pharmaceutical Catharanthus roseus leaf and root terpenoid indole alkaloids and their precursors in single runs by reversed-phase liquid chromatography. J AOAC Int 87(6):1287–1296

Drapeau D, Blanch HW, Wilke CR (1987) Ajmalicine, serpentine, and catharanthine accumulation in Catharanthus roseus bioreactor cultures. Planta Med 53(4):373–376

Düsman LT, Marinho Jorge TC, Souza MCD, Eberlin MN, Meurer EC, Bocca CC, Basso EA, Sarragiotto MH (2004) Monoterpene indole alkaloids from Palicourea crocea. J Nat Prod 67(11):1886–1888

Dutta A, Batra J, Pandey-Rai S, Singh D, Kumar S, Sen J (2005) Expression of terpenoid indole alkaloid biosynthetic pathway genes corresponds to accumulation of related alkaloids in Catharanthus roseus (L.) G. Don. Planta 220(3):376–383

El-Domyati FM, Abdel Azeiz AZ, Atef A, Elhalwagi A, Sabir JSM, Ramadan AM, Gadalla NO, Edris S, Baeshen MN, Hajrah NH, Al-Kordy MA (2014) Functional genomic analysis for overproduction of key terpenoid indole alkaloids in mature leaf of Catharanthus roseus. Life Sci J 11(10):1313–1320

El-Sayed M, Choi YH, Frederich M, Roytrakul S, Verpoorte R (2004) Alkaloid accumulation in Catharanthus roseus cell suspension cultures fed with stemmadenine. Biotechnol Lett 26(10):793–798

Favretto D, Piovan A, Filippini R, Caniato R (2001) Monitoring the production yields of vincristine and vinblastine in Catharanthus roseus from somatic embryogenesis. Semiquantitative determination by flow-injection electrospray ionization mass spectrometry. Rapid Commun Mass Spectrom 15(5):364–369

Fernández-Pérez F, Almagro L, Pedreño MA, Gomez Ros LV (2013) Synergistic and cytotoxic action of indole alkaloids produced from elicited cell cultures of Catharanthus roseus. Pharm Biol 51(3):304–310

Ferreres F, Pereira DM, Valentão P, Oliveira JM, Faria J, Gaspar L, Sottomayor M, Andrade PB (2010) Simple and reproducible HPLC–DAD–ESI-MS/MS analysis of alkaloids in Catharanthus roseus roots. J Pharm Biomed Anal 51(1):65–69

Gómez-Ariza JL, Lorenzo F, García-Barrera T (2005) Comparative study of atomic fluorescence spectroscopy and inductively coupled plasma mass spectrometry for mercury and arsenic multispeciation. Anal Bioanal Chem 382(2):485–492

Gorman M, Neuss N, Svoboda GH, Barnes AJ Jr, Cone NJ (1959) A note on the alkaloids of Vinca rosea Linn. (Catharanthus roseus G. Don.) II: catharanthine, lochnericine, vindolinine, and vindoline. J Am Pharm Assoc (Sci Ed) 48(4):256–257

Gupta MM, Singh DV, Tripathi AK, Pandey R, Verma RK, Singh S, Shasany AK, Khanuja SPS (2005) Simultaneous determination of vincristine, vinblastine, catharanthine, and vindoline in leaves of Catharanthus roseus by high-performance liquid chromatography. J Chromatogr Sci 43(9):450–453

Gustavsson SÅ, Samskog J, Markides KE, Långström B (2001) Studies of signal suppression in liquid chromatography–electrospray ionization mass spectrometry using volatile ion-pairing reagents. J Chromatogr A 937(1–2):41–47

Hanafy MS, Matter MA, Asker MS, Rady MR (2016) Production of indole alkaloids in hairy root cultures of Catharanthus roseus L. and their antimicrobial activity. S Afr J Bot 105:9–18

He L, Yang L, Tan R, Zhao S, Hu Z (2011a) Enhancement of vindoline production in suspension culture of the Catharanthus roseus cell line C20hi by light and methyl jasmonate elicitation. Anal Sci 27(12):1243–1243

He L, Yang L, Xiong A, Zhao S, Wang Z, Hu Z (2011b) Simultaneous quantification of four indole alkaloids in Catharanthus roseus cell line C20hi by UPLC-MS. Anal Sci 27(4):433–433

He L, Zhao S, Hu Z (2012) A novel cytochrome P450 gene from Catharanthus roseus cell line C20hi: cloning and characterization of expression. Acta Pharm Sin B 2(3):250–255

Hisiger S, Jolicoeur M (2007) Analysis of Catharanthus roseus alkaloids by HPLC. Phytochem Rev 6(2):207–234

Höfer R, Dong L, André F, Ginglinger JF, Lugan R, Gavira C, Grec S, Lang G, Memelink J, Van Der Krol S, Bouwmeester H (2013) Geraniol hydroxylase and hydroxygeraniol oxidase activities of the CYP76 family of cytochrome P450 enzymes and potential for engineering the early steps of the (seco) iridoid pathway. Metab Eng 20:221–232

Jeong WT, Lim HB (2018) A UPLC-ESI-Q-TOF method for rapid and reliable identification and quantification of major indole alkaloids in Catharanthus roseus. J Chromatogr B 1080:27–36

Jessome LL, Volmer DA (2006) Ion suppression: a major concern in mass spectrometry. LC GC North Am 24(5):498–510

Kaufmann A, Butcher P, Maden K, Walker S, Widmer M (2010) Comprehensive comparison of liquid chromatography selectivity as provided by two types of liquid chromatography detectors (high resolution mass spectrometry and tandem mass spectrometry): "where is the crossover point?". Anal Chim Acta 673(1):60–72

Knobloch KH, Bast G, Berlin J (1982) Medium- and light-induced formation of serpentine and anthocyanins in cell suspension cultures of Catharanthus roseus. Phytochemistry 21(3):591–594

Koel M, Kuhtinskaja M, Vaher M (2020) Extraction of bioactive compounds from Catharanthus roseus and Vinca minor. Sep Purif Technol 252:117438

Kohl W, Witte B, Höfle G (1982) Alkaloide aus Catharanthus roseus-Zellkulturen, III [1]/Alkaloids from Catharanthus roseus tissue cultures, III [1]. Z Naturforsch B 37(10):1346–1351

Kumar S, Bajpai V, Singh A, Kumar B (2018a) Identification, characterization and distribution of terpene indole alkaloids in ethanolic extracts of Catharanthus roseus using high-performance liquid chromatography/electrospray ionization quadrupole time-of-flight tandem mass spectrometry and the study of their geographical variation. Rapid Commun Mass Spectrom 32(4):319–332

Kumar S, Singh A, Kumar B, Singh B, Bahadur L, Lal M (2018b) Simultaneous quantitative determination of bioactive terpene indole alkaloids in ethanolic extracts of Catharanthus roseus (L.) G. Don by ultra high performance liquid chromatography–tandem mass spectrometry. J Pharm Biomed Anal 151:32–41

Kutney JP, Choi LS, Kolodziejczyk P, Sleigh SK, Stuart KL, Worth BR, Kurz WGW, Chatson KB, Constabel F (1980) Alkaloid production in Catharanthus roseus cell cultures: isolation and characterization of alkaloids from one cell line. Phytochemistry 19(12):2589–2595

Law B, Temesi D (2000) Factors to consider in the development of generic bioanalytical high-performance liquid chromatographic–mass spectrometric methods to support drug discovery. J Chromatogr B Biomed Sci Appl 748(1):21–30

Lee CW, Shuler ML (2000) The effect of inoculum density and conditioned medium on the production of ajmalicine and catharanthine from immobilized Catharanthus roseus cells. Biotechnol Bioeng 67(1):61–71

Liu DH, Ren WW, Cui LJ, Zhang LD, Sun XF, Tang KX (2011) Enhanced accumulation of catharanthine and vindoline in Catharanthus roseus hairy roots by overexpression of transcriptional factor ORCA2. Afr J Biotechnol 10(17):3260–3268

Liu J, Liu Y, Pan YJ, Zu YG, Tang ZH (2016a) Determination of alkaloids in Catharanthus roseus and Vinca minor by high-performance liquid chromatography–tandem mass spectrometry. Anal Lett 49(8):1143–1153

Liu Z, Wu HL, Li Y, Gu HW, Yin XL, Xie LX, Yu RQ (2016b) Rapid and simultaneous determination of five vinca alkaloids in Catharanthus roseus and human serum using trilinear component modeling of liquid chromatography–diode array detection data. J Chromatogr B 1026:114–123

Lucas D, Neal-Kababick J, Zweigenbaum J (2015) Characterization and quantitation of yohimbine and its analogs in botanicals and dietary supplements using LC/QTOF-MS and LC/QQQ-MS for determination of the presence of bark extract and yohimbine adulteration. J AOAC Int 98(2):330–335

Magagula NL, Mohanlall V, Odhav B (2012) Optimized thin layer chromatographic method for screening pharmaceutically valuable alkaloids of Catharanthus roseus (Madagascar Periwinkle). Int J Sci 2012(12):1–20

Maloney EM, Farnsworth NR, Blomster RN, Abraham DJ, Sharkey AG Jr (1965) Catharanthus lanceus VII. Isolation of tetrahydroalstonine, lochnerinine, and periformyline. J Pharm Sci 54(8):1166–1168

Mathias PI, Connor TH, B'Hymer C (2017) A review of high performance liquid chromatographic-mass spectrometric urinary methods for anticancer drug exposure of health care workers. J Chromatogr B 1060:316–324

McCalley DV (2002) Analysis of the cinchona alkaloids by high-performance liquid chromatography and other separation techniques. J Chromatogr A 967(1):1–19

Monforte-González M, Ayora-Talavera T, Maldonado-Mendoza IE, Loyola-Vargas VM (1992) Quantitative analysis of serpentine and ajmalicine in plant tissues of Catharanthus roseus and hyoscyamine and scopolamine in root tissues of Datura stramonium by thin layer chromatography-densitometry. Phytochem Anal 3(3):117–121

Moza BK, Trojanek J, Bose AK, Das KG, Funke P (1964) The structure of lochnericine and lochnerinine. Tetrahedron Lett 5(37):2561–2566

Mu F, Yang L, Wang W, Luo M, Fu Y, Guo X, Zu Y (2012) Negative-pressure cavitation extraction of four main vinca alkaloids from Catharanthus roseus leaves. Molecules 17(8):8742–8752

Mukhopadhyay S, Cordell GA (1981a) Catharanthus alkaloids. XXXV. Isolation of leurosidine N′b-oxide from Catharanthus roseus. J Nat Prod 44(5):611–613

Mukhopadhyay S, Cordell GA (1981b) Catharanthus alkaloids. XXXVI. Isolation of vincaleukoblastine (VLB) and periformyline from Catharanthus trichophyllus and pericyclivine from Catharanthus roseus. J Nat Prod 44(3):335–339

Mukhopadhyay S, Cordell GA, Sim GA, Cox PJ (1983a) Catharanthus alkaloids—XXXIX: trichophylline—a novel aspidosperma alkaloid from Catharanthus trichophyllus (Apocynaceae). Tetrahedron 39(22):3639–3643

Mukhopadhyay S, El-Sayed A, Handy GA, Cordell GA (1983b) Catharanthus alkaloids XXXVII. 16-Epi-Z-isositsirikine, a monomeric indole alkaloid with antineoplastic activity from Catharanthus roseus and Rhazya stricta. J Nat Prod 46(3):409–413

Naaranlahti T, Nordström M, Huhtikangas A, Lounasmaa M (1987) Determination of Catharanthus alkaloids by reversed-phase high-performance liquid chromatography. J Chromatogr A 410:488–493

Nagy-Turák A, Végh Z (1994) Extraction and in situ densitometric determination of alkaloids from Catharanthus roseus by means of overpressured layer chromatography on amino-bonded silica layers I. Optimization and validation of the separation system. J Chromatogr A 668(2):501–507

Nakagawa T, Shibukawa A, Shimono N, Kawashima T, Tanaka H, Haginaka J (1987) Retention properties of internal-surface reversed-phase silica: packing and recovery of drugs from human plasma. J Chromatogr B Biomed Sci Appl 420:297–311

Nawrocki J (1997) The silanol group and its role in liquid chromatography. J Chromatogr A 779(1–2):29–71

Neuss N, Boaz HE, Occolowitz JL, Wenkert E, Schell FM, Potier P, Kan C, Plat MM, Plat M (1973) The structure of vincarodine. Helv Chim Acta 56(7):2660–2666

Nobilis M, Kopecký J, Květina J, Svoboda Z, Pour M, Kuneš J, Holčapek M, Kolářová L (2003) Comparative biotransformation and disposition studies of nabumetone in humans and minipigs using high-performance liquid chromatography with ultraviolet, fluorescence and mass spectrometric detection. J Pharm Biomed Anal 32(4–5):641–656

O'Connor SE (2012) Strategies for engineering plant natural products: the iridoid-derived monoterpene indole alkaloids of Catharanthus roseus. Methods Enzymol 515:189–206. Academic Press.

O'Connor SE, Maresh JJ (2006) Chemistry and biology of monoterpene indole alkaloid biosynthesis. Nat Prod Rep 23(4):532–547

O'Gara JE, Walsh DP, Alden BA, Casellini P, Walter TH (1999) Systematic study of chromatographic behavior vs alkyl chain length for HPLC bonded phases containing an embedded carbamate group. Anal Chem 71(15):2992–2997

O'Keefe BR, Mahady GB, Gills JJ, Beecher CW, Schilling AB (1997) Stable vindoline production in transformed cell cultures of Catharanthus roseus. J Nat Prod 60(3):261–264

Papagiannopoulos M, Wollseifen HR, Mellenthin A, Haber B, Galensa R (2004) Identification and quantification of polyphenols in carob fruits (Ceratonia siliqua L.) and derived products by HPLC-UV-ESI/MS^n. J Agric Food Chem 52(12):3784–3791
Peebles CA, Hong SB, Gibson SI, Shanks JV, San KY (2006) Effects of terpenoid precursor feeding on Catharanthus roseus hairy roots over-expressing the alpha or the alpha and beta subunits of anthranilate synthase. Biotechnol Bioeng 93(3):534–540
Peraza-Sánchez SR, Gamboa-Angulo MM, Erosa-López C, Ramírez-Erosa I, Escalante-Erosa F, Peña-Rodríguez LM, Loyola-Vargas VM (1998) Production of 19(S)-epimisiline by hairy root cultures of Catharanthus roseus. Nat Prod Lett 11(3):217–224
Pluta PL, Crespi HL, Klein M, Blake MI, Studier MH, Katz JJ (1976) Biosynthesis of deuterated riboflavin: structure determination by NMR and mass spectrometry. J Pharm Sci 65(3):362–367
Rahim RA, Ahmad NH, Al Azzam KM, Mat I (2018) Determination and quantification of the vinblastine content in purple, red, and white Catharanthus roseus leaves using RP-HPLC method. Adv Pharm Bull 8(1):157–161
Renaudin JP (1984) Reversed-phase high-performance liquid chromatographic characteristics of indole alkaloids from cell suspension cultures of Catharanthus roseus. J Chromatogr A 291:165–174
Rijhwani SK, Shanks JV (1998) Effect of elicitor dosage and exposure time on biosynthesis of indole alkaloids by Catharanthus roseus hairy root cultures. Biotechnol Prog 14(3):442–449
Rischer H, Orešič M, Seppänen-Laakso T, Katajamaa M, Lammertyn F, Ardiles-Diaz W, Van Montagu MC, Inzé D, Oksman-Caldentey KM, Goossens A (2006) Gene-to-metabolite networks for terpenoid indole alkaloid biosynthesis in Catharanthus roseus cells. Proc Natl Acad Sci 103(14):5614–5619
Sagi S, Avula B, Wang YH, Khan IA (2016) Quantification and characterization of alkaloids from roots of Rauwolfia serpentina using ultra-high performance liquid chromatography-photo diode array-mass spectrometry. Anal Bioanal Chem 408(1):177–190
Schiel O, Witte L, Berlin J (1987) Geraniol-10-hydroxylase activity and its relation to monoterpene indole alkaloid accumulation in cell suspension cultures of Catharanthus roseus. Z Naturforsch C 42(9–10):1075–1081
Shams K, Nazif N, Azim N, Shafeek K, Missiry M, Ismail S, Nasr M (2009) Isolation and characterization of antineoplastic alkaloids from Catharanthus roseus L. Don. cultivated in Egypt. Afr J Tradit Complement Altern Med 6(2):118–122
Shanks JV, Bhadra R, Morgan J, Rijhwani S, Vani S (1998) Quantification of metabolites in the indole alkaloid pathways of Catharanthus roseus: implications for metabolic engineering. Biotechnol Bioeng 58(2–3):333–338
Shigemori H, Kagata T, Ishiyama H, Morah F, Ohsaki A, Kobayashi JI (2003) Naucleamides A–E, new monoterpene indole alkaloids from Nauclea latifolia. Chem Pharm Bull 51(1):58–61
Shokeen B, Sethy NK, Kumar S, Bhatia S (2007) Isolation and characterization of microsatellite markers for analysis of molecular variation in the medicinal plant Madagascar periwinkle (Catharanthus roseus (L.) G. Don). Plant Sci 172(3):441–451
Siddiqui MJA, Ismail Z, Saidan NH (2011) Simultaneous determination of secondary metabolites from Vinca rosea plant extractives by reverse phase high performance liquid chromatography. Pharmacogn Mag 7(26):92–96
Song KM, Park SW, Hong WH, Lee H, Kwak SS, Liu JR (1992) Isolation of vindoline from Catharanthus roseus by supercritical fluid extraction. Biotechnol Prog 8(6):583–586
Stöckigt J, Sheludko Y, Unger M, Gerasimenko I, Warzecha H, Stöckigt D (2002) High-performance liquid chromatographic, capillary electrophoretic and capillary electrophoretic–electrospray ionisation mass spectrometric analysis of selected alkaloid groups. J Chromatogr A 967(1):85–113
St-Pierre B, De Luca V (1995) A cytochrome P-450 monooxygenase catalyzes the first step in the conversion of tabersonine to vindoline in Catharanthus roseus. Plant Physiol 109(1):131–139
Tang Z, Yang L, Zu Y, Guo X (2009) Variations of vinblastine accumulation and redox state affected by exogenous H2O2 in Catharanthus roseus (L.) G. Don. Plant Growth Regul 57(1):15–20

Thakore D, Srivastava AK, Sinha AK (2015) Model based fed batch cultivation and elicitation for the overproduction of ajmalicine from hairy roots of Catharanthus roseus. Biochem Eng J 97:73–80

Theodoridis GA, Gika HG, Want EJ, Wilson ID (2012) Liquid chromatography–mass spectrometry based global metabolite profiling: a review. Anal Chim Acta 711:7–16

Tikhomiroff C, Jolicoeur M (2002) Screening of Catharanthus roseus secondary metabolites by high-performance liquid chromatography. J Chromatogr A 955(1):87–93

Uniyal GC, Bala S, Mathur AK, Kulkarni RN (2001) Symmetry C18 column: a better choice for the analysis of indole alkaloids of Catharanthus roseus. Phytochem Anal 12(3):206–210

Vay M, Sauter M, Mikus G, Burhenne J (2019) Quantification of microdosed oral yohimbine and its major metabolite in human plasma in the picogram range. Bioanalysis 11(16):1459–1467

Verma A, Hartonen K, Riekkola ML (2008) Optimisation of supercritical fluid extraction of indole alkaloids from Catharanthus roseus using experimental design methodology—comparison with other extraction techniques. Phytochem Anal 19(1):52–63

Verpoorte R, Niessen WMA (1994) Liquid chromatography coupled with mass spectrometry in the analysis of alkaloids. Phytochem Anal 5(5):217–232

Viju C, Agrawal SA, Jadhav AP (2023) Development and validation of a novel HPTLC method for quantitative estimation of vincamine from Catharanthus roseus Linn. Leaves extract. Indian Drugs 60(10):48

Wang CH, Wang GC, Wang Y, Zhang XQ, Huang XJ, Ye WC (2012a) Three new monomeric indole alkaloids from the roots of Catharanthus roseus. J Asian Nat Prod Res 14(3):249–255

Wang CH, Wang GC, Wang Y, Zhang XQ, Huang XJ, Zhang DM, Chen MF, Ye WC (2012b) Cytotoxic dimeric indole alkaloids from Catharanthus roseus. Fitoterapia 83(4):765–769

Wang Q, He J, Yan J, Wu D, Li H (2015) Spectroscopy and docking simulations of the interaction between lochnericine and bovine serum albumin. Luminescence 30(2):240–246

Wesołowska A, Grzeszczuk M, Wilas J, Kulpa D (2016) Gas chromatography-mass spectrometry (GC-MS) analysis of indole alkaloids isolated from Catharanthus roseus (L.) G. Don cultivated conventionally and derived from in vitro cultures. Not Bot Horti Agrobot Cluj-Napoca 44(1):100–106

Yamamoto K, Takahashi K, Mizuno H, Anegawa A, Ishizaki K, Fukaki H, Ohnishi M, Yamazaki M, Masujima T, Mimura T (2016) Cell-specific localization of alkaloids in Catharanthus roseus stem tissue measured with Imaging MS and Single-cell MS. Proc Natl Acad Sci 113(14):3891–3896

Yang L, Wang H, Zu YG, Zhao C, Zhang L, Chen X, Zhang Z (2011) Ultrasound-assisted extraction of the three terpenoid indole alkaloids vindoline, catharanthine and vinblastine from Catharanthus roseus using ionic liquid aqueous solutions. Chem Eng J 172(2–3):705–712

Zenk MH, El-Shagi H, Arens H, Stöckigt J, Weiler EW, Deus B (1977) Formation of the indole alkaloids serpentine and ajmalicine in cell suspension cultures of Catharanthus roseus. In: Plant tissue culture and its biotechnological application. Springer, Berlin, Heidelberg, pp 27–43

Zhang Y, Huang Z, Ding L, Yan H, Wang M, Zhu S (2010) Simultaneous determination of yohimbine, sildenafil, vardenafil and tadalafil in dietary supplements using high-performance liquid chromatography-tandem mass spectrometry. J Sep Sci 33(14):2109–2114

Zhang L, Gai Q, Yang L, Ma Y, Liu Y (2014) Simultaneous quantitative determination of five alkaloids in Catharanthus roseus by HPLC-ESI-MS/MS. Chin J Nat Med 12(10):786–793

Zhang A, Sun H, Yan G, Wang X (2017a) Recent developments and emerging trends of mass spectrometry for herbal ingredients analysis. TrAC Trends Anal Chem 94:70–76

Zhang Y, Sun Y, Mu X, Yuan L, Wang Q, Zhang L (2017b) Identification of metabolites of vindoline in rats using ultra-high performance liquid chromatography/quadrupole time-of-flight mass spectrometry. J Chromatogr B 1060:126–137

Zhao J, Zhu WH, Hu Q (2001) Enhanced catharanthine production in Catharanthus roseus cell cultures by combined elicitor treatment in shake flasks and bioreactors. Enzym Microb Technol 28(7–8):673–681

Zheng Z, Wu M (2004) Cadmium treatment enhances the production of alkaloid secondary metabolites in Catharanthus roseus. Plant Sci 166(2):507–514

Zhou H, Tai Y, Sun C, Pan Y (2005) Rapid identification of vinca alkaloids by direct-injection electrospray ionisation tandem mass spectrometry and confirmation by high-performance liquid chromatography–mass spectrometry. Phytochem Anal 16(5):328–333

Zhou P, Yang J, Zhu J, He S, Zhang W, Yu R, Zi J, Song L, Huang X (2015) Effects of β-cyclodextrin and methyl jasmonate on the production of vindoline, catharanthine, and ajmalicine in Catharanthus roseus cambial meristematic cell cultures. Appl Microbiol Biotechnol 99(17):7035–7045

Zschocke S, Liu JH, Stuppner H, Bauer R (1998) Comparative study of roots of Angelica sinensis and related umbelliferous drugs by thin layer chromatography, high-performance liquid chromatography, and liquid chromatography–mass spectrometry. Phytochem Anal 9(6):283–290

Zweigenbaum J, Lucas D, Neal-Kababick J (2016) Characterization and quantitation of yohimbine and analogues in botanicals and dietary supplements by LC/MS. Planta Med 82(05):45

Zweil AM, Ibrahim IA, Bekhit M (2019) Screening of available components of Egyptian Catharanthus roseus cultivars using HPLC-MS. J Chem Pharm Res 11(6):33–40

Bisindole Alkaloids: Identification, Characterization, and Isolation Aspects

6

Abstract

Bisindole alkaloids, also known as dimeric indole alkaloids, are composed of two monomeric indole units linked by C–C, C–O–C, or C–N bonds. These compounds are predominantly found in plant families such as Loganiaceae, Apocynaceae, and Rubiaceae. Compared to their monomeric counterparts, bisindole alkaloids exhibit significantly enhanced biological activity. *Catharanthus* species, particularly *C. roseus*, are a rich source of pharmacologically active bisindole alkaloids, including vinblastine and vincristine—two widely used anticancer agents. The near-simultaneous discovery of vinblastine remains a landmark achievement in alkaloid chemistry. Traditionally, the isolation, fractionation, and purification of bisindole alkaloids involve time-consuming, labor-intensive, and costly procedures. However, advanced analytical techniques such as HPTLC, HPLC, capillary electrophoresis–mass spectrometry (CE–MS), and HPLC–MS/MS have greatly improved sensitivity, specificity, and efficiency in both qualitative and quantitative analyses. This chapter provides a comprehensive overview of various modern analytical approaches employed for the detection and characterization of bisindole alkaloids, highlighting the benefits of using cutting-edge chromatographic and mass spectrometric methods over traditional techniques.

Keywords

Dimeric indole alkaloids · *Catharanthus* · Analytical methods · CE-MS

6.1 Introduction

Bisindole alkaloids, often referred to as dimeric indole alkaloids, comprise two similar or distinct monomeric indole alkaloid units linked by a C–C, C–O–C, or C–N bond and are typically found in plants belonging to the families Loganiaceae,

S. Kumar, B. Singh, *Catharanthus in Drug Discovery*,
https://doi.org/10.1007/978-981-95-5048-7_6

Apocynaceae, and Rubiaceae. *Catharanthus* species are recognized for their strong bisindole alkaloids. They typically exhibit greater potency regarding their biological action compared to their respective monomeric units. The bisindole alkaloids vinblastine and vincristine, discovered from *C. roseus*, serve as anticancer drugs. The nearly coincidental discovery of vinblastine by Noble et al. and Svoboda and colleagues is one of the most significant occurrences in alkaloid chemistry. Since its first discovery, vinblastine has been repeatedly isolated from *C. roseus* and has also been recovered from many other species. Vinblastine is utilized in the therapy of bladder and breast malignancies, as well as Hodgkin's disease, whereas vincristine is employed for leukemias and lymphomas. Vinorelbine is utilized in the management of non-small cell lung cancer, whereas vindesine plays a significant part in the therapy of acute lymphoblastic leukemia. Vinflunine received approval in 2010 and is currently undergoing Phase III trials for the treatment of solid tumors. In 1963, the FDA initially approved vincristine for the treatment of acute leukemia in children. Subsequently, vincristine has emerged as one of the most often utilized anticancer agents. Currently, vincristine is utilized in the treatment of both adult and pediatric malignancies, including Hodgkin's disease, acute leukemia, non-Hodgkin's lymphomas, neuroblastoma, rhabdomyosarcoma, Wilms' tumor, breast cancer, and small cell lung cancer. Vincristine is primarily utilized as a component in combination chemotherapy regimens. Currently, almost 50% of vincristine utilization is for the treatment of lymphomas, namely as part of the CHOP (cyclophosphamide, doxorubicin, vincristine, prednisone) regimen for first-line therapy of non-Hodgkin lymphomas. Notable therapeutic efficacy has been shown in certain patients; nonetheless, it is accompanied by considerable to severe toxicities, especially neurotoxicities (Brossi and Suffness 1990; Pan et al. 2016; Dhyani et al. 2022).

Since the last comprehensive review of monoterpenoid indole alkaloid biosynthesis, numerous focused studies have advanced our understanding of this complex field. Two major breakthroughs have significantly shaped recent progress: (i) the identification of strictosidine as the universal precursor of all monoterpenoid indole alkaloids, and (ii) the discovery that selected cell-free systems from *C. roseus* can generate a broad spectrum of alkaloid structures, including complex bisindoles. Vinblastine and vincristine, two potent bisindole alkaloids derived from *C. roseus*, are of immense pharmaceutical value but remain exceptionally expensive—approximately \$1 million and \$3.5 million per kilogram, respectively. This high cost stems from their extremely low natural abundance (around 0.0005% of the plant's dry weight). For example, isolating just 1 g of vinblastine requires one ton of dried leaves, while 530 kg of plant material is needed to obtain 1 kg of vincristine. The complexity of their extraction and purification—due to the presence of over 200 structurally similar molecules—further compounds the challenge. These factors, coupled with their proven efficacy in cancer treatment, have driven significant research into understanding their biosynthesis and developing alternative production methods such as cell cultures, metabolic engineering, and synthetic biology (Zhang et al. 2022; Nguyen et al. 2023). The process of isolation, fractionation, and purification of bisindole alkaloids by standard procedures is highly laborious, lengthy, costly, and time consuming than HPTLC, HPLC, and HPLC–MS/MS, which exhibit greater sensitivity and accuracy than traditional approaches. In this chapter, we have summarized a variety of analytical methods used for the qualitative and quantitative analysis of bisindole alkaloids, such as HPTLC/TLC, HPLC, CE–MS, HPLC–MS, and HPLC–MS/MS methods.

6.2 Qualitative Analysis of Bisindole Alkaloids

Researchers have used methods like HPTLC/TLC, HPLC, CE–MS, HPLC–MS, and HPLC–MS/MS to study complex mixtures, including extracts from *Catharanthus* plants, to separate, identify, and describe bisindole alkaloids, especially vinblastine, vincristine, vindesine, and vinorelbine. All these alkaloids are detected in positive ionization mode because the nitrogen atom gains a proton, creating $[M + H]^+$ ions.

6.2.1 Identification and Characterization of Bisindole Alkaloids Using Unit Mass Resolution Mass Spectrometry and Liquid Chromatography–Mass Spectrometry (LC–MS)

Over the past five decades, many researchers have conducted qualitative analyses on *Catharanthus* species, mostly concentrating on bisindole alkaloids such as vinblastine and vincristine, utilizing MS and HPLC–MS for qualitative assessment from various extracts, cell cultures, and plant sections. Previous research identified and characterized bisindole alkaloids, including vinblastine, vincristine, vindesine, vinorelbine, leurosidine, leurocristine, and anhydrovinblastine, at *m/z* 811, 825, 777, 777, 811, 825, and 793, respectively, in various extracts from plant parts such as leaves, stems, flowers, roots, and callus culture, utilizing positive HPLC–ESI-MS and HPLC–ESI–MS/MS techniques. The protonated adduct ion peak of vinblastine was identified at *m/z* 810 (M^+), with distinctive fragment ions at *m/z* 792 (M^+-H_2O), 779 (M^+-CH_3O), 751 (M^+-$COOCH_3$), and 692 (M^+-$(COOCH_3)_2$ in root callus culture (Miura et al. 1987; Trim et al. 2008; Chen et al. 2013). Jossang et al. documented $[M + H]^+$ ion at 811 for vinblastine, with product ions detected at *m/z* 355, 522, 542, 733, 751, and 793 (Figs. 6.1 and 6.2). Likewise, the purified fungal vincristine produced a protonated ion at *m/z* 825 $[M + H]^+$, and its *MS/MS* spectra exhibited fragment ions at *m/z* 766 and 807 (Jossang et al. 1998). Zhou et al. reported the development of a straightforward and expeditious method for identifying bisindole alkaloids from a crude extract of *C. roseus* via direct-injection ESI and MS/MS. Vindoline, vindolidine, vincristine, and vinblastine were discovered and described in a commercial extract of *C. roseus* with the aforementioned approach (Zhou et al. 2005). Pseudovinblastinediol, previously referred to as pseudovincaleukoblastinediol, has been extracted from *C. roseus*. The $[M + H]^+$ ion of pseudovinblastinediol was observed at *m/z* 768. Distinctive fragment ions at *m/z* 469 and 282, generally originating from the vindoline moiety, were displaced by 58 mass units to *m/z* 411 and 224, respectively. This change signifies the existence of a

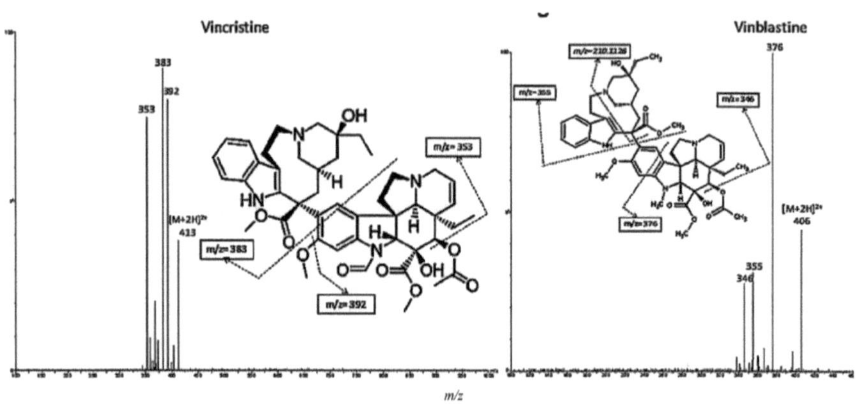

Fig. 6.1 MS/MS spectra of vinorelbine and vinblastine (Chen et al. 2013)

deacetoxyvindoline moiety as the dihydroindole component of the molecule. Subsequent mass spectral analyses of vinblastine indicated the presence of an additional hydroxyl group in the alicyclic segment of the indole moiety of pseudovinblastinediol relative to vinblastine. Pseudovinblastinediol produced a diacetate, with one acetate group located in the vindoline portion and the other in the indole segment, as demonstrated by an ion transition from *m/z* 371 to *m/z* 413 (Blaskó and Cordell 1990). Vingramine and methylvingramine were extracted from the seeds of *C. roseus*. The MS/MS spectral analysis revealed that both compounds possess a novel bisindole framework, comprising an indole alkaloid component B, characterized by the elimination of 5′,6′-ethylene, a C7′-C16′ linkage, a 14′-O-19′-tetrahydrofuran moiety, and an N-4′-isobutyramide substituent. Fragmentation mechanisms of both chemicals have not yet been reported (Jossang et al. 1998; Liu et al. 2013).

Similar to vinblastine, vinleurosine exhibited a $[M + H]^+$ ion at *m/z* 811 and analogous fragment ions generated during MS/MS analysis. Vinleurosine exhibited $[M + H]^+$ at *m/z* 809 in the MS spectrum. Significant fragment ions were detected at *m/z* 791 $[M\text{-}H_2O + H]^+$, *m/z* 748 $[M\text{-}C_3H_8O + H]^+$, 624, 527, 469, 353, and 224 in the MS2 spectrum. Vinblastine exhibited $[M + H]^+$ at *m/z* 825, with fragment ions detected at *m/z* 807 $[M\text{-}H_2O + H]^+$, *m/z* 765 $[M\text{-}C_3H_8O + H]^+$, and 704 in the MS and MS/MS spectra (Ali and Bashir 1984; Bashir 1983; Lin et al. 2015; Chen et al. 2013; Zhang et al. 2014; Abraham et al. 1969) (Figs. 6.1 and 6.2). Deacetoxyvinblastine exhibited a protonated adduct ion at *m/z* 752 $[M + H]^+$; nevertheless, a peak at *m/z* 766 was also detectable in the mass spectrum. The intermolecular methylation by the C-18′ methoxycarbonyl group resulted in a rise of 14 mass units in the molecular ion. The 18′-demethoxycarbonyl-3-hydraz ide derivative, which precludes intermolecular transmethylation, substantiated the nonexistence of the acetoxy group at C-4, displaying an *m/z* 493 fragment rather than the *m/z* 509 fragment typical of the ion possessing a hydroxyl at C-4 in the hydrazide derivative of vinblastine (Blaskó and Cordell 1990; Paci et al. 2003) (Fig. 6.3).

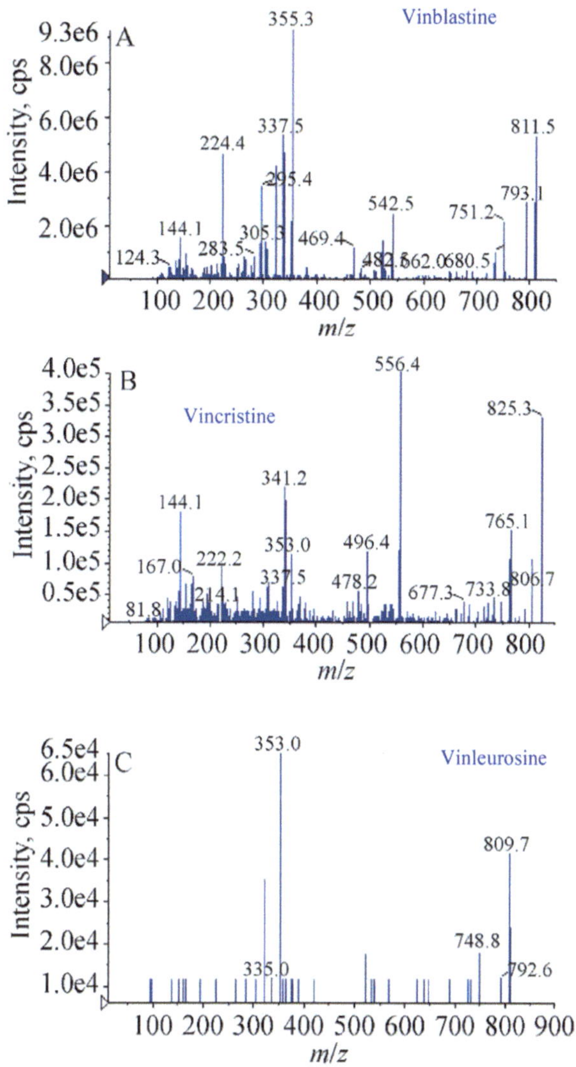

Fig. 6.2 MS/MS spectra of vinblastine, vincristine, and vinleurosine by ion trap MS

Catharine ([M + H]$^+$) exhibited significant peaks at *m/z* 822, 763, 735, 733, 622, 555, 554, 158, 144, 136, 135, 130, 122, 121, 108, and 107. The fragment ions indicated that catharine constituted one-half of this dimeric alkaloid derived from vindoline. The partial structure of catharine can be properly inferred from the mass spectral data. The *m/z* 367 denotes the molecular weight of the remaining portion of the dimeric alkaloid. The fragment ion at *m/z* 763 results from the removal of a carbomethoxy group. The *m/z* 822 to *m/z* 662 transitions, accompanied by a loss of 160 mass units, signify ABD cleavage resulting in the formation of fragment IIa. A

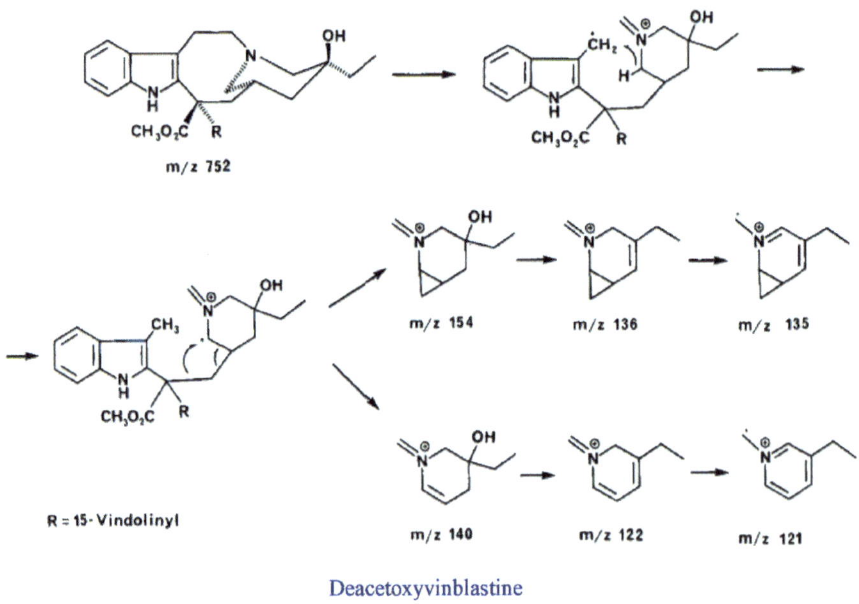

Fig. 6.3 Fragmentation pathways of deacetoxyvinblastine. (Source: Blaskó and Cordell 1990))

robust metastable ion for this transition was detected at *m/z* 533. The *m/z* 554 peak may correspond to ion (III), resulting from cleavage at E in II, analogous to the fundamental ion produced by vindoline at *m/z* 188. The transition from *m/z* 282 (IV) to *m/z* 222 (V) exhibits a faint metastable ion at *m/z* 174.8. Additional typical vindoline fragment ions were produced at *m/z* 135 (VI) and *m/z* 107 (VII), accompanied by a metastable peak at *m/z* 81.8, as well as at *m/z* 122 (I; III), arising from B, C, D cleavage (Fig. 6.4). Catharine comprises a vindoline and catharanthine moiety (Abraham et al. 1967; Chen et al. 2013).

Leurosidine and leurosidine-N-oxide were found and described from spectral data derived from *C. roseus*. Leurosidine is an isomer of vinblastine. The mass spectrum of leurosidine-N-oxide exhibited several ions characteristic of a vindoline moiety, notably the ion at *m/z* 262. The fragment ion at *m/z* 154, indicative of the alicyclic component of the velbanamine dimer, suggested the existence of a single oxygen atom in the piperidine ring (Mukhopadhyay and Cordell 1981a, b; Neuss et al. 1967. Vindogentianine was identified, characterized, and isolated from *C. roseus* (Tiong et al. 2015; Teoh and Das 2018). Leurosine is an exceedingly unstable alkaloid that decomposes when exposed to light, heat, or when left in the dark for several days. Consequently, it exhibits far more stability under these conditions than the sulfate salt. The composition of leurosine consists of two distinct components: vindoline and epoxyvelbanamine. The fragmentation mechanism of leurosine is illustrated in Fig. 6.5.

Leurosine exhibited the precursor ion ($[M + H]^+$) at *m/z* 808. The mass spectra of leurosine exhibited prominent fragment ions at *m/z* 120, 121, 122, and 135, which

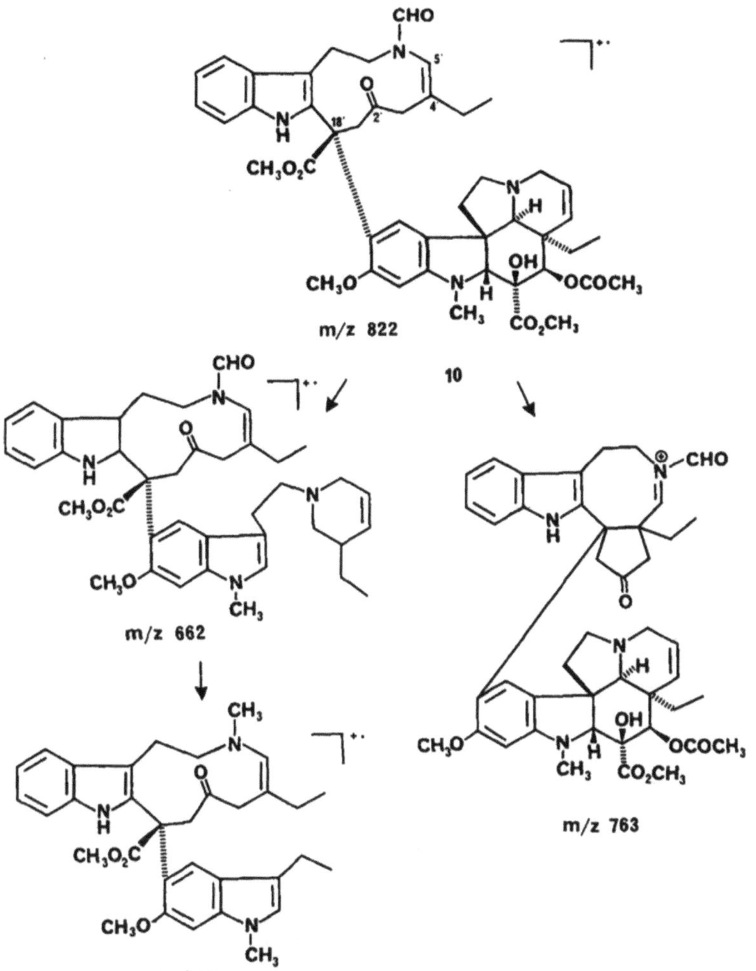

Fig. 6.4 Fragmentation pathways of catharine

are likewise observed in the spectrum of vindoline and vindoline-containing dimers. A noteworthy phenomenon was identified in the combination of different known vindoline (*m/z* 397) and catharanthine (*m/z* 355). The MS/MS spectra of the vindoline moiety comprised fragment ions at *m/z* 456, 397, 396, 296, 188, and 174. These ions were attached to the epoxyvelbanamine moiety, which is a modified homologue of catharanthine. Consequently, fragment ions at *m/z* 455, 396, 395, 295, 187, and 173 are anticipated to emerge in conjunction with the mass of a stable epoxyvelbanamine moiety. Consequently, the coupling of stable vindoline ions with the anticipated stable epoxyvelbanamine ions at *m/z* 353, 215, 214, 156, and 155 yielded diverse combinations that manifested as peaks in the mass spectrum. The

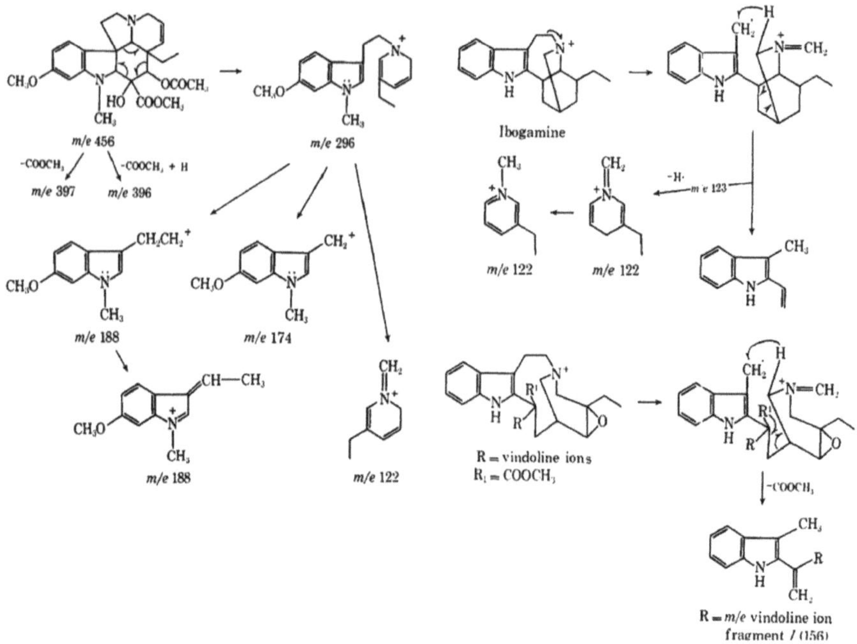

Fig. 6.5 Fragmentation pathways of leurosine

composition includes epoxyvelbanamine with a molecular ion at *m/z* 353 and a molecular formula $C_{21}H_{25}N_2O_3$, whereas vindoline has a molecular ion at *m/z* 455 ($C_{25}H_{31}N_2O_6$) (Abraham et al. 1969; Farnsworth et al. 1963; Tafur et al. 1975, 1976). The isolation and structural elucidation of catharanthamine were conducted in *C. roseus*. MS/MS analysis of catharanthamine revealed an *m/z* of 808 [M^+], producing fragment ions at 792, 766, 753, 752, 734, 648, 527, 469, 467, 451, 368, 296, 289, 270, 260, 256, 143, 135, and 121 (El-Sayed and Cordell 1981; Siddiqui et al. 2011; Jacobs et al. 2004). Leurosine-Nb-oxide exhibited a molecular ion peak at *m/z* 808 (M+) along with product ions at 670, 611, 610, 552, 551, 510, 509, 451, 450, 401, 388, 329, 152 (100%), 144, 135, 122, 121, and 107, consistent with the reported findings on leurosine (El-Sayed et al. 1980; Abraham et al. 1969). Leurosine exhibited a molecular ion at *m/z* 824. The principal fragment ions originate from *m/z* 808 and correspond to those of leurosine. Consequently, the cleavage of the vindolinyl ion results in a prominent epoxyvelbanamine ion at *m/z* 353, whereas fragment ions at *m/z* 121, 135, 143, and 282 define the vindoline moiety (Abraham et al. 1969). Isoleurosine and 17-desacetoxyleurosine were obtained from *C. roseus* (Svoboda et al. 1961, 1962; Cone et al. 1963) (Fig. 6.6). The mass spectrum of 17-desacetoxyvinblastine indicated a molecular weight of 752 and fragment ions at *m/z* 766 and 700. The fragmentation mechanism is illustrated in Scheme 5. Nonetheless, the mass spectra of many bisindole alkaloids exhibited peaks at M + 14 and M + 28 (De Bruyn et al. 1982; Zhang et al. 2013). 5′-Oxoleurosine has been extracted from *C. roseus*, and its structure has been determined using spectroscopic and spectrometric analyses.

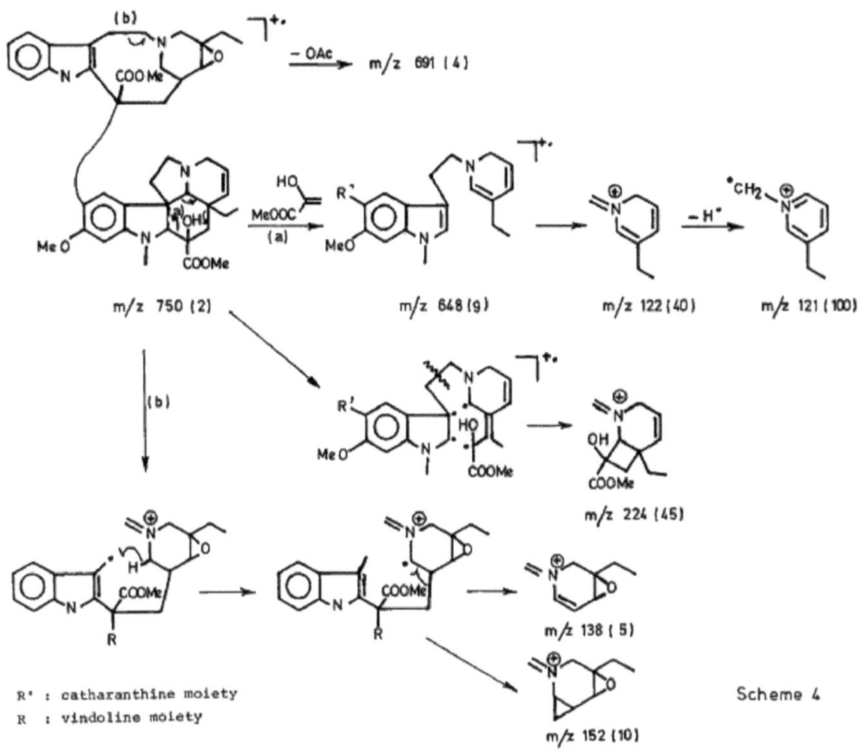

Fig. 6.6 Fragmentation pathways of 17-desacetoxyvinblastine

A molecular ion of 5′-oxoleurosine was detected at *m/z* 822, characterized as $C_{46}H_{54}N_4O_{10}$, indicating that the isolate possessed one additional oxygen atom and two fewer hydrogen atoms compared to leurosine. The mass spectrum exhibited a fragmentation pattern characteristic of a vindoline moiety, notably the ion at *m/z* 282. The differentiation between the positioning of the carbonyl group at C-5′ or C-19′ was established through a comparative analysis of the H-NMR spectra of 14 and that of 19′-oxoleurosine, which was produced from leurosine via oxidation with iodine under basic circumstances (Blaskó and Cordell 1990; Kutney et al. 1978). Vinleurosine was extracted from *C. roseus* (Creasey 1969; Abraham and Farnsworth 1969). The analysis revealed $[M + H]^+$ at *m/z* 809, with distinctive fragment ions detected at *m/z* 791, 749, 624, 527, 469, 353, and 224 (Chen et al. 2013; Zhang et al. 2014). Leurocolornbine exhibited a mass of 825 $[M + H]^+$ with the chemical formula $C_{46}H_{58}N_4O_{10}$. Fragment ions were detected at *m/z* 807 as a result of the loss of 18. Fragment ions at *m/z* 751/749 were observed as a result of the cleavage of the catharanthine molecule. Fragment ions at *m/z* 365–397 and 144–160 were identified as typical of vindoline and the indole ring. The vindoline component of 20 appears to be preserved, as shown by peaks at *m/z* 469, 282, 135, 122, and 121. Given the presence of typical indole fragments at *m/z* 143 and 144, it was inferred that the additional hydroxyl group resided in the alicyclic section of the indole moiety of the molecule (Blaskó and Cordell 1990).

Roseadine exhibited two prominent molecular ions at *m/z* 822 ($C_{47}H_{58}N_4O_9$) and 808 ($C_{46}H_{56}N_4O_9$). The *m/z* 808 corresponds to the isomer of leurosine. The ion at *m/z* 808 produced distinctive fragment ions at *m/z* 766, 765 (M-$COCH_3$), 749 (M-CO_2CH_3), and 648 (M^+-160). The ion at *m/z* 822 is ascribed to the incorporation of CH_2 into the molecular ion; these trans-methylation reactions are well documented in this class of compounds. Characteristic pieces of the vindoline moiety were detected at *m/z* 296, 282, 174, 135, 122, and 121, confirming the presence of this unit. Fragment ions at *m/z* 144, 143, and 156 are indicative of a substituted indole moiety in these compounds and further suggest that C-6 remains a methylene group. Fragment ions at *m/z* 527, 469, and 468 offered significant evidence pertaining to the alicyclic moiety. The high-resolution data indicated that the ion at *m/z* 468 contained a vindoline moiety along with an additional carbon atom, suggesting it was derived from the ion at *m/z* 527 through the loss of a methoxycarbonyl group. This implies that the vindoline unit was connected to the indole moiety via an extra carbon atom that possessed a methoxycarbonyl group. This cleavage reaction is atypical for the vinblastine alkaloid class. Rosemine exhibited an *m/z* of 810 as a molecular ion, attributable to the presence of fragment ions at *m/z* 751 and 355. Distinctive fragment ions from the vindoline nucleus were detected (*m/z* 282, 135, 122, 121, and 107). However, the mass spectrum mostly revealed an unsubstituted indole nucleus from an ion at *m/z* 610 (vindoline + 155 amu), with a base signal at *m/z* 154. This ion signifies the existence of a solitary hydroxy group within an alicyclic structure. Its characterization is partially derived from its spectrum features (El-Sayed et al. 1983; Blaskó and Cordell 1990; Siddiqui et al. 2011). Vincathicine was discovered in *C. roseus* and is isomeric with leurosine. Vincathicine produced fragment ions of the vindoline component as the dimeric alkaloids. It displayed fragment ions at *m/z* 649 (M-$C_6H_7O_5$), 469 ($C_{26}H_{33}N_2O_6$), 282 ($C_{14}H_{20}NO_5$), 135, and 122. The fragment ion at *m/z* 353 ($C_{21}H_{25}N_2O_3$) represents the molecular ion without the vindoline radical.

Vincathicine has been confirmed to be especially susceptible to intermolecular transmethylation in the mass spectrometer. At the minimum temperature, the transmethylation peak at *m/z* 822, together with the peaks at *m/z* 765 and 750, identified as M-C_2H_3O and M-$C_2H_2O_2$, respectively, was eradicated. A pronounced peak at *m/z* 366 ($C_{22}H_{26}N_2O_3$), which remains unchanged in the spectra of 16-acetylvincathicine and 17-deacetylvincathicine and appears at *m/z* 368 in the spectrum of dihydrovincathicine, originates from the altered velbanamine moiety (Tafur et al. 1975; Blaskó and Cordell 1990). Vinamidine was initially extracted from *C. roseus* by Svoboda and colleagues (Tafur et al. 1976; Wang et al. 2012; Nisar et al. 2016). Vincovaline, vincovalinine, and vincovalicine were detected at *m/z* 751, 749, and 735 in *Catharanthus* species (Blaskó and Cordell 1990; Dorman and Paschal 1976; Tafur et al. 1975; El-Sayed and Cordell 1981). Vindolicine is a distinctive alkaloid, extracted from *C. roseus*, whose structure represents the fusion of two vindoline units by condensation with a one-carbon moiety. While several bisindole alkaloids are recognized as products of the combination of two Aspidosperma units, vindolicine is the sole instance associated in this fashion (El-Sayed et al. 1983; Almagro et al. 2015; Forni et al. 2019). The protonated molecular ion $[M + H]^+$ was detected at *m/z* 925, along with its fragment ions at *m/z* 805, 749, 588, 463, and 383 (Chen et al. 2013).

6.2.2 Analysis of Bisindole Alkaloids by High Resolution Tandem Mass Spectrometry and Liquid Chromatography–High Resolution Tandem Mass Spectrometry

HR-MS/MS and HPLC–HR-MS/MS have emerged as indispensable methodologies for the structural characterization of compounds derived from complex matrices, such as plant extracts. HPLC–HR-MS/MS provides superior mass accuracy, sensitivity, and resolution. This method can resolve isobaric compounds and detect low-abundance molecules. Table 6.1 presents the details of MS and MS/MS ions. HPLC–HR-MS/MS provides exceptional chromatographic separation and extensive fragmentation patterns, enabling the detection, identification, and characterization of both targeted and non-targeted analyses of known and unknown compounds (Grata et al. 2009; Glauser et al. 2013; Xian et al. 2012). HPLC and UHPLC coupled with Orbitrap and time-of-flight (TOF), particularly TOF, are utilized for the structural characterization of bisindole alkaloids derived from *Catharanthus* species (Yamamoto et al. 2016; Jeong and Lim 2018; Kumar et al. 2018a). Several UPLC–Q-TOF methodologies have been documented for the structural characterization of bisindole alkaloids derived from the root, stem, and leaf tissues of *C. roseus*. The authors assert that these approaches exhibit greater sensitivity, speed, and reliability compared to ion trap mass spectrometry. Electrospray ionization–Q-TOF (ESI–Q-TOF) analysis facilitated the characterization of alkaloids and their fragment ions, providing enhanced accuracy and precision in identifying distinctive ions and fragmentation patterns. The alkaloid mixture, including vincristine and vinblastine, together with the extract solution, was separated on a C18 column using acetonitrile and water containing 0.1% formic acid and 10 mM ammonium acetate as the mobile phase over a duration of 16 min.

UPLC–Q-TOF facilitates the identification of six alkaloids by comparing retention durations, molecular ions, and fragmentation patterns with reference substances. A total of 16 supplementary indole alkaloids were provisionally discovered through comparison of chromatograms with chemical databases and literature sources (Jeong and Lim 2018; Zhang et al., 2017a, b) (Fig. 6.7).

The protonated ions of vinblastine and vincristine were identified at *m/z* 811.4278 and 825.4068, respectively. MS/MS spectra are presented in Figs. 6.8 and 6.9. Fragment ions at *m/z* 793.4177 and 807.3894 were detected as a result of H_2O loss, while product ions *m/z* 779.4014 and 793.3804 were generated from the loss of CH_3OH from *m/z* 811.4278 and 825.4068, respectively. Product ions at *m/z* 761.3909 and 775.3974 were generated via the elimination of H_2O from ions at *m/z* 779.4014 and 793.4177, respectively.

Both alkaloids yielded a common fragment at *m/z* 341.1495 due to loss of $C_{26}H_{35}N_2O_7$ and $C_{26}H_{33}N_2O_8$. Further loss of H_2O from *m/z* 341.1495 provided the ion at *m/z* 323.1760. Product ions at *m/z* 751.4108 and 765.3799 were formed due

Table 6.1 Spectrometric data of bisindole alkaloids in *Catharanthus* species

			Molecular wight					
RT	Compounds	Molecular formula	Calculated value	Observed value/ Error (ppm)		Fragment ions	Distribution	Reference
11.1	3,4-Anhydrovinblastine	$C_{46}H_{56}N_4O_8$	792.4098	793.4161	-1.3	707.3809 (22.7), 397.1777 (98.48)	Stem	Jeong and Lim (2018)
13.4	Catharine	$C_{46}H_{54}N_4O$	822.384	823.3913	0	763.3718 (9), 640.3128 (1), 457.2325 (1), 554.2645 (2), 367.1660 (2), 279.1563 (3), 255.1491 (3), 193.0893 (5), 144.0799 (6)		Jeong and Lim (2018)
26	Catharine	$C_{46}H_{54}N_4O_{10}$	823.3913	823.3918	0.6	805.3784 (5), 763.3643 (63), 745.3527 (5), 731.3392 (8), 661.3344 (5), 554.2603 (27), 522.2324 (6), 457.2339 (3), 397.2122 (4), 367.1619 (12), 339.1646 (10), 335.1390 (8), 255.1461 (50), 222.1098 (24), 144.0791 (100)	Root, stem, and leaves	Kumar et al. (2018a)
26.71	Catharine (isomer)	$C_{46}H_{54}N_4O_{10}$	823.3913	823.3912	−0.12	805.3784 (6), 763.3612 (59), 745.3512 (6), 731.3320 (11), 554.2603 (23), 522.2343 (8), 494.2310 (6), 367.1583 (14), 339.1621 (12), 307.1411 (14), 279.1442 (14), 255.1441 (48), 222.1088 (27), 144.0771 (100)	Root, stem, and leaves	Kumar et al. (2018a)

11.2	Vinblastine	$C_{46}H_{58}N_4O_9$	810.4204	811.426	−2.1	–	Root and leaf	Jeong and Lim (2018)
10.5	Vinblastine	$C_{46}H_{58}N_4O_9$	810.4204	811.4	<5	793.4 (29.4), 751.3 (36.6), 144.1 (100)	Root and leaf	Chen et al. (2013)
~4.3	Vinblastine	$C_{46}H_{58}N_4O_9$	810.4204	811.4272	−1.0	793.4 (28), 751.3 (20), 397.2 (100), 144.1 (51)	Root and leaf	Kumar et al. (2018a)
5.88	Vinblastine	$C_{46}H_{58}N_4O_9$	810.4204	811.4269	2.1	793.4174 (33.5), 751.3823 (22.3), 144.0800 (100)	Leaf and stem	Jeong and Lim (2018)
~3.0	Vinblastine	$C_{46}H_{58}N_4O_9$	810.4204	811.4	<5	793.3 (28), 397.2 (100), 144.1 (65)	C20hi cell line	He et al. (2011a, b)
~5.5	Vinblastine	$C_{46}H_{58}N_4O_9$	810.4204	811.4	<5	793.3 (45), 397.2 (100), 144.1 (52)	Leaf	Zweil et al. (2019)
~6.4	Vinblastine	$C_{46}H_{58}N_4O_9$	810.4204	811.4	<5	793.4 (30), 397.2 (100), 144.1 (68)	Leaf and stem	Koel et al. (2020)
~7.2	Vinblastine	$C_{46}H_{58}N_4O_9$	810.4204	811	–	793, 397	Leaf	Auriola et al. (1991)
—	Vinblastine	$C_{46}H_{58}N_4O_9$	810.4204	811	–	793, 751, 397	Not specified	Kumar et al. (2018a)
10.7	Vinblastine	$C_{46}H_{58}N_4O_9$	810.4204	811	–	Not specified	Leaf	Mu et al. (2012)

(continued)

Table 6.1 (continued)

			Molecular wight					
RT	Compounds	Molecular formula	Calculated value	Observed value/ Error (ppm)		Fragment ions	Distribution	Reference
12.96	Vinblastine	$C_{46}H_{58}N_4O_9$	811.4277	811.4278	−0.28	793.4177 (8), 751.4078 (40), 733.3873 (12), 675.3932 (6), 641.3179 (5), 586.3180 (5), 612.3008 (8), 586.3180 (5), 542.2942 (51), 522.2939 (28), 480.2623 (11), 464.2683 (11), 436.2590 (9), 401.1668 (10), 381.2208 (9), 353.1875 (28), 224.1283 (100)		Kumar et al. (2018a)
10.6	Vincristine	$C_{46}H_{56}N_4O_{10}$	824.3996	825.4045	−2.9	–	Leaf	Jeong and Lim (2018)
~4.1	Vincristine	$C_{46}H_{56}N_4O_{10}$	824.3996	825.4059	−0.8	793.4 (34), 751.3 (28), 397.2 (100), 144.1 (62)	Leaf	Kumar et al. (2018a)
6.14	Vincristine	$C_{46}H_{56}N_4O_{10}$	824.3996	825.4047	2.5	793.3824 (28.4), 397.1912 (92.6), 144.0803 (100)	Leaf	Jeong and Lim (2018)
~3.3	Vincristine	$C_{46}H_{56}N_4O_{10}$	824.3996	825.4	<5	793.4 (30), 397.2 (95), 144.1 (100)	C20hi cell line	He et al. (2011a, b)
~5.9	Vincristine	$C_{46}H_{56}N_4O_{10}$	824.3996	825.4	<5	793.4 (40), 397.2 (95), 144.1 (100)	Leaf	Zweil et al. (2019)
~6.8	Vincristine	$C_{46}H_{56}N_4O_{10}$	824.3996	825.4	<5	793.4 (33), 397.2 (98), 144.1 (100)	Leaf extract	Koel et al. (2020)

~7.9	Vincristine	$C_{46}H_{56}N_4O_{10}$	824.3996	825	–	793, 397, 144	Leaf	Auriola et al. (1991)
— (DI)	Vincristine	$C_{46}H_{56}N_4O_{10}$	824.3996	825	–	793, 397, 144	Not specified	Kumar et al. (2018a)
11.5	Vincristine	$C_{46}H_{56}N_4O_{10}$	824.3996	825	–	Not specified	Leaf	Mu et al. (2012)
26.02	Vincristine	$C_{46}H_{56}N_4O_{10}$	825.4069	825.4068	−0.12	807.3882 (5), 765.3778 (58), 747.3650 (4), 733.3548 (5), 705.3549 (4), 663.3478 (5), 556.2747 (100), 496.2553 (15), 341.1820 (32), 309.1567 (5), 281.1595 (12), 222.1091 (9), 194.0784 (5), 167.0672 (7), 144.0783 (15)		Kumar et al. (2018a)
8.09	Vindesine	$C_{43}H_{55}N_5O_7$	754.4174	754.4183	0.33	736.4102 (43), 709.3906 (28), 691.3856 (16), 651.3918 (19), 542.3015 (16), 522.3046 (9), 470.2268 (11), 412.2192 (76), 355.2013 (100), 337.1917 (58), 323.1714 (46), 305.1639 (22), 295.1798 (34), 225.1267 (22), 222.1264 (15), 144.0802 (23), 124.1116 (68)		Kumar et al. (2018a)
11.5	Vinleurosine	$C_{46}H_{56}N_4O_9$	808.4047	809.411	−1.2	749.3967 (16.9), 697.3619 (13.4), 353.1858 (14.9)	Root and leaf	Jeong and Lim (2018)

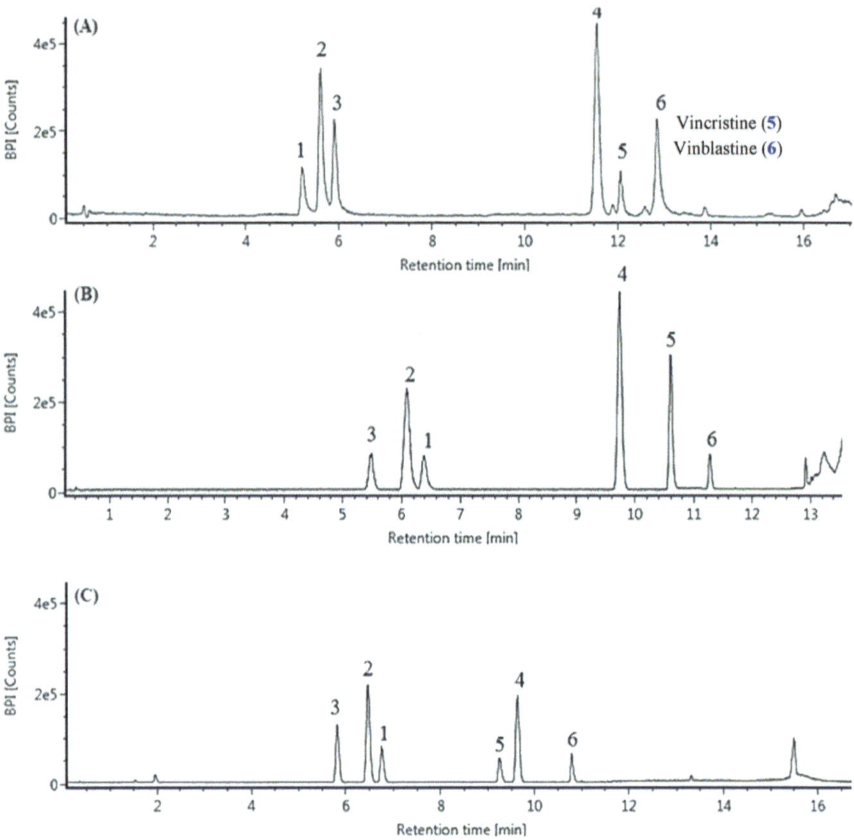

Fig. 6.7 Representative chromatograms of standard alkaloid mixture as a function of pH: (**a**) pH 4.5 (10 mM ammonium acetate with 0.1% acetic acid in water), (**b**) pH 3.3 (10 mM ammonium formate with 0.1% formic acid in water), and (**c**) pH 2.7 (0.1% formic acid in water). The peaks were identified as follows: (1) serpentine, (2) catharanthine, (3) ajmalicine, (4) vindoline, (5) vincristine, and (6) vinblastine

to the common loss of CH_3COOH from *m/z* 811.4278 and 825.4068, respectively. Further loss of CH_3OH from *m/z* 751.4108 and 765.3799 gave product ions at *m/z* 719.3903 and 733.354, respectively. Similarly, product ions at *m/z* 649.3807 and 663.3541 were formed due to loss of $C_4H_6O_3$ from product ions at *m/z* 751.4108 and 765.3799, respectively. Product ions at *m/z* 751.4108 and 765.3799 also produced ions at *m/z* 542.2007 and 556.2775, respectively, due to loss of $C_{11}H_{15}O_3$, and further loss of $HCOOCH_3$ gave product ions at *m/z* 482.2835 and 496.2554, respectively. Product ions at *m/z* 524.2945 and 538.2735 were formed due to loss of H_2O from ions at *m/z* 542.2007 and 556.2775, respectively. Similarly, product ions 464.2695 and 478.2478 were also formed due to loss of $HCOOCH_3$ from *m/z* 524.2945 and 538.2735, respectively (Kumar et al. 2018a, b).

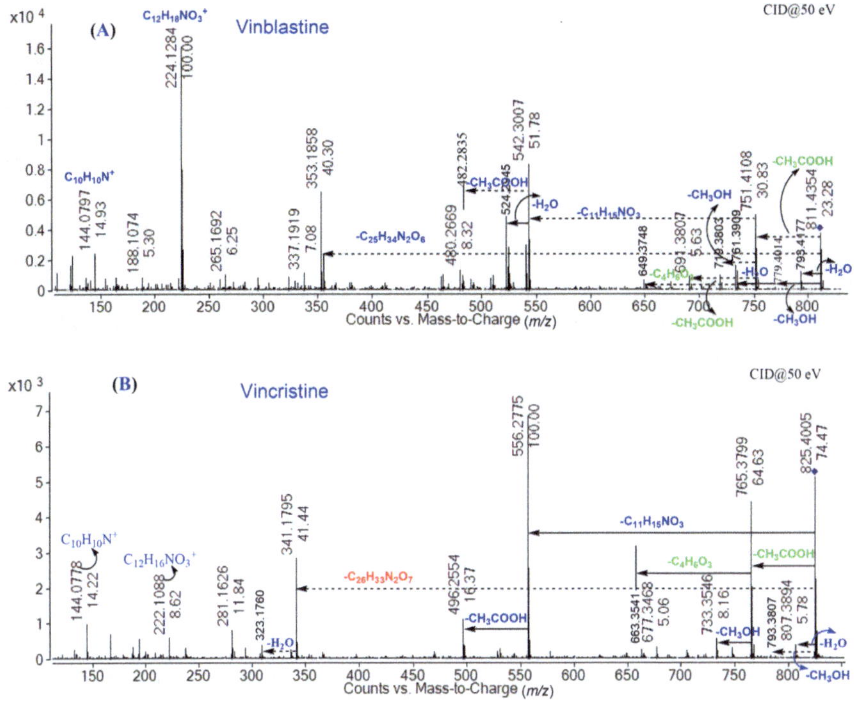

Fig. 6.8 HR-MS/MS spectra of vincristine and vinblastine

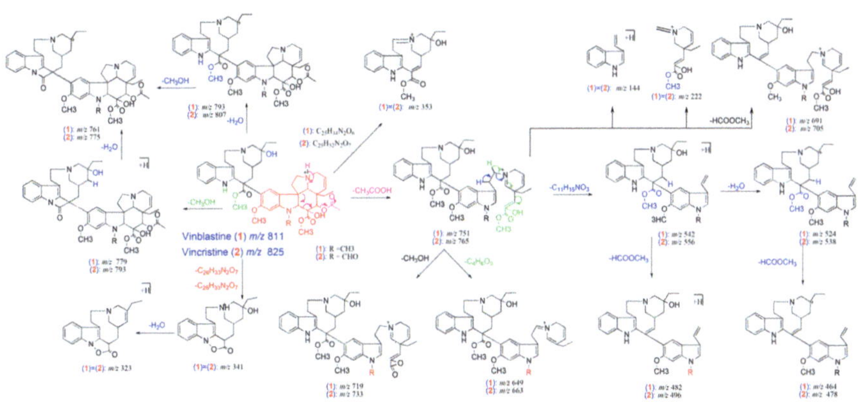

Fig. 6.9 MS/MS fragmentation pathways of vincristine and vinblastine by HPLC–HR-MS/MS

Product ions at *m/z* 691.3807 and 705.3648 were generated via the elimination of $HCOOCH_3$ from ions at *m/z* 751.4108 and 765.3799, respectively (Kumar et al. 2013, 2018a; Háda et al. 2013; Thimmaiah et al. 1990; Dubrovay et al. 2013; Chen et al. 2011; Zhang et al. 2014). The vinblastine-type N-oxide alkaloids, 17-desacetoxyvinblastine N′b-oxide and 20′-deoxyvinblastine N′b-oxide, were extracted from the leaves of *C. roseus*. The structures of 17-desacetoxyvinblastine N′b-oxide and 20′-deoxyvinblastine N′b-oxide were determined through the study of HR-ESI-MS spectroscopic data. The HR-ESI-MS spectra of 17-desacetoxyvinblastine N′b-oxide and 20′-deoxyvinblastine N′b-oxide exhibited quasi-molecular ions $[M + H]^+$ at *m/z* 767.3988 (molecular formula $C_{44}H_{56}N_4O_8$) and 809.4081 (molecular formula $C_{46}H_{58}N_4O_9$), respectively (Zhang et al. 2013). Vindesine was identified as a protonated ion at *m/z* 754.4174 $[M + H]^+$ in the MS spectrum, and its MS/MS spectrum exhibited product ions at *m/z* 736.4174, 709.3906, 412.2192, and 355.2013, corresponding to the losses of H_2O, $HCONH_2$, $C_{20}H_{26}N_2O_3$, and $C_{22}H_{29}N_3O_4$, respectively. The fragment ion at *m/z* 736.4174 produced product ions at *m/z* 719.3803, 651.3918, and 542.3015, corresponding to the losses of NH_3, $C_3H_3NO_2$, and $C_{10}H_{14}N_2O_2$, respectively. The elimination of H_2O from the ion at *m/z* 542.3015 produced the ion at *m/z* 524.2908. Sequential losses of H_2O were detected from the ion at *m/z* 709.3906, resulting in the ions at *m/z* 691.3856 and 673.3748. The ions at *m/z* 337.1917 and 323.1724 resulted from the loss of H_2O and CH_3OH, respectively, from the ion at *m/z* 355.2013. The additional loss of CH_3OH and H_2O from *m/z* 337.1917 and 323.1724, respectively, resulted in the formation of an ion at *m/z* 305.1639 (Kumar et al. 2018a; Kokkonen et al. 1990; Zhu et al. 2014; Maeda and Miwa 2013; Jungclas et al. 1982; Sottani et al. 1998; Ramírez et al. 1997; Chen et al. 2011; Zhang et al., 2017a, b) (Figs. 6.10 and 6.11).

Catharine (*m/z* 823 $[M + H]^+$) (Fig. 6.12) has been identified and described, exhibiting product ions at *m/z* 805 and 763 resulting from the loss of H_2O and CH_3COOH, respectively. The fragment ion at 763.3643 generated product ions at *m/z* 745, 731, 661, and 554, corresponding to the losses of H_2O, CH_3OH, $C_4H_6O_3$, and $C_{11}H_{14}NO_3$, respectively. The fragment ion at *m/z* 367 resulted from the breaking of the C-C link between the catharanthine and vindoline moieties. The resultant ions at *m/z* 457.2339 and 367.1619 produced *m/z* 397 and 335, respectively, following the elimination of CH_3COOH and CH_3OH. Additional product ions detected at *m/z* 255.222 and 144 are illustrated in Fig. 6.12 (Kumar et al. 2018a; Abraham et al. 1967).

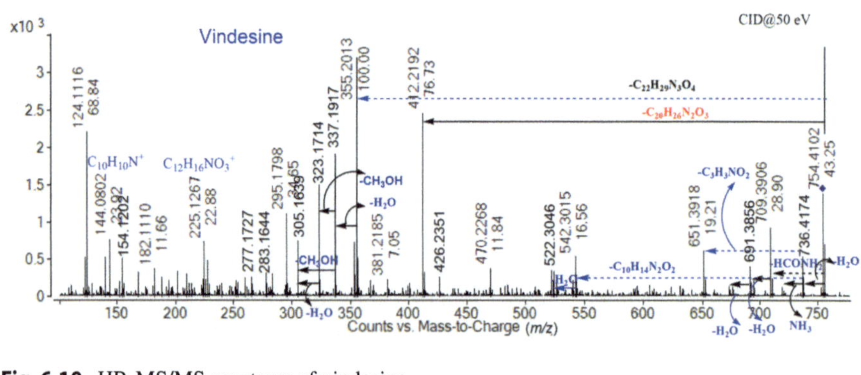

Fig. 6.10 HR-MS/MS spectrum of vindesine

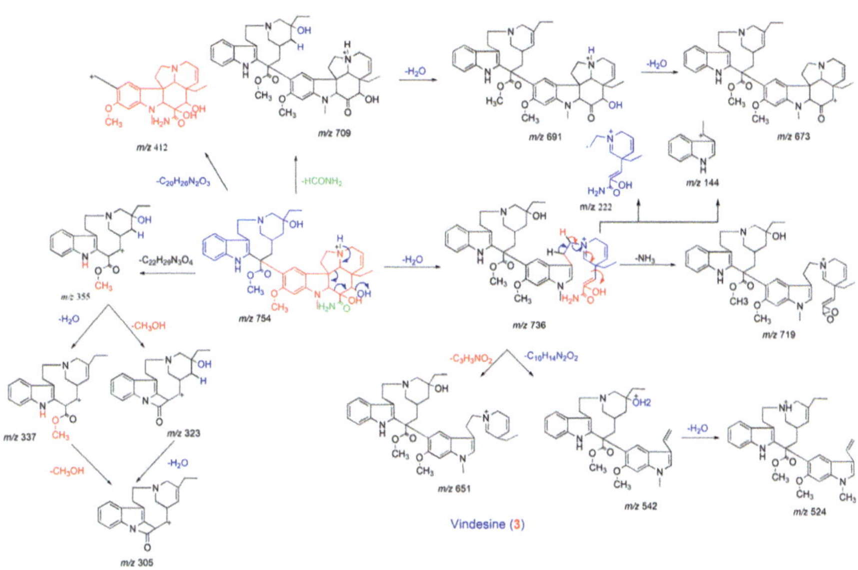

Fig. 6.11 Fragmentation pathways of vindesine

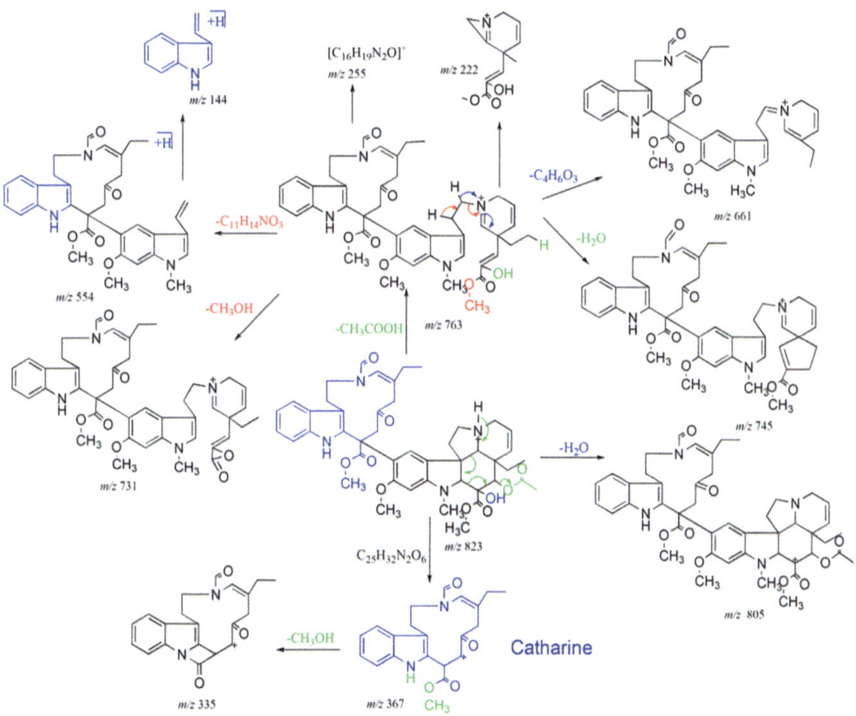

Fig. 6.12 Fragmentation pathways of catharine

Leurosine exhibits a protonated ion at *m/z* 808.4117 $[M + H]^+$ with the chemical formula $C_{46}H_{56}N_4O_9$ (Farnsworth et al. 1963; Tafur et al. 1976; Endo et al. 1987; Farnsworth 1972). Catharanthamin was extracted from the alkaloid fraction of *C. roseus* and purified by chromatographic techniques following the extraction of leurocristine. Its structure was elucidated using IR and NMR (Blaskó and Cordell 1990). Vingramine ($C_{46}H_{56}N_4O_8$) and methylvingramine ($C_{47}H_{58}N_4O_8$) were extracted from the seeds of *C. roseus*. Their structures were elucidated using HRFABMS and one- and two-dimensional NMR spectroscopy (Jossang et al. 1998; Zhang et al. 2013). 4′-Deoxyvinblastine was initially identified as an impurity in leurosine. Initially, 4′-deoxyvinblastine was thought to be isomeric with leurosine, leading to its designation as isoleurosine. High-resolution MS confirmed that isoleurosine has a chemical composition of $C_{46}H_{58}N_4O_8$, containing two additional hydrogen atoms and one fewer oxygen atom than leurosine. Subsequently, it also corresponds precisely with the natural product (Blaskó and Cordell 1990; Jacobs et al. 2004). An HPLC–MS technique was established, utilizing an Agilent Zorbax Eclipse XDB-C8 reversed-phase column (150 × 4.6 mm i.d.; 5 μm) at 25 °C. The separation employed isocratic elution with 0.1% triethylamine (solvent A) and methanol (solvent B) at a flow rate of 0.8 mL/min. Five alkaloids, including vincristine and vinblastine, were isolated and characterized in the commercial product. Fragment ions were detected as previously mentioned (Zhou et al. 2005). Figure 6.6 displays the total ion chromatogram and the extracted ion chromatogram of the detected alkaloids.

6.3 Quantitative Analysis of Bisindole Indole Alkaloids

The quantification of bisindole indole from different extracts of *Catharanthus* plant sections is crucial for optimizing extraction techniques, standardizing herbal products, and elucidating biochemical routes. Nonetheless, bisindole alkaloids, characterized by their intrinsic structural complexity and resemblance to other indole alkaloids, along with their naturally low abundance, present considerable analytical hurdles. HPLC, UPLC, HPLC–MS/MS, and UPLC–MS/MS are sophisticated hyphenated chromatographic and mass spectrometric methodologies.

6.3.1 Quantitative Analysis of Bisindole Indole Alkaloids by Thin-Layer Chromatography or High-Performance Thin-Layer Chromatography

In general, pre-coated silica gel 60 F254 plates serve as the stationary phase in conjunction with various combinations of the mobile phases chloroform, acetone, methanol, and formic acid. Dragendorff's reagent (in HCl) and Cerium Ammonium Sulfate (CAS) serve as detecting reagents at wavelengths of 254–366 nm, with maximal sensitivity typically at approximately 289 or 254 nm. The Rf values for vincristine and vinblastine were approximately 0.21–0.45 and 0.54–0.63, respectively. The extensive isolation of vinblastine and vincristine from *C. roseus* has garnered interest from pharmaceutical businesses, with several methods for isolating these alkaloids documented by Blaskó and Cordell in 1990. The phytochemical components of the ethanolic extract were also analyzed using TLC. Vinblastine and vincristine were extracted from *C. roseus* and assessed using HPLC and HPTLC in separate mixes. They were purified using TLC and HPLC and characterized via UV-Vis spectroscopy, ESI-MS, MS/MS, and 1H NMR. Isoleurosine was extracted from *C. roseus*, and its structure was revealed using UV and TLC analyses (Svoboda et al. 1961, 1962; Cone et al. 1963; Monforte-González et al. 1992; Tam et al. 1995; Toivonen et al. 1991; Blomster et al. 1967; Nagy-Turák and Végh 1994; DeLuca et al. 1986). Many scientists presented a systematic HPTLC approach for the quantitative analysis of vinblastine and vincristine utilizing overpressured layer chromatography (OPLC). A combination of vinblastine and vincristine was fractionated using HPTLC aluminum sheets coated with silica gel 60 F254, utilizing a mobile phase of hexane–dichloromethane–acetone–2-propanol (65:13:21:0.9 v/v) and in situ spectrodensitometry at 298 nm. The calibration curve was built solely for vinblastine, exhibiting linearity between 100 and 1000 ng. The LOD was reported as 50 ng, with a repeatability of 246.8 mg/kg vinblastine in the drug extract, exhibiting a standard deviation (SD) of 7.1 mg/kg and a relative standard deviation (RSD) of 2.9%. Vincristine and vinblastine were quantified in 80% methanol extracts of the dried aerial parts of *C. roseus* on Silica gel Merck 60 F254 TLC plates (10 × 10 cm) utilizing a mobile phase of benzene, methanol, and 25% ammonia (100:5:5:3). Detection is achieved using cerium ammonium sulfate (CAS) in 85% phosphoric acid, visualized under white light and UV light (366 nm). The detection limits, determined by TLC spot intensity with a 10 μL application, were reported as 0.0055% for vincristine and 0.05% for vinblastine. Rf values were established at 0.44–0.45 for vincristine and 0.63 for vinblastine (Magagula et al. 2012). Shams et al. (2009) analyzed alkaloid-rich fractions using Merck silica gel 60 F254 (20 × 20 cm) with a mobile phase consisting of ethyl acetate, benzene, ethyl alcohol, and 25% ammonia in a ratio of 100:5:5:3. Utilize Dragendorff's reagent, followed by 10% HCl spray, to observe alkaloids at wavelengths of 254, 289, 366, 400, and 500 nm. The retention factors of vinblastine and vincristine were measured at 0.54 and 0.21, respectively.

6.3.2 Quantitative Analysis by Liquid Chromatography and High-Performance Liquid Chromatography

The determination of bisindole alkaloids, including vinblastine and vincristine, is generally conducted on a C18 reversed-phase column utilizing liquid chromatography methods, particularly HPLC and UPLC, frequently integrated with detectors such as UV, DAD, and FLD (Fluorescence Detector). Chromophores on the aromatic rings are essential for UV and DAD detection or absorbance. Delocalized π electrons in the structure of alkaloids facilitates autofluorescence, enhancing their efficient detection with a diminished baseline background, particularly for those that are highly luminous, such as vincristine and vinblastine. Bisindole alkaloids are often more hydrophobic than their indole predecessors. It is advisable to establish two HPLC separation procedures, one for hydrophilic precursors and another for hydrophobic bisindole alkaloids. A more polar elution is conducted during the analysis of precursors in contrast to alkaloids. Table 6.2 presents the various strategies that have been developed. Some approaches exhibit significant similarity, differing primarily due to minor variations in equipment performance or particular laboratory research objectives. Multiple ways exist for constructing an HPLC separation procedure, and the following fundamental considerations must be taken into account. There is no consensus about the selection of the mobile phase in the analysis of TIAs. A successful separation can be accomplished utilizing a saline buffer with potassium phosphate or ammonium acetate as the aqueous phase, and methanol or acetonitrile as the organic phase. These approaches appear to be derived from Renaudin's earlier work (1984) on the separation of vindoline and vinblastine. The initial development of HPLC procedures with ion-pairing reagents was conducted by Naaranlahti et al. (1989) for vindoline, catharanthine, serpentine, vinblastine, vincristine, and ajmalicine. At a neutral pH, ion-exchange interactions and hydrogen bonding may transpire between dissociated silanol and protonated (or partially protonated) alkaloids, typically leading to peak tailing. TEA facilitates the competition of alkaloids for silanol sites. Typically, an aged column (presumably after processing several thousand samples) demonstrates diminished retention for alkaloids with low pKa (e.g., ajmalicine) while exhibiting enhanced retention for compounds with high pKa (such as tryptamine and serpentine) compared to a new column (Renaudin 1984). This behavior is attributed to the deterioration of the hydrophobic stationary phase caused by the uncapping of silanol groups after the breaking of the C18 alkyl chains. Amine modifier agents in the mobile phase influence interactions between the stationary phase and basic analytes (e.g., alkaloids), thereby altering the elution sequence.

Consequently, comparing the methodologies becomes laborious. Competitive amines can improve peak shape at neutral pH and diminish peak tailing. The pH of the mobile phase is a critical parameter, with acidic pH (pH 3–4) facilitating improved peak shape for basic alkaloids. At low pH, the majority of silanol groups remain undissociated, thereby restricting ion-exchange interactions, even when the alkaloids are protonated (McCalley 2002). To our knowledge, alkaline pH (pH 10) was exclusively examined by Barthe et al. (2002). Commercially available columns exist for usage at severe alkaline pH; however, it is important to note that older generation columns (e.g., C8, Phenyl, and C18) are unsuitable for high pH applications due to potential degradation of the stationary phase. In all instances, whether utilizing acidic or basic mobile phases, methodologies must adopt a pH significantly distant from the pKa of the target compounds, as minor fluctuations in mobile phase composition can alter the protonation state of the alkaloids, thereby affecting the elution profile and the consistency of retention times (Uniyal et al. 2001; McCalley 2002). Uniyal et al. (2001) noted that the elution time sequence altered from catharanthine, vincristine, vinblastine, and vindoline at pH 6 to vincristine, vindoline, catharanthine, and vinblastine at pH 7.5.

Table 6.2 Various analytes that have been developed from *Catharanthus*

S.no.	Instrument	Analytes	Column	Mobile phase	Detection	Elution program	Range of RT (min)	Flow rate (mL/min)	Plant part	Extract	Reference
1	HPLC	Ajmalicine, catharanthine, serpentine, and vinblastine	*μ*Bondapack C_{18} (300 × 3.9 mm, 10 μm) (Waters)	Methanol: 5 mM $(NH_4)_2HPO_4$ (pH 7.3) (71:29, v:v)	UV_{254}	Isocratic	12.00–19.00	1	Cell suspension cultures	Ethanolic	Renaudin (1984)
		Ajmalicine, catharanthine, tetrahydroalstonine, and vindoline,	*μ*Bondapack C_{18} (300 × 3.9 mm, 10 μm) (Waters)	Methanol: 5 mM $(NH_4)_2HPO_4$ (pH 7.3) (67:33, v:v)	UV_{254}	Isocratic	9.5–22	1	Hairy roots and cell suspensions	Aqueous solution with 001% acetic acid	Renaudin (1984)
		Ajmalicine, catharanthine, serpentine, vinblastine, vincristine, and vindoline	Spheri-5 C_{18} (220 × 4.6 mm, 5 μm)	Solvent A: 0.2% TEA in 25 mM CH_3COONH_4; solvent B: 0.2% TEA in a mix of methanol:acetonitrile (29:71, v:v)	PDA	Gradient	13–24.5		Cell suspension cultures	Ethanolic	Naaranlahti et al. (1987)
		Ajmalicine, catharanthine, serpentine, tabersonine, and tryptamine,	*μ*Bondapak C_{18} (300 × 3.9 mm, 10 μm)	Acetonitrile: 0.1 M CH_3COONH_4 (pH 7.2) (49:51, v:v)	UV_{280}	Gradient	2.0–18.0	1	Tissue cultures	Ethanolic	Auriola et al. (1989)
		Ajmalicine, catharanthine, serpentine, vinblastine, and vindoline,	*μ*Bondapack C_{18} (300 × 3.9 mm, 10 μm) (Waters)	Methanol: 5 mM $(NH_4)_2HPO_4$ (pH not mentioned) (67:33, v:v)	UV_{254}	Isocratic	8.5–59.20	0.8	Hairy roots and cell suspensions	Alkaloidal	Bhadra et al. (1993)
		Catharanthine, vinblastine, vindoline, and vincristine,	*μ*Bondapack C_{18} (300 × 3.9 mm, 10 μm) (Waters)	Acetonitrile: 0.1 M phosphate buffer: glacial acetic acid (pH 4.14) (38:62:0.3, v:v:v)	PDA	Isocratic	10.15–13.54	0.6	Leaves	90% Ethanolic	Singh et al. (2000)

(continued)

Table 6.2 (continued)

S.no.	Instrument	Analytes	Column	Mobile phase	Detection	Elution program	Range of RT (min)	Flow rate (mL/min)	Plant part	Extract	Reference
		Catharanthine, vinblastine, vincristine, and vindoline	Symmetry C_{18} (150 × 4 mm, 5 μm) (Waters)	Solvent A (pH 7.0): methanol: acetonitrile: 5 mM CH_3COONH_4: triethylamine (13:32:55:0.2, by volume); solvent B (pH 7.0): methanol: acetonitrile: 5 mM CH_3COONH_4: triethylamine (19:46:35:0.2, by volume)	PDA	Gradient	14.62–19.6		Cell suspension cultures and plant parts	–	Uniyal et al. (2001)
		Anhydrovinblastine, catharanthine, vindoline, vinflunine, and viorelbine	XTerra C_{18} (250 × 4.6 mm, 5 μm) (Waters)	Acetonitrile:water:boric acid (pH 10) (55:45:3.1, v:v:w)	UV_{21} 4 nm	Isocratic	6.0–20.8	1	–	–	Barthe et al. (2002)
		Catharanthine, serpentine, tabersonine, vinblastine, vindoline, and vincristine,	Zorbax XDB C_{18} (250 × 4.6 mm, 5 μm) (Agilent)	Solvent A: 5 mM KH_2PO_4 (pH 6); solvent B: acetonitrile	UV	Gradient	7.8–20.8	2	Hairy root cultures		Tikhomiroff and Jolicoeur (2002)
		Ajmalicine, catharanthine, loganine, serpentine, tryptamine, tryptophan, vinblastine, vincristine, and vindoline	Phenomenex Luna C_{18} (250 × 4.6 mm, 5 μm)		DAD		NA		Leaf and root		Digvijay et al. (2004)

		Ajmalicine, serpentine, vinblastine, and vindoline	C_{18} (125 × 4 mm, 5 μm)	0.025 M methanol–acetonitrile–ammonium acetate	$UV_{233, 254, 280}$	Gradient	–	1	Leaves, stems, and roots	Methanolic	Favali et al. (2004)
		Ajmalicine, serpentine,and tryptamine	Phenomenex Luna C_{18} (150 × 4.6 mm, 5 μm)	12% acetonitrile: 88% water +0.1% formic acid	UV_{254}	Isocratic	May-37	1	Cell suspension cultures	Methanolic	Wong et al. (2004)
		Catharanthine, serpentine, tabersonine, vindoline, vincristine, and vinblastine,	Zorbax XDB C_{18} (250 × 4.6 mm, 5 μm (Agilent)	Solvent A: 5 mM KH_2PO_4 (pH 6); solvent B: acetonitrile	UV	Gradient		2	Leaves	90% Ethanolic	Dutta et al. (2005)
		Catharanthine, vinblastine, vincristine, and vindoline	Chromolith performance C_{18} (100 × 4.6 mm)	21:79 (v/v) acetonitrile–0.1 M phosphate buffer containing 0.5% glacial acetic acid	UV_{254}	Gradient	6.81–26.93	1.2	Leaves	90% Ethanolic	Gupta et al. (2005)
		Vinblastine, vincristine, vindoline, and vindolidine,	Agilent Zorbax Eclipse XDB-C_8 (150 × 4.6 mm, 5 μm)	0.1% Triethylamine (solvent A) and methanol (solvent B)	UV_{280}	Gradient	10.0–40.0	0.8	Leaves	Commercial	Zhou et al. (2005)
		Catharanthine, vinblastine, vindoline, and vincristine	ODS C_{18} (250 × 4.6 mm) (Waters)	Water + 0.1% formic acid and B acetonitrile + 0.1% formic acid	UV_{220}			–	Leaves and roots	Methanolic	Guo et al. (2007)
		Vinblastine	HypersilBDS-C_{18} (150 × 4.6 mm, 5 μm) (Agilent)	5 mM $Na_2HPO_4 \times 2H_2O$ buffer (pH adjusted to 7.3 by H_3PO_4)–acetonitrile–methanol	UV_{218}	Gradient		1.5	Leaves	Alkaloidal	Verma et al. (2007)

(continued)

Table 6.2 (continued)

S.no.	Instrument	Analytes	Column	Mobile phase	Detection	Elution program	Range of RT (min)	Flow rate (mL/min)	Plant part	Extract	Reference
		Ajmalicine, cataranthine, vinblastine, Serpentine, vincristine, and vindoline	Bondpak C_{18} (1500 × 2.5 mm, 4.6 μm)	Acetonitrile and ammonium carbonate buffer pH 7.3 (1:1)	UV_{254}	Gradient	20.52–41.42		Leaves, root-callus, and roots		Ataei-Azimi et al. (2008)
		Ajamalicine, catharanthine, serpentine, vincristine, vinblastine, and vindoline	Phenomenex C_{18} (250 × 4.6 mm, 5¼ m)	0.01 M phosphate buffer (pH 4.9) and acetonitrile	UV_{220}	Gradient			*C. roseus*/ root, stem, leaf, flower, and seed	90% Ethanolic	Singh et al. (2000)
		Catharanthine, vinblastine, vincristine, and vindoline	Phenomenex Gemini C_{18} (150 × 2.0 mm, 5 μm) (Torrance, CA, USA)	10 mM ammonium acetate buffer (pH 5.0), acetonitrile and methanol	UV_{220}	Gradient		0.3	Leaves	Methanolic	Verma et al. (2008)
		Vincristine	Merck LiChro CART C_{18} (125 × 4.6 mm, 5 μm)	Acetonitrile: methanol; 3:1	UV_{220}	Isocratic		1	Leaf, root, and whole plant	Alkaloidal	Aslam et al. (2010a)
		Vincristine	Supelco ODS Hypersil, C_{18} (4.6 × 250 mm, 5 μm)	Methanol:water (50:50 v/v)	UV_{280}	Isocratic		0.5	Leaf, stem	Methanolic	Kalidass et al. (2009)
		Vinblastine and vincristine	Nova pack C_{18} (3.9 × 150 mm) (Waters)	Phosphate buffer solution of pH 6.5, acetonitril (55:45)	UV	Isocratic	8.05–11.83	6	*C. Roseus*	Methanolic, alkaloidal	Shams et al. (2009)

		A-3′,4′-anhydrovinblastine, catharanthine, vinblastine, and vindoline	Waters ODS C_{18} (250 × 94.6 mm, 5 μm)	A: Water : diethylamine, pH to 7.2 H_3PO_4, B: Methanol : acetonitrile = 4 : 1	UV_{220}	Gradient	–	1.5	Leaves	Methanolic	Tang et al. (2009)
		Vinblastine	Microsorb-MV C_{18} (150 × 4.6 mm, 5 μm)	Methanol and acetonitrile	UV_{240}	Gradient	7.02	0.2	Leaves	Aqueous	Ahmad et al. (2010)
		Vinblastine	Merck LiChro CART C_{18} (125 × 4 mm, 5 μm)	Acetonitrile and methanol, 3:1	UV_{255}	–		1	Leaves/leaf callus and in germinating embryos	Methanolic	Aslam et al. (2010a, b)
		Catharanthine, vinblastine, vindoline, and vincristine	Zorbax eclipse plus C_{18} (250 × 4.6 mm, 5 μm)	Methanol: acetonitrile: ammonium acetate buffer (25 mM) with 0.1% triethylamine (15:45:40 v/v)	UV	Isocratic	10.26–19.62	1	Leaves, stem, and flower	Methanol and methanol: water	Siddiqui et al. (2011)
		Catharanthine, vinblastine, and vindoline	HiQ sil-C_{18} (4.6 × 250 mm, 5 μm) (KYA TECH)	Methanol: acetonitrile (4:1, v/v, solution A) water:diethylamine (986:14, v/v) with phosphoric acid pH of 7.5 (solution B).	UV_{220}	–	10.0–26.0	1.5	Leaves	85% Ethanolic	Yang et al. (2011)
		Catharanthine, vinblastine, vincristine, and vindoline,	HiQ sil C_{18} (4.6 × 250 mm)	A: (methanol/acetonitrile), and solution B (water/diethylamine, pH 7.2 by H_3PO_4)	UV_{220}	Gradient	9.0, 11.9, 13.2, 20.2	2	Leaves	80% Ethanolic	Mu et al. (2012)
		Ajmalicine, catharanthine, serpentine, vinblastine, vindoline, and vincristine	C_{18} (Acquity, Waters, Manchester, UK)	98.9% Deionized water, 1.0% acetonitrile and 0.1% formic acid	UV	Isocratic		1	Leaves	Homogenized with acid water + 0.01% formic acid) 1:1 (m/v)	Andrade et al. (2013)

(continued)

Table 6.2 (continued)

S.no.	Instrument	Analytes	Column	Mobile phase	Detection	Elution program	Range of RT (min)	Flow rate (mL/min)	Plant part	Extract	Reference
		Vinblastine and vincristine	C_{18} (Waters)	5–95% Acetonitrile in water with 0.01% trifluoroacetic acid	$UV_{254,220}$	Gradient		0.5	Endophytic fungus fusarium oxysporum	Alkaloidal	Kumar et al. (2013)
		Catharanthine, tryptamine, vinblastine, and vindoline	Himpak VP-ODSC_{18} (150 × 4.6 mm)	Methanol, acetonitrile, and 5 mm diammonium hydrogen phosphate (3:4:3)	UV_{220}	Isocratic		1	Suspension culture	Methanolic	Pandiangan et al. (2013)
		Anhydravinblastine, catharanthine, vinblastine, vincristine, and vindoline	Kromasil C_{18} (250 × 4.6 mm, 5 μm)	Water:diethylamine (986:14, v/v) with phosphoric acid to a pH of 7.5 (solution A) and methanol:acetonitrile (4:1, v/v, solution B)	UV_{220}			2	Leaves	80% Ethanolic	Luo et al. (2014)
		Vinblastine and vincristine	Octadecylsilane C_{18} (250 × 4.6 mm, 5 μm (Supelco)	Acetonitrile:water 50:50	UV	Isocratic	1.229–1.474	2	Leaves	Methanolic	Saravanan and Karthi (2014)
		Vinblastine, vincristine, and catharanthine	C_{18} (250 × 4 mm, 5 μm)	55% Sodium phosphate buffer (5 mmol/L, pH 6), 45% acetonitrile)	UV	Gradient	NA	0.7	Root	Methanolic	Hanafy et al. (2016)
		Vinblastine	Agilent Zorbax Eclipse XDB-C_{18}	Ammonium acetate (20 mM) 70:30 acetonitrile, at 30	UV_{254}	Gradient		0.2	Leaves	Carbon dioxide and ethanol	Falcão et al. (2017)
		Ajmalicine, catharanthine, vinblastine, vincristine, and vindoline,	Diamonsil C_{18} (250 × 4.6 mm, 5 μm)	Water:diethylamide (v/v), [pH 7.2], methanol:acetonitrile = 4:1 (v/v)	UV_{220}	Gradient	NA	1.5	Leaf and root	Methanolic	Hashemabadi et al. (2018)

		Ajmalicine, catharanthine, serpentine, vinblastine, vincristine, and vindoline	C_{18} (50 × 2.1 mm, 1.7 µm)	Water (solvent A) and acetonitrile (solvent B) containing 0.1% formic acid ammonium formate	UV	Gradient	5.47–11.25	0.4	Root, stem, and leaf	70% Methanolic	Jeong et al. (2018)
		Catharanthine, vinblastine, vincristine, and vindoline	Eclipse XDB-C18 (4.6 × 250 mm, 5 µm) (Agilent Technologies)	A and TFA–water–acetonitrile (0.01:5:95) as solvent B	UV_{254}	Gradient		1	CMC suspension culture	Methanolic	Moon et al. (2018)
		Vinblastine	Microsorb—MV C_{18} (250 × 4.6 mm, 5 µm)	Methanol–phosphate buffer (5 mM, pH 6.0)–acetonitrile	UV_{254}	Gradient	20	2	Leaves	Aqueous	Rahim et al. (2018)
		Vincristine	Zorbax SB-C_{18} (2.1 × 50 mm, 1.8 µm)	Methanol–15 mmol/L ammonium acetate containing 0.02% formic acid (65: 35, V/V)	UV_{254}			1	Aerial parts	Carbon dioxide and ethanol	Karimi and Raofie (2019)
3	HPLC-MS	Catharanthine, vinblastine, vincristine, vindoline, vindolidine, vindolinine, and 19S-vindolinine	Hypersil ODS (200 × 4.6 mm, 5 µm) (Agilent)	Solvent A: 25 mM CH_3COONH_4 in methanol CH_3COONH_4 in water; solvent B: 25 mM	MS	Gradient	14.8–21.3	1	Leaf	Isopropyl alcoholic	Chu et al. (1997)
		Vinblastine and vincristine	Zorbax Bonus C_{18} (150 × 2.1 mm, 5 µm) (Agilent)	Solvent A: 20 mM CH_3COONH_4; solvent B: acetonitrile	MS	Gradient	18–21.5	0.2	Aerial parts and roots	Methanolic	Choi et al. (2002)
		Vinblastine, vincristine, vindoline, and vindolidine,	Direct flow injection	Methanol	MS		N/A	–	Whole plant	Commercial	Zhou et al. (2005)
		Vinblastine and vincristine	Xterra MS C_{18} (4.6 × 150 mm, 5 µm) (Waters)	10 mM ammonium acetate (pH 10) and acetonitrile (55:45)	MS	Isocratic		1	Leaves suspension cultures	Ethanolic	Rischer et al. (2006)

(continued)

Table 6.2 (continued)

S.no.	Instrument	Analytes	Column	Mobile phase	Detection	Elution program	Range of RT (min)	Flow rate (mL/min)	Plant part	Extract	Reference
		Vincristine, vinblastine, and vinorelbine	Phenomenex Polar-RP (250 × 2.0 mm, 4 μm)	Solvent A (46%) + 0.1% TFA/water; solvent B (54%) + 1:1 methanol: acetonitrile	MS	Isocratic	5–14.40	0.2	Plasma		Schmidt et al. (2006)
		Ajmalicine, ajmalicine (isomer), catharanthine, serpentine, serpentine (isomer), tabersonine, vindolinine, and 19-S-vindolinine	Luna C_{18} (250 × 4.6 mm, 5 μm) (Phenomenex, United Kingdom)	Acetonitrile (A) and acetic acid 1% (B)	MS	Gradient	10.0–40.0	1	Root	Methanol:water (1:1)	Ferreres et al. (2010)
		Vinblastine	Merck LiChro CART C_{18} (125 × 94 mm, 5 μm)	Acetonitrile:methanol (3:1)	UV_{255}	Gradient		1	Embryos/tissues	Chloroform	Aslam et al. (2011)
		Ajmalicine, catharanthine, serpentine, vinblastine, vincristine, and vindoline	C_{18} (Acquity, Waters, Manchester, UK)	98.9% Deionized water, 1.0% acetonitrile and 0.1% formic acid	MS	Isocratic		1	Leaves	Homogenized with acid water + 0.01% formic acid) 1:1 (m/v)	Andrade et al. (2013)
		Ajmalicine, catharanthine, vinblastine, vindoline, and vinleurosine	DL-C_{18} (250 × 4.6 mm, 5.0 μm) (Japan)	Acetonitrile (A) and 10 mmol ammonium acetate in water (B)	UV, MS	Gradient		1–0.5	Leave	Alkaloidal	Chen et al. (2013)

		Catharanthine, serpentine, taberosinine, vinblastine, vincristine, vindoline, vindolidine, vindolinine, and 19-S-vindolinine	BEH C_{18} (50 mm × 2.1 mm, 1.7 μm)	0.1% formic acid (A) and acetonitrile (B)	MS	Gradient	1.74–4.47	0.5	Leaf	Methanol:water (1:1)	El-Domyati et al. (2014)
		Catharanthine, vinblastine, vincristine, and vindoline, vinleurosine	Agilent Eclipse C_{18} (150 × 4.6 mm, 5 μm)	Methanol mmol/L ammonium acetate containing 0.02% formic acid (65:35, V/V)	MS (MRM)	Gradient	2.51–2.97	1	Leaves, stems, and roots	Water–0.1% H_2SO_4 methanol	Zhang et al. (2014)
		Catharanthine and vindoline	C_{18} (2.1 × 50 mm, 3.5 μm)	Acetonitrile–0.1% formic acid in water	MS	Gradient		0.4	Plasma		Lin et al. (2015)
		Catharanthine, vinblastine, vincristine, and vindoline	Phenomenex Gemini C_{18} (250 × 4.6 mm, 5 μm)	Methanol/acetonitrile/10 mM ammonium acetate (15:40:45, v/v/v)	MS	Isocratic		1	Root-callus and roots	75% Ethanolic	Zhang et al. (2015)
		Catharanthine, serpentine, tabersonine, tryptamine, vinblastine, vincamine, vindoline, and yohimbine	Cosmosil C_{18} (4.6 × 250 mm, 5 μm), WondaSil C_{18} (200 × 4.6 mm, 5 μm)	(A) 0.05 mol/L ammonium acetate and (B) acetonitrile, acetonitrile/0.2% formic acid in water, 37:63, v/v	MS	Gradient	4.88–36.37	1	Leaves and human serum samples	Water:methanol (50:50, v/v) and alkaloidal	Liu et al. (2016a, b)
		16 alkaloids	Eclipse XDB-C_{18} (150 × 4.6 mm, 5 μm) (Agilent)	25 mM ammonium acetate and acetonitrile	LTQ Orbitrap	Isocratic at 20:80		0.25	Stem tissue		Yamamoto et al. (2016)
		Vinblastine	Zorbax XDB C_{18} (150 × 4 mm, 5 μm)	Methanol and nano pure water with 0.01% acetic acid	MS	Isocratic	4.91–5.2	1	Leaves	90% Ethanolic	Ayob et al. (2017)

(continued)

Table 6.2 (continued)

S.no.	Instrument	Analytes	Column	Mobile phase	Detection	Elution program	Range of RT (min)	Flow rate (mL/min)	Plant part	Extract	Reference
		Ajmalicine, serpentine, vinblastine and vincristine,	VP-ODS C_{18} (200 × 150 mm, 5.0 μm)	(A) 0.04% HoAc–H_2O, (B) 0.04% HoAc–MeCN	MS	–		0.5	Leaves	70% Methanolic	Zhang et al. (2017a, b)
5	UHPLC-MS	Ajmaline, ajmalicine, serpentine, reserpine, vinblastine, vincristine, vindesine, vindoline, vinpocetine, and yohimbine,	ACQUITY UPLC BEHTM C_{18} (2.1 × 50 mm, 1.7 μm)	Acetonitrile, methanol, and formic acid	MS	Gradient	2.09–5.92	0.5	Leaf, stem, and root	Ethanolic	Kumar et al. (2018b)
6	CE-MS	Catharanthine, vinblastine, and vindoline	(6500 × 50 μm) (Yongnian, Hebei, China)	20 mM ammonium acetate and 1.5% acetic acid	MS	Directly injected	13–17	0.4	Stem	Methanolic	Chen et al. (2011)

Tikhomiroff and Jolicoeur (2002) achieved an identical elution sequence using a phosphate buffer at pH 6 with acetonitrile as the mobile phase, as reported by Uniyal et al. (2001) at pH 7.5. This discrepancy may be attributed to Uniyal et al. (2001) employing TEA and an alternative column, leading to a variation in the silanol activity of the column packing. Barthe et al. (2002) discovered that altering the mobile phase pH from 5 to 10 changed the elution sequence of vindoline and catharanthine. Consequently, a prudent amalgamation of buffer and mobile phase pH can be advantageous when formulating a separation approach targeting certain peaks. For example, modifying the mobile phase pH to 4.7 rather than 3.9 facilitated the separation of strictosidine and nonpolar alkaloids from a Tabernaemontana divaricata extract (Dagnino et al. 1995). Bhadra et al. (1993) and Rijhwani and Shanks (1998) successfully isolated up to five alkaloids in a single run for *C. roseus* metabolites. Van der Heijden et al. (1987) employed a previously established isocratic HPLC method for *Tabernaemontana* species. Moreover, the separation of alkaloids can be significantly enhanced with the application of gradient elution (Naaranlahti et al. 1987; Tikhomiroff and Jolicoeur 2002). It facilitates effective separation without prolonging HPLC run time. Certain co-eluting peaks cannot be precisely quantified by UV, yet they were differentiated using an ESI-MS detector in single ion recording mode, relying on the mass of the protonated molecule. The examination of a crude extract from *C. roseus* hairy roots was undertaken by Tikhomiroff and Jolicoeur (2002) and established the protocol of gradient technique for isolation of compounds. All alkaloids are eluted within a 23-min run. This review illustrates the efficacy of a gradient method for the separation of intricate combinations of *C. roseus* alkaloids, enabling the simultaneous isolation of up to 12 recognized compounds in a single run.

A method utilizing NPCE followed by RP-HPLC was developed and optimized for the extraction and quantification of vindoline, catharanthine, vincristine, and vinblastine from *C. roseus* leaves. Sixty mesh particles were utilized, along with 80% ethanol, a negative pressure of −0.075 MPa, a solid-to-liquid ratio of 1:20, an extraction duration of 30 min, and three extraction cycles. Under these optimal circumstances, the extraction yields of vindoline, catharanthine, vincristine, and vinblastine are 0.5783, 0.2843, 0.018, and 0.126 mg/g DW, respectively. The extraction yields are comparable to those obtained from the established ultrasonic extraction method and exceed the yields from maceration extraction and heated reflux extraction. The method asserts that NPCE–RP-HPLC is a superior option for the extraction and quantification of vinca alkaloids in both pilot- and industrial-scale applications (Mu et al. 2012). Vinblastine and vincristine were isolated using TLC and HPLC. One liter of culture filtrate produced 76 mg of vinblastine and 67 mg of vincristine (Jossang et al. 1998).

The identification and quantification of 3′,4′-anhydrovinblastine in HPLC were conducted through coinjection and by comparing retention duration, UV spectra, and peak areas with those of an external standard. The peak associated with 3′,4′-anhydrovinblastine was previously found via 1H-NMR (Sottomayor et al. 1998; Kutney et al. 1979, 1982; Andrews et al. 1995). The antineoplastic alkaloids, vinblastine and vincristine, were isolated using vacuum liquid chromatography on a silica gel and aluminum oxide (1:1) mixed bed, followed by a charcoal column, and ultimately purified through centrifugally accelerated radial chromatography (Shams et al. 2009). Leurocolornbine was extracted from *C. roseus* using gradient pH purification, subsequently followed by preparative HPLC separation. The separation of vinblastine and vincristine was conducted using a reversed-phase C18 Microsorb-MV column (250 × 4.6 mm, 5 μm) at ambient temperature. The mobile phase comprised methanol, phosphate buffer (5 mM, pH 6.0), and acetonitrile, employing a gradient elution with varying proportions at a flow rate of 2.0 mL min^{-1}, with detection at 254 nm. The HPLC method was employed to quantify vinblastine in the purple, red, and white forms of *C. roseus* leaves. The separation was accomplished in just 8 min. The peak confirmation was conducted based on the retention periods and UV spectra of the reference material. The method was validated concerning linearity, accuracy, recovery, LOD, and LOQ. The results indicated that the purple variety yields vinblastine concentrations 1.2 and 1.5 times greater than those of the white and pink kinds, respectively. The results derived from various cultivars are therefore beneficial for vinblastine production from the *C. roseus* plant (Kalidass et al. 2009; Thakore et al. 2013; Aslam et al. 2010a, b; Favali et al. 2004; Verma et al. 2007; Rahim et al. 2018; Falcão et al. 2017; Karimi and Raofie 2019).

6.4 Quantitative Analysis of Bisindole Alkaloids High-Performance Liquid Chromatography–Mass Spectrometry

Numerous enhancements can be realized to augment HPLC as a more potent analytical technology, including HPLC–MS and UPLC–MS. The purpose of enhancing HPLC techniques should be to streamline method scale-up processes and augment the capability for incorporating more advanced MS detectors. The determination of molecular weight for each peak is particularly beneficial for analyzing plant or cell extracts using UV and fluorescence spectra. The application of MS techniques, such as TSP-MS, ESI-MS (electrospray ionization mass spectrometry), or MALDI-TOF, can improve the identification of low-abundance chemicals and impure peaks with more precision than UV spectra. As previously noted, the silanol activity of HPLC packing necessitates the use of an acidic mobile phase to mitigate peak tailing and guarantee batch-to-batch repeatability. While amine modifiers may be incorporated into the mobile phase as masking agents, their application is inadvisable for ESI-MS detection due to the potential for ion suppression, which significantly diminishes instrument sensitivity (Gustavsson et al. 2001; Antignac et al. 2005). The introduction of columns incorporating a carbamate internal group in the bonded phase appears to have resolved this issue. Peak tailing was found to be diminished for basic compounds in comparison to conventional C18 alkyl packing (O'Gara et al. 1999). The selection of several MS interfaces for alkaloid analysis has been examined (Verpoorte and Niessen 1994). Hallard developed an LC-ESI-MS method for the measurement of 10 alkaloids and 19 precursors. Chu et al. (1997) employed an HPLC separation system integrated with an ESI interface to identify vincristine and vinblastine in plant extracts, achieving a detection limit of 0.2 μg mL^{-1}. The sensitivity for these chemicals was 2–3 times greater in ESI-MS compared to UV detection. Moreover, ESI-MS enhances sensitivity and facilitates the identification of various chemicals that cannot currently be discerned by a UV detector due to the close similarity of the UV spectra of certain secondary metabolites. Furthermore, as the elucidation of numerous alkaloid structures is accomplished using 1H-NMR, the integration of this detection method with the separation capabilities of HPLC could prove to be highly effective. Bobzin et al. (2000) proposed this method for identifying marine alkaloids as inhibitors of an enzyme associated with type II diabetes. HPLC–NMR was employed for the evaluation of plant extracts (Bringmann et al. 1999, 2000; Zhao et al. 1999).

Vinblastine and vinleurosine alkaloids were found using direct injection ion trap-mass spectrometry (IT-MS) to obtain MS spectra. The quantification of these alkaloids was achieved by liquid chromatography (LC) utilizing UV and MS detection methods. Figure 6.7 displays a TIC of active alkaloids in *C. roseus* and an EIC of five standard analytes (B) and the sample (C) obtained via LC–MS in positive ion mode. The analytes yielded strong indications for the protonated molecular ions $[M + H]^+$ and their associated product ions. Two approaches were employed to assess various validation parameters, including repeatability, LOD, calibration range, and recovery. MS exhibited elevated selectivity and sensitivity for the quantification of alkaloids in positive ionization mode. Following the optimization of the methodologies, the separation, identification, and quantification of the five constituents in *C. roseus* were thoroughly achieved using HPLC with UV and MS detection. Vinblastine and vinleurosine concentrations were measured at 7.52 and 2.10 μg/g, as well as 6.02 and 1.85 μg/g, utilizing LC–UV and LC–MS, respectively (Chen et al. 2013). Vinblastine, vincristine, and vinleurosine were chromatographically isolated on a C18 HPLC column. The mobile phase consisted of methanol with 15 nmol L^{-1} ammonium acetate and 0.02% formic acid in a 65:35 (V/V) ratio. The measurement of these alkaloids was performed using the multiple reaction monitoring (MRM) technique. This approach was confirmed, demonstrating that the intra- and inter-day precision and accuracy of the five alkaloids ranged from 1.2% to 11.5% (RSD%) and from −10.9% to 10.5% (RE%). The recovery rates of the alkaloids in the samples ranged from 79.9% to 91.5%.

Vinblastine, vincristine, and vinleurosine exhibited stability at room temperature for 2 hours, at 4 °C for 12 hours, and at −20 °C for 2 weeks. The established approach was effectively utilized to quantify the five alkaloids in three portions of three batches of *C. roseus*, utilizing small samples obtained from three districts in China. Jeong et al. devised a sensitive, quick, and innovative ultra-performance liquid chromatography–quadrupole time-of-flight (UPLC–Q-TOF) MS method for the quantification of indole alkaloids (vincristine, vinblastine, ajmalicine, catharanthine, serpentine, and vindoline) in *C. roseus*. The alkaloids were isolated using a C18 column with a mobile phase of acetonitrile and water, supplemented with 0.1% formic acid and 10 mM ammonium acetate, achieving effective resolution in 13 min. The ultra-performance liquid chromatography-quadrupole time-of-flight (UPLC–Q-TOF) mass spectrometry method was validated in accordance with ICH guidelines, yielding exceptional linearity ($R^2 > 0.9988$), a LOD ranging from 1 to 10 ng/mL, a limit of quantification from 3 to 30 ng/mL, as well as intraday and inter-day precision, and extraction recovery rates between 92.8% and 104.1% for all components. The validated UPLC–Q-TOF method was utilized for the quantification of extracts from the root, stem, and leaves of *C. roseus*. The concentrations of bisindole alkaloids (vincristine and vinblastine) were elevated in the aerial portions, but the concentrations of mono-indole alkaloids (ajmalicine, catharanthine, serpentine, and vindoline) were elevated in the roots (Fig. 6.8) (Jeong et al. 2018). Vinblastine, vincristine, vinleurosine, vindoline, and catharanthine were separated using a C_{18} HPLC column with a mobile phase consisting of methanol and 15 nmol/L ammonium acetate containing 0.02% formic acid at a ratio of 65:35 (V/V) (EIC depicted in Fig. 6.9). The measurement of these alkaloids was conducted using the Multiple Reaction Monitoring (MRM) method. This procedure was validated, and the results met the objectives of the study. The intra- and inter-day precision and accuracy of the five alkaloids ranged from 1.2% to 11.5% (RSD%) and from −10.9% to 10.5% (RE%). The recovery rates of the five alkaloids in the samples ranged from 79.9% to 91.5%. The five analytes exhibited stability at room temperature for 2 hours, at 4 °C for 12 hours, and at −20 °C for 2 weeks. The HPLC–ESI-MS/MS technique enables the concurrent quantification of five significant alkaloids in trace samples of *C. roseus*. The concentration of vinleurosine was documented as 2.4–17.6 μg/g, DW in the leaf, stem, and root of *C. roseus* specimens obtained from three distinct regions of China (Zhang et al. 2014).

An efficient, sensitive, and reproducible UHPLC–ESI-MS/MS method was developed and validated for the simultaneous quantification of anticancer agents (vincristine, vinblastine, vindesine), antihypertensive compounds (ajmaline, ajmalicine, reserpine), an aphrodisiac (yohimbine), a sedative (serpentine), a dietary supplement (vinpocetine), and a precursor of vinblastine (vindoline) in crude extracts of *C. roseus* utilizing UHPLC coupled with electrospray ionization hybrid triple quadrupole-linear ion trap mass spectrometry (UHPLC–ESI-QqQLIT-MS/MS) in multiple reaction monitoring (MRM) mode. All analytes exhibited strong signal strength in positive ionization mode. The calibration curves exhibited linearity within the concentration range of 0.5–1000 ng/mL, with correlation values (R^2) exceeding 0.9931. The LOD and LOQ varied from 0.039 to 0.583 ng/mL and from 0.118 to 1.767 ng/mL, respectively. The intraday precision (0.2–2.7% RSD), inter-day precision (0.4–2.9% RSD), and stability (0.7–3.4%) were deemed acceptable. The recoveries varied from 99.63% to 104.30%, with a %RSD of less than or equal to 3.03%, demonstrating the method's high accuracy. The validated methodology was employed for the concurrent quantification of terpene indole alkaloids in ethanolic extracts from 39 samples of *C. roseus* gathered from five distinct locations in India. Serpentine was identified as one of the most prevalent components in all samples, as previously reported. The findings demonstrated that this approach was appropriate for the study of specific alkaloids from crude extracts of *C. roseus*.

The established UHPLC–ESI-MS/MS technique is applicable for the separation and quantification of alkaloids from *C. roseus* and its herbal derivatives, owing to its precision, reliability, and reproducibility. The efficacy and effectiveness of the designed and validated UHPLC–ESI-MS/MS method were assessed through its application in the quantitative analysis of 10 TIAs in ethanolic extracts from 39 samples of *C. roseus* leaf, stem, and roots. Serpentine was identified as the predominant analyte among all targets. The greatest concentration of serpentine was recorded in the Uttar Pradesh stem sample C-5S (50.074 mg/g), succeeded by the Bihar stem sample CR-3R (49.885 mg/g) and the Arunachal Pradesh root sample C-44R (49.005 mg/g). Serpentine content (19.701 mg/g) was identified as the most abundant compound in the aqueous extract of *C. roseus*, as reported in a previous study utilizing HPLC–DAD–ESI-MS/MS analysis. Ajmalicine was also detected in significant quantities in root samples C-11R (17.70 mg/g), C-44R (16.88 mg/g), and CR-3R (15.76 mg/g). Similarly, a significant concentration of vincristine was noted in CR-4R (6.52 mg/g), followed by C-41L (0.502 mg/g) and CR-2L (0.478 mg/g), whereas vinblastine was predominantly found in C-44R (1.514 mg/g), followed by CR-4R (1.38 mg/g) and CR-2R (1.296 mg/g). CR-2L had a high vindoline concentration of 19.46 mg/g, followed by CR-3L at 18.44 mg/g. In contrast, samples C-2R (2.42 mg/g), C-5R (2.26 mg/g), and C-44R (2.16 mg/g) demonstrated comparable yohimbine content. Vindesine was identified in concentrations ranging from 1.544 mg/g in C-5R to 3.22 mg/g in C-2R. Minimal quantities of ajmaline (<0.14 mg/g) were identified in all samples. Vinpocetine and reserpine were identified in concentrations of less than 0.0204 mg/g and 0.0534 mg/g, respectively. The anticancer drugs vincristine and vinblastine were evenly distributed in the Jammu and Kashmir and Arunachal Pradesh samples. Vincristine was predominantly detected in elevated concentrations in the leaf and root samples. The anticancer drugs vincristine and vinblastine were evenly distributed in the root and leaf samples collected from Jammu and Kashmir, Uttar Pradesh and Arunachal Pradesh. *C. roseus* is a significant source of serpentine, similar to the *Rauwolfia* species (Kumar et al. 2018b).

References

Abraham DJ, Farnsworth NR (1969) Structure eucidation and chemistry of Catharanthus -alkaloids III: structure of leurosine, an active anticancer alkaloid. J Pharm Sci 58(6):694–698

Abraham DJ, Farnsworth NR, Blomster RN, Rhodes RE (1967) Structure elucidation and chemistry of Catharanthus alkaloids II. Isolation and partial structure of catharine, a dimeric indole alkaloid from C. zanceus and C. roseus. J Pharm Sci 56(3):401–403

Abraham DJ, Farnsworth NR, Loub WD, Blomster RN (1969) Structure elucidation and chemistry of Catharanthus alkaloids IV. Structures of hoerhammericine and hoerhammerinine. J Org Chem 34(6):1575–1576

Ahmad NH, Rahim RA, Mat I (2010) Catharanthus roseus aqueous extract is cytotoxic to Jurkat leukaemic T-cells but induces the proliferation of normal peripheral blood mononuclear cells. Trop Life Sci Res 21(2):101–113

Ali I, Bashir M (1984) Isolation and structural studies on the alkaloids in flowers of Catharanthus roseus. J Nat Prod 47(3):554–555

Almagro L, Fernández-Pérez F, Pedreño M (2015) Indole alkaloids from Catharanthus roseus: bioproduction and their effect on human health. Molecules 20(2):2973–3000

Andrade SAL, Malik S, Sawaya ACHF, Bottcher A, Mazzafera P (2013) Association with arbuscular mycorrhizal fungi influences alkaloid synthesis and accumulation in Catharanthus roseus and Nicotiana tabacum plants. Acta Physiol Plant 35(3):867–880

Andrews CW, Wisowaty J, Davis AO, Crouch RC, Martin GE (1995) Molecular modeling, NMR spectroscopy, and conformational analysis of 3′,4′-anhydrovinblastine. J Heterocyclic Chem 32(3):1011–1017

Antignac JP, Brosseaud A, Gaudin-Hirret I, André F, Le Bizec B (2005) Analytical strategies for the direct mass spectrometric analysis of steroid and corticosteroid phase II metabolites. Steroids 70(3):205–216

Aslam J, Khan SH, Siddiqui ZH, Fatima Z, Maqsood M, Bhat MA, Nasim SA, Ilah A, Ahmad IZ, Khan SA, Mujib A (2010a) Catharanthus roseus (L.) G. Don. An important drug: its applications and production. Pharm Glob 4(12):1–16

Aslam J, Mujib A, Fatima Z, Sharma MP (2010b) Variations in vinblastine production at different stages of somatic embryogenesis, embryo, and field-grown plantlets of Catharanthus roseus (L.) G. Don, as revealed by HPLC. In Vitro Cell Dev Biol – Plant 46(4):348–353

Aslam J, Mujib A, Sharma MP (2011) Influence of freezing and non-freezing temperature on somatic embryogenesis and vinblastine production in Catharanthus roseus (L.) G. Don. Acta Physiol Plant 33(2):473–480

Ataei-Azimi A, Hashemloian BD, Ebrahimzadeh H, Majd A (2008) High in vitro production of anticancer indole alkaloids from periwinkle (Catharanthus roseus) tissue culture. Afr J Biotechnol 7(16):2834–2839

Auriola S, Ranta VP, Naaranlahti T, Lapinjoki SP (1989) Thermospray liquid chromatography–mass spectrometric analysis of Catharanthus alkaloids. J Chromatogr A 474(1):181–185

Auriola S, Naaranlahti T, Lapinjoki SP (1991) Determination of Catharanthus roseus alkaloids by high-performance liquid chromatography–isotope dilution thermospray–mass spectrometry. J Chromatogr A 554(1–2):227–231

Ayob FW, Simarani K, Zainal Abidin N, Mohamad J (2017) First report on a novel N igrospora sphaerica isolated from C atharanthus roseus plant with anticarcinogenic properties. Microb Biotechnol 10(4):926–932

Barthe L, Ribet JP, Pélissou M, Degude MJ, Fahy J, Duflos A (2002) Optimization of the separation of Vinca alkaloids by nonaqueous capillary electrophoresis. J Chromatogr A 968(1–2):241–250

Bashir M (1983) Isolation of new alkaloids from Catharanthus roseus. Planta Med 49(10):124–125

Bhadra R, Vani S, Shanks JV (1993) Production of indole alkaloids by selected hairy root lines of Catharanthus roseus. Biotechnol Bioeng 41(5):581–592

Blaskó G, Cordell GA (1990) Isolation, structure elucidation, and biosynthesis of the bisindole alkaloids of Catharanthus. In: Cordell GA (ed) The alkaloids: chemistry and pharmacology, vol 37. Academic Press, San Diego, pp 1–76

Blomster RN, Farnsworth NR, Abraham DJ (1967) Catharanthus alkaloids X. Isolation of lanceine and vinosidine from Catharanthus lanceus roots. J Pharm Sci 56(2):284–286

Bobzin SC, Yang S, Kasten TP (2000) Application of liquid chromatography–nuclear magnetic resonance spectroscopy to the identification of natural products. J Chromatogr B Biomed Sci Appl 748(1):259–267

Bringmann G, Messer K, Wohlfarth M, Kraus J, Dumbuya K, Rückert M (1999) HPLC−CD on-line coupling in combination with HPLC−NMR and HPLC−MS/MS for the determination of the full absolute stereostructure of new metabolites in plant extracts. Anal Chem 71(14):2678–2686

Bringmann G, Wohlfarth M, Heubes M (2000) Observation of exchangeable protons by high-performance liquid chromatography–nuclear magnetic resonance spectroscopy and high-performance liquid chromatography–electrospray ionization mass spectrometry: a useful tool for the hyphenated analysis of natural products. J Chromatogr A 904(2):243–249

Brossi A, Suffness M (eds) (1990) The alkaloids: antitumor bisindole alkaloids from Catharanthus roseus (L.), vol 37. Academic Press, San Diego

Chen Q, Li N, Zhang W, Chen J, Chen Z (2011) Simultaneous determination of vinblastine and its monomeric precursors vindoline and catharanthine in Catharanthus roseus by capillary electrophoresis–mass spectrometry. J Sep Sci 34(20):2885–2892

Chen Q, Zhang W, Zhang Y, Chen J, Chen Z (2013) Identification and quantification of active alkaloids in Catharanthus roseus by liquid chromatography–ion trap mass spectrometry. Food Chem 139(1–4):845–852

Choi YH, Yoo KP, Kim J (2002) Supercritical fluid extraction and liquid chromatography-electrospray mass analysis of vinblastine from Catharanthus roseus. Chem Pharm Bull 50(9):1294–1296

Chu IH, Bodnar JA, Bowman RN, White EL (1997) Determination of vincristine and vinblastine in Catharanthus roseus plants by high performance liquid chromatography/electrospray ionization mass spectrometry. J Liq Chromatogr Relat Technol 20(8):1159–1174

Cone NJ, Miller R, Neuss N (1963) Alkaloids of Vinca rosea Linn. (Catharanthus roseus G. Don) XV: analysis of vinca alkaloids by thin-layer chromatography. J Pharm Sci 52(7):688–692

Creasey WA (1969) Biochemical effects of the vinca alkaloids—IV: studies with vinleurosine. Biochem Pharmacol 18(1):227–232

Dagnino D, Schripsema J, Verpoorte R (1995) Terpenoid indole alkaloid biosynthesis and enzyme activities in two cell lines of Tabernaemontana divaricata. Phytochemistry 39(2):341–349

De Bruyn A, De Taeye L, Simonds R, Verzele M, De Pauw C (1982) Alkaloids from Catharanthus roseus. Isolation and identification of 17-Desacetoxyvinblastine and 17-Desacetoxyleurosine. Bull Soc Chim Belg 91(1):75–85

DeLuca V, Balsevich J, Tyler RT, Eilert U, Panchuk BD, Kurz WG (1986) Biosynthesis of indole alkaloids: developmental regulation of the biosynthetic pathway from tabersonine to vindoline in Catharanthus roseus. J Plant Physiol 125(1–2):147–156

Dhyani P, Quispe C, Sharma E, Bahukhandi A, Sati P, Attri DC, Szopa A, Sharifi-Rad J, Docea AO, Mardare I, Calina D (2022) Anticancer potential of alkaloids: a key emphasis to colchicine, vinblastine, vincristine, vindesine, vinorelbine and vincamine. Cancer Cell Int 22(1):206

Digvijay S, Shashi PR, Suchi S, Kumar RS, Raghavendra M, Sushil K (2004) Simultaneous quantification of some pharmaceutical Catharanthus roseus leaf and root terpenoid indole alkaloids and their precursors in single runs by reversed-phase liquid chromatography. J AOAC Int 87(6):1287–1296

Dorman DE, Paschal JW (1976) 13C NMR spectroscopy: comparison of the spectra of some dimeric Catharanthus alkaloids and their derivatives. Org Magn Reson 8(8):413–418

Dubrovay Z, Háda V, Béni Z, Szántay C Jr (2013) NMR and mass spectrometric characterization of vinblastine, vincristine and some new related impurities – part I. J Pharm Biomed Anal 84:293–308

Dutta A, Batra J, Pandey-Rai S, Singh D, Kumar S, Sen J (2005) Expression of terpenoid indole alkaloid biosynthetic pathway genes corresponds to accumulation of related alkaloids in Catharanthus roseus (L.) G. Don. Planta 220(3):376–383

El-Domyati FM, Abdel Azeiz AZ, Atef A, Elhalwagi A, Sabir JSM, Ramadan AM, Gadalla NO, Edris S, Baeshen MN, Hajrah NH, Al-Kordy MA (2014) Functional genomic analysis for overproduction of key terpenoid indole alkaloids in mature leaf of Catharanthus roseus. Life Sci J 11(10):1313–1320

El-Sayed A, Cordell GA (1981) Catharanthus alkaloids. XXXIV. Catharanthamine, a new antitumor bisindole alkaloid from Catharanthus roseus. J Nat Prod 44(3):289–293

El-Sayed A, Handy GA, Cordell GA (1980) Catharanthus alkaloids XXXIII. 21′-oxo-leurosine from Catharanthus roseus (Apocynaceae). J Nat Prod 43(1):157–161

El-Sayed A, Handy GA, Cordell GA (1983) Catharanthus Alkaloids, XXXVIII. Confirming structural evidence and antineoplastic activity of the bisindole alkaloids leurosine-N′b-oxide (pleurosine), roseadine, and vindolicine from Catharanthus roseus. J Nat Prod 46(4):517–527

Endo T, Goodbody A, Misawa M (1987) Alkaloid production in root and shoot cultures of Catharanthus roseus. Planta Med 53(5):479–482

Falcão MA, Scopel R, Almeida RN, do Espirito Santo AT, Franceschini G, Garcez JJ, Vargas RM, Cassel E (2017) Supercritical fluid extraction of vinblastine from Catharanthus roseus. J Supercrit Fluids 129:9–15

Farnsworth NR (1972) Anomalous isolation of an active antitumor alkaloid from a fraction of Catharanthus lanceus devoid of anticancer activity. J Pharm Sci 61(11):1840–1841

Farnsworth NR, Loub WD, Blomster RN (1963) Studies on Catharanthus lanceus (Vinca lancea) I. Isolation of leurosine, perivine, and yohimbine. J Pharm Sci 52(11):1114

Favali MA, Musetti R, Benvenuti S, Bianchi A, Pressacco L (2004) Catharanthus roseus L. plants and explants infected with phytoplasmas: alkaloid production and structural observations. Protoplasma 223(1):45–51

Ferreres F, Pereira DM, Valentão P, Oliveira JM, Faria J, Gaspar L, Sottomayor M, Andrade PB (2010) Simple and reproducible HPLC–DAD–ESI-MS/MS analysis of alkaloids in Catharanthus roseus roots. J Pharm Biomed Anal 51(1):65–69

Forni C, Facchiano F, Bartoli M, Pieretti S, Facchiano A, D'Arcangelo D, Norelli S, Valle G, Nisini R, Beninati S, Tabolacci C (2019) Beneficial role of phytochemicals on oxidative stress and age-related diseases. Biomed Res Int 2019:8748253

Glauser G, Veyrat N, Rochat B, Wolfender JL, Turlings TC (2013) Ultra-high pressure liquid chromatography–mass spectrometry for plant metabolomics: a systematic comparison of high-resolution quadrupole-time-of-flight and single stage Orbitrap mass spectrometers. J Chromatogr A 1292:151–159

Grata E, Guillarme D, Glauser G, Boccard J, Carrupt PA, Veuthey JL, Rudaz S, Wolfender JL (2009) Metabolite profiling of plant extracts by ultra-high-pressure liquid chromatography at elevated temperature coupled to time-of-flight mass spectrometry. J Chromatogr A 1216(30):5660–5668

Guo XR, Yang L, Yu JH, Tang ZH, Zu YG (2007) Alkaloid variations in Catharanthus roseus seedlings treated by different temperatures in short term and long term. J For Res 18(4):313–315

Gupta MM, Singh DV, Tripathi AK, Pandey R, Verma RK, Singh S, Shasany AK, Khanuja SPS (2005) Simultaneous determination of vincristine, vinblastine, catharanthine, and vindoline in leaves of Catharanthus roseus by high-performance liquid chromatography. J Chromatogr Sci 43(9):450–453

Gustavsson SÅ, Samskog J, Markides KE, Långström B (2001) Studies of signal suppression in liquid chromatography–electrospray ionization mass spectrometry using volatile ion-pairing reagents. J Chromatogr A 937(1–2):41–47

Háda V, Dubrovay Z, Lakó-Futó Á, Galambos J, Gulyás Z, Aranyi A, Szántay C Jr (2013) NMR and mass spectrometric characterization of vinblastine, vincristine and some new related impurities—part II. J Pharm Biomed Anal 84:309–322

Hanafy MS, Matter MA, Asker MS, Rady MR (2016) Production of indole alkaloids in hairy root cultures of Catharanthus roseus L. and their antimicrobial activity. S Afr J Bot 105:9–18

Hashemabadi D, Sabzevari F, Kaviani B, Ansari MH (2018) Organic N-fertilizer, rhizobacterial inoculation and fungal compost improve nutrient uptake, plant growth and the levels of vindoline, ajmalicine, vinblastine, catharanthine and total alkaloids in Catharanthus roseus L. Folia Hortic 30(2):203–213

He L, Yang L, Tan R, Zhao S, Hu Z (2011a) Enhancement of vindoline production in suspension culture of the Catharanthus roseus cell line C20hi by light and methyl jasmonate elicitation. Anal Sci 27(12):1243–1243

He L, Yang L, Xiong A, Zhao S, Wang Z, Hu Z (2011b) Simultaneous quantification of four indole alkaloids in Catharanthus roseus cell line C20hi by UPLC-MS. Anal Sci 27(4):433–433

Jacobs DI, Snoeijer W, Hallard D, Verpoorte R (2004) The Catharanthus alkaloids: pharmacognosy and biotechnology. Curr Med Chem 11(5):607–628

Jeong WT, Lim HB (2018) A UPLC-ESI-Q-TOF method for rapid and reliable identification and quantification of major indole alkaloids in Catharanthus roseus. J Chromatogr B 1080:27–36

Jossang A, Fodor P, Bodo B (1998) A new structural class of bisindole alkaloids from the seeds of Catharanthus roseus: vingramine and methylvingramine. J Org Chem 63(21):7162–7167

Jungclas H, Danigel H, Schmidt L, Dellbrügge J (1982) Combined liquid chromatography time-of-flight mass spectrometry: an application of 252Cf fission fragment induced desorption mass spectrometry. Org Mass Spectrom 17(10):499–502

Kalidass C, Mohan VR, Daniel A (2009) Effect of auxin and cytokinin on vincristine production by callus cultures of Catharanthus roseus L. (Apocynaceae). Trop Subtrop Agroecosyst 12(2):283–288

Karimi M, Raofie F (2019) Micronization of vincristine extracted from Catharanthus roseus by expansion of supercritical fluid solution. J Supercrit Fluids 146:172–179

Koel M, Kuhtinskaja M, Vaher M (2020) Extraction of bioactive compounds from Catharanthus roseus and Vinca minor. Sep Purif Technol 252:117438

Kokkonen P, Schröder E, Niessen WMA, Tjaden UR, Van Der Greef J (1990) Important parameters in liquid chromatography—continuous-flow fast atom bombardment mass spectrometry. J Chromatogr A 511:35–47

Kumar A, Patil D, Rajamohanan PR, Ahmad A (2013) Isolation, purification and characterization of vinblastine and vincristine from endophytic fungus Fusarium oxysporum isolated from Catharanthus roseus. PLoS One 8(9):e71805

Kumar S, Bajpai V, Singh A, Kumar B (2018a) Identification, characterization and distribution of terpene indole alkaloids in ethanolic extracts of Catharanthus roseus using high-performance liquid chromatography/electrospray ionization quadrupole time-of-flight tandem mass spectrometry and the study of their geographical variation. Rapid Commun Mass Spectrom 32(4):319–332

Kumar S, Singh A, Kumar B, Singh B, Bahadur L, Lal M (2018b) Simultaneous quantitative determination of bioactive terpene indole alkaloids in ethanolic extracts of Catharanthus roseus (L.) G. Don by ultra high performance liquid chromatography–tandem mass spectrometry. J Pharm Biomed Anal 151:32–41

Kutney JP, Balsevich J, Honda T, Liao PH, Thieelier HP, Worth BR (1978) Total synthesis of indole and dihydroindole alkaloids. XVI. Derivatives of vinblastine and vincristine: change of functionality in the vindoline unit. Can J Chem 56(19):2560–2566

Kutney JP, Badger RA, Beck JF, Bosshardt H, Matough FS, Ridaura-Sanz VE, So YH, Sood RS, Worth BR (1979) Dihydropyridines in synthesis and biosynthesis. I. Secodine and precursors of dehydrosecodine. Can J Chem 57(3):289–299

Kutney JP, Aweryn B, Choi LS, Kolodziejczyk P, Kurz WG, Chatson KB, Constabel F (1982) Alkaloid production in Catharanthus roseus cell cultures XI. Biotransformation of 3′,4′-anhydrovinblastine to other bisindole alkaloids. Helv Chim Acta 65(4):1271–1278

Lin C, Cai J, Yang X, Hu L, Lin G (2015) Liquid chromatography mass spectrometry simultaneous determination of vindoline and catharanthine in rat plasma and its application to a pharmacokinetic study. Biomed Chromatogr 29(1):97–102

Liu L, Cao JX, Yao YC, Xu SP (2013) Progress of pharmacological studies on alkaloids from Apocynaceae. J Asian Nat Prod Res 15(2):166–184

Liu J, Liu Y, Pan YJ, Zu YG, Tang ZH (2016a) Determination of alkaloids in Catharanthus roseus and Vinca minor by high-performance liquid chromatography–tandem mass spectrometry. Anal Lett 49(8):1143–1153

Liu Z, Wu HL, Li Y, Gu HW, Yin XL, Xie LX, Yu RQ (2016b) Rapid and simultaneous determination of five vinca alkaloids in Catharanthus roseus and human serum using trilinear component modeling of liquid chromatography–diode array detection data. J Chromatogr B 1026:114–123

Luo M, Yang LQ, Yao XH, Mu FS, Zhang DY, Song ZY, Qiao Q, Fu YJ, Zu YG (2014) Optimization of enzyme-assisted negative pressure cavitation extraction of five main indole alkaloids from Catharanthus roseus leaves and its pilot-scale application. Sep Purif Technol 125:66–73

Maeda S, Miwa Y (2013) Multicomponent high-performance liquid chromatography/tandem mass spectrometry analysis of ten chemotherapeutic drugs in wipe samples. J Chromatogr B 921:43–48

Magagula NL, Mohanlall V, Odhav B (2012) Optimized thin layer chromatographic method for screening pharmaceutically valuable alkaloids of Catharanthus roseus (Madagascar Periwinkle). Int J Sci 2012(12):1–20

McCalley DV (2002) Analysis of the cinchona alkaloids by high-performance liquid chromatography and other separation techniques. J Chromatogr A 967(1):1–19
Miura Y, Hirata K, Kurano N (1987) Isolation of vinblastine in callus culture with differentiated roots of Catharanthus roseus (L.) G. Don. Agric Biol Chem 51(2):611–614
Monforte-González M, Ayora-Talavera T, Maldonado-Mendoza IE, Loyola-Vargas VM (1992) Quantitative analysis of serpentine and ajmalicine in plant tissues of Catharanthus roseus and hyoscyamine and scopolamine in root tissues of Datura stramonium by thin layer chromatography-densitometry. Phytochem Anal 3(3):117–121
Moon SH, Pandurangan M, Kim DH, Venkatesh J, Patel RV, Mistry BM (2018) A rich source of potential bioactive compounds with anticancer activities by Catharanthus roseus cambium meristematic stem cell cultures. J Ethnopharmacol 217:107–117
Mu F, Yang L, Wang W, Luo M, Fu Y, Guo X, Zu Y (2012) Negative-pressure cavitation extraction of four main vinca alkaloids from Catharanthus roseus leaves. Molecules 17(8):8742–8752
Mukhopadhyay S, Cordell GA (1981a) Catharanthus alkaloids. XXXV. Isolation of leurosidine N′b-oxide from Catharanthus roseus. J Nat Prod 44(5):611–613
Mukhopadhyay S, Cordell GA (1981b) Catharanthus alkaloids. XXXVI. Isolation of vincaleukoblastine (VLB) and periformyline from Catharanthus trichophyllus and pericyclivine from Catharanthus roseus. J Nat Prod 44(3):335–339
Naaranlahti T, Nordström M, Huhtikangas A, Lounasmaa M (1987) Determination of Catharanthus alkaloids by reversed-phase high-performance liquid chromatography. J Chromatogr A 410:488–493
Naaranlahti T, Ranta VP, Jarho P, Nordström M, Lapinjoki SP (1989) Electrochemical detection of indole alkaloids of Catharanthus roseus in high-performance liquid chromatography. Analyst 114(10):1229–1231
Nagy-Turák A, Végh Z (1994) Extraction and in situ densitometric determination of alkaloids from Catharanthus roseus by means of overpressured layer chromatography on amino-bonded silica layers I. Optimization and validation of the separation system. J Chromatogr A 668(2):501–507
Neuss N, Huckstep LL, Cone NJ (1967) Vinca alkaloids. XXIX. Structure of leurosidine (vinrosidine, VRD), an oncolytic alkaloid from Vinca rosea Linn. (Catharanthus roseus G. Don). Tetrahedron Lett 8(9):811–816
Nguyen DQ, Nguyen NL, Nguyen VT, Nguyen THG, Do TTT, Nguyen TH, Dung DH, Nguyen TKL, Nguyen QH, Le TT, Nguyen HH (2023) Isolation and identification of vincristine and vinblastine producing endophytic fungi from Catharanthus roseus (L.) G. Don. Russ J Plant Physiol 70(8):188
Nisar A, Mamat AS, Hatim MI, Aslam MS, Syarhabil M (2016) An updated review on Catharanthus roseus: phytochemical and pharmacological analysis. Indian Res J Pharm Sci 9:631–653
O'Gara JE, Walsh DP, Alden BA, Casellini P, Walter TH (1999) Systematic study of chromatographic behavior vs alkyl chain length for HPLC bonded phases containing an embedded carbamate group. Anal Chem 71(15):2992–2997
Paci A, Mercier L, Bourget P (2003) Identification and quantitation of antineoplastic compounds in chemotherapeutic infusion bags by use of HPTLC: application to the vinca-alkaloids. J Pharm Biomed Anal 30(5):1603–1610
Pan Q, Mustafa NR, Tang K, Choi YH, Verpoorte R (2016) Monoterpenoid indole alkaloids biosynthesis and its regulation in Catharanthus roseus: a literature review from genes to metabolites. Phytochem Rev 15(2):221–250
Pandiangan D, Tilaar W, Nainggolan N, Wahyudi L (2013) Relations between catharanthine content enhancement with the other associated secondary metabolites in Catharanthus roseus cell culture that treated tryptophan. Int J Sci Res 4(1):2208–2212
Rahim RA, Ahmad NH, Al Azzam KM, Mat I (2018) Determination and quantification of the vinblastine content in purple, red, and white Catharanthus roseus leaves using RP-HPLC method. Adv Pharm Bull 8(1):157–161
Ramírez J, Ogan K, Ratain MJ (1997) Determination of vinca alkaloids in human plasma by liquid chromatography/atmospheric pressure chemical ionization mass spectrometry. Cancer Chemother Pharmacol 39(4):286–290

Renaudin JP (1984) Reversed-phase high-performance liquid chromatographic characteristics of indole alkaloids from cell suspension cultures of Catharanthus roseus. J Chromatogr A 291:165–174
Rijhwani SK, Shanks JV (1998) Effect of elicitor dosage and exposure time on biosynthesis of indole alkaloids by Catharanthus roseus hairy root cultures. Biotechnol Prog 14(3):442–449
Rischer H, Orešič M, Seppänen-Laakso T, Katajamaa M, Lammertyn F, Ardiles-Diaz W, Van Montagu MC, Inzé D, Oksman-Caldentey KM, Goossens A (2006) Gene-to-metabolite networks for terpenoid indole alkaloid biosynthesis in Catharanthus roseus cells. Proc Natl Acad Sci 103(14):5614–5619
Saravanan S, Karthi S (2014) HPLC analysis for methanolic extract of Catharanthus roseus under elevated CO_2. World J Pharm Pharm Sci 3(10):683–693
Schmidt MS, Huang R, Classon RJ, Murry DJ (2006) Determination of vincristine in infant plasma by liquid chromatography–atmospheric pressure chemical ionization–mass spectroscopy. J Pharm Biomed Anal 41(2):540–543
Shams K, Nazif N, Azim N, Shafeek K, Missiry M, Ismail S, Nasr M (2009) Isolation and characterization of antineoplastic alkaloids from Catharanthus roseus L. Don. cultivated in Egypt. Afr J Tradit Complement Altern Med 6(2):118–122
Siddiqui MJA, Ismail Z, Saidan NH (2011) Simultaneous determination of secondary metabolites from Vinca rosea plant extractives by reverse phase high performance liquid chromatography. Pharmacogn Mag 7(26):92–96
Singh DV, Maithy A, Verma RK, Gupta MM, Kumar S (2000) Simultaneous determination of Catharanthus alkaloids using reversed phase high performance liquid chromatography. J Liq Chromatogr Relat Technol 23(4):601–607
Sottani C, Turci R, Perbellini L, Minoia C (1998) Liquid–liquid extraction procedure for trace determination of cyclophosphamide in human urine by high-performance liquid chromatography tandem mass spectrometry. Rapid Commun Mass Spectrom 12(16):1063–1068
Sottomayor M, Lopez-Serrano M, DiCosmo F, Ros Barceló A (1998) Purification and characterization of α-3′, 4′-anhydrovinblastine synthase (peroxidase-like) from Catharanthus roseus (L.) G. Don. FEBS Lett 428(3):299–303
Svoboda GH, Gorman M, Neuss N, Barnes AJ Jr (1961) Alkaloids of Vinca rosea Linn. (Catharanthus roseus G. Don.). VIII. Preparation and characterization of new minor alkaloids. J Pharm Sci 50(5):409–413
Svoboda GH, Gorman M, Barnes AJ Jr, Oliver AT (1962) Alkaloids of Vinca rosea Linn. (Catharanthus roseus G. Don.). XII. Preparation and characterization of trace alkaloids. J Pharm Sci 51(6):518–523
Tafur S, Jones WE, Dorman DE, Logsdon EE, Svoboda GH (1975) Alkaloids of Vinca rosea L. (Catharanthus roseus G. Don) XXXVI: isolation and characterization of new dimeric alkaloids. J Pharm Sci 64(12):1953–1957
Tafur SS, Occolowitz JL, Elzey TK, Paschal JW, Dorman DE (1976) Alkaloids of Vinca rosea (Catharanthus roseus G. Don). XXXVII. Structure of vincathicine. J Org Chem 41(6):1001–1005
Tam MN, Nikolova-Damyanova B, Pyuskyulev B (1995) Quantitative thin layer chromatography of indole alkaloids. II. Catharanthine and vindoline. J Liq Chromatogr Relat Technol 18(5):849–858
Tang Z, Yang L, Zu Y, Guo X (2009) Variations of vinblastine accumulation and redox state affected by exogenous H2O2 in Catharanthus roseus (L.) G. Don. Plant Growth Regul 57(1):15–20
Teoh SL, Das S (2018) Phytochemicals and their effective role in the treatment of diabetes mellitus: a short review. Phytochem Rev 17(5):1111–1128
Thakore D, Srivastava AK, Sinha AK (2013) Yield enhancement strategies for enhancement of indole alkaloids in hairy root cultures of Catharanthus roseus. Int J Chem Eng Appl 4(3):153–156
Thimmaiah KN, Horton JK, Qian XD, Beck WT, Houghton JA, Houghton PJ (1990) Structural determinants of phenoxazine type compounds required to modulate the accumulation of vinblastine and vincristine in multidrug-resistant cell lines. Cancer Commun 2(7):249–259

Tikhomiroff C, Jolicoeur M (2002) Screening of Catharanthus roseus secondary metabolites by high-performance liquid chromatography. J Chromatogr A 955(1):87–93
Tiong SH, Looi CY, Arya A, Wong WF, Hazni H, Mustafa MR, Awang K (2015) Vindogentianine, a hypoglycemic alkaloid from Catharanthus roseus (L.) G. Don (Apocynaceae). Fitoterapia 102:182–188
Toivonen L, Ojala M, Kauppinen V (1991) Studies on the optimization of growth and indole alkaloid production by hairy root cultures of Catharanthus roseus. Biotechnol Bioeng 37(7):673–680
Trim PJ, Henson CM, Avery JL, McEwen A, Snel MF, Claude E, Marshall PS, West A, Princivalle AP, Clench MR (2008) Matrix-assisted laser desorption/ionization-ion mobility separation-mass spectrometry imaging of vinblastine in whole body tissue sections. Anal Chem 80(22):8628–8634
Uniyal GC, Bala S, Mathur AK, Kulkarni RN (2001) Symmetry C18 column: a better choice for the analysis of indole alkaloids of Catharanthus roseus. Phytochem Anal 12(3):206–210
Van der Heijden R, Lamping PJ, Out PP, Wijnsma R, Verpoorte R (1987) High-performance liquid chromatographic determination of indole alkaloids in a suspension culture of Tabernaemontana divaricata. J Chromatogr A 396:287–295
Verma A, Laakso I, Seppänen-Laakso T, Huhtikangas A, Riekkola ML (2007) A simplified procedure for indole alkaloid extraction from Catharanthus roseus combined with a semi-synthetic production process for vinblastine. Molecules 12(7):1307–1315
Verma A, Hartonen K, Riekkola ML (2008) Optimisation of supercritical fluid extraction of indole alkaloids from Catharanthus roseus using experimental design methodology—comparison with other extraction techniques. Phytochem Anal 19(1):52–63
Verpoorte R, Niessen WMA (1994) Liquid chromatography coupled with mass spectrometry in the analysis of alkaloids. Phytochem Anal 5(5):217–232
Wang CH, Wang GC, Wang Y, Zhang XQ, Huang XJ, Ye WC (2012) Three new monomeric indole alkaloids from the roots of Catharanthus roseus. J Asian Nat Prod Res 14(3):249–255
Wong PL, Royce AJ, Lee-Parsons CW (2004) Improved ajmalicine production and recovery from Catharanthus roseus suspensions with increased product removal rates. Biochem Eng J 21(3):253–258
Xian F, Hendrickson CL, Marshall AG (2012) High resolution mass spectrometry. Anal Chem 84(2):708–719
Yamamoto K, Takahashi K, Mizuno H, Anegawa A, Ishizaki K, Fukaki H, Ohnishi M, Yamazaki M, Masujima T, Mimura T (2016) Cell-specific localization of alkaloids in Catharanthus roseus stem tissue measured with Imaging MS and Single-cell MS. Proc Natl Acad Sci 113(14):3891–3896
Yang L, Wang H, Zu YG, Zhao C, Zhang L, Chen X, Zhang Z (2011) Ultrasound-assisted extraction of the three terpenoid indole alkaloids vindoline, catharanthine and vinblastine from Catharanthus roseus using ionic liquid aqueous solutions. Chem Eng J 172(2–3):705–712
Zhang WK, Xu JK, Tian HY, Wang L, Zhang XQ, Xiao XZ, Li P, Ye WC (2013) Two new vinblastine-type N-oxide alkaloids from Catharanthus roseus. Nat Prod Res 27(20):1911–1916
Zhang L, Gai Q, Yang L, Ma Y, Liu Y (2014) Simultaneous quantitative determination of five alkaloids in Catharanthus roseus by HPLC-ESI-MS/MS. Chin J Nat Med 12(10):786–793
Zhang W, Yang J, Zi J, Zhu J, Song L, Yu R (2015) Effects of adding vindoline and MeJA on production of vincristine and vinblastine, and transcription of their biosynthetic genes in the cultured CMCs of Catharanthus roseus. Nat Prod Commun 10(12):2095–2096
Zhang A, Sun H, Yan G, Wang X (2017a) Recent developments and emerging trends of mass spectrometry for herbal ingredients analysis. TrAC Trends Anal Chem 94:70–76
Zhang Y, Sun Y, Mu X, Yuan L, Wang Q, Zhang L (2017b) Identification of metabolites of vindoline in rats using ultra-high performance liquid chromatography/quadrupole time-of-flight mass spectrometry. J Chromatogr B 1060:126–137
Zhang J, Hansen LG, Gudich O, Viehrig K, Lassen LM, Schrübbers L, Adhikari KB, Rubaszka P, Carrasquer-Alvarez E, Chen L, D'ambrosio V (2022) A microbial supply chain for production of the anti-cancer drug vinblastine. Nature 609(7926):341–347

Zhao Y, Nookandeh A, Schneider B, Sun X, Schmitt B, Stöckigt J (1999) Lignans from Torreya jackii identified by stopped-flow high-performance liquid chromatography–nuclear magnetic resonance spectroscopy. J Chromatogr A 837(1–2):83–91

Zhou H, Tai Y, Sun C, Pan Y (2005) Rapid identification of vinca alkaloids by direct-injection electrospray ionisation tandem mass spectrometry and confirmation by high-performance liquid chromatography–mass spectrometry. Phytochem Anal 16(5):328–333

Zhu RH, Cai HL, Jiang ZP, Xu P, Dai LB, Peng WX (2014) Validated HILIC–MS/MS assay for determination of vindesine in human plasma: application to a population pharmacokinetic study. J Pharm Biomed Anal 96:31–36

Zweil AM, Ibrahim IA, Bekhit M (2019) Screening of available components of Egyptian Catharanthus roseus cultivars using HPLC-MS. J Chem Pharm Res 11(6):33–40

Phenolics: Identification, Characterization, Isolation, and Pharmacology in Genus *Catharanthus*

7

Abstract

Phenolic compounds are important secondary metabolites in plants, recognized for their wide-ranging biological activities. In *Catharanthus* species, phenolics complement alkaloids in the plant's therapeutic profile. This communication focuses on the identification, characterization, isolation, and pharmacological significance of phenolics in *Catharanthus* plant member species. Techniques like high-performance liquid chromatography (HPLC), liquid chromatography-mass spectrometry (LC-MS), and nuclear magnetic resonance (NMR) have advanced the understanding of phenolic acids and flavonoids. Such compounds demonstrate antioxidant, anti-inflammatory, antimicrobial, and anticancer properties, enhancing the medicinal value of *Catharanthus* species. Investigating their roles and potential synergistic effects with other phytochemicals may pave the way for new therapeutic discoveries, underscoring the need for more focused research on these bioactive constituents.

Keywords

Phenolics · Characterization · *Catharanthus* · Bioactives

7.1 Introduction

Catharanthus species contain a lot of bioactive phenolic compounds, in addition to alkaloids like vincristine and vinblastine. These chemicals are very important for the medicinal effects of the leaves, stems, roots, seeds, and fruits of this species. These phenolic compounds have been extensively researched for their advantageous effects on human health, encompassing antioxidant qualities and their potential in the prevention of chronic disorders such as inflammation, cardiovascular diseases, cancer, and diabetes. The most common types are phenolic acids and

S. Kumar, B. Singh, *Catharanthus in Drug Discovery*,
https://doi.org/10.1007/978-981-95-5048-7_7

flavonoids. There are both soluble conjugated (glycosides) and insoluble versions. Phenolic acids are usually found in forms that do not dissolve, while flavonoids are usually found as glycosides, which means they have one or more sugar groups connected to them by hydroxyl groups (O-glycosides) or carbon–carbon bonds (C-glycosides). In the early 1980s, sophisticated analytical methods, especially mass spectrometry (MS), were created to measure free, soluble conjugated, and insoluble bound phenolics with great accuracy. Research indicates that insoluble-bound phenolics demonstrate markedly superior antioxidant activity relative to their free and soluble conjugated forms. This underscores the significance of these phenolic compounds in augmenting the antioxidant activity of *Catharanthus* species and their potential in addressing oxidative stress-related disorders (Cheynier 2012; Tsimogiannis and Oreopoulou 2019; Mutha et al. 2021). Flavonoids and phenolic acids exhibit a vast array of structural diversity, including polymeric forms. They are responsible for the principal organoleptic characteristics of plant-based foods and beverages, particularly their color and flavor profiles, and they significantly contribute to the nutritional attributes of fruits and vegetables. Carotenoids are the most important and common natural colors, and are tetrapyrrole derivatives of phenolic chemicals that are found in plants. Flavonoids are among the most common types of plant phenolics. A total of 8000 flavonoid variants have been found to date (Treutter 2006; Wu et al. 2023; Saini et al. 2024). It is important to have quick and dependable ways to analyze crude extracts so that bioactive components may be found, identified, and defined for structure–activity relationships and food quality control.

Modern mass spectrometric techniques have become indispensable tools for the analysis of flavonoids in plant materials and food products. These advanced methods provide detailed structural information, even from trace amounts of both pure compounds and complex mixtures. Techniques such as high-performance liquid chromatography (HPLC), ultra-performance liquid chromatography (UPLC), HPLC coupled with mass spectrometry (HPLC–MS), and tandem mass spectrometry (HPLC–MS/MS) are widely utilized for the identification and quantification of natural compounds. The mass spectrometer functions as a highly sensitive and universal detector, capable of identifying extremely low concentrations while also delivering valuable insights into molecular mass and structure. Numerous phenolic compounds have already been characterized from *Catharanthus* species using HPLC–MS (Steinmann and Ganzera 2011; Cetinkaya et al. 2025). This chapter focuses on the application of HPTLC, HPLC, UPLC, and HPLC–MS techniques in the analysis of flavonoids, phenolic acids, and proanthocyanidins.

7.2 Qualitative Analysis of Phenolic Acids, Flavonoids, and Anthocyanins in *Catharanthus* Species

High-performance liquid chromatography coupled with diode array detection (HPLC–DAD) and electrospray ionization tandem mass spectrometry (HPLC–ESI-MS/MS) are widely employed analytical techniques for the detection, identification, and characterization of phenolic acids, flavonoids, and anthocyanins in various plant extracts and preparations. These methods enable high sensitivity and selectivity in profiling phenolic compounds. Although phenolic compounds can be detected in both positive and negative ionization modes, the negative mode is generally preferred due to enhanced ionization efficiency for most analytes. In these modes, deprotonated ions $[M - H]^-$ and protonated ions $[M + H]^+$ are commonly observed. Numerous studies have reported the use of C18 reversed-phase columns with mobile phases containing 0.1% formic acid, methanol, and acetonitrile, employed in either isocratic or gradient elution systems. Detection wavelengths in UV and DAD modes typically range from 235 to 350 nm (Stefova et al. 2003; Bligh et al. 2013).

7.2.1 MS and HPLC–MS/MS Analysis of Flavonoids

High-performance liquid chromatography coupled with mass spectrometry (HPLC/MS) is a well-established, reliable, and highly sensitive analytical technique widely used for the qualitative and quantitative analysis of phenolic acids, flavonoids, and isoflavonoids in complex plant extracts. This technique enables detailed structural elucidation of these bioactive compounds, making it especially valuable in phytochemical research. One of the key advantages of HPLC/MS lies in its ability to distinguish between structurally similar compounds, such as isomeric flavonoid glycosides. Studies focusing on the structural characterization of flavonoids have revealed that the position of sugar substitution on the flavonoid backbone plays a critical role in determining the fragmentation behavior of flavonoid O-glycosides during MS analysis. Specifically, the formation of the aglycone ion (commonly referred to as Y_0^-) and the radical aglycone ion ($[Y_0 - H]^{-\cdot}$) is essential for differentiating between isomeric forms. The radical aglycone ion provides particularly informative fragmentation patterns, aiding in accurate structural identification. This phenomenon has been well demonstrated in the case of genistein-7-O-glucoside, where the $[Y_0 - H]^{-\cdot}$ ion was observed in the product ion spectrum using high-resolution time-of-flight mass spectrometry (Cuyckens et al. 2001; Cuyckens and Claeys 2004; March and Brodbelt 2008) (Table 7.1).

Domon and Costello (1988) came up with the standard naming scheme for glycol conjugates, which was used to name and describe the detected fragment ions. This system uses the letters ${}^{k,l}X_j^n$, Y_j^n, and Z_j^n to show the fragment ions. These particular ions correspond to pieces that preserve the flavonoid aglycone molecule subsequent to glycosidic bond cleavage.

In this notation, j is the number of interglycosidic bonds that have been broken, counting from the aglycone end of the molecule.

- n shows where the oligosaccharide is bonded to the aglycone structure.
- k and l show the exact places where the carbohydrate rings break apart, giving us a lot of information about the fragmentation mechanism.

Table 7.1 Details of chromatographic and spectrometric methods by HPLC, UPLC, HPLC–MS, and PLC–MS/MS

Compound	Precursor m/z $[M + H]^+$	Product ion (m/z)	Wavelength	Chromatographic column	Mobile phase	Injection volume (μL)	Flow rate (mL/min)	Run time (min)	Retention (min)	LOD/LOQ (ng/mL)	Linearity range (ng/mL)	Recovery/precision (%)	Reference
3,5-Dicaffeoylquinic acid	515	353, 191, 179	320	Spherisorb ODS-2, 250 × 4.6 mm, 5μm	0.1% formic acid (A) + methanol (B), gradient	20	0.5	55	26.3	–	–	–	Ferreres et al. (2008)
3-Feruloylquinic acid	367	191, 173	320	Spherisorb ODS-2, 250 × 4.6 mm, 5μm	0.1% formic acid (A) + methanol (B), gradient	20	0.5	55	22.9	–	–	–	Ferreres et al. (2008)
3-p-Coumaroylquinic acid	337	191, 173	320	Spherisorb ODS-2, 250 × 4.6 mm, 5μm	0.1% formic acid (A) + methanol (B), gradient	20	0.5	55	16.2	–	–	–	Ferreres et al. (2008)
4,5-Dicaffeoylquinic acid	515	353, 191, 179	320	Spherisorb ODS-2, 250 × 4.6 mm, 5μm	0.1% formic acid (A) + methanol (B), gradient	20	0.5	55	27.6	–	–	–	Ferreres et al. (2008)

5-Feruloylquinic acid	367	191, 173	320	Spherisorb ODS-2, 250 × 4.6 mm, 5µm	0.1% formic acid (A) + methanol (B), gradient	20	0.5	55	21.8	–	–	–	Ferreres et al. (2008)
5-p-Coumaroylquinic acid	337	191, 173	320	Spherisorb ODS-2, 250 × 4.6 mm, 5µm	0.1% formic acid (A) + methanol (B), gradient	20	0.5	55	18.5	–	–	–	Ferreres et al. (2008)
Acropyrone	–	–	235, 254, 280, 340	Eurospher-10 C18	Water (pH 2, formic acid adjusted): methanol (linear gradient)	100	–	–					Akpotu et al. (2017)
Chlorogenic acid	353	191, 179	320	Spherisorb ODS-2, 250 × 4.6 mm, 5µm	0.1% formic acid (A) + methanol (B), gradient	20	0.5	55	11.6	–	–	–	Ferreres et al. (2008)

(continued)

Table 7.1 (continued)

Compound	Precursor m/z $[M + H]^+$	Product ion (m/z)	Wavelength	Chromatographic column	Mobile phase	Injection volume (µL)	Flow rate (mL/min)	Run time (min)	Retention (min)	LOD/LOQ (ng/mL)	Linearity range (ng/mL)	Recovery/precision (%)	Reference
Citreoisocoumarin	–	–	235, 254, 280, 340	Eurospher-10 C18 (125 × 4 mm)	Water (pH 2, formic acid adjusted): methanol (linear gradient)	100	–	–					Akpotu et al. (2017)
Citreoisocoumarinol	–	–	235, 254, 280, 340	Eurospher-10 C18	Water (pH 2, formic acid adjusted): methanol (linear gradient)	100	–	–					Akpotu et al. (2017)
Cladosporin	–	–	235, 254, 280, 340	Eurospher-10 C18	Water (pH 2, formic acid adjusted): methanol (linear gradient)	100	–	–					Akpotu et al. (2017)

Gallic acid	–	–	254	Hypersil gold-C18 (250 × 4.6 mm, 5μm)	Acetonitrile: methanol: 3% formic acid (50:50:0.3 v/v/v)	10	0.5	–	6.388 (std); 6.429–6.465 (samples)	–	–	–	Nisar (2017)
Hydroxyemodin	–	–	235, 254, 280, 340	Eurospher-10 C18	Water (pH 2, formic acid adjusted): methanol (linear gradient)	100	–	–					Akpotu et al. (2017)
Isorhamnetin-O-glucoside	477	315	350	Spherisorb ODS-2, 250 × 4.6 mm, 5μm	0.1% formic acid (A) + methanol (B), gradient	20	0.5	55	~34–38	–	–	–	Ferreres et al. (2008)
Kaempferol	–	–	254	Athena C18 250 × 4.6 mm	Phosphate buffer (pH 5.8): acetonitrile (55:45 v/v)	20	1	–	9.373 (std), 9.463 (sample)	–	–	–	Rao and Ahmed (2014)

(continued)

Table 7.1 (continued)

Compound	Precursor m/z $[M + H]^+$	Product ion (m/z)	Wavelength	Chromatographic column	Mobile phase	Injection volume (μL)	Flow rate (mL/min)	Run time (min)	Retention (min)	LOD/LOQ (ng/mL)	Linearity range (ng/mL)	Recovery/precision (%)	Reference
Kaempferol-O-glucoside	447	285	350	Spherisorb ODS-2, 250 × 4.6 mm, 5μm	0.1% formic acid (A) + methanol (B), gradient	20	0.5	55	~32–36	–	–	–	Ferreres et al. (2008)
Methyl 2-(4-hydroxyphenyl) acetate	–	–	235, 254, 280, 340	Eurospher-10 C18	Water (pH 2, formic acid adjusted): methanol (linear gradient)	100	–	–					Akpotu et al. (2017)
Nigricinol	–	–	235, 254, 280, 340	Eurospher-10 C18	Water (pH 2, formic acid adjusted): methanol (linear gradient)	100	–	–					Akpotu et al. (2017)

Quercetin	–	–	254	Athena C18 250 × 4.6 mm	Phosphate buffer (pH 5.8): acetonitrile (55:45 v/v)	20	1	–	6.093 (std), 6.160 (sample)	–	–	–	Rao and Ahmed (2014)
Quercetin	–	–	254	Hypersil gold-C18 (250 × 4.6 mm, 5μm)	Acetonitrile: methanol: 3% formic acid (50:50:0.3 v/v/v)	10	0.5	–	6.691 (std); 6.683–6.788 (samples)	–	–	–	Nisar (2017)
Quercetin-O-glucoside	463	301	350	Spherisorb ODS-2, 250 × 4.6 mm, 5μm	0.1% formic acid (A) + methanol (B), gradient	20	0.5	55	~33–37	–	–	–	Ferreres et al. (2008)
Questinol	–	–	235, 254, 280, 340	Eurospher-10 C18	Water (pH 2, formic acid adjusted): methanol (linear gradient)	100	–	–	Not specified	Not provided	Not provided	Not provided	Akpotu et al. (2017)

(continued)

Table 7.1 (continued)

Compound	Precursor m/z $[M + H]^+$	Product ion (m/z)	Wavelength	Chromatographic column	Mobile phase	Injection volume (μL)	Flow rate (mL/min)	Run time (min)	Retention (min)	LOD/LOQ (ng/mL)	Linearity range (ng/mL)	Recovery/precision (%)	Reference
Rutin	–	–	254	Athena C18 250 × 4.6 mm	Phosphate buffer (pH 5.8): acetonitrile (55:45 v/v)	20	1	–	2.357 (std), 2.420 (sample)	–	–	–	Rao and Ahmed (2014)
Rutin	–	–	254	Hypersil gold-C18 (250 × 4.6 mm, 5μm)	Acetonitrile: methanol: 3% formic acid (50:50:0.3 v/v/v)	10	0.5	–	4.239 (std); 4.142–4.223 (samples)	–	–	–	Nisar (2017)

So, ions like ${}^{k,l}X_j^n$, Y_j^n, and Z_j^n are helpful because they show which parts of the glycosidic chain are missing and which parts are still linked to the aglycone. This systematic naming scheme makes it easier to figure out the structure of flavonoid glycosides, especially when looking at complicated fragmentation spectra (Fig. 7.2).

We used the naming system that Domon and Costello came up with to figure out the aglycone-derived ions in the MS fragmentation spectra. In this context, the aglycone ions of kaempferol, quercetin, and isorhamnetin were detected at m/z 285, 300/301, and 315, respectively, usually showing up as very common or even base peaks in their MS^2 or MS^3 spectra. Four flavonoid glycosides were found and structurally described in the methanolic extract of hairy root cultures of *C. roseus*. Among these, 3′,4′-Di-O-methylquercetin-7-O-[(4″→13‴)-2‴,6‴,10‴,14‴-tetramethylhexadec-13‴-ol-14‴-enyl]-β-D-glucopyranoside has a unique fragment ion at m/z 329 (Y_0^-) and retro-Diels–Alder (RDA) fragment ions at m/z 222 and 178 (Fig. 7.1). In the same way, 4′-O-Methylkaempferol-3-O-[(4″ → 13‴)-2‴,6‴,10‴,14‴-tetramethylhexadecan-13‴-olyl]-β-D-glucopyranoside produced significant ions at m/z 268 and 252 (Fig. 7.3). The butin derivatives, 3′,4′-Di-O-methylbutin-7-O-[(6″→1‴)-3‴,11‴-dimethyl-7‴-methylenedodeca-3‴,10‴-dienyl]-β-D-glucopyranoside and 4′-O-Methylbutin-7-O-[(6″→1‴)-3‴,11‴-dimethyl-7‴-hydroxymethylenedodecanyl]-β-D-glucopyranoside exhibited a comparable fragmentation process, yielding aglycone-derived ions at m/z 283 and 269, respectively, in addition to RDA fragments at m/z 208 and 150 (Figs. 7.2, 7.3, and 7.4).

HPLC–MS and tandem MS (MS/MS) techniques have been extensively utilized for the identification and structural characterization of phenolic compounds found in the seeds, stems, leaves, and petals of *C. roseus*. To get the best results, chemical mixtures or crude extracts were separated using a reversed-phase C18 column (RP-C18, 250 × 4 mm, 5μm; LiChroCART) and a LiChroCART guard column

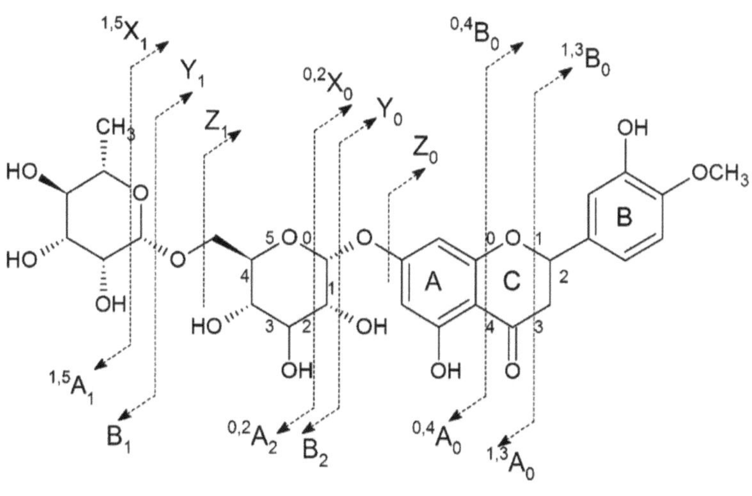

Fig. 7.1 MS/MS fragmentation nomenclature of flavonoids

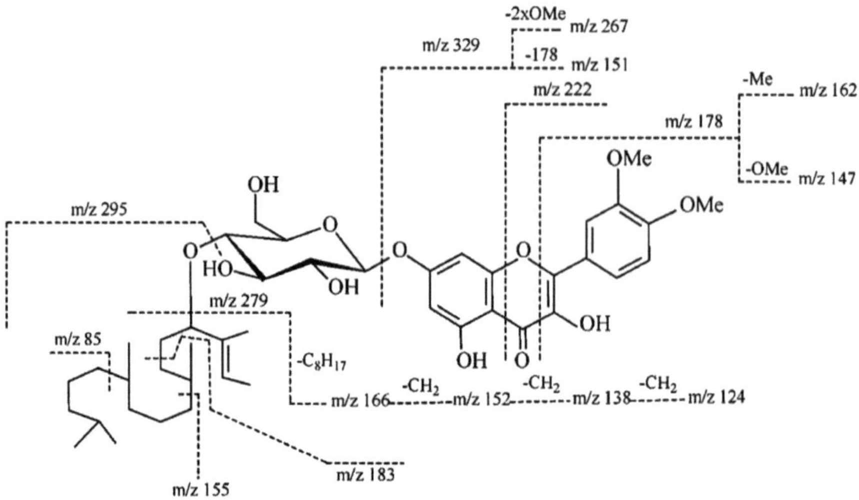

Fig. 7.2 Fragmentation pathways of 3′,4′-di-*O*-methylquercetin-7-*O*-[(4″ → 13‴)-2‴,6‴,10‴, 14‴-tetramethylhexadec-13‴- ol-14‴-enyl]-*â*-D-glucopyranoside (Chung et al. 2009)

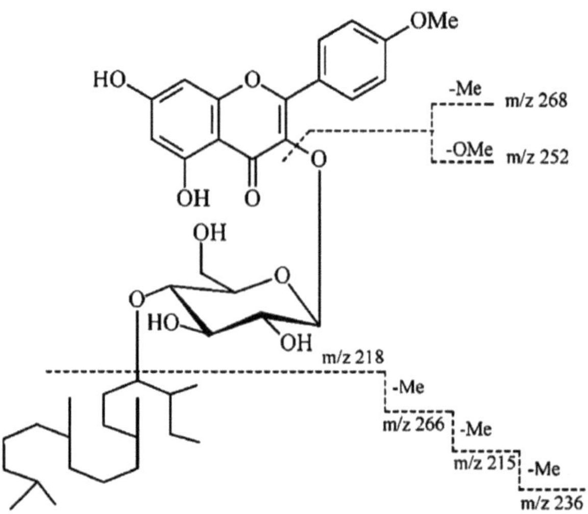

Fig. 7.3 Fragmentation pathways of 4′-*O*-methylkaempferol-3-*O*-[(4″ → 13‴)- 2‴,6‴,10‴, 14‴-tetramethylhexadecan-13‴-olyl]-*â*-D-glucopyranoside

(4 × 4 mm). A mobile phase made up of 1% acetic acid in water (solvent A) and acetonitrile (solvent B) was used for the chromatographic separation. The gradient program started with 12% B and went up to 20% B over the course of 30 min. HPLC–DAD-ESI-MS/MS was used to keep an eye on the separation profiles that came out, and Fig. 7.5 shows the total ion chromatograms (TICs) that were made from extracts of (a) stems, (b) leaves, (c) seeds, and (d) petals of *C. roseus*.

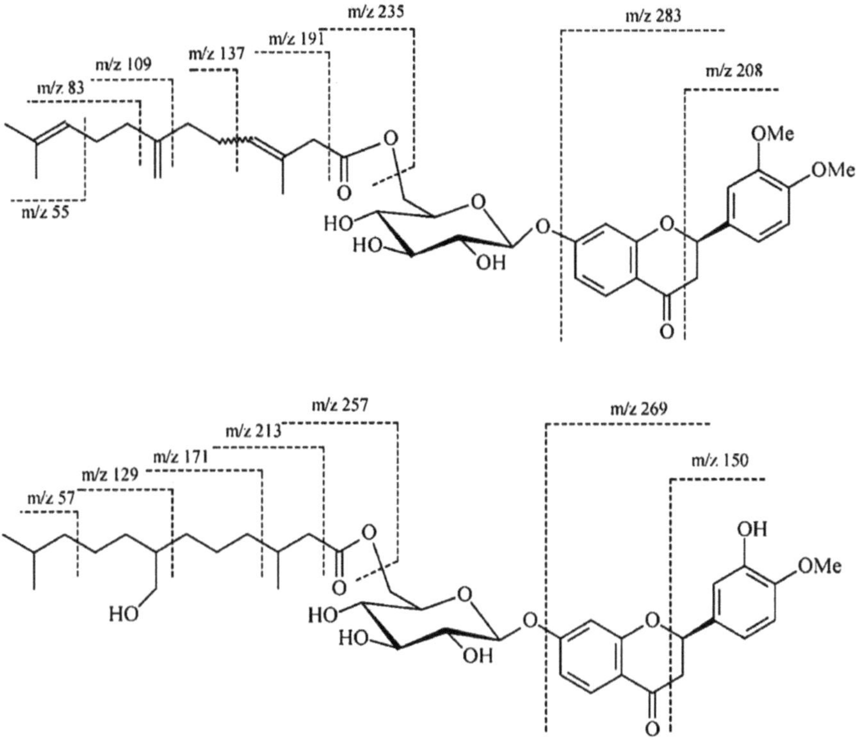

Fig. 7.4 MS/MS fragmentation pathways of 3′,4′-di-O-methylbutin-7-O-[(6″→1‴)-3‴,11‴-dimethyl-7‴-methylenedodeca-3‴,10‴-dienyl]-β-D-glucopyranoside and 4′-O-methylbutin-7-O-[(6″→1‴)-3‴,11‴-dimethyl-7‴-hydroxymethylenedodecanyl]-β-D-glucopyranoside, highlighting key cleavage sites and characteristic ion formations

Ferreres and colleagues have thoroughly characterized flavonoid glycosides in *C. roseus* via HPLC-DAD-ESI-MS/MS. Ferreres et al. (2010) identified deprotonated molecular ions $[M-H]^-$ at m/z 739 and 755 in vacuolar extracts, corresponding to kaempferol-3-*O*-(6-*O*-rhamnosyl-hexoside)-7-*O*-hexoside. Ferreres et al. (2011) subsequently reported that this molecule generated fragment ions at m/z 285 and 300, resulting from the elimination of glucose moieties. Rhamnosyl (146 Da) and galactosyl (164 Da) residue losses were also noted, exhibiting distinctive diagnostic fragmentations in accordance with prior studies (Carbone et al. 2004; Chen et al. 2005; Zhao et al. 2020). Ferreres et al. (2008) found 18 phenolic compounds in hydroalcoholic extracts of *C. roseus* seeds, stems, leaves, and petals in a previous study (Fig. 7.6). These were a mix of kaempferol, quercetin, and isorhamnetin glycosides, such as: kaempferol-3-*O*-(2,6-di-*O*-rhamnosyl-galactoside)-7-*O*-hexoside, kaempferol-3-*O*-(6-*O*-rhamnosyl-galactoside)-7-*O*-galactoside, kaempferol-3-*O*-(6-*O*-rhamnosyl-galactoside)-7-*O*-glucoside, quercetin-3-*O*-(2,6-di-*O*-rhamnosyl-galactoside), quercetin-3-*O*-(2,6-di-*O*-rhamnosyl-glucoside),

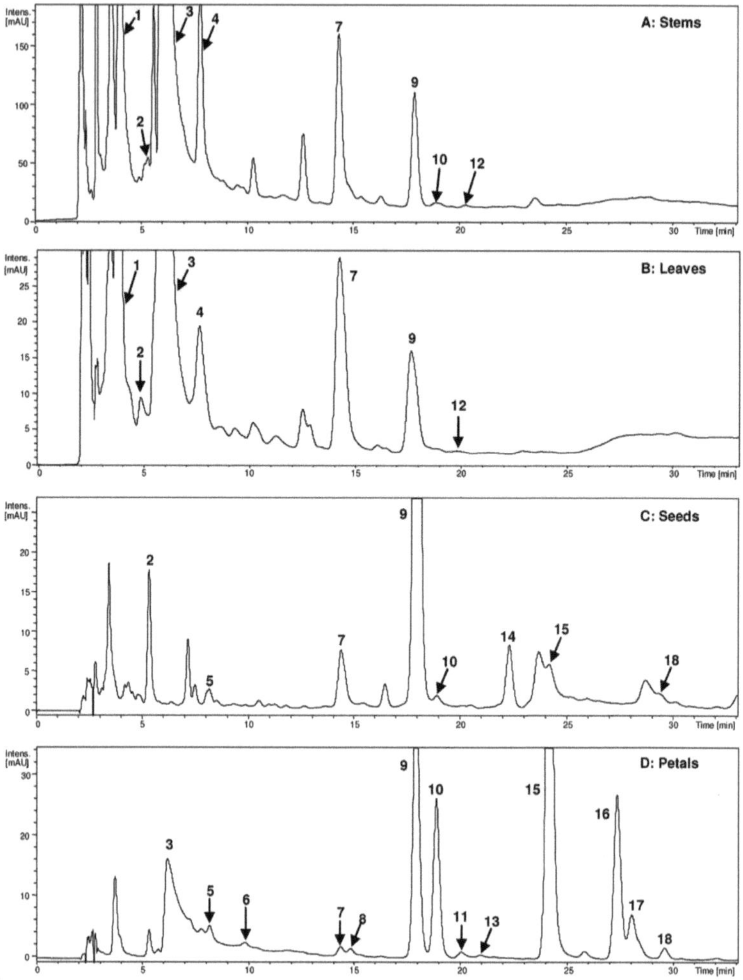

Fig. 7.5 Total ion chromatogram of extracts of *C. roseus* by HPLCDAD-ESI-MS/MS ((**a**)—stems, (**b**)—leaves, (**c**)—seeds, (**d**)—petals)

kaempferol-3-*O*-(2,6-di-*O*-rhamnosyl-glucoside), quercetin-3-*O*-(6-*O*-rhamnosyl-galactoside), quercetin-3-*O*-(6-*O*-rhamnosyl-glucoside), isorhamnetin-3-*O*-(2,6-di-O-rhamnosyl-galactoside), isorhamnetin-3-*O*-(2,6-di-*O*-rhamnosyl-glucoside), More mono- and diglycosylated versions of kaempferol and isorhamnetin. The distinct deprotonated aglycone ions were reliably detected at m/z 285 (kaempferol), m/z 300/301 (quercetin), and m/z 315 (isorhamnetin), typically appearing as the base peaks in MS^2 or MS^3 investigations. Quercetin-3-*O*-(2,6-di-*O*-rhamnosyl-galactoside) (m/z 755) and kaempferol-3-*O*-(2,6-di-O-rhamnosyl-galactoside) (m/z 739) were two of the most common flavonoids in stems and leaves. Their MS^2 spectra had strong base peaks at m/z 300 and 285, which confirmed that they were triglycosides made up of two rhamnose

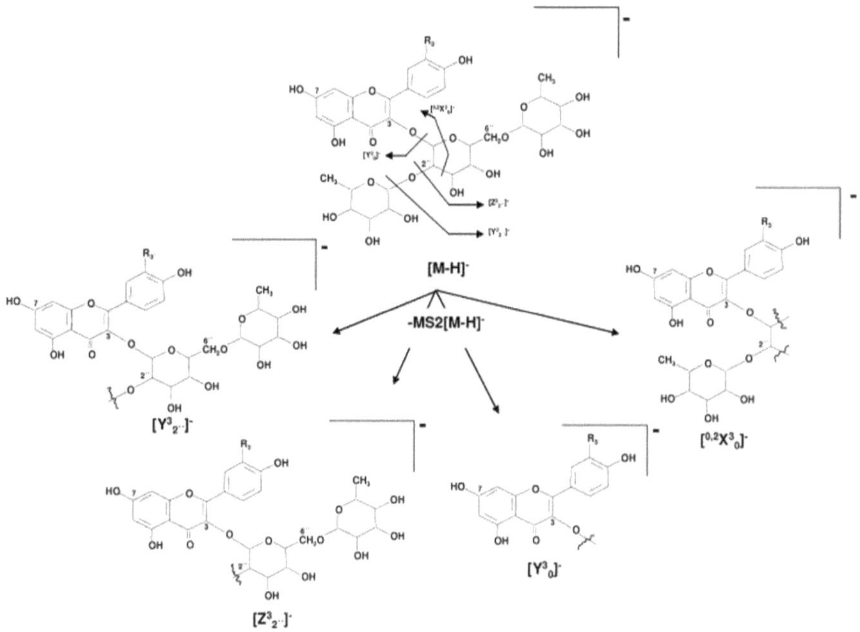

Fig. 7.6 M/S Fragmentation pathways flavonol-3-O-(2,6-di-O-rhamnosyl-hexoside) derivatives. (Quercetin-3-O-(2,6-di-O-rhamnosyl-galactoside) and quercetin-3-O-(2,6-di-O-rhamnosyl-glucoside) ($R^{3'}$ = OH), kaempferol-3-O-(2,6-di-O-rhamnosyl-galactoside) and kaempferol-3-O-(2,6-di-O-rhamnosyl-glucoside) ($R^{3'}$ = H) and isorhamnetin-3-O-(2,6-di-O-rhamnosyl-galactoside) and isorhamnetin-3-O-(2,6-di-O-rhamnosyl-glucoside) ($R^{3'}$ = OCH_3)

and one hexose unit. The predominance of the aglycone fragment indicates the attachment of the triglycoside to a solitary phenolic hydroxyl group. Six diglycosides, specifically rhamnohexoside derivatives of quercetin, kaempferol, and isorhamnetin, were also identified: quercetin-3-*O*-(6-*O*-rhamnosyl-galactoside), quercetin-3-*O*-(6-*O*-rhamnosyl-glucoside), kaempferol-3-*O*-(6-*O*-rhamnosyl-galactoside), kaempferol-3-*O*-(6-*O*-rhamnosyl-glucoside), isorhamnetin-3-O-(6-O-rhamnosyl-galactoside), and sorhamnetin-3-*O*-(6-*O*-rhamnosyl-glucoside). The MS^2 fragmentation spectra showed mostly deprotonated aglycone ions at m/z 285, 300/301, or 315, which show a 1→6 interglycosidic connection. Three glycosides with extra glycosylation at the 7 position were also spoken about: kaempferol-3-*O*-(2,6-di-*O*-rhamnosyl-galactoside)-7-*O*-hexoside, kaempferol-3-*O*-(6-*O*-rhamnosyl-galactoside)-7-*O*-galactoside, and kaempferol-3-*O*-(6-*O*-rhamnosyl-galactoside)-7-*O*-glucoside. These exhibited MS^2 spectra characterized by a neutral loss of 162 Da, signifying hexose removal. In MS^3, the fragmentation patterns changed: kaempferol-3-*O*-(2,6-di-*O*-rhamnosyl-galactoside)-7-*O*-hexoside acted like triglycosides, whereas kaempferol-3-*O*-(6-*O*-rhamnosyl-galactoside)-7-*O*-galactoside and its glucoside counterpart acted like diglycosides. These probably come from the seeds of kaempferol-3-*O*-(2,6-di-*O*-rhamnosyl-galactoside) and the petals of kaempferol-3-*O*-(6-*O*-rhamnosyl-galactoside/-glucoside). Interestingly, syringetin-3-*O*-robinoside (3,5,7,4′-tetrahydroxy-3′,5′-dimethoxyflavone-3-*O*-(6-*O*-

rhamnosyl-galactoside)), which has been found in *C. roseus* stems before, was not found in our tests. Extracted ion chromatograms for m/z 653 $[M - H]^-[M - H]^-$ and the matching syringetin aglycone ion at m/z 345 validated its absence in the analyzed materials (Ferreres et al. 2008).

Three anthocyanins were tentatively identified at m/z 317, 331, and 345 as petunidin, malvidin, and hirsutidin, respectively. The MS/MS spectrum at *m/z* 345 showed fragment ions at m/z 330, 329, 313, 312, and 301. These ions were caused by the loss of CH_3, CH_4 (or H + CH_3), CH_3OH, (CH_3OH + H), and CO_2 from the molecular ion. The ions at *m/z* 284 formed when CH*3*OH, H, and CO were lost one after the other, or when CO was lost. Petunidin 3-O-glucoside, malvidin 3-O-glucoside, and hirsutidin 3-*O*-glucoside had *m/z* values of 479, 493, and 507, respectively. They all lost a glucose moiety at 162 ($C_6H_{10}O_5$) and produced fragment ions at *m/z* 317, 331, and 345, which correspond to petunidin, malvidin, and hirsutidin (Vukics and Guttman 2010; Stobiecki 2000). Petunidin 3-*O*-(6-*O*-p-coumaroyl) glucoside, malvidin 3-*O*-(6-*O*-p-coumaroyl) glucoside, and hirsutidin 3-*O*-(6-*O*-p-coumaroyl) glucoside all had $[M]^+$ at m/z 625, 639, and 653. The loss of p-coumaroyl resulted in fragment ions at *m/z* 479, 493, and 507, derived from *m/z* 625, 639, and 653, with a corresponding glucose loss observed at *m/z* 317, 331, and 345 in petunidin 3-*O*-(6-*O*-p-coumaroyl) glucoside, malvidin 3-*O*-(6-*O*-p-coumaroyl) glucoside, and hirsutidin 3-*O*-(6-*O*-p-coumaroyl) glucoside, respectively (Piovan et al. 1998).

7.2.2 Analysis of Phenolic Acids by MS/LC-MS

The phenic acid 3-*O*-caffeoylquinic acid exhibited a deprotonated molecular ion at *m/z* 353 $[M - H]^-$. The MS^2 spectra exhibited distinctive fragment ions at *m/z* 191 and 179. Similarly, 4-O-caffeoylquinic acid has a precursor ion at *m/z* 353 $[M - H]^-$, along with fragment ions at m/z 191 (relative intensity 80), 179 (60), and 173. The deprotonated ion of 5-*O*-caffeoylquinic acid appeared at *m/z* 353 $[M - H]^-$, with MS/MS analysis revealing fragment ions at *m/z* 191 and 179. These signals were reliably identified in both stem and leaf extracts (Ferreres et al. 2008).

7.3 Quantitative Analysis of Flavonoids in *Catharanthus* Species

7.3.1 Quantitative Analysis of Phenolic Acids and Flavonoids by HPLC

All phenolics and flavonoid aglycones have at least one aromatic ring, which causes the absorption shift toward longer wavelengths in UV light. The A-ring of flavonoids accounts for the initial maximum detected in the 240–285 nm range, whereas the subsequent maximum, noted at 300–550 nm, arises from the substitution pattern and conjugation of the C-ring (Leite et al. 2001; Cristea et al. 2003; Seal 2016). The UV detector at a specific wavelength is yielding information that remains somewhat limited. Consequently, the use of legitimate standards is advised to acquire additional information from chromatography. Diode array detectors (DADs) are highly advantageous since they furnish a comprehensive record of the UV-Vis spectrum for each molecule. Each type of phenolic chemical exhibited a distinct spectrum in the UV-Vis region due to its unique characteristics. This distinctive spectrum enabled the identification and characterization of compounds from plant extracts through comparison with standards based on retention time and UV spectrum (Gomes et al. 2017; Park et al. 2016). When two compounds, such as isomers or superimposable entities, exhibit nearly identical retention durations and UV spectra in a chromatogram, advanced detectors capable of providing extensive structural information, such as LC-MS, are necessary (Hsu et al. 2017; Zohra et al. 2019). Due to the vast array of flavonoid glycosides present in nature, absolute quantification without authentic standards is frequently unfeasible unless they are commercially accessible. Nonetheless, the sugar moieties serve as inadequate chromophores and exert minimal influence on the UV spectra; it is customary to quantify glycosides utilizing standards of the equivalent aglycone or a comparable compound. Most phenolic acids are commercially accessible, allowing for their accurate identification by the comparison of retention times and UV spectra with reference compounds (Seal 2016; Kumar 2017; Russo et al. 2018). Hydroxycinnamic acids exhibit absorption in two regions of the UV spectrum, with a peak at 225–235 nm and two closely situated peaks at 290–330 nm (Stelzner et al. 2019; Harbaum et al. 2007). The dual absorption in this area results from the existence of *cis*- and *trans*- isomers, with the ratio of these two forms influencing the final spectrum. In an alkaline environment, all three maxima experience a bathochromic change. The many esters of the same acid have analogous spectra, irrespective of the alcohol moiety (quinic acid, sugar, or tartaric acid) (Pereira et al. 2009). The UV spectrum of benzoic acids is more affected by the degree of hydroxylation of the benzene moiety than that of hydroxycinnamic acids. The methylation of hydroxyl groups exerts a negligible influence on the UV spectrum. Only dihydroxylated benzoic acids exhibit two maxima (Yu et al. 2001; Barros et al. 2009; Pereira et al. 2009). In *C. roseus*, multiple studies have documented the occurrence of hydroxycinnamic acids esterified with quinic acid, specifically 3-O-caffeoylquinic, 4-O-caffeoylquinic, and 5-O-caffeoylquinic acids, as determined by HPLC with a UV detector calibrated at 254 nm. The distribution of these chemicals is contingent upon the source of the samples, including stems and leaves. These chemicals were not discovered in the seeds and petals. The tissue-specific distribution of caffeoylquinic acids in leaves and stems correlates with their established role as protectants against herbivores and pathogens, a function especially pertinent to the organs involved in vegetative growth (stems and leaves). The concentrations of rutin and quercetin in flower petals are 29.46% and 6.51%, respectively, while the concentrations of rutin, gallic acid, and quercetin in leaves are 25.16%, 8.57%, and 10.47%, respectively. The quantities of rutin, gallic acid, and quercetin in the stem were recorded as 13.02%, 5.89%, and 7.47%, respectively. Quercetin (13.49%) was exclusively identified in the root. The most phenolic concentration was observed in flower petals, specifically 194.50 mg GAE/g (Nisar 2017). A study revealed that HPLC method for the detection and quantification of glucosides, revealing that cell cultures of *Vanilla planifolia* Andrews produced the highest concentration of glucovanillin compared to those of *Ilex dumosa* Reissek and *Catharanthus roseus*. The maximum yield of glucovanillin was achieved from *V. planifolia* cells supplied with 6.6 mM vanillin and collected after 12 h. The yield was 3.28 mM glucovanillin, representing 49.7%. This glucoside was predominantly stored within the cells. Piovan described the synthesis of anthocyanin in *C. roseus* flowers from both field-grown and somatic embryogenesis-regenerated plants and cell cultures. The anthocyanins were characterized as the 3-O-glucosides and the 3-O-(6-O-p-coumaroyl) glucosides of hirsutidin, malvidin, and petunidin, respectively, both under in vivo and in vitro conditions. The impact of environmental factors on in vitro anthocyanin accumulation is delineated by the HPLC method (Piovan and Filippini 2007).

7.3.2 Quantitative Analysis of Phenolic Acids and Flavonoids by MS/LC-MS

Analysis of phenolic composition from *C. roseus* was performed using HPLC–ESI-MS/MS. The HPLC–MS methods have been developed by different researchers using a C18 column with the mobile phase acetonitrile (Ferreres et al. 2008). To improve the separation and resolution, buffers were used, which yielded better results (Dai et al. 2016). The distribution of these metabolites within the different plant tissues obtained from leaves, stems, seeds, and flowers were studied. Fifteen glycosides of quercetin, isorhamnetin, and kaempferol were detected as the majority constituents. The two main flavonoids kaempferol-3-*O*-(6-*O*-rhamnosyl-galactoside)-7-*O*-glucoside and quercetin-3-*O*-(2,6-di-O-rhamnosyl-glucoside were reported in *C. roseus* leaves and stems (Nishibe et al. 1996; Brun et al. 1999). Kaempferol-3-*O*-(2,6-di-*O*-rhamnosyl-galactoside) was abundant in petals, with quercetin-3-*O*-(2,6-di-*O*-rhamnosyl-galactoside) in the petals.

Another compound with a similar fragmentation is isorhamnetin-3-*O*-(2,6-di-*O*-rhamnosyl-galactoside) and isorhamnetin-3-*O*-(2,6-di-*O*-rhamnosyl-galactoside), detected only in trace amounts in the stems and leaves (Cui et al. 2000; Pereira et al. 2005). Quercetin-3-*O*-(2,6-di-*O*rhamnosyl-glucoside) was reported solely in the petals and in trace amounts. Kaempferol-3-*O*-(2,6-di-*O*-rhamnosyl-glucoside) was reported in higher amounts in the stems, seeds, and petals, whereas compound isorhamnetin-3-*O*-(2,6-di-*O*-rhamnosyl-glucoside) was found in the seeds. Kaempferol-3-*O*-(6-*O*-rhamnosyl-galactoside) and isorhamnetin-3-*O*-(6-*O*-rhamnosyl-glucoside) were reported in small amounts in the seeds, while petals contained most abundant amounts of compounds quercetin-3-*O*-(6-*O*-rhamnosyl-galactoside), quercetin-3-*O*-(6-*O*-rhamnosyl-glucoside), kaempferol-3-*O*-(6-*O*-rhamnosyl-galactoside); kaempferol-3-*O*-(6-*O*-rhamnosyl-glucoside); isorhamnetin-3-*O*-(6-*O*-rhamnosyl-galactoside); isorhamnetin-3-*O*-(6-*O*-rhamnosyl-glucoside (Steinmann and Ganzera 2011; Simirgiotis et al. 2012). Kaempferol-3-*O*-(2,6-di-*O*-rhamnosyl-galactoside)-7-*O*-hexoside was the most abundant in seeds. High concentrations of anthocyanins (0.001–0.50μmol/g fr. wt) were reported in all cells present in a medium-induced culture. The aglycones of the anthocyanins were identified as petunidin, malvidin, and hirsutidin (Knobloch et al. 1982). The total anthocyanin contents and the relative abundance of the molecular ions of the six glucosides, detected in the ESI mass spectrum of the anthocyanin extracts, are quoted. The anthocyanin content in the flowers from both plant types was comparable and approximately 1.5 times higher than those obtained from cell suspensions. Hirsutidin 3-*O*-(6-*O*-*p*-coumaroyl) glucoside was the major pigment (49.3–70 relative abundance); the 3-*O*-(6-*O*-*p*-coumaroyl) glucosides were more abundant than the simple glucosides. The extracts from both embryogenic plant flowers and cell cultures contained the malvidin derivatives in higher amounts than the petunidin ones (Filippini et al. 2003). Kaempferol-3-*O*-(2,6-di-*O*-rhamnosyl-galactoside)-7-*O*-hexoside was detected at 292.3 mg/kg dry basis and 52.7 mg/kg dry basis in seeds and leaves, respectively. The kaempferol-3-*O*-(2,6-di-*O*-rhamnosyl-galactoside) in seeds, stems, leaves and petals detected were 2714.2, 190.8, 8.5, 8, and 120.8 mg/kg, respectively. Similarly, quercetin-3-*O*-(2,6-di-*O*-rhamnosyl-galactoside) was detected at 582.7, 190.5, 310.9, 1027.9 in seeds, stems, leaves, and petals, respectively (Pereira et al. 2013; Mustafa and Verpoorte 2007). Petunidin-3-O-glucoside, malvidin-3-O-glucoside, hirsutidin-3-O-glucoside, petunidin-3-*O*-(6-*O*-p-coumaroyl) glucoside, malvidin-3-O-(6-O-p-coumaroyl) glucoside, and hirsutidin-3-*O*-(6-*O*-*p*-coumaroyl) glucoside were reported at 3.5, 1.5, 28, 11.5, 6.2, and 49.3 mg/g fresh weight, respectively (Piovan et al. 1998).

References

Akpotu MO, Eze PM, Abba CC, Umeokoli BO, Nwachukwu CU, Okoye FB, Esimone CO (2017) Antimicrobial activities of secondary metabolites of endophytic fungi isolated from Catharanthus roseus. J Health Sci 7(1):15–22

Barros L, Dueñas M, Ferreira IC, Baptista P, Santos-Buelga C (2009) Phenolic acids determination by HPLC–DAD–ESI/MS in sixteen different Portuguese wild mushrooms species. Food Chem Toxicol 47(6):1076–1079

Bligh SW, Ogegbo O, Wang ZT (2013) Flavonoids by HPLC. In: Ramawat KG, Mérillon JM (eds) Natural products. Springer, Berlin, pp 2107–2144

Brun G, Dijoux MG, David B, Mariotte AM (1999) A new flavonol glycoside from Catharanthus roseus. Phytochemistry 50:167–169

Carbone V, Montoro P, de Tommasi N, Pizza C (2004) Analysis of flavonoids from Cyclanthera pedata fruits by liquid chromatography/electrospray mass spectrometry. J Pharm Biomed Anal 34(2):295–304

Cetinkaya A, Yayla S, Hurkul MM, Ozkan SA (2025) Comprehensive review on chromatographic analysis of flavonoids in fruits. J Chromatogr Open 7:100209

Chen LJ, Song H, Du QZ, Li J, Ito Y (2005) Analysis of flavonoids in the extracts from the seeds of Oroxylum indicum using high-speed countercurrent chromatography/mass spectrometry. J Liq Chromatogr Relat Technol 28(10):1549–1555

Cheynier V (2012) Phenolic compounds: from plants to foods. Phytochem Rev 11(2):153–177

Chung IM, Ahmad A, Ali M, Lee OK, Kim MY, Kim JH, Yoon DY, Peebles CA, San KY (2009) Flavonoid glucosides from the hairy roots of Catharanthus roseus. J Nat Prod 72(4):613–620

Cristea D, Bareau I, Vilarem G (2003) Identification and quantitative HPLC analysis of the main flavonoids present in weld (Reseda luteola L.). Dyes Pigments 57(3):267–272

Cui M, Song F, Zhou Y, Liu Z, Liu S (2000) Rapid identification of saponins in plant extracts by electrospray ionization multi-stage tandem mass spectrometry and liquid chromatography/tandem mass spectrometry. Rapid Commun Mass Spectrom 14(14):1280–1286

Cuyckens F, Claeys M (2004) Mass spectrometry in the structural analysis of flavonoids. J Mass Spectrom 39(1):1–15

Cuyckens F, Rozenberg R, de Hoffmann E, Claeys M (2001) Structure characterization of flavonoid O-diglycosides by positive and negative nano-electrospray ionization ion trap mass spectrometry. J Mass Spectrom 36(11):1203–1210

Dai Y, Rozema E, Verpoorte R, Choi YH (2016) Application of natural deep eutectic solvents to the extraction of anthocyanins from Catharanthus roseus with high extractability and stability replacing conventional organic solvents. J Chromatogr A 1434:50–56

Domon B, Costello CE (1988) Structure elucidation of glycosphingolipids and gangliosides using high-performance tandem mass spectrometry. Biochemistry 27(5):1534–1543

Ferreres F, Pereira DM, Valentão P, Andrade PB, Seabra RM, Sottomayor M (2008) New phenolic compounds and antioxidant potential of Catharanthus roseus. J Agric Food Chem 56(21):9967–9974

Ferreres F, Pereira DM, Valentão P, Oliveira JM, Faria J, Gaspar L, Sottomayor M, Andrade PB (2010) Simple and reproducible HPLC–DAD–ESI-MS/MS analysis of alkaloids in Catharanthus roseus roots. J Pharm Biomed Anal 51(1):65–69

Ferreres F, Figueiredo R, Bettencourt S, Carqueijeiro I, Oliveira J, Gil-Izquierdo A, Pereira DM, Valentão P, Andrade PB, Duarte P, Barcelo AR (2011) Identification of phenolic compounds in isolated vacuoles of the medicinal plant Catharanthus roseus and their interaction with vacuolar class III peroxidase: an H_2O_2 affair? J Exp Bot 62(8):2841–2854

Filippini R, Caniato R, Piovan A, Cappelletti EM (2003) Production of anthocyanins by Catharanthus roseus. Fitoterapia 74(1–2):62–67

Gomes SV, Portugal LA, dos Anjos JP, de Jesus ON, de Oliveira EJ, David JP, David JM (2017) Accelerated solvent extraction of phenolic compounds exploiting a Box-Behnken design and quantification of five flavonoids by HPLC-DAD in Passiflora species. Microchem J 132:28–35

Harbaum B, Hubbermann EM, Wolff C, Herges R, Zhu Z, Schwarz K (2007) Identification of flavonoids and hydroxycinnamic acids in Pak Choi varieties (Brassica campestris L. ssp. chinensis var. communis) by HPLC–ESI-MS^n and NMR and their quantification by HPLC–DAD. J Agric Food Chem 55(20):8251–8260

Hsu BY, Lin SW, Inbaraj BS, Chen BH (2017) Simultaneous determination of phenolic acids and flavonoids in Chenopodium formosanum Koidz. (djulis) by HPLC-DAD-ESI–MS/MS. J Pharm Biomed Anal 132:109–116

Knobloch KH, Bast G, Berlin J (1982) Medium- and light-induced formation of serpentine and anthocyanins in cell suspension cultures of Catharanthus roseus. Phytochemistry 21(3):591–594

Kumar BR (2017) Application of HPLC and ESI-MS techniques in the analysis of phenolic acids and flavonoids from green leafy vegetables (GLVs). J Pharm Anal 7(6):349–364

Leite JPV, Rastrelli L, Romussi G, Oliveira AB, Vilegas JH, Vilegas W, Pizza C (2001) Isolation and HPLC quantitative analysis of flavonoid glycosides from Brazilian beverages (Maytenus ilicifolia and M. aquifolium). J Agric Food Chem 49(8):3796–3801

March R, Brodbelt J (2008) Analysis of flavonoids: tandem mass spectrometry, computational methods, and NMR. J Mass Spectrom 43(12):1581–1617

Mustafa NR, Verpoorte R (2007) Phenolic compounds in Catharanthus roseus. Phytochem Rev 6(2–3):243–258

Mutha RE, Tatiya AU, Surana SJ (2021) Flavonoids as natural phenolic compounds and their role in therapeutics: an overview. Futur J Pharm Sci 7(1):1–25

Nisar A (2017) Identification of flavonoids (quercetin, gallic acid and rutin) from Catharanthus roseus plant parts using deep eutectic solvent. Recent Adv Biol Med 3(1277):1–6

Nishibe S, Takenaka T, Fujikawa T, Yasukawa K, Takido M, Morimitsu Y, Hirota A, Kawamura T, Noro Y (1996) Bioactive phenolic compounds from Catharanthus roseus and Vinca minor. Nat Med 50:378–383

Park JS, Kim IS, Rehman SU, Na CS, Yoo HH (2016) HPLC determination of bioactive flavonoids in Hovenia dulcis fruit extracts. J Chromatogr Sci 54(2):130–135

Pereira CA, Yariwake JH, McCullagh M (2005) Distinction of the C-glycosylflavone isomer pairs orientin/isoorientin and vitexin/isovitexin using HPLC-MS exact mass measurement and in-source CID. Phytochem Anal 16(5):295–301

Pereira DM, Ferreres F, Oliveira J, Valentão P, Andrade PB, Sottomayor M (2009) Targeted metabolite analysis of Catharanthus roseus and its biological potential. Food Chem Toxicol 47(6):1349–1354

Pereira DM, Valentão P, Sottomayor M, Ferreres F, Andrade PB (2013) Phenolic compounds in Catharanthus roseus. In: Natural products: phytochemistry, botany and metabolism of alkaloids, phenolics and terpenes. Springer, Berlin, pp 2093–2106

Piovan A, Filippini R (2007) Anthocyanins in Catharanthus roseus in vivo and in vitro: a review. Phytochem Rev 6(2–3):235–242

Piovan A, Filippini R, Favretto D (1998) Characterization of the anthocyanins of Catharanthus roseus (L.) G. Don in vivo and in vitro by electrospray ionization ion trap mass spectrometry. Rapid Commun Mass Spectrom 12(7):361–367

Rao AS, Ahmed MF (2014) Simultaneous determination of phenolic compounds in Catharanthus roseus leaves by HPLC method. Int J Pharm Sci Res 5(3):977–981

Russo M, Fanali C, Tripodo G, Dugo P, Muleo R, Dugo L, De Gara L, Mondello L (2018) Analysis of phenolic compounds in different parts of pomegranate (Punica granatum) fruit by HPLC-PDA-ESI/MS and evaluation of their antioxidant activity: application to different Italian varieties. Anal Bioanal Chem 410(15):3507–3520

Saini N, Anmol A, Kumar S, Wani AW, Bakshi M, Dhiman Z (2024) Exploring phenolic compounds as natural stress alleviators in plants-a comprehensive review. Physiol Mol Plant Pathol 133:e102383

Seal T (2016) Quantitative HPLC analysis of phenolic acids, flavonoids and ascorbic acid in four different solvent extracts of two wild edible leaves, Sonchus arvensis and Oenanthe linearis of North-Eastern region in India. J Appl Pharm Sci 6(2):157–166

Simirgiotis MJ, Silva M, Becerra J, Schmeda-Hirschmann G (2012) Direct characterisation of phenolic antioxidants in infusions from four Mapuche medicinal plants by liquid chromatography with diode array detection (HPLC–DAD) and electrospray ionisation tandem mass spectrometry (HPLC–ESI–MS). Food Chem 131(1):318–327

Stefova M, Stafilov T, Kulevanova S (2003) HPLC analysis of flavonoids. In: Encyclopedia of chromatography, pp 183–195 CRC Press, Taylor & Francis, New York

Steinmann D, Ganzera M (2011) Recent advances on HPLC/MS in medicinal plant analysis. J Pharm Biomed Anal 55(4):744–757

Stelzner J, Roemhild R, Garibay-Hernández A, Harbaum-Piayda B, Mock HP, Bilger W (2019) Hydroxycinnamic acids in sunflower leaves serve as UV-A screening pigments. Photochem Photobiol Sci 18(7):1649–1659

Stobiecki M (2000) Application of mass spectrometry for identification and structural studies of flavonoid glycosides. Phytochemistry 54(3):237–256

Treutter D (2006) Significance of flavonoids in plant resistance: a review. Environ Chem Lett 4(3):147–157

Tsimogiannis D, Oreopoulou V (2019) Classification of phenolic compounds in plants. In: Polyphenols in plants. Academic Press, London, pp 263–284

Vukics V, Guttman A (2010) Structural characterization of flavonoid glycosides by multi-stage mass spectrometry. Mass Spectrom Rev 29(1):1–16

Wu J, Lv S, Zhao L, Gao T, Yu C, Hu J, Ma F (2023) Advances in the study of the function and mechanism of the action of flavonoids in plants under environmental stresses. Planta 257(6):108

Yu J, Vasanthan T, Temelli F (2001) Analysis of phenolic acids in barley by high-performance liquid chromatography. J Agric Food Chem 49(9):4352–4358

Zhao X, Zhang S, Liu D, Yang M, Wei J (2020) Analysis of flavonoids in Dalbergia odorifera by ultra-performance liquid chromatography with tandem mass spectrometry. Molecules 25(2):389

Zohra T, Ovais M, Khalil AT, Qasim M, Ayaz M, Shinwari ZK (2019) Extraction optimization, total phenolic, flavonoid contents, HPLC-DAD analysis and diverse pharmacological evaluations of Dysphania ambrosioides (L.) Mosyakin & Clemants. Nat Prod Res 33(1):136–142

Volatile Compounds: Identification, Characterization, Isolation, and Pharmacology in Genus *Catharanthus*

8

Abstract

Catharanthus species are rich sources of volatile compounds with significant pharmacological potential. This study focuses on the identification, characterization, and isolation of volatile constituents across different *Catharanthus* species using advanced analytical techniques such as Gas chromatography-Mass spectrometry (GC-MS). The volatile profile includes terpenoids, alkaloids, and aromatic compounds contributing to the plant's bioactivity. Their pharmacological relevance is assessed, highlighting antimicrobial, antioxidant, and anticancer properties. Understanding these volatiles aids in drug discovery and therapeutic applications. The comprehensive analysis provides insights into the biochemical diversity of *Catharanthus*, supporting its use in traditional medicine and the development of novel natural product-based pharmaceuticals.

Keywords

Volatile bioactives · GC–MS · *Catharanthus*

8.1 Introduction

Phytochemicals with low molecular weights (less than 300 Da) are called volatile compounds. These components are evaporating more easily at ambient temperature (Azodanlou et al. 1999). Some volatile chemicals spread to the olfactory epithelium, dissolve in mucus, and may bind to olfactory receptors to create a smell. The smell of olive oil comes from aldehydes, esters, alcohols, ketones, hydrocarbons, and furans. Some volatile chemicals are still unidentified volatile molecules. The C6 and C5 volatile molecules are the most common ones found in virgin olive oils. Most virgin olive oils in Europe have hexanal, trans-2-hexenal, hexan-1-ol, and 3-methylbutan-1-ol in them. A research of extra-virgin olive oils from Italy, Spain,

S. Kumar, B. Singh, *Catharanthus in Drug Discovery*,
https://doi.org/10.1007/978-981-95-5048-7_8

and Morocco found that Italian oils include a lot of C6 volatile compounds, although their scent is not very fruity. Moroccan extra virgin olive oils had a lot of fruity esters, as ethyl isobutyrate, ethyl butyrate, ethyl 2-methylbutyrate, ethyl 3-methylbutyrate, and ethyl cyclohexylcarboxylate. The volatile elements with a high concentration are not always the main sources of smell. Major or minor volatile molecules are important for the quality of olive oil. The volatile molecules in olive oil that are below their sensory threshold do not directly contribute to the aroma. They could help us understand how the volatiles are made and broken down. Lastly, they add a lot to the smell and could be useful quality identifiers. This part has C5 carbonyl compounds, pentenols, hydrocarbons, and some other chemicals that are not made from fatty acid conversions (Plumb 1992; Effah et al. 2019). The volatile composition of olive oil is affected by the cultivar, the geographic area, the ripeness of the fruit, and the methods and parameters used to process it. Various cultivars of fruit cultivated in the same environment make oils with various volatile constituents. The same is true for Dragon fruit of the same cultivar growing in different parts of the world. It is known that the fragrance chemicals in oil increase as the fruit ripens, but only to a certain extent. Aside from the state of the fruit at harvest, variations in the postharvest treatment of the fruit and oil result in distinct volatile profiles (Ranjan et al. 2023; Bajo-Fernández et al. 2024). GC-MS is one of the most flexible and commonly used technology platforms in current metabolomic research. The newest GC–MS and LC–MS/MS methods are better than immunoassays when it comes to selectivity and limits of quantification. They also show good linearity even at low concentrations. GC–MS is another way to analyze things that has been around for 40 years. GC–MS is utilized for the analysis of steroid hormone metabolites and their precursors (Lisec et al. 2006; Sosa et al. 2016; Hadi and Hameed 2017).

8.2 Qualitative and Quantitative Analysis of Volatile Compounds by Gas Chromatography–Mass Spectrometry in *Catharanthus* Species

It has been said that volatile chemicals in the *Catharanthus* species are quite complicated combinations. The qualitative and quantitative study of these substances was conducted using a high-resolution analytical approach, such as GC–MS, due to their unique characteristics. Before GC–MS analysis, a fractionation step is needed to make sure the sample does not include any nonvolatile parts. We used GC–MS to look at the oils we got from *C. roseus*. Volatile components recovered during the water distillation of leaves and industrial marcs (plant wastes) from *C. roseus*. There are more than 170 chemicals in the leaves and industrial marcs (plant residues) of *C. roseus*, such as alkanes, alcohol, aldehydes, ketones, fatty acids, fatty acid esters, terpenoids, and phenylpropanoids (Table 8.1 and Fig. 8.1). Some chemicals have been found in *C. roseus* for the first time, and most of them have not been found in the Apocynaceae family before. The most important part was made up of fatty acids and fatty acid esters (84.8% for leaves and 51.4% for marcs). Alkanes (0.8%/0.6%),

Table 8.1 Volatile compounds identified by GC-MS, their bioactivities and their isolated sources

S. no.	Compound name	Class	Activity	Species/part	Reference
1.	Tetradecane	Alkanes	Unknown	*C. roseus*/ leaves and marcs	Brun et al. (2001), Lawal et al. (2015), Pandey-Rai et al. (2006)
2.	Pentadecane	Alkanes	Cytotoxic	*C. roseus*/ leaves and marcs	Brun et al. (2001), Pandey-Rai et al. (2006), Bruno et al. (2015)
3.	Heptadecane	Alkanes	Unknown	*C. roseus*/ leaves and marcs	Brun et al. (2001), Pandey-Rai et al. (2006), Lawal et al. (2015)
4.	Octadecane	Alkanes	Unknown	*C. roseus*/ leaves and marcs	Brun et al. (2001), Pandey-Rai et al. (2006)
5.	1-(*E*)-1-Hexenyl-cyclohexanol	Alcohols	Unknown	*C. roseus*/ leaves and marcs	Brun et al. (2001)
6.	Heptanal	Aldehydes	Unknown	*C. roseus*/ leaves and marcs	Brun et al. (2001), Pandey-Rai et al. (2006)
7.	Octanal	Aldehydes	Unknown	*C. roseus*/ leaves and marcs	Brun et al. (2001)
8.	(*E*)-2-Octenal	Aldehydes	Unknown	*C. roseus*/ leaves and marcs	Brun et al. (2001)
9.	Decanal	Aldehydes	Antimicrobial	*C. roseus*/ leaves and marcs	Brun et al. (2001), Pandey-Rai et al. (2006), Mahboubi and Feizabadi (2009)

(continued)

Table 8.1 (continued)

S. no.	Compound name	Class	Activity	Species/part	Reference
10.	(*E*)-2-Decenal	Aldehydes	Nematicidal, synergistic	*C. roseus*/ leaves and marcs	Brun et al. (2001), Caboni et al. (2012), Ntalli et al. (2016)
11.	(*E*)-2-Nonenal	Aldehydes	Unknown	*C. roseus*/ leaves and marcs	Brun et al. (2001)
12.	(*E*,*Z*)-2,4-Decadienal	Aldehydes	Nematicidal, synergistic	*C. roseus*/ leaves and marcs	Caboni et al. (2012), Brun et al. (2001), Ntalli et al. (2016)
13.	(*E*,*E*)-2,4-Decadienal	Aldehydes	Nematicidal, synergistic	*C. roseus*/ leaves and marcs	Brun et al. (2001), Caboni et al. (2012), Ntalli et al. (2016)
14.	(*E*)-2-Undecenal	Aldehydes	Unknown	*C. roseus*/ leaves and marcs	Brun et al. (2001)
15.	Pentadecanal	Aldehydes	Unknown	*C. roseus*/ leaves and marcs	Brun et al. (2001), Pandey-Rai et al. (2006)
16.	6-Methyl-5-hepten-2-one	Ketones	Unknown	*C. roseus*/ leaves and marcs	Brun et al. (2001)
17.	Heptenone	Ketones	Unknown	*C. roseus*/ leaves and marcs	Brun et al. (2001)
18.	2,2,6-Trimethylcyclohexanone	Ketones	Unknown	*C. roseus*/ leaves and marcs	Brun et al. (2001)

19.	(*E*,*Z*)-Octa-3,5-dien-2-one	Ketones	Unknown	*C. roseus*/ leaves and marcs	Brun et al. (2001), Pandey-Rai et al. (2006)
20.	(*E*,*E*)-Octa-3,5-dien-2-one	Ketones	Unknown	*C. roseus*/ leaves and marcs	Brun et al. (2001)
21.	6-Methyl-3,5-heptadien-2-one	Ketones	Unknown	*C. roseus*/ leaves and marcs	Brun et al. (2001)
22.	Pentadecane-2-one	Ketones	Unknown	*C. roseus*/ leaves and marcs	Brun et al. (2001)
23.	Caproic acid	Fatty acids and esters	Unknown	*C. roseus*/ leaves and marcs	Brun et al. (2001)
24.	Methyl heptanoate	Fatty acids and esters	Unknown	*C. roseus*/ leaves and marcs	Brun et al. (2001)
25.	Methyl caprylate	Fatty acids and esters	Unknown	*C. roseus*/ leaves and marcs	Brun et al. (2001)
26.	Caprylic acid	Fatty acids and esters	Antibacterial, bactericidal, synergistic, antimicrobial	*C. roseus*/ leaves and marcs	Brun et al. (2001)

(continued)

Table 8.1 (continued)

S. no.	Compound name	Class	Activity	Species/part	Reference
27.	Capric acid	Fatty acids and esters	Anti-bacterial, anti-inflammatory, microbicidal	*C. roseus*/ leaves and marcs	Brun et al. (2001), Huang et al. (2014), Thormar et al. (2006), Chadeganipour and Haims (2001), Antifungal activities; Kato (1981)
28.	Methyl undecanoate	Fatty acids and esters	Progestational activity	*C. roseus*/ leaves and marcs	Brun et al. (2001), Attardi et al. (2006)
29.	Methyl laurate	Fatty acids and esters	Unknown	*C. roseus*/ leaves and marcs	Brun et al. (2001)
30.	Lauric acid	Fatty acids and esters	Antifungal; antimicrobial activity, antibacterial	*C. roseus*/ leaves and marcs	Brun et al. (2001), Walters et al. (2003), Řiháková et al. (2001), Yang et al. (2009a), Dayrit (2015), Nitbani et al. (2016), Anzaku et al. (2017)
31.	Tridecanoic acid	Fatty acids and esters	Antioxidant	*C. roseus*/ leaves and marcs	Brun et al. (2001), Pandey-Rai et al. (2006), Henry et al. (2002)
32.	Methyl myristate	Fatty acids and esters	Unknown	*C. roseus*/ leaves and marcs	Brun et al. (2001)
33.	Myristic acid	Fatty acids and esters	Unknown	*C. roseus*/ leaves and marcs, *C. roseus*/ leaves and flowers	Brun et al. (2001), Pandey-Rai et al. (2006)

34.	Methyl pentadecanoate	Fatty acids and esters	Unknown	*C. roseus*/ leaves and marcs	Brun et al. (2001)
35.	Pentadecanoic acid	Fatty acids and esters	Unknown	*C. roseus*/ leaves and marcs	Brun et al. (2001)
36.	Methyl 9-tridecenoate	Fatty acids and esters	Unknown	*C. roseus*/ leaves and marcs	Brun et al. (2001)
37.	Methyl palmitate	Fatty acids and esters	Anti-inflammatory	*C. roseus*/ leaves and marcs	Brun et al. (2001), Saeed et al. (2012)
38.	Palmitic acid	Fatty acids and esters	Antitumor, protonophoric, bactericidal	*C. roseus*/ leaves and marcs	Brun et al. (2001), Harada et al. (2002), Shinohara et al. (1995), Ivanova et al. (2017), Pandey-Rai et al. (2006)
39.	α-Isophorone	Terpenoids	Unknown	*C. roseus*/ leaves and marcs	Brun et al. (2001)
40.	(*Z*)-linalool oxide	Terpenoids	Unknown	*C. roseus*/ leaves and marcs	Brun et al. (2001)
41.	(*E*)-linalool oxide	Terpenoids	Unknown	*C. roseus*/ leaves and marcs	Brun et al. (2001)

(continued)

Table 8.1 (continued)

S. no.	Compound name	Class	Activity	Species/part	Reference
42.	Safranal	Terpenoids	Antitumor, anticonvulsant, anxiolytic, hypnotic, antidepressant, dermatologic, effects on sexual behavior; respiratory, antidotal, antihypertensive, anti-ischemia, antinociceptive, anti-inflammatory effects, antimicrobial, antioxidant, DNA interactions, genoprotective, toxicology, anticancer	*C. roseus*/ leaves and marcs	Brun et al. (2001), Samarghandian et al. (2014), Hosseinzadeh and Talebzadeh (2005), Hosseinzadeh and Shariaty (2007), Hosseinzadeh and Noraei (2009), Kanakis et al. (2007), Rezaee and Hosseinzadeh (2013), Zhang et al. (2018)
43.	β-Cyclocitral	Terpenoids	Unknown	*C. roseus*/ leaves and flowers	Brun et al. (2001)
44.	β-Pulegone	Terpenoids	Unknown	*C. roseus*/ leaves and flowers	Brun et al. (2001), Pandey-Rai et al. (2006)
45.	Geranylacetone	Terpenoids	Antimicrobial	*C. roseus*/ leaves and flowers	Brun et al. (2001), Pandey-Rai et al. (2006), Bonikowski et al. (2015)
46.	Germacrene D	Terpenoids	Antibacterial, antioxidant	*C. roseus*/ leaves and flowers	Brun et al. (2001), Juteau et al. (2002)
47.	Dihydroactinidiolide	Terpenoids	Antibacterial, anticancer	*C. roseus*/ leaves and flowers	Brun et al. (2001)
48.	(E,E)-farnesyl acetate	Terpenoids	Unknown	*C. roseus*/ leaves and flowers	Brun et al. (2001)
49.	Hexahydrofarnesylacetone	Terpenoids	Unknown	*C. roseus*/ leaves and flowers	Brun et al. (2001)

50.	Farnesylacetone	Terpenoids	Vasodilatation effect	*C. roseus*/leaves and flowers	Brun et al. (2001), Park et al. (2008)
51.	Isophytol	Terpenoids	Unknown	*C. roseus*/leaves and flowers	Brun et al. (2001)
52.	Edulane II	Compounds with 13 carbons	Unknown	*C. roseus*/leaves and flowers	Brun et al. (2001)
53.	Theaspirane B	Compounds with 13 carbons	Unknown	*C. roseus*/leaves and flowers	Brun et al. (2001)
54.	Theaspirane A	Compounds with 13 carbons	Unknown	*C. roseus*/leaves and flowers	Brun et al. (2001)
55.	(*E*)-*β*-Damascenone	Compounds with 13 carbons	Antispasmodic	*C. roseus*/leaves and flowers	Brun et al. (2001), Pongprayoon et al. (1992)
56.	(*E*)-*α*-ionone	Compounds with 13 carbons	Unknown	*C. roseus*/leaves and flowers	Brun et al. (2001)
57.	(*E*)-*β*-Ionone-5,6-epoxide	Compounds with 13 carbons	Unknown	*C. roseus*/leaves and flowers	Brun et al. (2001)
58.	(*E*)-*β*-Ionone	Compounds with 13 carbons	Unknown	*C. roseus*/leaves and flowers	Brun et al. (2001)

(continued)

Table 8.1 (continued)

S. no.	Compound name	Class	Activity	Species/part	Reference
59.	Megastigmatrienone	Compounds with 13 carbons	Unknown	*C. roseus*/ leaves and flowers	Brun et al. (2001)
60.	(*E*,*E*)-Pseudoionone	Compounds with 13 carbons	Unknown	*C. roseus*/ leaves and flowers	Brun et al. (2001)
61.	(*E*)-Anethole	Phenylpropanoids	Fumigant, acaricidal, antimicrobial, antioxidant, anti-inflammatory gastroprotector, antimicrobial, antifungal activity, antihelmintic insecticidal, secretolytic expectorant effects, spasmolytic effect, antinociceptive, anti-inflammatory activity, gastroprotective activity, sedative activity, reproductive toxicity, estrogenic effects,	*C. roseus*/ leaves and flowers	Brun et al. (2001), Chang and Ahn (2002), Senra et al. (2013), Esfandyari-Manesh et al. (2013), Freire et al. (2005), Marinov and Valcheva-Kuzmanova (2015)
62.	Eugenol	Phenylpropanoids	Antimicrobial, anti-inflammatory, analgesic, antioxidant, anticancer	*C. roseus*/ leaves and flowers	Brun et al. (2001), Kamatou et al. (2012), Raja et al. (2015), Abdullah et al. (2018)
63.	4-Methylbenzaldehyde	Miscellaneous compounds	Unknown	*C. roseus*/ leaves and flowers	Brun et al. (2001)
64.	Ethylbenzaldehyde	Miscellaneous compounds	Unknown	*C. roseus*/ leaves and flowers	Brun et al. (2001)
65.	Methyl salicylate	Miscellaneous compounds	Antioxidant, toxicity, antimicrobial, antioxidant	*C. roseus*/ leaves and flowers	Brun et al. (2001), Sayyari et al. (2011), Davison et al. (1961), Oloyede (2016)

66.	2,6,6-Trimethylcyclohexen-1-acetaldehyde	Miscellaneous compounds	Unknown	*C. roseus*/ leaves and flowers	Brun et al. (2001)
67.	α-Pinene	Terpenoids	Anti-inflammatory, chondroprotective	*C. roseus*/ leaves and flowers	Lawal et al. (2015), Rufino et al. (2014), Chalchat et al. (2000)
68.	β-Pinene	Terpenoids	Anti-inflammatory, chondroprotective	*C. roseus*/ leaves and flowers	Lawal et al. (2015), Rufino et al. (2014), Chalchat et al. (2000)
69.	Limonene	Terpenoids	Antimicrobial, anticancer, toxicity, antiparasitic	*C. roseus*/ leaves and flowers	Lawal et al. (2015), Erasto and Viljoen (2008), Negro et al. (2016)
70.	β-Phellandrene	Terpenoids	Unknown	*C. roseus*/ leaves and flowers	Lawal et al. (2015)
71.	(*Z*)-β-Ocimene	Terpenoids	Unknown	*C. roseus*/ leaves and flowers	Lawal et al. (2015)
72.	γ-Terpinene	Terpenoids	Acaricidal	*C. roseus*/ leaves and flowers	Lawal et al. (2015), Li and Liu (2009)
73.	Trans-Sabinene hydrate	Terpenoids	Unknown	*C. roseus*/ leaves and flowers	Lawal et al. (2015)
74.	Isopulegol	Terpenoids	Toxicologic, dermatologic, antifungal, anti-inflammatory, antioxidant, analgesic, gastroprotective	*C. roseus*/ leaves and flowers	Lawal et al. (2015), Bhatia et al. (2008), Silva et al. (2009)

(continued)

Table 8.1 (continued)

S. no.	Compound name	Class	Activity	Species/part	Reference
75.	(*E*,*Z*)-2,6-Nonadienal	Terpenoids	Unknown	*C. roseus*/ leaves and flowers	Lawal et al. (2015)
76.	Terpinen-4-ol	Terpenoids	Antifungal, anticonvulsant, larvicidal, cancers	*C. roseus*/ leaves and flowers	Lawal et al. (2015), Morcia et al. (2012), Govindarajan et al. (2016), Shapira et al. (2016)
77.	α-Terpineol	Terpenoids	Anticonvulsant, cardiovascular; antidiarrheal, antibacterial, antimicrobial	*C. roseus*/ leaves and flowers	Lawal et al. (2015), dos Santos Negreiros et al. (2019), Zhou et al. (2014)
78.	*n*-Dodecane	Alkane	Unknown	*C. roseus*/ leaves and flowers	Lawal et al. (2015)
79.	Iso-Dihydro carveol	Terpenoids	Unknown	*C. roseus*/ leaves and flowers	Lawal et al. (2015)
80.	Geraniol	Terpenoids	Antimicrobial, antioxidant, anti-inflammatory, vascular effects, antifungal, antibacterial, anti-candida, prostate cancer	*C. roseus*/ leaves and flowers	Lawal et al. (2015), Pandey-Rai et al. (2006), Chen and Viljoen (2010), Pereira et al. (2015), Barros et al. (2009), Sharma et al. (2016), Kim et al. (2011)
81.	Perilla aldehyde	Terpenoids	Unknown	*C. roseus*/ leaves and flowers	Lawal et al. (2015)
82.	2,6,11-Trimethyl dodecane	Alkane	Unknown	*C. roseus*/ leaves and flowers	Lawal et al. (2015)

83.	*n*-Tridecane	Alkane	Unknown	*C. roseus*/ leaves and flowers	Lawal et al. (2015)
84.	*α*-Terpinyl acetate	Terpenoids	Unknown	*C. roseus*/ leaves and flowers	Lawal et al. (2015)
85.	Neryl acetate	Alkane	Unknown	*C. roseus*/ leaves and flowers	Lawal et al. (2015)
86.	Methyl eugenol	Phenypropene	Unknown	*C. roseus*/ leaves and flowers	Lawal et al. (2015)
87.	*Trans*-*α*-Bergamotene	Terpenoids	Unknown	*C. roseus*/ leaves and flowers	Lawal et al. (2015)
88.	Dodecyl alcohol	Alcohol	Unknown	*C. roseus*/ leaves and flowers	Lawal et al. (2015)
89.	2,6-bis(1,1-Dimethylethyl)-2,5-cyclohexadiene-1,4-dione	Terpenoids	Unknown	*C. roseus*/ leaves and flowers	Lawal et al. (2015)
90.	*n*-Pentadecane	Alkane	Unknown	*C. roseus*/ leaves and flowers	Lawal et al. (2015)
91.	(*E*,*E*)-*α*-Farnesene	Terpenoids	Unknown	*C. roseus*/ leaves and flowers	Lawal et al. (2015)

(continued)

Table 8.1 (continued)

S. no.	Compound name	Class	Activity	Species/part	Reference
92.	β-Bisabolene	Terpenoids	Synergistic, bactericidal	*C. roseus*/ leaves and flowers	Lawal et al. (2015), Nascimento et al. (2007)
93.	2,4-bis(1,1-Dimethylethyl) phenol	Phenol	Unknown	*C. roseus*/ leaves and flowers	Lawal et al. (2015), Padmavathi et al. (2014)
94.	δ-Cadinene	Terpenoids	Unknown	*C. roseus*/ leaves and flowers	Lawal et al. (2015)
95.	Elemol	Isopropanol	Unknown	*C. roseus*/ leaves and flowers	Lawal et al. (2015)
96.	Davanone	Terpenoids	Unknown	*C. roseus*/ leaves and flowers	Lawal et al. (2015)
97.	*n*-Hexadecene	Terpenoids	Unknown	*C. roseus*/ leaves and flowers	Lawal et al. (2015)
98.	Widdrol	Terpenoids	Antifungal	*C. roseus*/ leaves and flowers	Lawal et al. (2015), Nuñez et al. (2006)
99.	Cedrol	Terpenoids	Hair growth promoting, antiseptic, anti-inflammatory, antispasmodic, tonic, astringent, diuretic, sedative, insecticidal, antifungal	*C. roseus*/ leaves and flowers	Lawal et al. (2015), Zhang et al. (2016)
100.	Dodecyl acetate	Fatty acids and esters	Unknown	*C. roseus*/ leaves and flowers	Lawal et al. (2015)

101.	3-Phenyldodecane	Alkane	Unknown	*C. roseus*/ leaves and flowers	Lawal et al. (2015)
102.	Isoamyl laurate	Fatty acids and esters	Unknown	*C. roseus*/ leaves and flowers	Lawal et al. (2015)
103.	Farnesyl acetone	Terpene ketone	Unknown	*C. roseus*/ leaves and flowers	Lawal et al. (2015)
104.	Hexadecanol	Fatty alcohol	Unknown	*C. roseus*/ leaves and flowers	Lawal et al. (2015)
105.	*n*-Hexadecanoic acid	Fatty acids and esters	Cytotoxic	*C. roseus*/ leaves and flowers	Lawal et al. (2015), Ravi and Krishnan (2017)
106.	Linoleic acid methyl ester	Fatty acids and esters	Unknown	*C. roseus*/ leaves and flowers	Lawal et al. (2015)
107.	Linolenic acid methyl ester	Fatty acids and esters	Unknown	*C. roseus*/ leaves and flowers	Lawal et al. (2015)
108.	Linoleic acid	Fatty acids and esters	Antibacterial, desaturation, antioxidant	*C. roseus*/ leaves and flowers	Lawal et al. (2015), Dilika et al. (2000), Peluffo et al. (1976)
109.	Oleic acid ethyl ester	Fatty acids and esters	Unknown	*C. roseus*/ leaves and flowers	Lawal et al. (2015)

(continued)

Table 8.1 (continued)

S. no.	Compound name	Class	Activity	Species/part	Reference
110.	Linolenic acid ethyl ester	Fatty acids and esters	Unknown	*C. roseus*/ leaves and flowers	Lawal et al. (2015)
111.	Stearic acid	Fatty acids and esters		*C. roseus*/ leaves and flowers	Lawal et al. (2015), Pandey-Rai et al. (2006)
112.	(*Z*)-9-Tricosene	Alkene	Bio-pesticide	*C. roseus*/ leaves and flowers	Lawal et al. (2015), Uebel et al. (1976)
113.	Dehydroabietic acid	Terpenoids	Unknown	*C. roseus*/ leaves and flowers	Lawal et al. (2015)
114.	α-Tocopherol	Methylated phenol	Gene-regulatory, antioxidant	*C. roseus*/ leaves and flowers	Lawal et al. (2015), Rimbach et al. (2010), Huang et al. (1996)
115.	Dotriacontane	Alkane	Unknown	*C. roseus*/ leaves and flowers	Lawal et al. (2015)
116.	Lanosterol	Terpenoids	Unknown	*C. roseus*/ leaves and flowers	Lawal et al. (2015)
117.	α-Amyrin	Terpenoids	Unknown	*C. roseus*/ leaves and flowers	Lawal et al. (2015)
118.	γ-Sitosterol	Terpenoids	Antidiabetic	*C. roseus*/ leaves and flowers	Lawal et al. (2015), Balamurugan et al. (2011)

119.	Tetracontane	Alkanes	Unknown	*C. roseus*/ leaves and flowers	Pandey-Rai et al. (2006)
120.	Tridecane	Alkanes	Unknown	*C. roseus*/ leaves and flowers	Pandey-Rai et al. (2006)
121.	Hexadecane	Alkanes	Unknown	*C. roseus*/ leaves and flowers	Pandey-Rai et al. (2006), Brun et al. (2001)
122.	Nonadecane	Alkanes	Unknown	*C. roseus*/ leaves and flowers	Pandey-Rai et al. (2006)
123.	Eicosane	Alkanes	Unknown	*C. roseus*/ leaves and flowers	Pandey-Rai et al. (2006)
124.	Heneicosene	Alkanes	Unknown	*C. roseus*/ leaves and flowers	Pandey-Rai et al. (2006)
125.	Heneicosane	Alkanes	Unknown	*C. roseus*/ leaves and flowers	Pandey-Rai et al. (2006)
126.	Docosene	Alkanes	Unknown	*C. roseus*/ leaves and flowers	Pandey-Rai et al. (2006)
127.	Docosane	Alkanes	Unknown	*C. roseus*/ leaves and flowers	Pandey-Rai et al. (2006)

(continued)

Table 8.1 (continued)

S. no.	Compound name	Class	Activity	Species/part	Reference
128.	Tricosane	Alkanes	Unknown	*C. roseus*/ leaves and flowers	Pandey-Rai et al. (2006)
129.	Tetracosane	Alkanes	Unknown	*C. roseus*/ leaves and flowers	Pandey-Rai et al. (2006)
130.	9-Octadecen-1-ol	Alcohols	Unknown	*C. roseus*/ leaves and flowers	Pandey-Rai et al. (2006)
131.	Eicosan-1-ol	Alcohols	Unknown	*C. roseus*/ leaves and flowers	Pandey-Rai et al. (2006)
132.	(*E*,*E*)-2,4-Hexadienal	Aldehydes	Unknown	*C. roseus*/ leaves and flowers	Pandey-Rai et al. (2006)
133.	(*E*,*E*)-2,4-Heptadienal	Aldehydes	Unknown	*C. roseus*/ leaves and flowers	Pandey-Rai et al. (2006), Brun et al. (2001)
134.	(*E*)-Octenal	Aldehydes	Unknown	*C. roseus*/ leaves and flowers	Pandey-Rai et al. (2006)
135.	Nonanal	Aldehydes	Antidiarrhoeal	*C. roseus*/ leaves and flowers	Pandey-Rai et al. (2006), Brun et al. (2001), Zavala-Sanchez et al. (2002)
136.	(*E*)-2-Decenal	Aldehydes	Nematicidal	*C. roseus*/ leaves and flowers	Pandey-Rai et al. (2006), Caboni et al. (2012)

137.	(*E*)-Decenal	Aldehydes	Nematicidal	*C. roseus*/ leaves and flowers	Pandey-Rai et al. (2006), Caboni et al. (2012)
138.	Undecanal	Aldehydes	Unknown	*C. roseus*/ leaves and flowers	Pandey-Rai et al. (2006)
139.	(*E*,*E*)-2,4-Dodecadienal	Aldehydes	Unknown	*C. roseus*/ leaves and flowers	Pandey-Rai et al. (2006)
140.	Dodecanal	Aldehydes	Unknown	*C. roseus*/ leaves and flowers	Pandey-Rai et al. (2006), Lawal et al. (2015)
141.	Tetradecanal	Aldehydes	Unknown	*C. roseus*/ leaves and flowers	Pandey-Rai et al. (2006), Lawal et al. (2015)
142.	Hexadecanal	Aldehydes	Unknown	*C. roseus*/ leaves and flowers	Pandey-Rai et al. (2006), Lawal et al. (2015)
143.	2-methyl 4-heptanone	Ketones	Unknown	*C. roseus*/ leaves and flowers	Pandey-Rai et al. (2006)
144.	6-methyl hept-5-ene-2-one	Ketones	Unknown	*C. roseus*/ leaves and flowers	Pandey-Rai et al. (2006)
145.	3-Octen-2-one	Ketones	Unknown	*C. roseus*/ leaves and flowers	Pandey-Rai et al. (2006), Brun et al. (2001)

(continued)

Table 8.1 (continued)

S. no.	Compound name	Class	Activity	Species/part	Reference
146.	Nonanoic acid	Fatty acids and esters	Unknown	*C. roseus*/ leaves and flowers	Pandey-Rai et al. (2006), Brun et al. (2001)
147.	Undecanoic acid	Fatty acids and esters	Unknown	*C. roseus*/ leaves and flowers	Pandey-Rai et al. (2006)
148.	Dodecanoic acid	Fatty acids and esters	Unknown	*C. roseus*/ leaves and flowers	Pandey-Rai et al. (2006)
149.	Pentadecanoic acid	Fatty acids and esters	Unknown	*C. roseus*/ leaves and flowers	Pandey-Rai et al. (2006)
150.	Hexatrienoic acid methyl ester	Fatty acids and esters	Unknown	*C. roseus*/ leaves and flowers	Pandey-Rai et al. (2006)
151.	Hexadecenoic acid	Fatty acids and esters	Unknown	*C. roseus*/ leaves and flowers	Pandey-Rai et al. (2006)
152.	Hexadecyl acetate	Fatty acids and esters	Unknown	*C. roseus*/ leaves and flowers	Pandey-Rai et al. (2006)
153.	Octadecenoic acid	Fatty acids and esters	Unknown	*C. roseus*/ leaves and flowers	Pandey-Rai et al. (2006)
154.	Myrcene	Terpenoids	Unknown	*C. roseus*/ leaves and flowers	Pandey-Rai et al. (2006)

155.	1,8-cineole	Terpenoids	Repellent, toxicant, grain protectant; antimicrobial, toxicity, feeding deterrence, antiinflammatory, antinociceptive, locomotor, anti-infectious bronchitis virus (ibv), antitumor, antinociceptive	*C. roseus*/ leaves and flowers	Pandey-Rai et al. (2006), Obeng-Ofori et al. (1997), Tripathi et al. (2001), Vuuren and Viljoen (2007), Santos and Rao (2000), Kovar et al. (1987), Yang et al. (2010), Murata et al. (2013), Hendry et al. (2009), Liapi et al. (2007)
156.	Linalool	Terpenoids	Antimicrobial, anti-inflammatory, anticancer, antioxidant, anticancer, toxic, repellent, arvicidal, ecotoxicity, antidepressant, hepatoprotective	*C. roseus*/ leaves and flowers	Pandey-Rai et al. (2006), Lawal et al. (2015), Brun et al. (2001), Kamatou and Viljoen (2008), Seol et al. (2016), Herman et al. (2016), Sun et al. (2015), Tabari et al. (2017)
157.	Citronellol	Terpenoids	Anticonvulsant, antihyperalgesic, orofacial antinociceptive	*C. roseus*/ leaves and flowers	Pandey-Rai et al. (2006), Santos et al. (2018)
158.	Nerol	Terpenoids	Unknown	*C. roseus*/ leaves and flowers	Pandey-Rai et al. (2006)
159.	Neral	Terpenoids	Anti-inflammatory	*C. roseus*/ leaves and flowers	Pandey-Rai et al. (2006), Lawal et al. (2015), Brun et al. (2001), Liao et al. (2015)
160.	γ-Eudesmol	Terpenoids	Unknown	*C. roseus*/ leaves and flowers	Pandey-Rai et al. (2006)
161.	α-Eudesmol	Terpenoids	Unknown	*C. roseus*/ leaves and flowers	Pandey-Rai et al. (2006)

(continued)

Table 8.1 (continued)

S. no.	Compound name	Class	Activity	Species/part	Reference
162.	(*Z*,*Z*)-Farnesol	Terpenoids	Unknown	*C. roseus*/ leaves and flowers	Pandey-Rai et al. (2006)
163.	(*E*,*E*)-Farnesol	Terpenoids	Unknown	*C. roseus*/ leaves and flowers	Pandey-Rai et al. (2006), Brun et al. (2001)
164.	(*E*,*Z*)-Farnesol	Terpenoids	Unknown	*C. roseus*/ leaves and flowers	Pandey-Rai et al. (2006)
165.	Dihydrofarnesyl acetone	Terpenoids	Unknown	*C. roseus*/ leaves and flowers	Pandey-Rai et al. (2006)
166.	Farnesyl acetone	Terpenoids	Unknown	*C. roseus*/ leaves and flowers	Pandey-Rai et al. (2006)
167.	Phytol	Terpenoids	Antianxiety, cytotoxic, metabolism-modulating, antioxidant, autophagy, apoptosis-inducing, anti-nociceptive, anti-inflammatory, immune-modulating, and antimicrobial	*C. roseus*/ leaves and flowers	Pandey-Rai et al. (2006), Lawal et al. (2015), Islam et al. (2018)
168.	Benzaldehyde	Miscellaneous compounds	Unknown	*C. roseus*/ leaves and flowers	Pandey-Rai et al. (2006), Brun et al. (2001), Kochi et al. (1980)
169.	2-Hydroxy 5-methoxy p-cymene	Miscellaneous compounds	Unknown	*C. roseus*/ leaves and flowers	Pandey-Rai et al. (2006)
170.	Di-ter-butyl p-cresol	Miscellaneous compounds	Unknown	*C. roseus*/ leaves and flowers	Pandey-Rai et al. (2006)

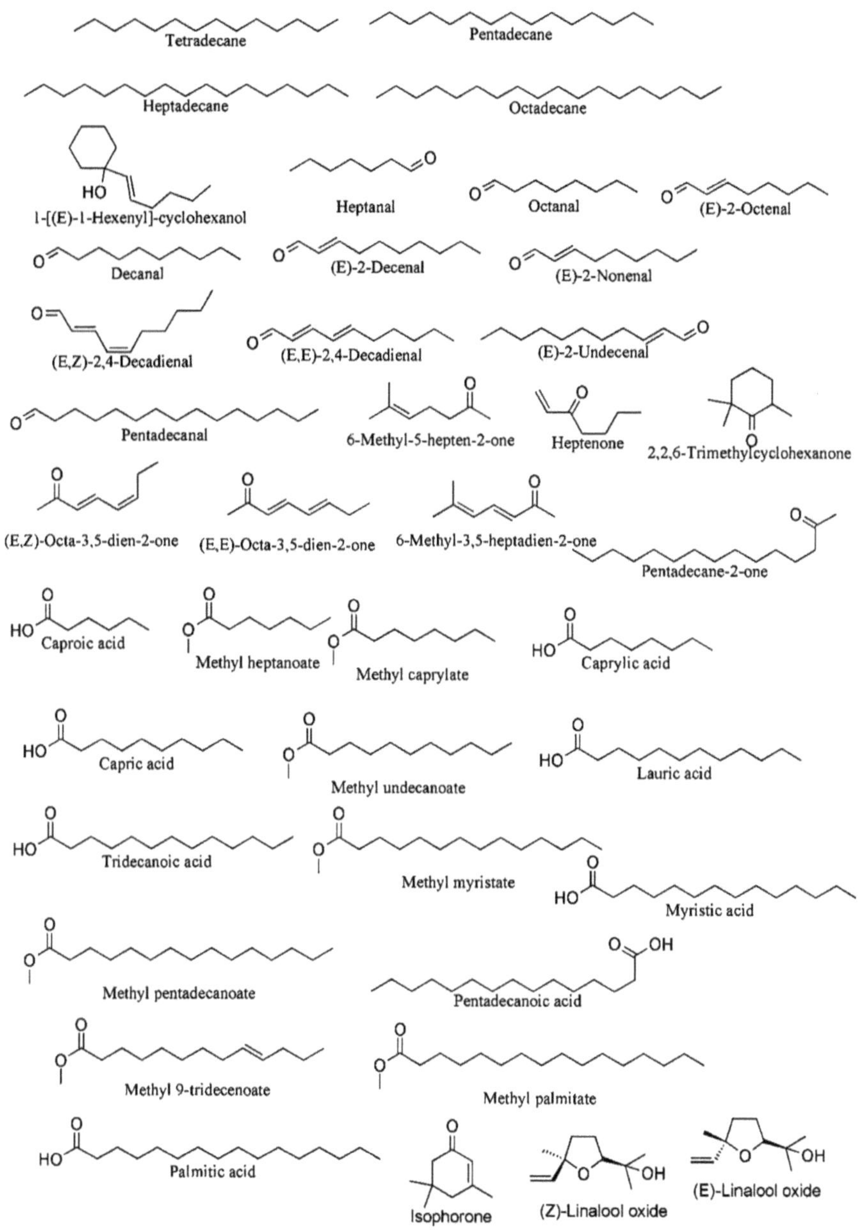

Fig. 8.1 Volatile compounds identified by GC-MS in *Catharanthus* species

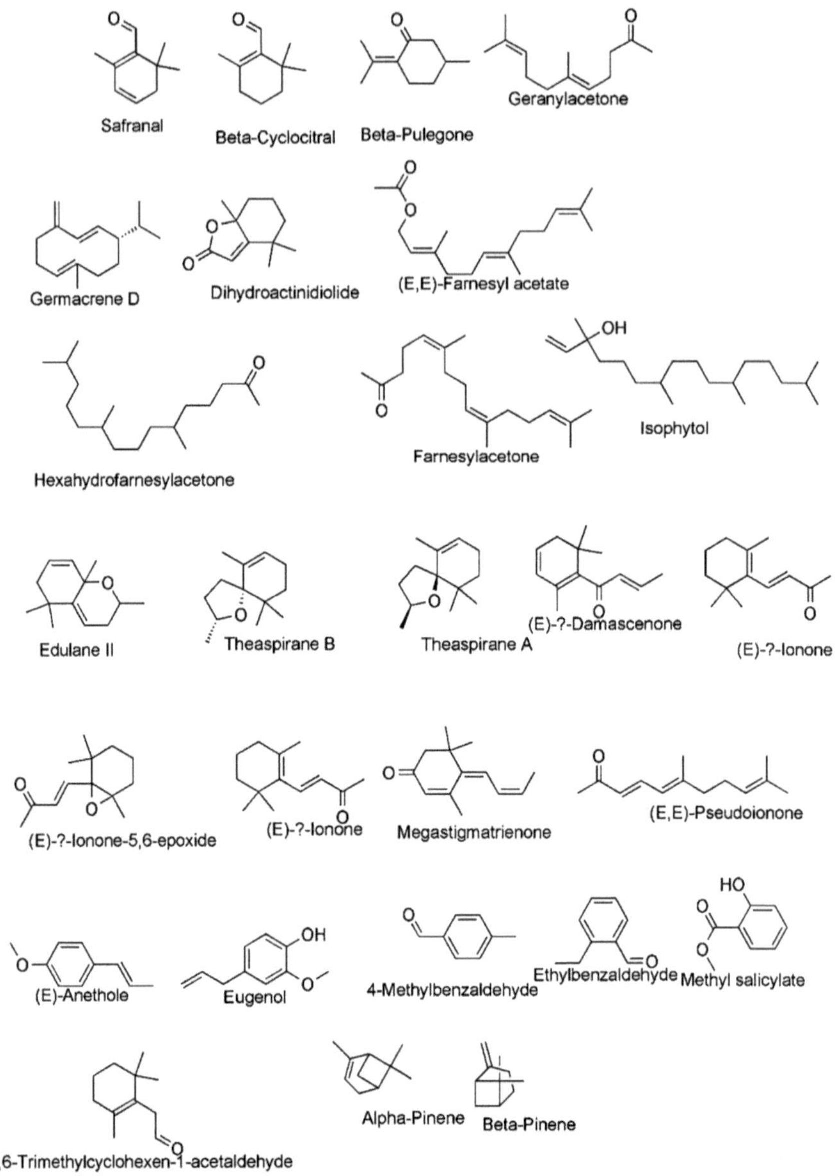

Fig. 8.1 (continued)

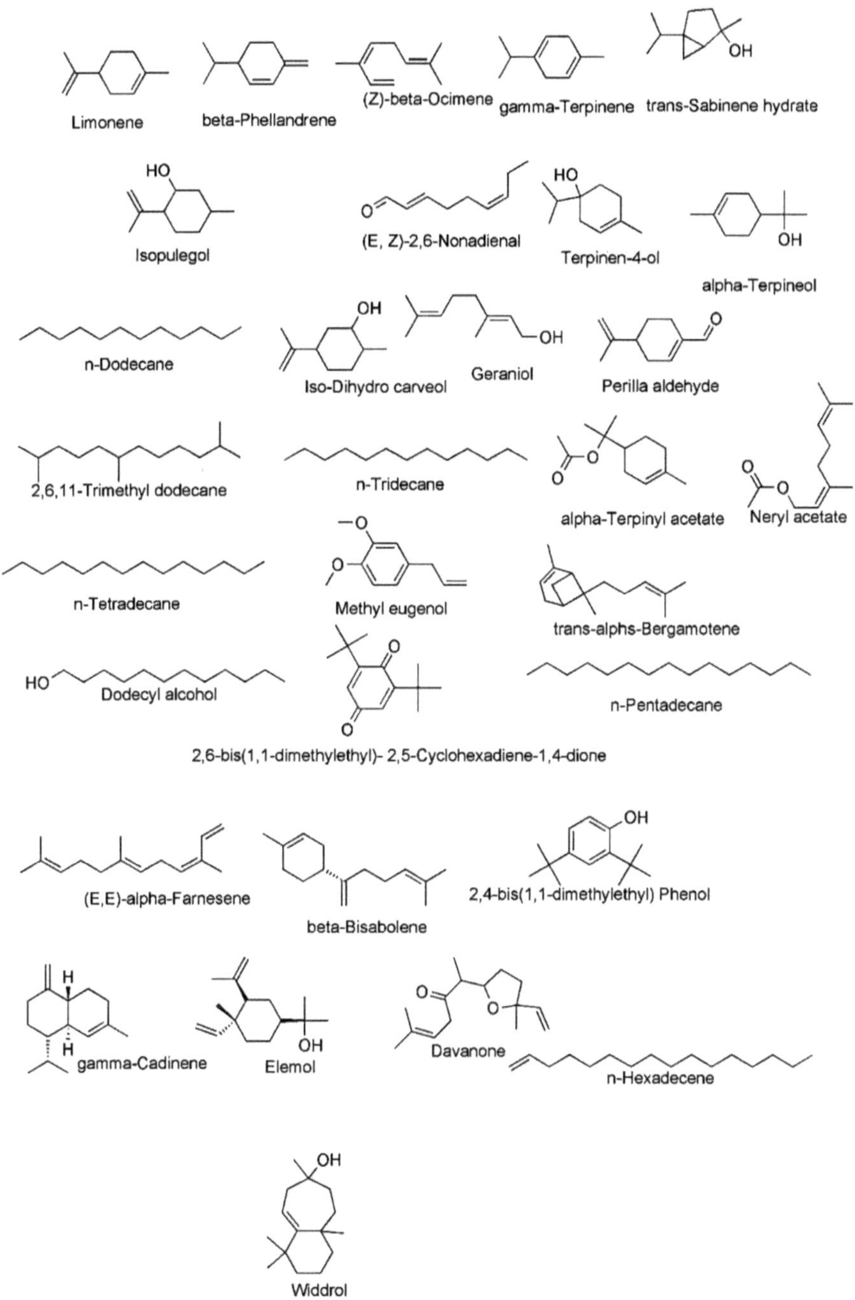

Fig. 8.1 (continued)

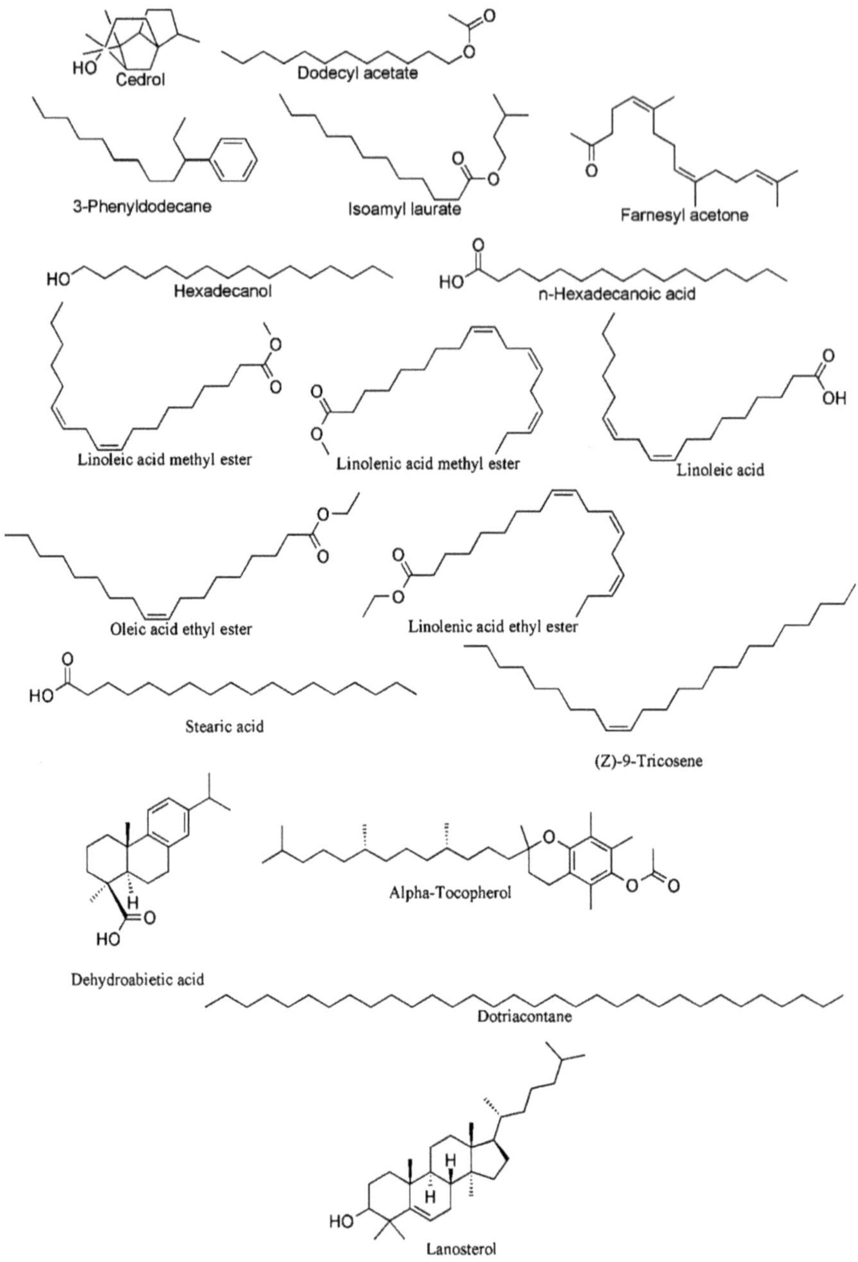

Fig. 8.1 (continued)

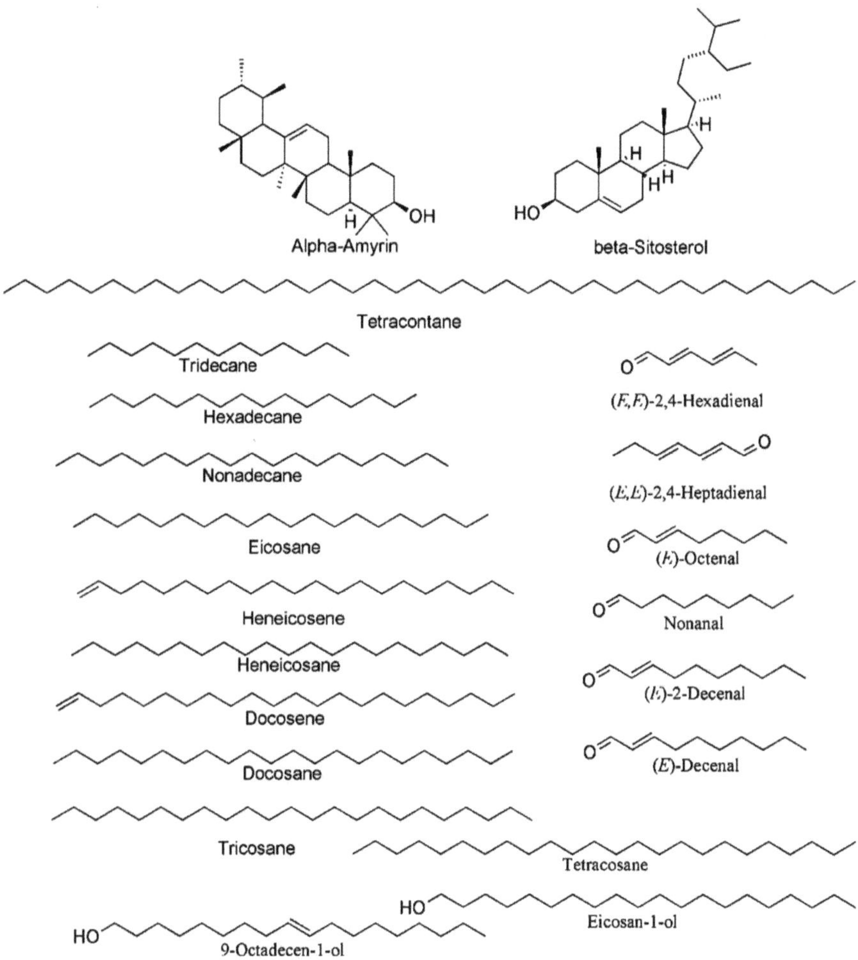

Fig. 8.1 (continued)

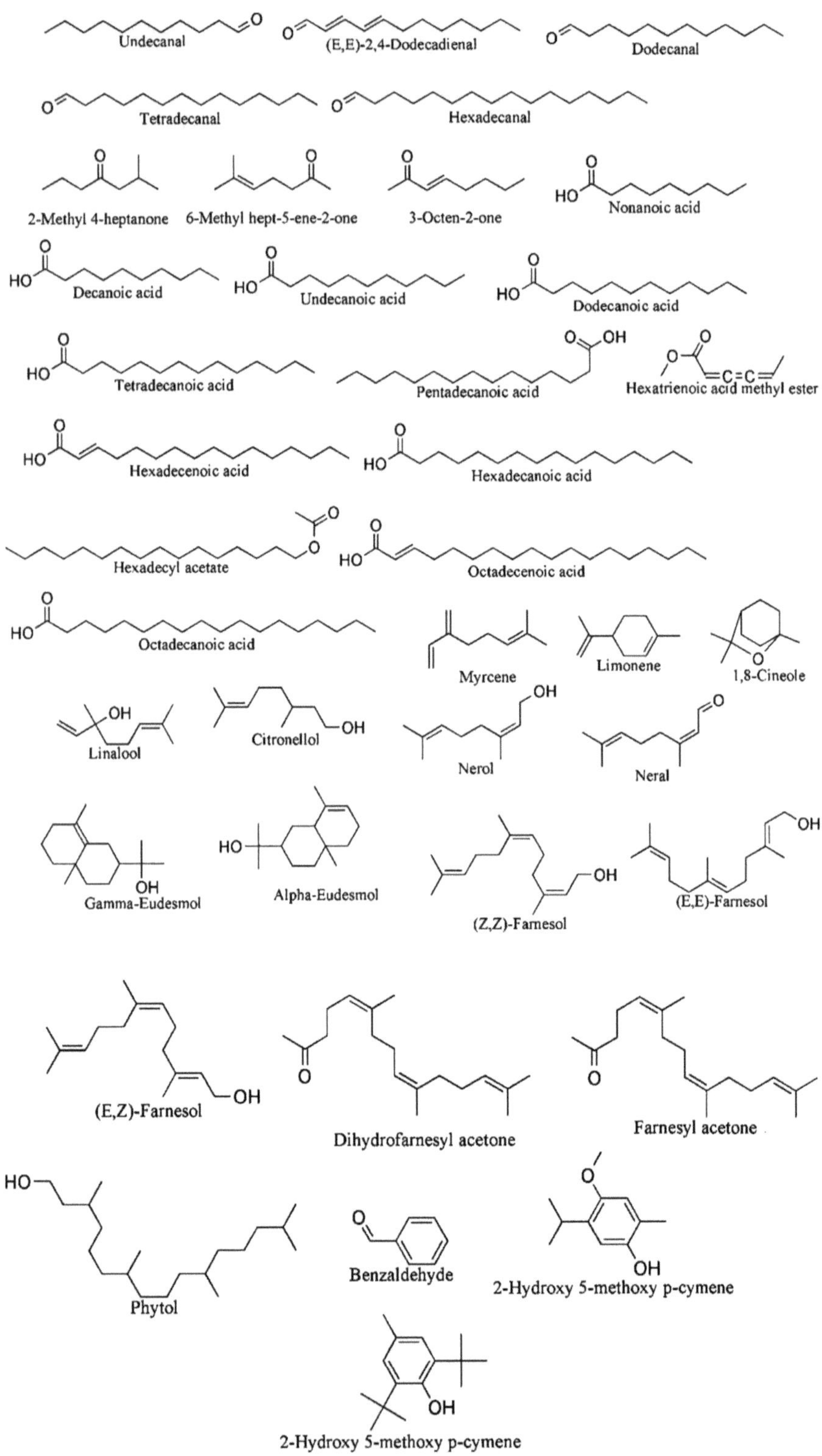

Fig. 8.1 (continued)

alcohols (0.2%), aldehydes (1.3%/7.8%), ketones (0.5%/4.2%), phenylpropanoids (0.2%), terpenoids (6.1%/15.9%), compounds with 13 carbons (2.6%/6.7%), and other compounds (0.3%/0.9%) were also found. After extracting dry leaves by refluxing with methanol, these chemicals were reported. The marc distillation product has fewer fatty parts than the leaves. So, this might have made the terpenoids and their derivatives in the marcs look like they were more than they really were. Linalool, (Z)- and (E)-linalool oxides, neral, geranial, isophorone, and pugelone were discovered. It is interesting that di-norterpenoids, such as edulane II and two spiro ethers, theaspiranes A and B, were only found in the leaves (Brun et al. 2001). Lawal et al. (2015) revealed that volatile compounds constituted 91.2%–99.2% of the total oil content of identified compounds in *C. roseus*. The pink flower *C. roseus* has mostly monoterpene hydrocarbons (39.0%), diterpenes (33.4%), and fat derivatives (15.3%). In leaf oil, fat derivatives (76.1%) are the most common type of chemical.

The pink flower *C. roseus* of leaf oil contained primarily linolenic acid ethyl ester (43.9%), stearic acid (10.6%), phytol (7.3%), and hexadecanoic acid (6.8%). Other important molecules were 2,6-bis (1,1-dimethylethyl)-2,5-cyclohexadiene-1,4-dione (6.2%), oleic acid ethyl ester (5.8%), and linoleic acid (5.6%). But the flower did have limonene (34.1%), phytol (29.6%), and linolenic acid ethyl ester (14.0%). The floral oil also has methyl eugenol (5.5%) and farnesyl acetone (5.1%) as important parts. There were clearly no sesquiterpene hydrocarbons in any of the pink flower *C. roseus* oils. The flower oil of the white flower *C. roseus* had a lot of monoterpene hydrocarbons (41.6%), oxygenated hydrocarbons (12.3%), and alkanes (22.3%). The main types of chemicals in the leaf oil include monoterpene hydrocarbons (23.2%), oxygenated hydrocarbons (30.2%), alkanes (10.5%), and fat derivatives (15.5%). Limonene (23.2%), dodecyl alcohol (9.8%), geraniol (7.3%), and citral (7.0%) were the primary parts of the leaf oil of white flower *C. roseus*. Limonene (37.2%) and dotriacontane (16.1%) were the most common parts of the floral oil (Lawal et al. 2015). The detected components accounted for 91.3%, whereas the composition of approximately 8.7% of the leaf oil remained uncharacterized. The leaf oil's main ingredients were (E,E)-2,4-hexadienal (7.7%), citronellol (7.9%), geraniol (7.9%), di-ter-butyl-p-cresol (4.7%), pentadecanal (6.6%), hexadecenoic acid (4.9%), palmitic acid (4.9%), and phytol (6.4%). These eight compounds made up 54.0% of the oil. The oil included about 1.5–3.5% of the following compounds: (E)-2-octenal (2.3%), (E,Z)-octadien-2-one (1.7%); neral (2.2%), geranial (1.6%), nonanoic acid (3.3%), (Z,Z)-farnesol (3.2%), (E,Z)-farnesol (1.8%), octadecanoic acid (1.5%), and tetracosane (1.7%). The remaining detected components were present in lesser amounts. The majority of the flower essential oil (99.4%) was made up of 41 known compounds. Heneicosane (20.8%) and tricosane (37.9%) were the two main parts of the oil, making up 58.7% of it. The other main parts were: 1.6% pentadecane, 1.9% heptadecane, 1.5% (Z,Z) farnesol, 2.0% octadecane, 2.8% nonadecane, 1.9% eicosane, 3.1% 9-octadecen-1-ol, 2.7% phytol, 4.3% docosane, 1.7% eicosane-1-ol, and 6.1% tetracosane. It is interesting to observe that the primary components (≥3.5%) of the leaf essential oils were distinct from those of the flower and leaf oils, even though the two oils shared

24 compounds. The two volatile oils have very varied amounts of different types of compounds: For leaf oil, terpenoids (35.9%) are more common than alcohols (20.3%), fatty acids and esters (18.2%), alkanes (6.6%), and ketones (3.2%). For floral oil, alkanes (82.9%) are more common than terpenoids (6.7%), alcohols (5.9%), fatty acids and esters (3.9%), and ketones (not found). Some other interesting things were that the two oils had palmitic and stearic acids, which made them slightly waxy, and that they contained a lot of phytol, with 2.7% in flower oil and 6.4% in leaf oil. Comparing the leaf oils of *C. roseus* from New Delhi (India) and France11 indicated that they were very different. The oil from France was said to have a lot of palmitic acid (64.9%), methyl palmitate (7.2%), and hexahydrofarnesyl acetone (4.0%). The oil from New Delhi in this investigation had low levels of palmitic acid (4.9%), and we did not find the molecules methyl palmitate and hexahydrofarnesyl acetone. The New Delhi oil had a lot of pentadecanal, citronellol, neral, and geraniol, while the French oil only had low levels of these substances. The French oil did not have the chemicals 2-hydroxy, 5-methoxy-p-cymene, hexadecenoic acid, and phytol that were found in the New Delhi oil. The oil from France included β-ionone, αionone, (E)-β-damascenone, edulane II, and theaspiranes A and B, whereas the oil from New Delhi did not. The discrepancies between the two oils could be because they used different types of *C. roseus* as raw material to extract the essential oil, because the plants were grown in distinct agroclimatic settings, or because the genotype–environment interactions were different. The French oil was made from dried leaves, whereas the New Delhi oil was made from fresh leaves. This might be another reason why the oils from the two countries are different (Pandey-Rai et al. 2006).

References

Abdullah ML, Hafez MM, Al-Hoshani A, Al-Shabanah O (2018) Anti-metastatic and anti-proliferative activity of eugenol against triple negative and HER2 positive breast cancer cells. BMC Complement Altern Med 18(1):321

Anzaku AA, Akyala JI, Juliet A, Obianuju EC (2017) Antibacterial activity of lauric acid on some selected clinical isolates. Ann Clin Lab Res 5:1–5

Attardi BJ, Hild SA, Reel JR (2006) Dimethandrolone undecanoate: a new potent orally active androgen with progestational activity. Endocrinology 147(6):3016–3026

Azodanlou R, Darbellay C, Luisier JL, Villettaz JC, Amadò R (1999) A new concept for the measurement of total volatile compounds of food. Z Lebensm Unters Forsch A 208(4):254–258

Bajo-Fernández M, Souza-Silva ÉA, Barbas C, Rey-Stolle MF, García A (2024) GC-MS-based metabolomics of volatile organic compounds in exhaled breath: applications in health and disease. Front Mol Biosci 10:1295955

Balamurugan R, Duraipandiyan V, Ignacimuthu S (2011) Antidiabetic activity of γ-sitosterol isolated from Lippia nodiflora L. in streptozotocin induced diabetic rats. Eur J Pharmacol 667(1–3):410–418

Barros LA, Yamanaka AR, Silva LE, Vanzeler MLA, Braum DT, Bonaldo J (2009) In vitro larvicidal activity of geraniol and citronellal against Contracaecum sp. (Nematoda: Anisakidae). Braz J Med Biol Res 42(10):918–920

Bhatia SP, McGinty D, Letizia CS, Api AM (2008) Fragrance material review on isopulegol. Food Chem Toxicol 46(11):S185–S189

Bonikowski R, Świtakowska P, Kula J (2015) Synthesis, odour evaluation and antimicrobial activity of some geranyl acetone and nerolidol analogues. Flavour Fragr J 30(3):238–244
Brun G, Bessière JM, Dijoux-Franca MG, David B, Mariotte AM (2001) Volatile components of Catharanthus roseus (L.) G. Don (Apocynaceae). Flavour Fragr J 16(2):116–119
Bruno F, Castelli G, Migliazzo A, Piazza M, Galante A, Verde VL, Calderone S, Nucatolo G, Vitale F (2015) Cytotoxic screening and in vitro evaluation of pentadecane against Leishmania infantum promastigotes and amastigotes. J Parasitol 101(6):701–706
Caboni P, Ntalli NG, Aissani N, Cavoski I, Angioni A (2012) Nematicidal activity of (E,E)-2,4-decadienal and (E)-2-decenal from Ailanthus altissima against Meloidogyne javanica. J Agric Food Chem 60(4):1146–1151
Chadeganipour M, Haims A (2001) Antifungal activities of pelargonic and capric acid on Microsporum gypseum. Mycoses 44(3–4):109–112
Chalchat JC, Chiron F, Garry RP, Lacoste J, Sautou V (2000) Photochemical hydroperoxidation of terpenes. Antimicrobial activity of α-pinene, β-pinene and limonene hydroperoxides. J Essent Oil Res 12(1):125–134
Chang KS, Ahn YJ (2002) Fumigant activity of (E)-anethole identified in Illicium verum fruit against Blattella germanica. Pest Manag Sci 58(2):161–166
Chen W, Viljoen AM (2010) Geraniol—a review of a commercially important fragrance material. S Afr J Bot 76(4):643–651
Davison C, Smith PK, Zimmerman EF (1961) On the metabolism and toxicity of methyl salicylate. J Pharmacol Exp Ther 132(2):207–211
Dayrit FM (2015) The properties of lauric acid and their significance in coconut oil. J Am Oil Chem Soc 92(1):1–15
Dilika F, Bremner PD, Meyer JJM (2000) Antibacterial activity of linoleic and oleic acids isolated from Helichrysum pedunculatum: a plant used during circumcision rites. Fitoterapia 71(4):450–452
dos Santos Negreiros P, da Costa DS, da Silva VG, de Carvalho Lima IB, Nunes DB, de Melo Sousa FB, Araújo TDSL, Medeiros JVR, Dos Santos RF, Oliveira RDCM (2019) Antidiarrheal activity of α-terpineol in mice. Biomed Pharmacother 110:631–640
Effah E, Holopainen JK, McCormick AC (2019) Potential roles of volatile organic compounds in plant competition. Perspect Plant Ecol Evol Syst 38:58–63
Erasto P, Viljoen AM (2008) Limonene—a review: biosynthetic, ecological and pharmacological relevance. Nat Prod Commun 3(7):1193–1202
Esfandyari-Manesh M, Ghaedi Z, Asemi M, Khanavi M, Manayi A, Jamalifar H, Atyabi F, Dinarvand R (2013) Study of antimicrobial activity of anethole and carvone loaded PLGA nanoparticles. J Pharm Res 7(4):290–295
Freire RS, Morais SM, Catunda-Junior FEA, Pinheiro DC (2005) Synthesis and antioxidant, anti-inflammatory and gastroprotector activities of anethole and related compounds. Bioorg Med Chem 13(13):4353–4358
Govindarajan M, Rajeswary M, Hoti SL, Benelli G (2016) Larvicidal potential of carvacrol and terpinen-4-ol from the essential oil of Origanum vulgare (Lamiaceae) against mosquito vectors. Res Vet Sci 104:77–82
Hadi MY, Hameed IH (2017) Uses of Gas Chromatography-Mass Spectrometry (GC-MS) technique for analysis of bioactive chemical compounds of Lepidium sativum: a review. Res J Pharm Technol 10(11):4039–4042
Harada H, Yamashita U, Kurihara H, Fukushi E, Kawabata J, Kamei Y (2002) Antitumor activity of palmitic acid found as a selective cytotoxic substance in a marine red alga. Anticancer Res 22(5):2587–2590
Hendry ER, Worthington T, Conway BR, Lambert PA (2009) Antimicrobial efficacy of eucalyptus oil and 1,8-cineole alone and in combination with chlorhexidine digluconate against microorganisms grown in planktonic and biofilm cultures. J Antimicrob Chemother 64(6):1219–1225
Henry GE, Momin RA, Nair MG, Dewitt DL (2002) Antioxidant and cyclooxygenase activities of fatty acids found in food. J Agric Food Chem 50(8):2231–2234

Herman A, Tambor K, Herman A (2016) Linalool affects the antimicrobial efficacy of essential oils. Curr Microbiol 72(2):165–172
Hosseinzadeh H, Noraei NB (2009) Anxiolytic and hypnotic effect of Crocus sativus aqueous extract and its constituents, crocin and safranal, in mice. Phytother Res 23(6):768–774
Hosseinzadeh H, Shariaty VM (2007) Anti-nociceptive effect of safranal, a constituent of Crocus sativus (saffron), in mice. Pharmacology 2:498–503
Hosseinzadeh H, Talebzadeh F (2005) Anticonvulsant evaluation of safranal and crocin from Crocus sativus in mice. Fitoterapia 76(7–8):722–724
Huang SW, Hopia A, Schwarz K, Frankel EN, German JB (1996) Antioxidant activity of α-tocopherol and trolox in different lipid substrates: bulk oils vs oil-in-water emulsions. J Agric Food Chem 44(2):444–452
Huang WC, Tsai TH, Chuang LT, Li YY, Zouboulis CC, Tsai PJ (2014) Anti-bacterial and anti-inflammatory properties of capric acid against Propionibacterium acnes: a comparative study with lauric acid. J Dermatol Sci 73(3):232–240
Islam MT, Ali ES, Uddin SJ, Shaw S, Islam MA, Ahmed MI, Shill MC, Karmakar UK, Yarla NS, Khan IN, Billah MM (2018) Phytol: a review of biomedical activities. Food Chem Toxicol 121:82–94
Ivanova EP, Nguyen SH, Guo Y, Baulin VA, Webb HK, Truong VK, Wandiyanto JV, Garvey CJ, Mahon PJ, Mainwaring DE, Crawford RJ (2017) Bactericidal activity of self-assembled palmitic and stearic fatty acid crystals on highly ordered pyrolytic graphite. Acta Biomater 59:148–157
Juteau F, Masotti V, Bessiere JM, Dherbomez M, Viano J (2002) Antibacterial and antioxidant activities of Artemisia annua essential oil. Fitoterapia 73(6):532–535
Kamatou GP, Viljoen AM (2008) Linalool – a review of a biologically active compound of commercial importance. Nat Prod Commun 3(7):1183–1192
Kamatou GP, Vermaak I, Viljoen AM (2012) Eugenol—from the remote Maluku Islands to the international marketplace: a review of a remarkable and versatile molecule. Molecules 17(6):6953–6981
Kanakis CD, Tarantilis PA, Tajmir-Riahi HA, Polissiou MG (2007) Crocetin, dimethylcrocetin, and safranal bind human serum albumin: stability and antioxidative properties. J Agric Food Chem 55(3):970–977
Kato N (1981) Antimicrobial activity of fatty acids and their esters against a film-forming yeast in soy sauce. J Food Saf 3(2):121–126
Kim SH, Bae HC, Park EJ, Lee CR, Kim BJ, Lee S, Park HH, Kim SJ, So I, Kim TW, Jeon JH (2011) Geraniol inhibits prostate cancer growth by targeting cell cycle and apoptosis pathways. Biochem Biophys Res Commun 407(1):129–134
Kochi M, Takeuchi S, Mizutani T, Mochizuki K, Matsumoto Y, Saito Y (1980) Antitumor activity of benzaldehyde. Cancer Treat Rep 64(1):21–23
Kovar KA, Gropper B, Friess D, Ammon HPT (1987) Blood levels of 1,8-cineole and locomotor activity of mice after inhalation and oral administration of rosemary oil. Planta Med 53(4):315–318
Lawal OA, Ogunwande IA, Ibirogba AE, Layode OM, Opoku AR (2015) Chemical constituents of essential oils from Catharanthus roseus (L.) G. Don grown in Nigeria. J Essent Oil Bear Plants 18(1):57–63
Li GX, Liu ZQ (2009) Unusual antioxidant behavior of α- and γ-terpinene in protecting methyl linoleate, DNA, and erythrocyte. J Agric Food Chem 57(9):3943–3948
Liao PC, Yang TS, Chou JC, Chen J, Lee SC, Kuo YH, Ho CL, Chao LKP (2015) Anti-inflammatory activity of neral and geranial isolated from fruits of Litsea cubeba lour. J Funct Foods 19:248–258
Liapi C, Anifantis G, Chinou I, Kourounakis AP, Theodosopoulos S, Galanopoulou P (2007) Antinociceptive properties of 1,8-cineole and β-pinene, from the essential oil of Eucalyptus camaldulensis leaves, in rodents. Planta Med 73(12):1247–1254
Lisec J, Schauer N, Kopka J, Willmitzer L, Fernie AR (2006) Gas chromatography mass spectrometry–based metabolite profiling in plants. Nat Protoc 1(1):387–396

Mahboubi M, Feizabadi MM (2009) Antimicrobial activity of Ducrosia anethifolia essential oil and main component, decanal against methicillin-resistant and methicillin-susceptible Staphylococcus aureus. J Essent Oil Bear Plants 12(5):574–579

Marinov V, Valcheva-Kuzmanova S (2015) Review on the pharmacological activities of anethole. Scr Sci Pharm 2(2):14–19

Morcia C, Malnati M, Terzi V (2012) In vitro antifungal activity of terpinen-4-ol, eugenol, carvone, 1,8-cineole (eucalyptol) and thymol against mycotoxigenic plant pathogens. Food Addit Contam: Part A 29(3):415–422

Murata S, Shiragami R, Kosugi C, Tezuka T, Yamazaki M, Hirano A, Yoshimura Y, Suzuki M, Shuto K, Ohkohchi N, Koda K (2013) Antitumor effect of 1,8-cineole against colon cancer. Oncol Rep 30(6):2647–2652

Nascimento AM, Brandao MG, Oliveira GB, Fortes IC, Chartone-Souza E (2007) Synergistic bactericidal activity of Eremanthus erythropappus oil or β-bisabolene with ampicillin against Staphylococcus aureus. Antonie Van Leeuwenhoek 92(1):95–100

Negro V, Mancini G, Ruggeri B, Fino D (2016) Citrus waste as feedstock for bio-based products recovery: review on limonene case study and energy valorization. Bioresour Technol 214:806–815

Nitbani FO, Siswanta D, Solikhah EN (2016) Isolation and antibacterial activity test of lauric acid from crude coconut oil (Cocos nucifera L.). Proc Chem 18:132–140

Ntalli N, Oplos C, Michailidis M, Thanasenaris A, Kontea D, Caboni P, Tsiropoulos NG, Menkissoglu-Spiroudi U, Adamski Z (2016) Strong synergistic activity and egg hatch inhibition by (E,E)-2,4-decadienal and (E)-2-decenal in Meloidogyne species. J Pest Sci 89(2):565–579

Nuñez YO, Salabarria IS, Collado IG, Hernández-Galán R (2006) The antifungal activity of widdrol and its biotransformation by Colletotrichum gloeosporioides (Penz.) Penz. & Sacc. and Botrytis cinerea Pers.: Fr. J Agric Food Chem 54(20):7517–7521

Obeng-Ofori D, Reichmuth CH, Bekele J, Hassanali A (1997) Biological activity of 1,8-cineole, a major component of essential oil of Ocimum kenyense (Ayobangira) against stored product beetles. J Appl Entomol 121(1–5):237–243

Oloyede GK (2016) Toxicity, antimicrobial and antioxidant activities of methyl salicylate dominated essential oils of Laportea aestuans (Gaud). Arab J Chem 9:840–845

Padmavathi AR, Abinaya B, Pandian SK (2014) Phenol, 2,4-bis (1,1-dimethylethyl) of marine bacterial origin inhibits quorum sensing mediated biofilm formation in the uropathogen Serratia marcescens. Biofouling 30(9):1111–1122

Pandey-Rai S, Mallavarapu GR, Naqvi AA, Yadav A, Rai SK, Srivastava S, Singh D, Mishra R, Kumar S (2006) Volatile components of leaves and flowers of periwinkle Catharanthus roseus (L.) G. Don from New Delhi. Flavour Fragr J 21(3):427–430

Park BG, Kwon SC, Park GM, Ham J, Shin WS, Lee S (2008) Vasodilatation effect of farnesylacetones, active constituents of Sargassum siliquastrum, on the basilar and carotid arteries of rabbits. Bioorg Med Chem Lett 18(24):6324–6326

Peluffo RO, Nervi AM, Brenner RR (1976) Linoleic acid desaturation activity of liver microsomes of essential fatty acid deficient and sufficient rats. Biochim Biophys Acta Lipids Lipid Metab 441(1):25–31

Pereira FDO, Mendes JM, Lima IO, Mota KSDL, Oliveira WAD, Lima EDO (2015) Antifungal activity of geraniol and citronellol, two monoterpene alcohols, against Trichophyton rubrum involves inhibition of ergosterol biosynthesis. Pharm Biol 53(2):228–234

Plumb RH Jr (1992) The importance of volatile organic compounds as a disposal site monitoring parameter. In: Groundwater contamination and analysis at hazardous waste sites. Marcel Dekker, New York, pp 173–197

Pongprayoon U, Baeckström P, Jacobsson U, Lindström M, Bohlin L (1992) Antispasmodic activity of β-damascenone and E-phytol isolated from Ipomoea pes-caprae. Planta Med 58(1):19–21

Raja MRC, Srinivasan V, Selvaraj S, Mahapatra SK (2015) Versatile and synergistic potential of eugenol: a review. Pharm Anal Acta 6(5):367

Ranjan S, Roy C, Sinha SK (2023) Gas chromatography–mass spectrometry (GC-MS): a comprehensive review of synergistic combinations and their applications in the past two decades. J Anal Sci Appl Biotechnol 5(2):72–85

Ravi L, Krishnan K (2017) Cytotoxic potential of N-hexadecanoic acid extracted from Kigelia pinnata leaves. Asian J Cell Biol 12(1):20–27

Rezaee R, Hosseinzadeh H (2013) Safranal: from an aromatic natural product to a rewarding pharmacological agent. Iran J Basic Med Sci 16(1):12–26

Řiháková Z, Plocková M, Filip V, Šmidrkal J (2001) Antifungal activity of lauric acid derivatives against Aspergillus niger. Eur Food Res Technol 213(6):488–490

Rimbach G, Moehring J, Huebbe P, Lodge JK (2010) Gene-regulatory activity of α-tocopherol. Molecules 15(3):1746–1761

Rufino AT, Ribeiro M, Judas F, Salgueiro L, Lopes MC, Cavaleiro C, Mendes AF (2014) Anti-inflammatory and chondroprotective activity of (+)-α-pinene: structural and enantiomeric selectivity. J Nat Prod 77(2):264–269

Saeed NM, El-Demerdash E, Abdel-Rahman HM, Algandaby MM, Al-Abbasi FA, Abdel-Naim AB (2012) Anti-inflammatory activity of methyl palmitate and ethyl palmitate in different experimental rat models. Toxicol Appl Pharmacol 264(1):84–93

Samarghandian S, Shoshtari ME, Sargolzaei J, Hossinimoghadam H, Farahzad JA (2014) Anti-tumor activity of safranal against neuroblastoma cells. Pharmacogn Mag 10(2):419–426

Santos FA, Rao VSN (2000) Antiinflammatory and antinociceptive effects of 1,8-cineole, a terpenoid oxide present in many plant essential oils. Phytother Res 14(4):240–244

Santos PL, Matos JPS, Picot L, Almeida JR, Quintans JS, Quintans-Junior LJ (2018) Citronellol, a monoterpene alcohol with promising pharmacological activities—a systematic review. Food Chem Toxicol 123:459–469

Sayyari M, Babalar M, Kalantari S, Martínez-Romero D, Guillén F, Serrano M, Valero D (2011) Vapour treatments with methyl salicylate or methyl jasmonate alleviated chilling injury and enhanced antioxidant potential during postharvest storage of pomegranates. Food Chem 124(3):964–970

Senra TDOS, Zeringóta V, de Oliveira Monteiro CM, Calmon F, Maturano R, Gomes GA, Faza A, de Carvalho MG, Daemon E (2013) Assessment of the acaricidal activity of carvacrol, (E)-cinnamaldehyde, trans-anethole, and linalool on larvae of Rhipicephalus microplus and Dermacentor nitens (Acari: Ixodidae). Parasitol Res 112(4):1461–1466

Seol GH, Kang P, Lee HS, Seol GH (2016) Antioxidant activity of linalool in patients with carpal tunnel syndrome. BMC Neurol 16(1):17

Shapira S, Pleban S, Kazanov D, Tirosh P, Arber N (2016) Terpinen-4-ol: a novel and promising therapeutic agent for human gastrointestinal cancers. PLoS One 11(6):e0156540

Sharma Y, Khan LA, Manzoor N (2016) Anti-Candida activity of geraniol involves disruption of cell membrane integrity and function. J Mycol Médicale 26(3):244–254

Shinohara Y, Unami A, Teshima M, Nishida H, van Dam K, Terada H (1995) Inhibitory effect of Mg^{2+} on the protonophoric activity of palmitic acid. Biochim Biophys Acta Bioenerg 1228(2–3):229–234

Silva MIG, Moura BA, de Aquino Neto MR, da Rocha Tomé A, Rocha NFM, de Carvalho AMR, Macêdo DS, Vasconcelos SMM, de Sousa DP, de Barros Viana GS, de Sousa FCF (2009) Gastroprotective activity of isopulegol on experimentally induced gastric lesions in mice: investigation of possible mechanisms of action. Naunyn Schmiedeberg's Arch Pharmacol 380(3):233–245

Sosa AA, Bagi SH, Hameed IH (2016) Analysis of bioactive chemical compounds of Euphorbia lathyrus using gas chromatography-mass spectrometry and Fourier-transform infrared spectroscopy. Int J Pharmacogn Phytochem Res 8(5):109–126

Sun XB, Wang SM, Li T, Yang YQ (2015) Anticancer activity of linalool terpenoid: apoptosis induction and cell cycle arrest in prostate cancer cells. Trop J Pharm Res 14(4):619–625

Tabari MA, Youssefi MR, Maggi F, Benelli G (2017) Toxic and repellent activity of selected monoterpenoids (thymol, carvacrol and linalool) against the castor bean tick, Ixodes ricinus (Acari: Ixodidae). Vet Parasitol 245:86–91

Thormar H, Hilmarsson H, Bergsson G (2006) Stable concentrated emulsions of the 1-monoglyceride of capric acid (monocaprin) with microbicidal activities against the food-borne bacteria Campylobacter jejuni, Salmonella spp., and Escherichia coli. Appl Environ Microbiol 72(1):522–526

Tripathi AK, Prajapati V, Aggarwal KK, Kumar S (2001) Toxicity, feeding deterrence, and effect of activity of 1,8-cineole from Artemisia annua on progeny production of Tribolium castanaeum (Coleoptera: Tenebrionidae). J Econ Entomol 94(4):979–983

Uebel EC, Sonnet PE, Miller RW (1976) House fly sex pheromone: enhancement of mating strike activity by combination of (Z)-9-tricosene with branched saturated hydrocarbons. Environ Entomol 5(5):905–908

Vuuren SV, Viljoen AM (2007) Antimicrobial activity of limonene enantiomers and 1,8-cineole alone and in combination. Flavour Fragr J 22(6):540–544

Walters DR, Walker RL, Walker KC (2003) Lauric acid exhibits antifungal activity against plant pathogenic fungi. J Phytopathol 151(4):228–230

Yang D, Pornpattananangkul D, Nakatsuji T, Chan M, Carson D, Huang CM, Zhang L (2009a) The antimicrobial activity of liposomal lauric acids against Propionibacterium acnes. Biomaterials 30(30):6035–6040

Yang M, Sun J, Lu Z, Chen G, Guan S, Liu X, Jiang B, Ye M, Guo DA (2009b) Phytochemical analysis of traditional Chinese medicine using liquid chromatography coupled with mass spectrometry. J Chromatogr A 1216(11):2045–2062

Yang Z, Wu N, Fu Y, Yang G, Wang W, Zu Y, Efferth T (2010) Anti-infectious bronchitis virus (IBV) activity of 1,8-cineole: effect on nucleocapsid (N) protein. J Biomol Struct Dyn 28(3):323–330

Zavala-Sanchez MA, Pérez-Gutiérrez S, Pérez-González C, Sanchez-Saldivar D, Arias-García L (2002) Antidiarrhoeal activity of nonanal, an aldehyde isolated from Artemisia ludoviciana. Pharm Biol 40(4):263–268

Zhang L, Wei G, Liu Y, Zu Y, Gai Q, Yang L (2016) Antihyperglycemic and antioxidant activities of total alkaloids from Catharanthus roseus in streptozotocin-induced diabetic rats. J For Res 27(1):167–174

Zhang D, Li X, Hu Y, Jiang H, Wu Y, Ding Y, Yu K, He H, Xu J, Sun L, Qian F (2018) Tabersonine attenuates lipopolysaccharide-induced acute lung injury via suppressing TRAF6 ubiquitination. Biochem Pharmacol 154:183–192

Zhou H, Tao N, Jia L (2014) Antifungal activity of citral, octanal and α-terpineol against Geotrichum citri-aurantii. Food Control 37:277–283

2D and 3D Structures of Bioactive Compounds of *Catharanthus*: Salient Physicochemical Properties

9

Abstract

The two-dimensional (2D) and three-dimensional (3D) structures of bioactive compounds from *Catharanthus*, such as vincristine, vinblastine, and ajmalicine, provide crucial insights into their pharmacological actions. These structures reveal functional groups, stereochemistry, and molecular conformations essential for biological activity. The key physicochemical properties include molecular weight, lipophilicity (log P), hydrogen bond donors/acceptors, and topological polar surface area (TPSA), which influence solubility, permeability, and drug-likeness. 3D modeling aids in understanding target binding interactions and optimizing drug design. Structure–activity relationship (SAR) and quantitative structure–activity relationship (QSAR) studies further highlight the role of spatial orientation and electronic distribution in modulating therapeutic efficacy and bioavailability of *Catharanthus*-derived compounds.

Keywords

2D · 3D · Hydrogen bond · *Catharanthus* · Bioactives

9.1 Introduction

Two-dimensional (2D) and three-dimensional (3D) structural representations are critical for understanding critical insights about the molecule. 2D structural representation provides information about atoms and their connectivity and is useful for preparing chemical databases, such as compound libraries for structural elucidation. By looking at the structure, we can guess how polar, soluble, and reactive a compound is, as well as use quantitative structure–activity relationship (QSAR) modeling. Stereochemistry is understood by 3D structural representation. It is critical to have a visual representation of molecular interactions with biological targets (e.g.,

S. Kumar, B. Singh, *Catharanthus in Drug Discovery*,
https://doi.org/10.1007/978-981-95-5048-7_9

receptor binding, enzyme inhibition). SBDD and LBDD in 3D provide docking studies, pharmacophore modeling, and non-covalent interactions (e.g., hydrogen bonds, van der Waals forces, and π–π stacking). In this chapter, we covered 2D and 3D structures and the features reported in *Catharanthus* species (Putz et al. 2016; Bahia et al. 2023).

9.2 Classification of Bioactive Compounds

Bioactive phytochemicals are naturally occurring compounds from medicinal plants that alter living organisms' physiological and biological processes. Despite their small or trace quantities, they significantly influence the biological process. Table 9.1 reports the class, subclass, and examples of phytochemicals. The main compounds found in *Catharanthus* species are phenolic compounds (polyphenols) and alkaloids, especially monoterpene indole. *C. roseus* is the most studied species of this genus, and different phytochemicals reported from this species are presented in Table 9.2.

9.3 Monoterpene Indole Alkaloids

MIAs are a subclass of diverse and large naturally occurring alkaloids, and they are bonded (fused) together with indole and monoterpene ($C_{10}H_{16}$). MIAs are isolated from Apocynaceae, Loganiaceae, and Rubiaceae families. A total of 103 MIAs (**1–103**) with their 2D and 3D structures are summarized in Fig. 9.1a–g. Most of the compounds were isolated from *C. roseus*. MIAs consist of a complex tetracyclic and pentacyclic frameworks. Their 2D and 3D structures often show more than five stereogenic centers. The fused rings and steric hindrance make the 3D structures more rigid with nonplanar conformations. A rigid and nonplanar conformation makes it more specific and suitable for bonding with a target acceptor.

9.4 Bisindole Alkaloids

Two MIAs unites, such as catharanthine, vindoline, or tabersonine, either same or different in nature, are covalently bonded through C–C, C–N, and C–O linkages to form bisindole alkaloids. In *Catharanthus* species, C–C bridges are the most common. These structures show high 3D complexity due to fused ring systems. Bisindole alkaloids have four nitrogen atoms, which make them susceptible for interaction with protein receptors. A total of 35 bisindole alkaloids (**108–142**) are summarized in Fig. 9.2a–c.

Table 9.1 Classification and key bonding characteristics of phytochemicals reported from *Catharanthus* species

S. no.	Class of compounds	Subclass	Key bonding feature	Examples
1.	Phenolic compounds (polyphenols)	Flavonoids	C6–C3–C6 backbone (two aromatic rings linked by a 3-carbon bridge)	Flavone–Apigenin and luteolin Flavanone–quercetin and kaempferol Flavanone–Naringenin and hesperidin Flavanol–Catechin Anthocyanin–Cyanidin and delphinidin Isoflavone–Genistein and daidzein
		Phenolic acids	One aromatic ring with hydroxyl and carboxylic groups	Caffeic acid and gallic acid
		Tannins	C–C bonds (4 → 8, 4 → 6), ester bonds	Tannic acid and epigallocatechin gallate
		Lignans	Two aromatic rings connected by a 2-carbon ethylene bridge (C6–C2–C6)	Secoisolariciresinol and podophyllotoxin
		Stilbenes		Resveratrol
2.	Alkaloids	Pyridine and piperidine	Six-membered nitrogen-containing rings	Nicotine and coniine
		Tropane	Bicyclic structure with nitrogen	Atropine, scopolamine, and cocaine
		Quinoline	Benzene fused with pyridine	Quinine and quinidine
		Isoquinoline	Benzene fused with pyridine at different position	Morphine, papaverine, and berberine
		Indole	Benzene fused with a pyrrole ring	Reserpine, vinblastine, and psilocybin
		Imidazole	Five-membered ring with two nitrogens	Pilocarpine and histamine
		Purine	Fused imidazole and pyrimidine rings	Caffeine and theobromine
		Steroidal alkaloids	Steroid backbone with nitrogen	Solanine and tomatine
		Terpenoid alkaloids	Terpene-derived skeletons	Aconitine, taxol

(continued)

Table 9.1 (continued)

S. no.	Class of compounds	Subclass	Key bonding feature	Examples
3.	Terpenoids (isoprenoids)	Monoterpenes	Head-to-tail (1′ → 4) linkage of isoprene units	Limonene
		Sesquiterpenes	Head-to-tail isoprene linkages	Farnesol and humulene
		Diterpenes	Head-to-tail linkages four isoprene units	Taxol
4.		Triterpenes	Six isoprene unit bonded C–C bond at C4–C4′	Lupeol, ursolic acid, oleanolic acid, and betulinic acid
		Carotenoids		β-Carotene and lycopene
5.	Steroids and phytosterols		Four fused rings (C6–C6–C6–C5)	β-Sitosterol and stigmasterol

9.5 Phenolics

Phenolics and polyphenols are a large and diverse class of secondary metabolites. All phenolic compounds have one or more aromatic rings with a hydroxyl group (OH). Their classification and characteristic linkages are described in Table 4.1. A total of 32 flavones (143–175), including flavonols (**143–159**), flavanones (**160**, **161**), and anthocyanidins and anthocyanins (**163–174**). Phenolic acids (**175–183**), are also reported in *Catharanthus* species (Fig. 9.3a–c).

9.6 Volatile Organic Compounds (VOCs)

A total of 157 volatile compounds, including alkanes (**184–198**), alkenes (**199–215**), aldehydes (**216–242**), ketones (**243–265**), alcohols (266–292), ethers (**293–295**, **335–339**), alcohol-ester (**296**), volatile triterpenes (**297–299**), carboxylic acids (**300–314**), and esters (**315–334**), are reported in *Catharanthus* species (Fig. 9.4a–f).

Table 9.2 Various classes of phytochemical reported from *C. roseus*

S. no.	Alkaloids Sub-class	Compound name	Mol. weight	Mol. formula
1.	MIAs	Vincadifformine	338.44	$C_{21}H_{26}N_2O_2$
2.		Minovincinine	354.40	$C_{21}H_{26}N_2O_3$
3.		Echitoveniline	548.64	$C_{31}H_{36}N_2O_7$
4.		Echitovenidine	436.54	$C_{26}H_{32}N_2O_4$
5.		Echitovenine	396.48	$C_{23}H_{28}N_2O_4$
6.		Tabersonine	336.43	$C_{21}H_{24}N_2O_2$
7.		11-Hydroxytabersonine	352.43	$C_{21}H_{24}N_2O_3$
8.		11-Methoxytabersonine	366.46	$C_{22}H_{26}N_2O_3$
9.		Nor 4-hydroxytabersonine	336.43	$C_{21}H_{24}N_2O_2$
10.		16-Methoxytabersonine	366.46	$C_{22}H_{26}N_2O_3$
11.		16-Hydroxytabersonine	336.43	$C_{21}H_{24}N_2O_3$
12.		19-Hydroxy-11-methoxytabersonine	382.46	$C_{22}H_{26}N_2O_4$
13.		19-Acetoxy-11-hydroxytabersonine	382.46	$C_{22}H_{26}N_2O_4$
14.		19-Acetoxy-11-methoxytabersonine	382.46	$C_{22}H_{26}N_2O_4$
15.		Minovincine	382.46	$C_{22}H_{26}N_2O_4$
16.		Lochnericine	352.43	$C_{21}H_{24}N_2O_3$
17.		19(*S*)-Epimisiline	368.42	$C_{21}H_{24}N_2O_4$
18.		Lochnerinine	382.46	$C_{22}H_{26}N_2O_4$
19.		Horhammericine	368.17	$C_{21}H_{24}N_2O_4$
20.		Horhammerinine	368.17	$C_{21}H_{24}N_2O_4$
21.		Lochneridine	382.19	$C_{22}H_{26}N_2O_4$
22.		20-*epi*-Lochneridine	382.19	$C_{22}H_{26}N_2O_4$
23.		Akuammicine	322.40	$C_{20}H_{22}N_2O_2$
24.		Vinervine	338.41	$C_{20}H_{22}N_2O_3$
25.		Cathaphylline	368.42	$C_{21}H_{24}N_2O_4$
26.		Tubotaiwine (dihydrocondylocarpine)	324.42	$C_{20}H_{24}N_2O_2$
27.		Vindoline	456.53	$C_{25}H_{32}N_2O_6$
28.		Vindorosine	426.5	$C_{24}H_{30}N_2O_5$
29.		Deacetylvindoline	414.50	$C_{23}H_{30}N_2O_5$
30.		4-*O*-Deacetyl-N,-demethylvindolin	400.47	$C_{22}H_{28}N_2O_5$
31.		4-*O*-Deacetylvindorosine	426.50	$C_{24}H_{30}N_2O_5$
32.		N,-Demethyldeacetylvindorosine	400.47	$C_{22}H_{28}N_2O_5$
33.		16-*O*-Demethylvindoline	428.52	$C_{24}H_{32}N_2O_5$
34.		16-Methoxy-2,3-dihydro-3-hydroxytabersonine	384.47	$C_{22}H_{28}N_2O_4$
35.		Bannucine	539.63	$C_{29}H_{34}N_4O_6$
36.		Vindolidine	426.51	$C_{24}H_{30}N_2O_5$
37.		Vincoline	368.43	$C_{21}H_{24}N_2O_4$
38.		Kitraline	382.45	$C_{22}H_{26}N_2O_4$
39.		Dihydrovindoline	458.50	$C_{25}H_{34}N_2O_6$
40.		Venalstonine	336.40	$C_{21}H_{24}N_2O_2$

(continued)

Table 9.2 (continued)

S. no.	Alkaloids Sub-class	Compound name	Mol. weight	Mol. formula
41.		Epivindolinine	336.42	$C_{21}H_{24}N_2O_2$
42.		5(*S*)-Hydroxy-14,15-dihydrovindolinine	354.44	$C_{21}H_{26}N_2O_3$
43.		Vindolinine-N_b-oxide (9-Epivindolinine-N_b-oxide)	352.42	$C_{21}H_{24}N_2O_3$
44.		Venalstonidine	352.40	$C_{21}H_{24}N_2O_3$
45.		11-Methoxy-2,16-di-hydrotabersonine	368.46	$C_{22}H_{28}N_2O_3$
46.		2,3-Dihydrotabersonine	338.40	$C_{21}H_{26}N_2O_2$
47.		Cathovaline	426.50	$C_{24}H_{30}N_2O_5$
48.		Deacetylcathovaline	384.46	$C_{22}H_{28}N_2O_4$
49.		Hydroxycathovaline	442.50	$C_{24}H_{30}N_2O_6$
50.		Vinpocetine	350.50	$C_{22}H_{26}N_2O_2$
51.		Vincamine	338.44	$C_{21}H_{26}N_2O_2$
52.		Trichophylline	350.45	$C_{22}H_{26}N_2O_2$
53.		Catharanthine	354.44	$C_{21}H_{26}N_2O_3$
54.		Vincarodine	366.41	$C_{21}H_{22}N_2O_4$
55.		Pleiocarpamine	322.40	$C_{20}H_{22}N_2O_2$
56.		Coronaridine	338.40	$C_{21}H_{26}N_2O_2$
57.		Stemmadenine	354.44	$C_{21}H_{26}N_2O_3$
58.		Dehydrosecodine	336.40	$C_{21}H_{24}N_2O_2$
59.		Akuammiline	394.50	$C_{23}H_{26}N_2O_4$
60.		10-Hydroxy-deacetylakuammiline	310.39	$C_{19}H_{22}N_2O_2$
61.		Desacetylakuammiline	324.41	$C_{20}H_{24}N_2O_2$
62.		10-*O*-methylsarpagine (Lochnerine)	324.40	$C_{20}H_{24}N_2O_2$
63.		Sarpagine	310.40	$C_{19}H_{22}N_2O_2$
64.		Perivine	338.40	$C_{20}H_{22}N_2O_3$
65.		Pericyclivine	322.40	$C_{20}H_{22}N_2O_2$
66.		Vellosiminol	294.40	$C_{19}H_{22}N_2O$
67.		21-Hydroxycyclolochnerine	340.41	$C_{20}H_{24}N_2O_3$
68.		10-Hydroxydeformodihydropseudoakuammigine 10-*O*-*L*-arabinopyranoside	516.58	$C_{27}H_{36}N_2O_8$
69.		10-Hydroxycathafoline 10-*O*-larabinopyranoside	516.58	$C_{27}H_{36}N_2O_8$
70.		Ajmaline	326.40	$C_{20}H_{26}N_2O_2$
71.		Norseredamine	324.40	$C_{20}H_{24}N_2O_2$
72.		Tetraphyllicine	308.40	$C_{20}H_{24}N_2O$
73.		10-Hydroxynortetraphyllicine	324.41	$C_{20}H_{24}N_2O$
74.		Pseudoyohimbine	354.40	$C_{21}H_{26}N_2O_3$
75.		Yohimbine	354.40	$C_{21}H_{26}N_2O_3$
76.		Reserpine	608.70	$C_{33}H_{40}N_2O_9$
77.		3,4,5,6-tetradehydroyohimbine	351.41	$C_{21}H_{23}N_2O_3^+$
78.		Ajmalicine	352.40	$C_{21}H_{24}$N2O3
79.		3-Isoajmalicine	352.40	$C_{21}H_{24}N_2O_3$
80.		Akuammigine	352.40	$C_{21}H_{24}N_2O_3$
81.		Tetrahydroalstonine	352.40	$C_{21}H_{24}N_2O_3$
82.		19-Epiajmalicine	352.40	$C_{21}H_{24}N_2O_3$

(continued)

Table 9.2 (continued)

S. no.	Alkaloids Sub-class	Compound name	Mol. weight	Mol. formula
83.		3-Iso-19-epi-ajmalicine	352.42	$C_{21}H_{24}N_2O_3$
84.		3-Epiajmalicine	352.42	$C_{21}H_{24}N_2O_3$
85.		Reserpiline	412.50	$C_{23}H_{28}N_2O_5$
86.		7-Hydroxyindolenineajmalicine	368.42	$C_{21}H_{24}N_2O_4$
87.		Alstonine	348.40	$C_{21}H_{20}N_2O_3$
88.		Cathenamine	350.40	$C_{21}H_{22}N_2O_3$
89.		Pseudoindoxyl ajmalicine	368.42	$C_{21}H_{24}N_2O_4$
90.		Strictosidine lactam	498.50	$C_{26}H_{30}N_2O_8$
91.		Mitraphylline	368.40	$C_{21}H_{24}N_2O_4$
92.		Serpentine	349.40	$C_{21}H_{21}N_2O_3^+$
93.		Vincoside	498.50	$C_{26}H_{30}N_2O_8$
94.		Dihydrositsirikine	356.50	$C_{21}H_{28}N_2O_3$
95.		Sịtsirikine	354.40	$C_{21}H_{26}N_2O_3$
96.		Isositsirikine (16*R*,19*E*)	354.40	$C_{21}H_{26}N_2O_3$
97.		Isositsirikine (16*R*,19*Z*)	354.40	$C_{21}H_{26}N_2O_3$
98.		*O*-Acetylstemmadenin (strychnos)	334.40	$C_{21}H_{22}N_2O_2$
99.		Vallesiachotamine	350.40	$C_{21}H_{22}N_2O_3$
100.		17-Isovallesiachotamine	350.40	$C_{21}H_{22}N_2O_3$
101.		Geissoschizine	352.40	$C_{21}H_{24}N_2O_3$
102.		Strictosidine	530.60	$C_{27}H_{34}N_2O_9$
103.		Anthirine	296.40	$C_{19}H_{24}N_2O$
104.		*β*-Carboline	168.19	$C_{11}H_8N_2$
105.		*n*,*n*-dimethyltryptamine	188.27	$C_{12}H_{16}N_2$
106.		Tryptamine,N_b-acetyl	202.25	$C_{12}H_{14}N_2O$
107.		Iochrovicine	285.33	$C_{17}H_{19}NO_3$
108.		Vinblastine	811.00	$C_{46}H_{58}N_4O_9$
109.		Vincristine	825.00	$C_{46}H_{56}N_4O_{10}$
110.		4′-Deoxyvinblastine	795.00	$C_{46}H_{58}N_4O_8$
111.		Demethylvinblastine	796.90	$C_{45}H_{56}N_4O_9$
112.		Deoxyvincaleukoblastine B	795.00	$C_{46}H_{58}N_4O_8$
113.		Pseudovinblastinediol	768.93	$C_{44}H_{56}N_4O_8$
114.		3′,4′-Anhydrovinblastine	793.00	$C_{46}H_{56}N_4O_8$
115.		Deacetoxyvinblastine (Desacetoxyvinblastine)	752.90	$C_{44}H_{56}N_4O_7$
116.		17-Deacetoxyvinblastine $N_b^?$ -oxide	769.94	$C_{44}H_{57}N_4O_8^+$
117.		20′-Deoxyvinblastine $N_b^?$ -oxide	827.98	$C_{46}H_{59}N_4O_{10}^+$
118.		Vindesine	753.90	$C_{43}H_{55}N_5O_7$
119.		Catharine	822.90	$C_{46}H_{54}N_4O_{10}$
120.		Leurosidine	811.00	$C_{46}H_{58}$N4O9
121.		Leurosidine-*N*-oxide	842.97	$C_{46}H_{58}N_4O_{11}^+$
122.		Leurosine	810.98	$C_{46}H_{56}N_4O_9$

(continued)

Table 9.2 (continued)

S. no.	Alkaloids Sub-class	Compound name	Mol. weight	Mol. formula
123.		Leurosine-N_b-oxide (Pleurosine)	824.97	$C_{46}H_{56}N_4O_{10}$
124.		Isoleurosine	795.00	$C_{46}H_{58}N_4O_8$
125.		Deacetoxyleurosine (Desacetoxyleurosine)	750.92	$C_{44}H_{54}N_4O_7$
126.		5′-Oxoleurosine	822.90	$C_{46}H_{54}N_4O_{10}$
127.		19′-Oxoleurosine	822.90	$C_{46}H_{54}N_4O_{10}$
128.		Vinleurosine	796.97	$C_{46}H_{56}N_4O_9$
129.		Leurocolornbine	826.97	$C_{46}H_{58}N_4O_{10}$
130.		Roseadine	809.00	$C_{46}H_{56}N_4O_9$
131.		Vincathicine	809.00	$C_{46}H_{56}N_4O_9$
132.		Catharinine (Vinamidine)	812.98	$C_{46}H_{56}N_4O_{10}$
133.		Vincovaline	810.97	$C_{46}H_{58}N_4O_9$
134.		Vincovalinine	750.92	$C_{44}H_{54}N_4O_7$
135.		Vincovalicine	825.96	$C_{46}H_{57}N_4O_{10}$
136.		Vincadioline	770.95	$C_{44}H_{58}N_4O_8$
137.		Vindolicine	924.08	$C_{52}H_{65}N_3O_{12}$
138.		Catharanthamine	809.00	$C_{46}H_{56}N_4O_9$
139.		Vingramine	792.95	$C_{46}H_{56}N_4O_8$
140.		Vinorelbine	778.90	$C_{45}H_{54}N_4O_8$
141.		Vinrosidine	810.98	$C_{46}H_{58}N_4O_9$
142.		Methylvingramine	806.98	$C_{47}H_{58}N_4O_8$
	Flavonoid			
143.		Quercetin	302.23	$C_{15}H_{10}O_7$
144.		Quercetin-3-*O*-arabinoside	434.30	$C_{20}H_{18}O_{11}$
145.		Rutin	610.50	$C_{27}H_{30}O_{16}$
146.		Quercetin-3-*O*-(2,6-di-*O*-rhamnosyl) galactoside	756.65	$C_{33}H_{40}O_{20}$
147.		Quercetin-3-*O*-(2,6-di-*O*-rhamnosyl)glucoside	756.65	$C_{33}H_{40}O_{20}$
148.		Quercetin-3-*O*-(6-*O*-rhamnosyl)galactoside	610.51	$C_{27}H_{30}O_{16}$
149.		Kaempferol	286.24	$C_{15}H_{10}O_6$
150.		Kaempferol-3-*O*-(6-*O*-rhamnosyl) glucoside	594.52	$C_{27}H_{30}O_{15}$
151.		Kaempferol-3-*O*-(2,6-di-*O*-rhamnosyl) galactoside	740.66	$C_{33}H_{40}O_{19}$
152.		Kaempferol-3-*O*-(2,6-di-*O*-rhamnosyl-glucoside	756.66	$C_{33}H_{40}O_{20}$
153.		Isorhamnetin-3-*O*-(2,6-di-*O*-rhamnosyl) galactoside	770.70	$C_{34}H_{42}O_{20}$
154.		Isorhamnetin-3-*O*-(2,6-di-*O*-rhamnosyl) glucoside	786.70	$C_{34}H_{42}O_{21}$
155.		Isorhamnetin-3-*O*-(6-*O*-rhamnosyl) glucoside	624.54	$C_{28}H_{32}O_{16}$
156.		Isorhamnetin-3-*O*-(6-*O*-rhamnosyl) galactoside	624.54	$C_{28}H_{32}O_{16}$
157.		Syringetin-3-*O*-robinobioside	654.57	$C_{29}H_{34}O_{17}$

(continued)

Table 9.2 (continued)

S. no.	Alkaloids Sub-class	Compound name	Mol. weight	Mol. formula
158.		3′,4′-di-*O*-methylquercetin-7-*O*-[(4″ → 3‴)-2‴,6‴,10‴,14‴-tetramethylhexadec-13‴-ol-14‴-enyl]-*β*-*D*-glucopyranoside	756.92	$C_{42}H_{60}O_{12}$
159.		4′-*O*-methylkaempferol-3-*O*-[(4″ → 3‴)-2‴,6‴,10‴,14‴-tetramethylhexadecan-13‴-olyl]-β-D-glucopyranoside	728.91	$C_{41}H_{60}O_{11}$
160.		3′,4′-di-*O*-methylbutin-7-*O*-[(6″ → 1‴)-3‴,11‴-dimethyl-7‴-methylenedodeca-3‴,10‴-dienyl]-*β*-*D*-glucopyranoside	680.78	$C_{38}H_{48}O_{11}$
161.		4′-*O*-methylbutin-7-*O*-[(6″ → 1‴)-3‴,11‴-dimethyl-7‴-hydroxymethylenedo-decanyl]-*β*-D-glucopyranoside	688.80	$C_{37}H_{52}O_{12}$
162.		Pelargonidin	271.24	$C_{15}H_{11}O_5^+$
163.		Cyanidin	287.24	$C_{15}H_{11}O_6^+$
164.		Peonidin	301.72	$C_{16}H_{13}O_6$
165.		Malvidin	331.30	$C_{17}H_{15}O_7^+$
166.		Petunidin	352.72	$C_{16}H_{13}ClO_7$
167.		Hirsutidin	345.30	$C_{18}H_{17}O_7^+$
168.		Delphinidin	338.69	$C_{15}H_{11}ClO_7$
169.		Petunidin-3-*O*-glucoside	479.11	$C_{22}H_{23}O_{12}$
170.		Malvidin-3-*O*-glucoside	493.43	$C_{23}H_{25}O_{12}^+$
171.		Hirsutidin-3-*O*-glucoside	507.46	$C_{24}H_{27}O_{12}^+$
172.		Petunidin-3-*O*-(6-*O*-*p*-coumaroyl) glucoside	625.55	$C_{31}H_{29}O_{14}^+$
173.		Hirsutidin-3-*O*-(6-*O*-*p*-coumaroyl) glucoside	653.60	$C_{33}H_{33}O_{14}^+$
174.		Malvidin-3-*O*-(6-*O*-*p*-coumaroyl) glucoside	639.57	$C_{32}H_{31}O_{14}^+$
	Phenolic acid			
175.		Gallic acid	170.12	$C_7H_6O_5$
176.		Ferulic acid	194.18	$C_{10}H_{10}O_4$
177.		Caffeic acid	180.16	$C_9H_8O_4$
178.		Benzoic acid	122.12	$C_7H_6O_2$
179.		Vanillic acid	168.15	$C_8H_8O_4$
180.		Coniferyl aldehyde	178.18	$C_{10}H_{10}O_3$
181.		4-*O*-Caffeoylquinic acid	354.31	$C_{16}H_{18}O_9$
182.		5-*O*-Caffeoylquinic acid	354.31	$C_{16}H_{18}O_9$
183.		3-*O*-Caffeoylquinic acid	354.31	$C_{16}H1_8O_9$
	Volatildeile compound			
184.		*n*-Dodecane	170.33	$C_{12}H_{26}$
185.		*n*-Tridecane	184.36	$C_{13}H_{28}$
186.		*n*-Tetradecane	198.39	$C_{14}H_{30}$

(continued)

Table 9.2 (continued)

S. no.	Alkaloids Sub-class	Compound name	Mol. weight	Mol. formula
187.		*n*-Pentadecane	212.41	$C_{15}H_{32}$
188.		Heptadecane	240.47	$C_{17}H_{36}$
189.		Octadecane	254.50	$C_{18}H_{38}$
190.		Nonadecane	268.53	$C_{19}H_{40}$
191.		Eicosane	282.55	$C_{20}H_{42}$
192.		Heneicosane	296.57	$C_{21}H_{44}$
193.		2,6,11-Trimethyl dodecane	212.41	$C_{15}H_{32}$
194.		Docosane	310.60	$C_{22}H_{46}$
195.		Tricosane	324.62	$C_{23}H_{48}$
196.		Tetracosane	338.65	$C_{24}H_{50}$
197.		Dotriacontane	450.88	$C_{32}H_{66}$
198.		Tetracontane	563.09	$C_{40}H_{82}$
199.		*n*-Hexadecene	224.43	$C_{16}H_{32}$
200.		Heneicosene	294.56	$C_{21}H_{42}$
201.		Docosene	308.59	$C_{22}H_{44}$
202.		(*Z*)-9-Tricosene	322.61	$C_{23}H_{46}$
203.		(*Z*)-*β*-Ocimene	136.23	$C_{10}H_{16}$
204.		Myrcene	136.23	$C_{10}H_{16}$
205.		(*E*,*E*)-*α*-Farnesene	204.35	$C_{15}H_{24}$
206.		3-Phenyldodecane	246.43	$C_{18}H_{30}$
207.		Limonene	136.23	$C_{10}H_{16}$
208.		*β*-Phellandrene	136.23	$C_{10}H_{16}$
209.		*γ*-Terpinene	136.23	$C_{10}H_{16}$
210.		*β*-Bisabolene	204.35	$C_{15}H_{24}$
211.		Germacrene D	204.35	$C_{15}H_{24}$
212.		*γ*-Cadinene	204.35	$C_{15}H_{24}$
213.		*Trans*-*α*-Bergamotene	204.35	$C_{15}H_{24}$
214.		*α*-Pinene	136.23	$C_{10}H_{16}$
215.		*β*-Pinene	136.23	$C_{10}H_{16}$
216.		Heptanal	114.19	$C_7H_{14}O$
217.		Octanal	128.21	$C_8H_{16}O$
218.		Nonanal	142.24	$C_9H_{18}O$
219.		Decanal	156.26	$C_{10}H_{20}O$
220.		Undecanal	170.29	$C_{11}H_{22}O$
221.		Dodecanal	184.32	$C_{12}H_{24}O$
222.		Tetradecanal	212.41	$C_{14}H_{28}O$
223.		Pentadecanal	226.45	$C_{15}H_{30}O$
224.		Hexadecanal	240.47	$C_{16}H_{32}O$
225.		(*E*)-2-Octenal	126.20	$C_8H_{14}O$
226.		(*E*)-2-Nonenal	140.22	$C_9H_{16}O$
227.		(*E*)-2-Decenal	154.25	$C_{10}H_{18}O$

(continued)

Table 9.2 (continued)

S. no.	Alkaloids Sub-class	Compound name	Mol. weight	Mol. formula
228.		(*E*)-2-Undecenal	168.27	$C_{11}H_{20}O$
229.		(*E*,*E*)-2,4-Hexadienal	96.13	C_6H_8O
230.		(*E*,*E*)-2,4-Heptadienal	110.16	$C_7H_{10}O$
231.		(*E*,*E*)-2,4-Decadienal	152.23	$C_{10}H_{16}O$
232.		(*E*,*E*)-2,4-Dodecadienal	180.29	$C_{12}H_{20}O$
233.		(*E*,*Z*)-2,4-Decadienal	152.23	$C_{10}H_{16}O$
234.		(*E*, *Z*)-2,6-Nonadienal	138.20	$C_9H_{14}O$
235.		Neral	152.23	$C_{10}H_{16}O$
236.		6,6-Trimethylcyclohexen-1-acetaldehyde	166.26	$C_{11}H_{18}O$
237.		Perilla aldehyde	150.22	$C_{10}H_{14}O$
238.		*β*-Cyclocitra	152.23	$C_{10}H_{16}O$
239.		Safranal	150.22	$C_{10}H_{14}O$
240.		Benzaldehyde	106.12	C_7H_6O
241.		4-Methylbenzaldehyde	120.15	C_8H_8O
242.		Ethylbenzaldehyde	134.17	$C_9H_{10}O$
243.		2-methyl 4-heptanone	128.21	$C_8H_{16}O$
244.		Pentadecane-2-one	226.40	$C_{15}H_{30}O$
245.		Hexahydrofarnesylacetone	268.48	$C_{18}H_{36}O$
246.		Heptenone	112.17	$C_7H_{12}O$
247.		3-Octen-2-one	126.20	$C_8H_{14}O$
248.		Isophorone	138.21	$C_9H_{14}O$
249.		*β*-Pulegone	152.23	$C_{10}H_{16}O$
250.		(*E*)-Ionone-5,6-epoxide	208.30	$C_{13}H_{20}O_2$
251.		(*E*)-*α*-ionone	192.30	$C_{13}H_{20}O$
252.		(*E*,*E*)-Octa-3,5-dien-2-one	124.18	$C_8H_{12}O$
253.		(E,Z)-Octa-3,5-dien-2-one	124.18	$C_8H_{12}O$
254.		6-Methyl-3,5-heptadien-2-one	124.18	$C_8H_{12}O$
255.		(*E*)-*β*-ionone	192.30	$C_{13}H_{20}O$
256.		(*E*,*E*)-Pseudoionone	192.30	$C_{13}H_{20}O$
257.		Megastigmatrienone	190.28	$C_{13}H_{18}O$
258.		(*E*)-Damascenone	190.28	$C1_3H_{18}O$
259.		2,6-bis(1,1-dimethylethyl)-2,5-Cyclohexadiene-1,4-dione	220.31	$C_{14}H_{20}O_2$
260.		6-Methyl hept-5-ene-2-one	126.20	$C_8H_{14}O$
261.		2,2,6-Trimethylcyclohexanone	140.23	$C_9H_{16}O$
262.		6-Methyl-5-hepten-2-one	126.20	$C_8H_{14}O$
263.		Geranylacetone	194.31	$C_{13}H_{22}O$
264.		Farnesylacetone	262.44	$C_{18}H_{30}O$
265.		Davanone	236.35	$C_{15}H_{24}O_2$
266.		Dodecyl alcohol	186.34	$C_{12}H_{26}O$
267.		Hexadecanol	242.44	$C_{16}H_{34}O$
268.		Eicosan-1-ol	298.55	$C_{20}H_{42}O$

(continued)

Table 9.2 (continued)

S. no.	Alkaloids Sub-class	Compound name	Mol. weight	Mol. formula
269.		Phytol	296.54	$C_{20}H_{40}O$
270.		Citronellol	156.27	$C_{10}H_{20}O$
271.		9-Octadecen-1-ol	268.48	$C_{18}H_{36}O$
272.		Nerol	154.25	$C_{10}H_{18}O$
273.		Geraniol	154.25	$C_{10}H_{18}O$
274.		(*E*,*E*)-Farnesol	222.37	$C_{15}H_{26}O$
275.		(*Z*,*Z*)-Farnesol	222.37	$C_{15}H_{26}O$
276.		(*E*,*Z*)-Farnesol	222.37	$C_{15}H_{26}O$
277.		*Iso*-Dihydro carveol	154.25	$C_{10}H_{18}O$
278.		Isopulegol	154.25	$C_{10}H_{18}O$
279.		Linalool	154.25	$C_{10}H_{18}O$
280.		Isophytol	296.54	$C_{20}H_{40}O$
281.		1-[(*E*)-1-Hexenyl]-cyclohexanol	182.30	$C_{12}H_{22}O$
282.		*Trans*-Sabinene hydrate	154.25	$C_{10}H_{18}O$
283.		α-Eudesmol	222.37	$C_{15}H_{26}O$
284.		γ-Eudesmol	222.37	$C_{15}H_{26}O$
285.		Terpinen-4-ol	154.25	$C_{10}H_{18}O$
286.		α-Terpineol	154.25	$C_{10}H_{18}O$
287.		Cedrol	222.37	$C_{15}H_{26}O$
288.		Widdrol	222.37	$C_{15}H_{26}O$
289.		Elemol	222.37	$C_{15}H_{26}O$
290.		2,6-di-tert-butyl-p-cresol	220.35	$C_{15}H_{24}O$
291.		2,4-bis(1,1-Dimethylethyl) phenol	206.32	$C_{14}H_{22}O$
292.		2-Hydroxy 5-methoxy p-cymene	166.22	$C_{10}H_{14}O_2$
293.		(*E*)-Anethole	148.20	$C_{10}H_{12}O$
294.		Eugenol	164.20	$C_{10}H_{12}O_2$
295.		Methyl eugenol	178.23	$C_{11}H_{14}O_2$
296.		Methyl salicylate	152.15	$C_8H_8O_3$
297.		Lanosterol	426.72	$C_{30}H_{50}O$
298.		β-Sitosterol	414.72	$C_{29}H_{50}O$
299.		α-Amyrin	426.72	$C_{30}H_{50}O$
300.		Caproic acid	116.16	$C_6H_{12}O_2$
301.		Caprylic acid	144.21	$C_8H_{16}O_2$
302.		Nonanoic acid	158.24	$C_9H_{18}O_2$
303.		Capric acid	172.27	$C_{10}H_{20}O_2$
304.		Undecanoic acid	186.29	$C_{11}H_{22O2}$
305.		Lauric acid	200.32	$C_{12}H_{24}O_2$
306.		Tridecanoic acid	214.35	$C_{13}H_{26}O_2$
307.		Myristic acid	228.37	$C_{14}H_{28}O_2$
308.		Pentadecanoic acid	242.40	$C_{15}H_{30}O_2$
309.		Palmitic acid	256.43	$C_{16}H_{32}O_2$
310.		Stearic acid	284.48	$C_{18}H_{36}O_2$

(continued)

Table 9.2 (continued)

S. no.	Alkaloids Sub-class	Compound name	Mol. weight	Mol. formula
311.		Hexadecenoic acid	254.41	$C_{16}H_{30}O_2$
312.		Linoleic acid	280.45	$C_{18}H_{32}O_2$
313.		Octadecenoic acid	282.46	$C_{18}H_{34}O_2$
314.		Dehydroabietic acid	302.45	$C_{20}H_{30}O_2$
315.		Methyl heptanoate	144.21	$C_8H_{16}O_2$
316.		Methyl caprylate	158.24	$C_9H_{18}O_2$
317.		Methyl undecanoate	200.32	$C_{12}H_{24}O_2$
318.		Methyl myristate	242.40	$C_{15}H_{30}O_2$
319.		Methyl pentadecanoate	256.43	$C_{16}H_{32}O_2$
320.		Methyl palmitate	270.46	$C_{17}H_{34}O_2$
321.		Methyl 9-tridecenoate	226.36	$C_{14}H2_6O_2$
322.		Hexatrienoic acid methyl ester	292.46	$C_{19}H_{32}O_2$
323.		Linoleic acid methyl ester	294.48	$C_{19}H_{34}O_2$
324.		Linolenic acid methyl ester	292.46	$C_{19}H_{32}O_2$
325.		Oleic acid ethyl ester	310.52	$C_{20}H_{38}O_2$
326.		Linolenic acid ethyl ester	320.52	$C_{21}H_{36}O_2$
327.		Isoamyl laurate	270.46	$C_{17}H_{34}O_2$
328.		Dodecyl acetate	228.37	$C_{14}H_{28}O_2$
329.		Hexadecyl acetate	284.48	$C_{18}H_{36}O_2$
330.		Neryl acetate	196.29	$C_{12}H_{20}O_2$
331.		(*E*,*E*)-farnesyl acetate	264.41	$C_{17}H_{28}O_2$
332.		α-Terpinyl acetate	196.29	$C_{12}H_{20}O_2$
333.		α-Tocopherol	430.71	$C_{29}H_{50}O_2$
334.		Dihydroactinidiolide	180.24	$C_{11}H_{16}O_2$
335.		(*Z*)-Linalool oxide	170.25	$C_{10}H_{18}O_2$
336.		(*E*)-Linalool oxide	170.25	$C_{10}H_{18}O_2$
337.		Theaspirane B	194.31	$C_{13}H_{22}O$
338.		Theaspirane A	194.31	$C_{13}H_{22}O$
339.		Edulane II	204.35	$C_{15}H_{24}$
340.		1,8-Cineole	154.25	$C_{10}H_{18}O$

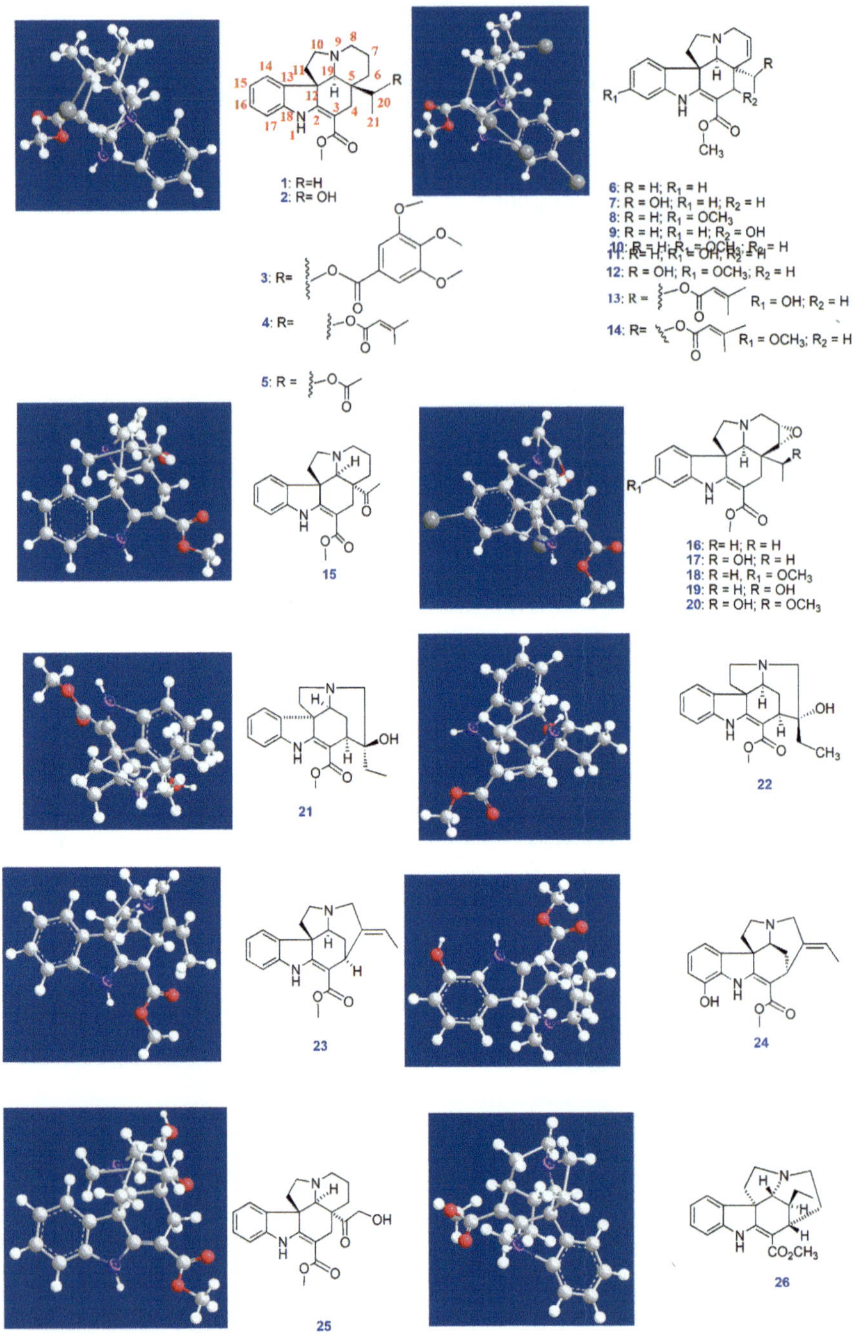

Fig. 9.1 (**a**) Two-dimensional and three-dimensional chemical structures (**1–26**) of representative monoterpene indole alkaloids reported in *Catharanthus* species. (**b**) Two-dimensional and three-dimensional chemical structures (**27–45**) of representative monoterpene indole alkaloids reported in *Catharanthus* species. (**c**) Two-dimensional and three-dimensional chemical structures (**46–61**)

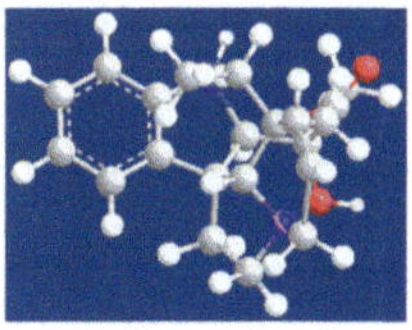

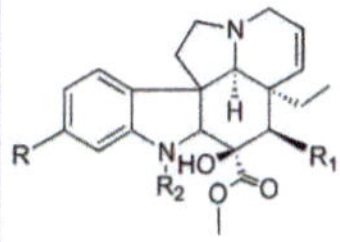

27: R = OCH_3; R_1 = $OCOCH_3$; R_2 =
28: R = H; R_1 = $OCOCH_3$; R_2 = CH_3
29: R = OCH_3; R_1 = $OCOCH_3$; R_2 =
30: R = OCH_3; R_1 = OH; R_2 = H
31: R = H; R_1 = OH; R_2 = CH_3
32: R = OH; R_1 = OH; R_2 = CH_3
33: R = OCH_3; R_1 = $OCOCH_3$; R_2 = H
34: R = OCH_3; R_1 = OH; R_2 = H

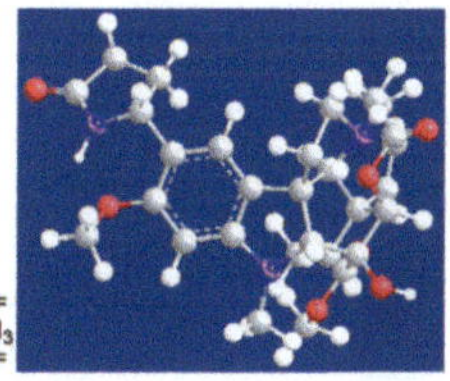

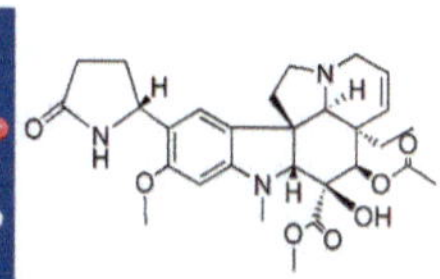

35

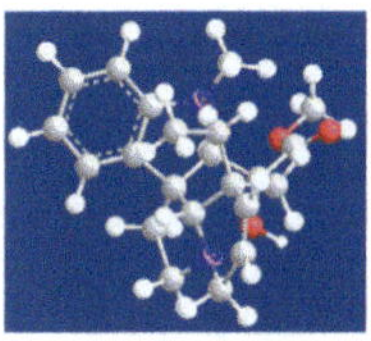

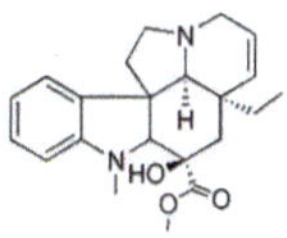

36

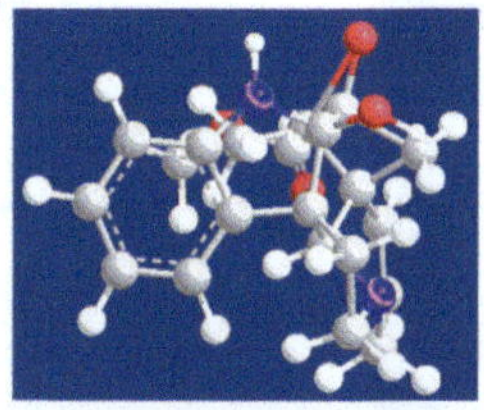

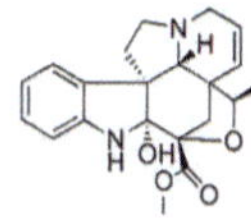

37

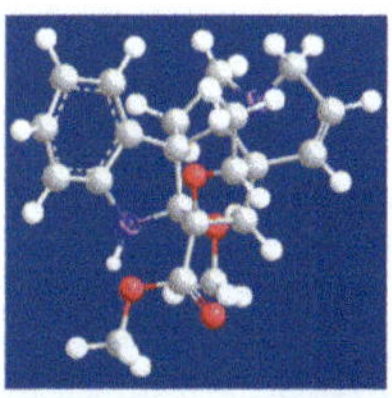

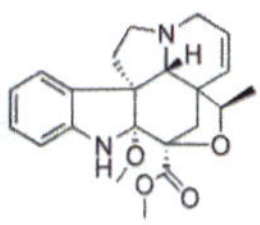

38

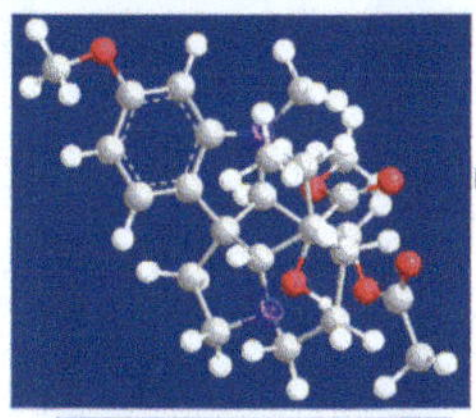

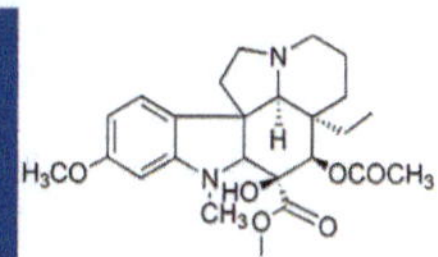

39

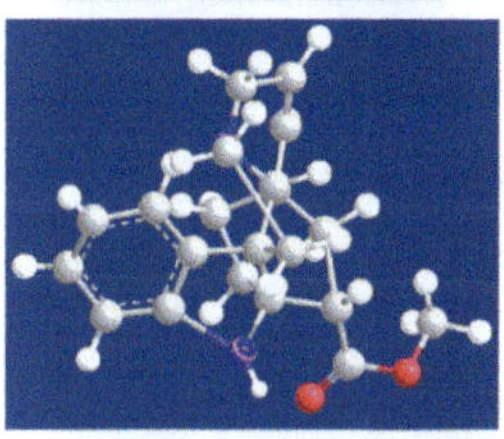

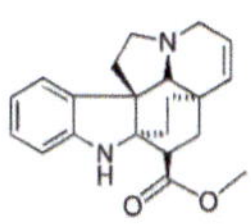

40

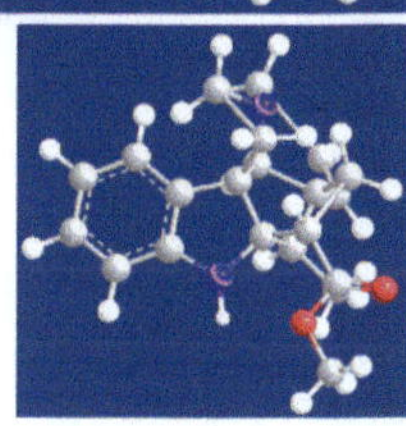

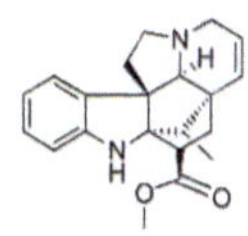

41

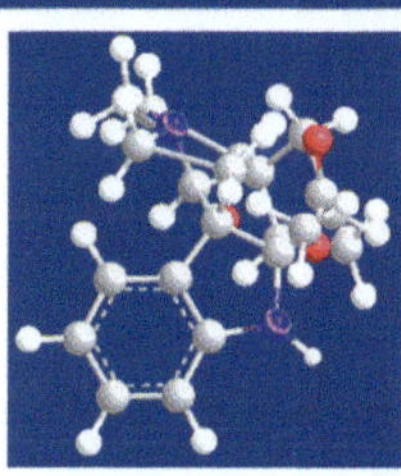

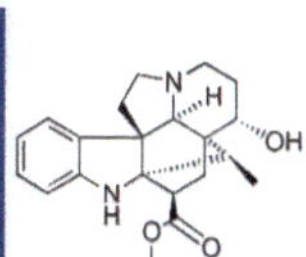

42

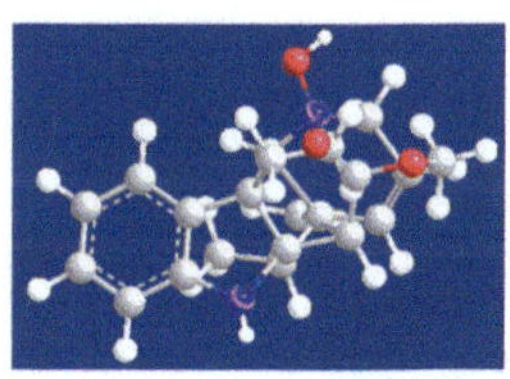

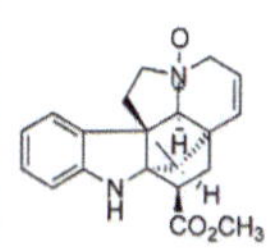

43

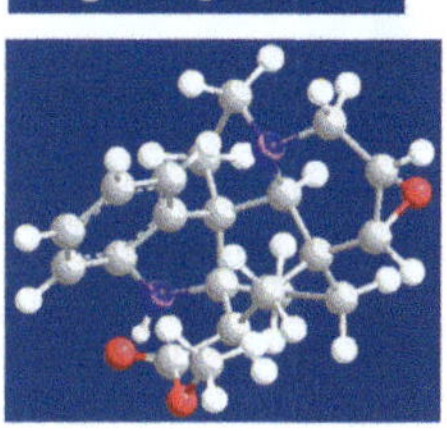

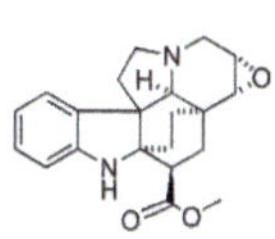

44

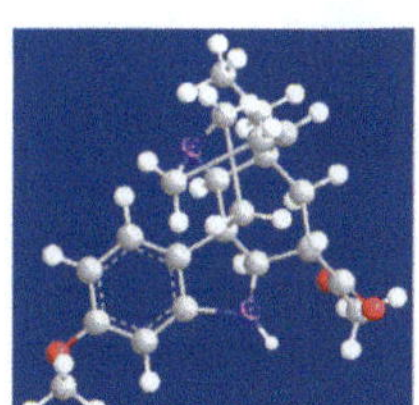

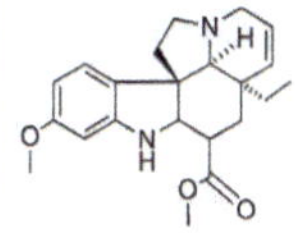

45

Fig. 9.1 (continued) of representative monoterpene indole alkaloids reported in *Catharanthus* species. (**d**) Two-dimensional and three-dimensional chemical structures (**62–73**) of representative monoterpene indole alkaloids reported in *Catharanthus* species. (**e**) Two-dimensional and three-dimensional chemical structures (**74–83**) of representative monoterpene indole alkaloids reported in

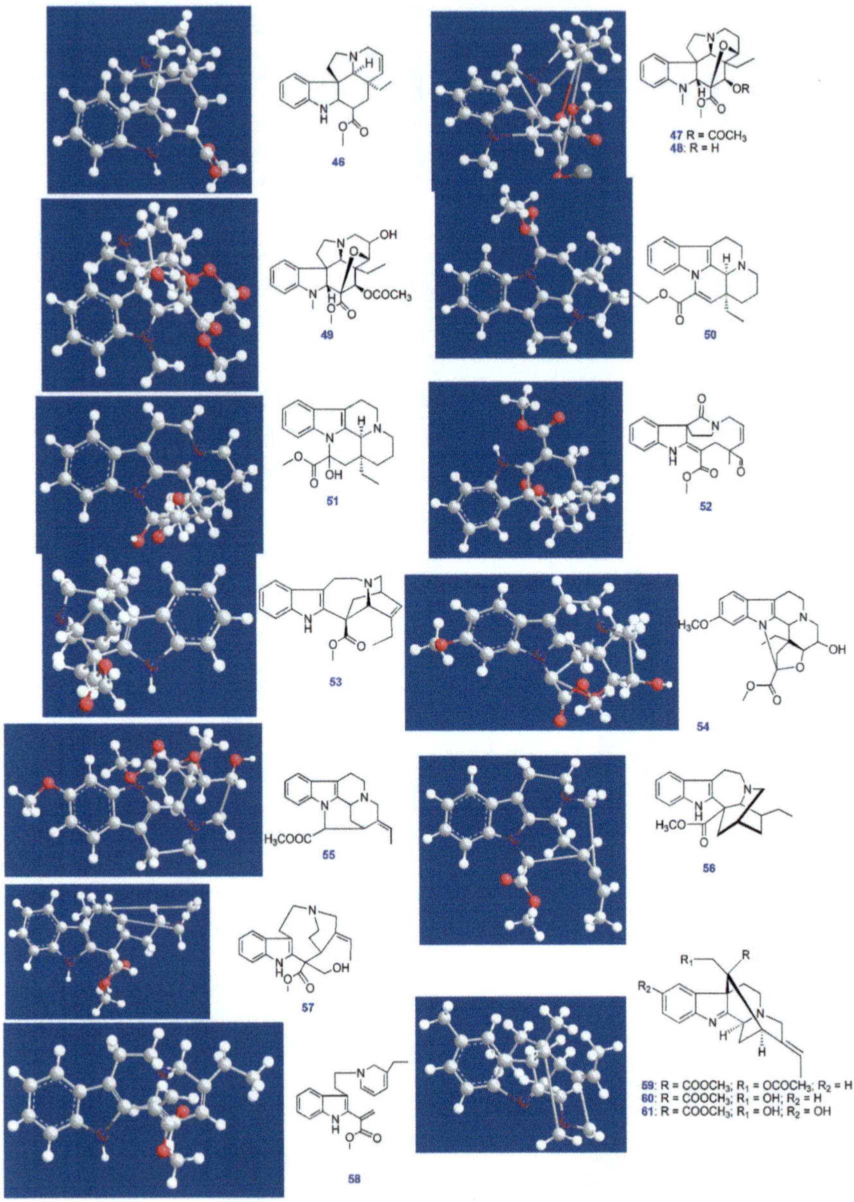

Fig. 9.1 (continued) *Catharanthus* species. (**f**) Two-dimensional and three-dimensional chemical structures (**84–93**) of representative monoterpene indole alkaloids reported in *Catharanthus* species. (**g**) Two-dimensional and three-dimensional chemical structures (**94–103**) and other alkaloids (**104–107**) of representative monoterpene indole alkaloids reported in *Catharanthus* species

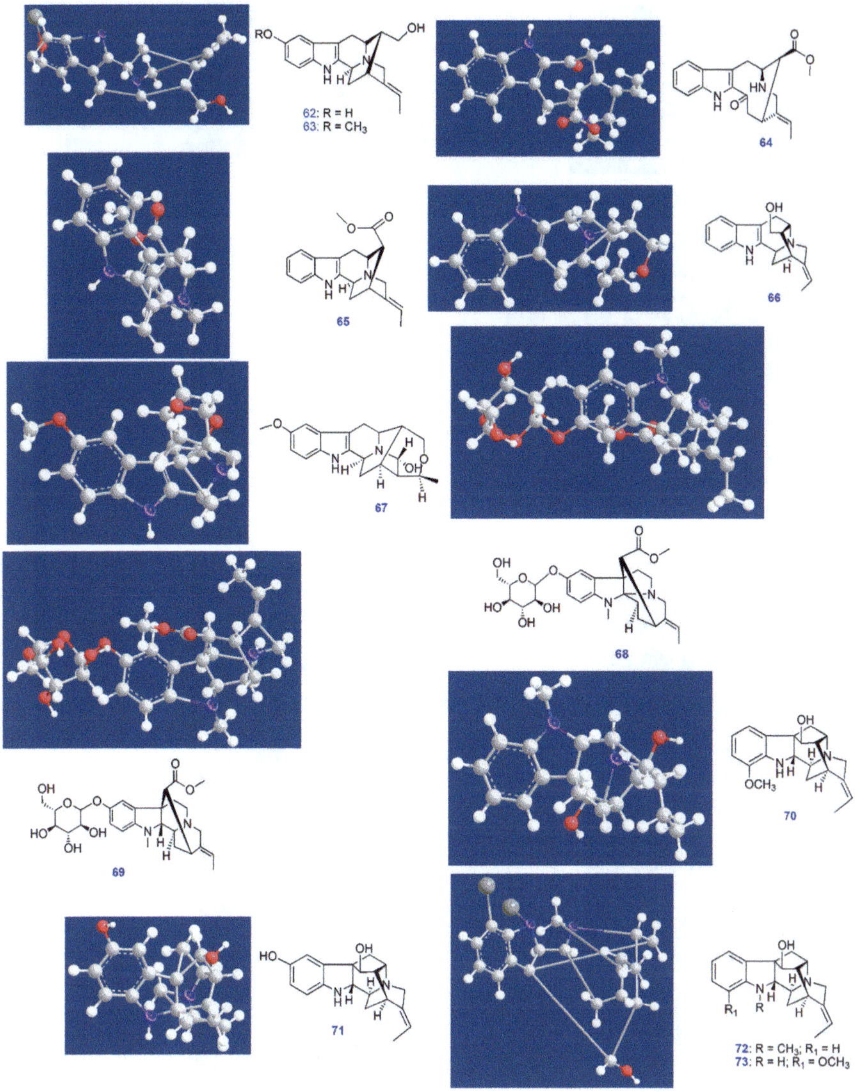

Fig. 9.1 (continued)

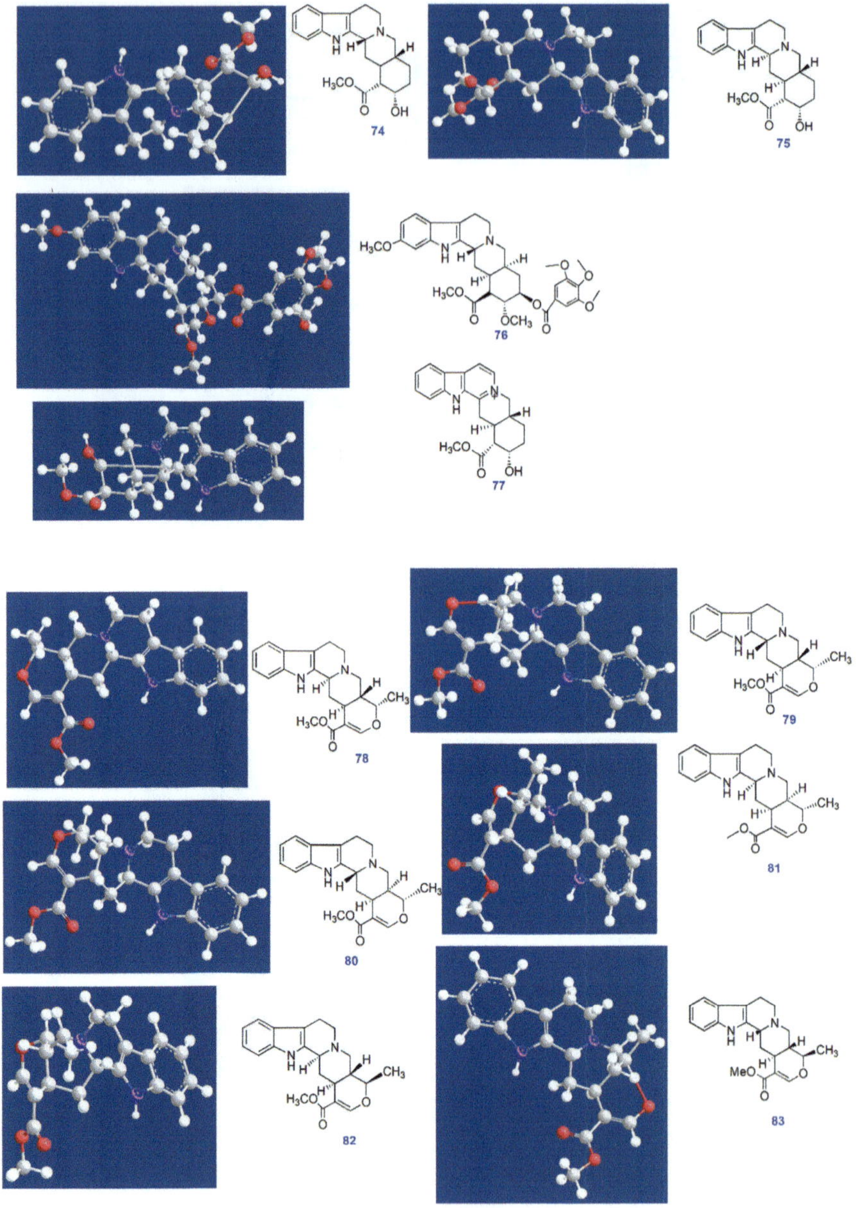

Fig. 9.1 (continued)

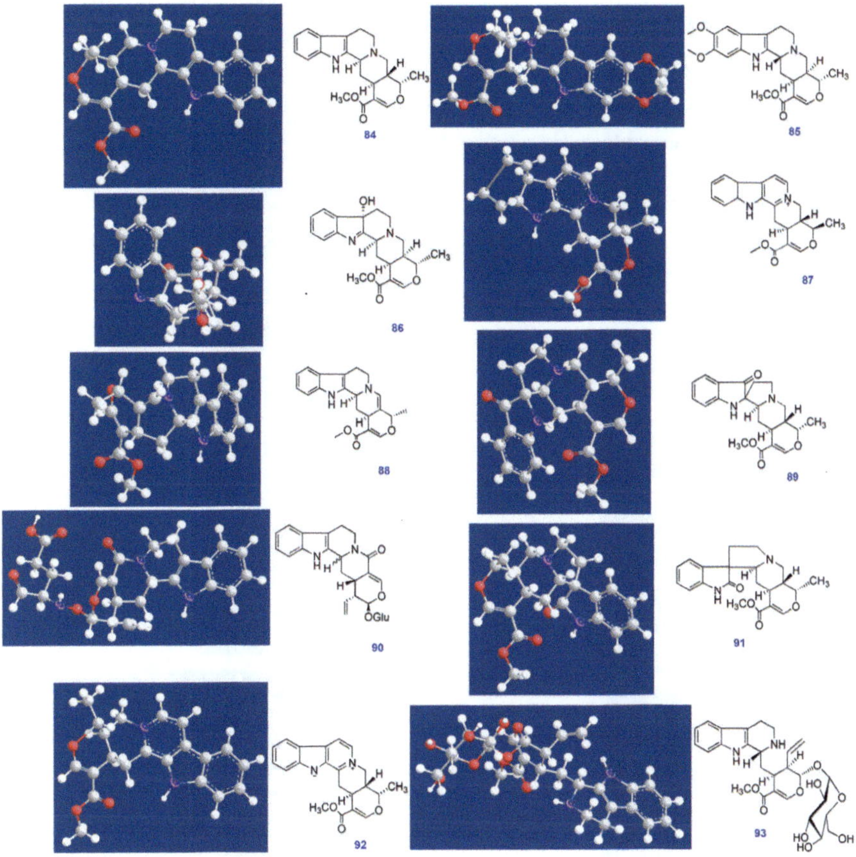

Fig. 9.1 (continued)

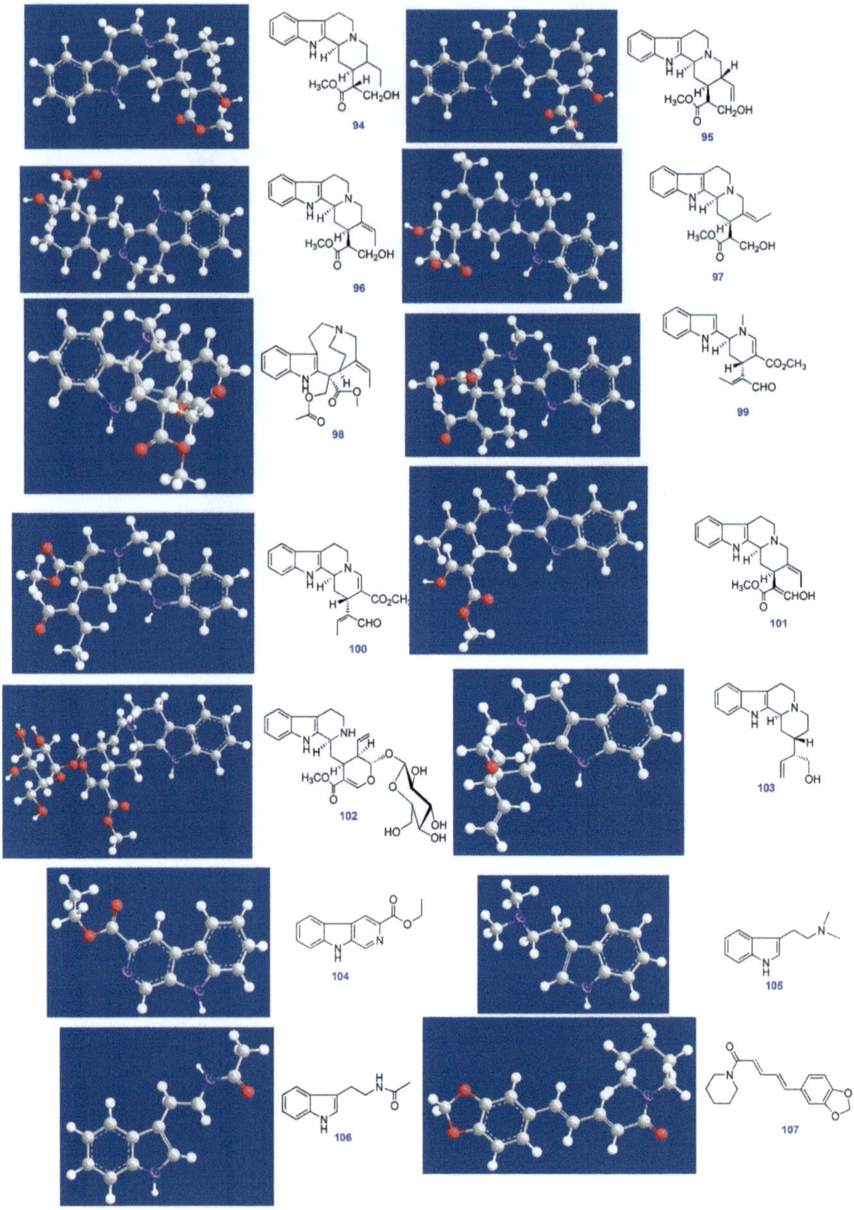

Fig. 9.1 (continued)

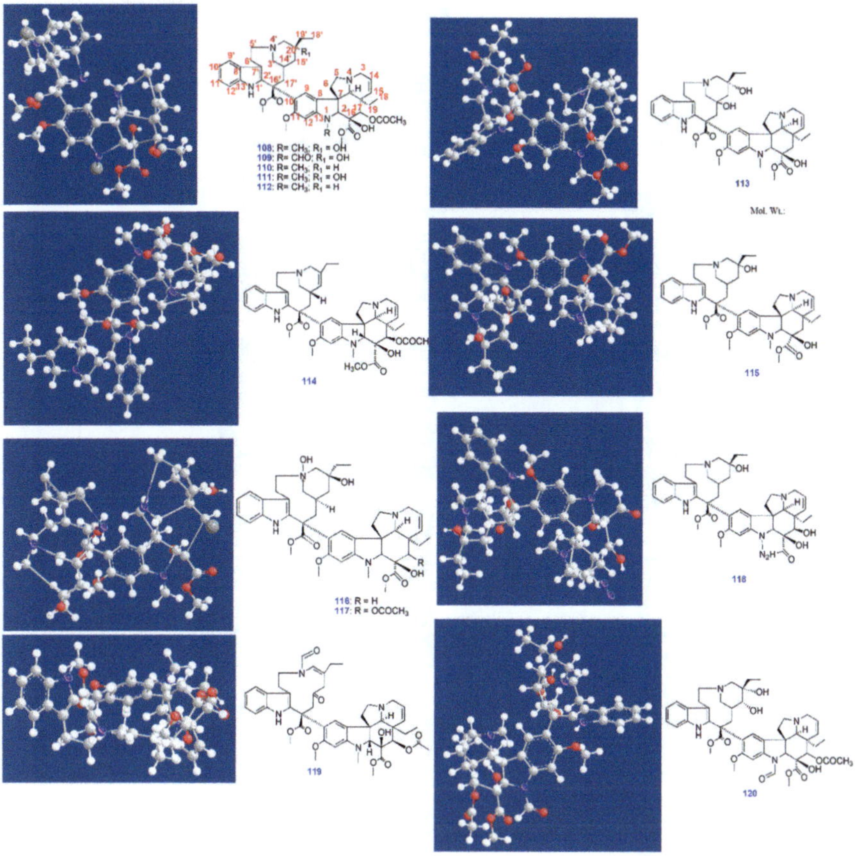

Fig. 9.2 (**a**) Two-dimensional and three-dimensional chemical structures (**108–120**) of representative bisindole alkaloids reported in *Catharanthus* species. (**b**) Two-dimensional and three-dimensional chemical structures (**121–130**) of representative bisindole alkaloids reported in *Catharanthus* species. (**c**) Two-dimensional and three-dimensional chemical structures (**131–142**) of representative bisindole alkaloids reported in *Catharanthus* species

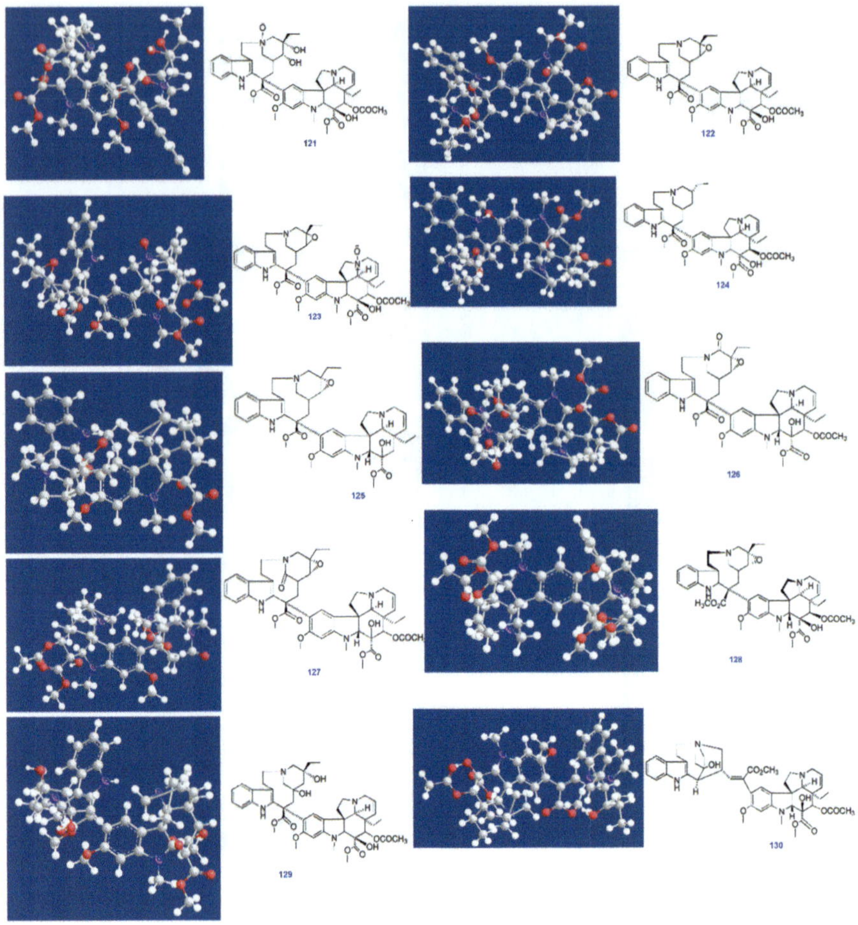

Fig. 9.2 (continued)

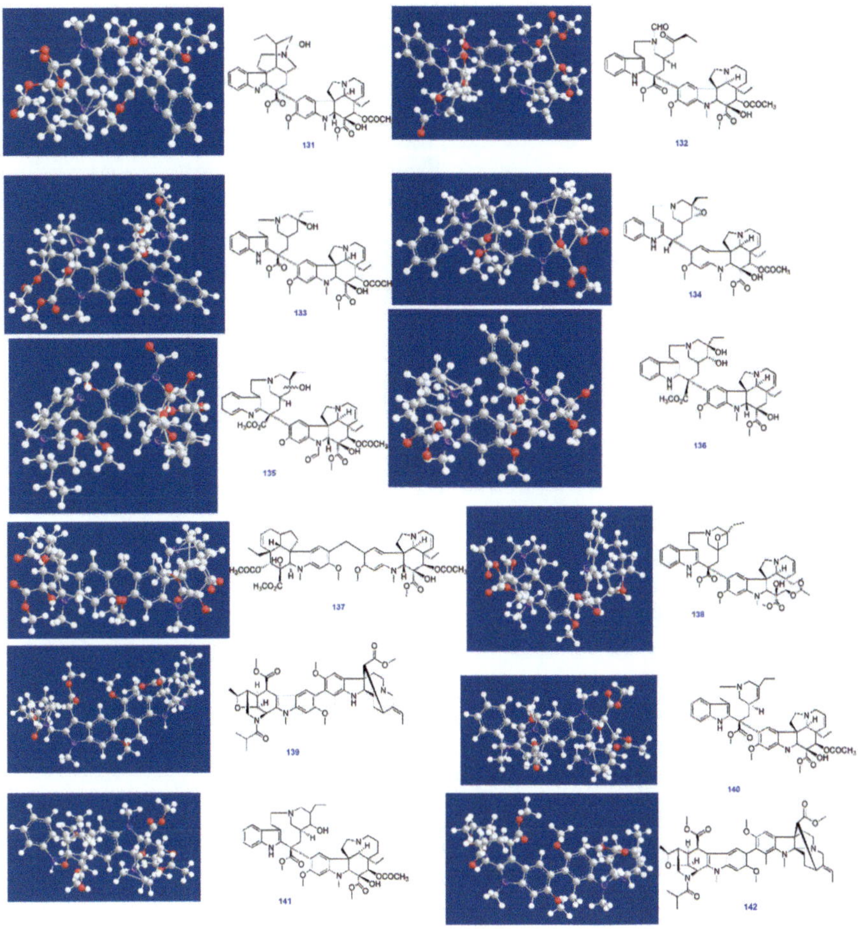

Fig. 9.2 (continued)

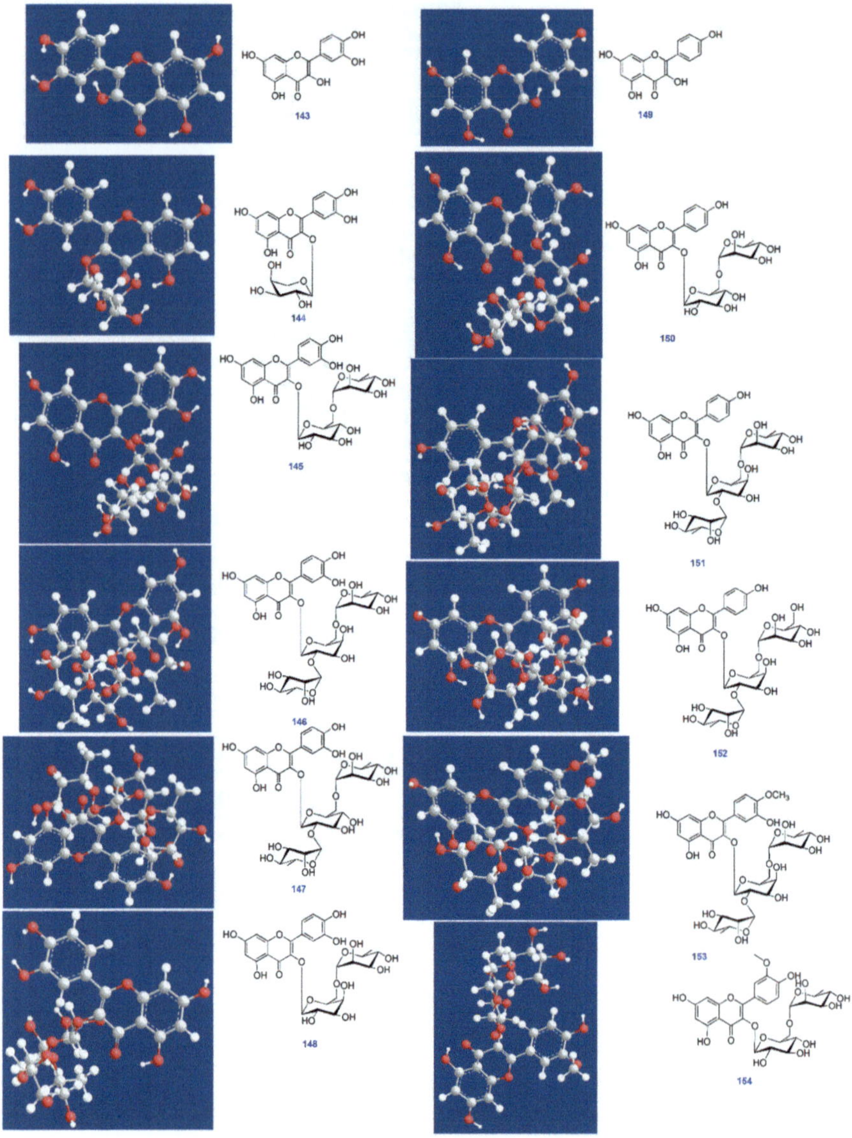

Fig. 9.3 (**a**) Two-dimensional and three-dimensional chemical structures (**143–154**) of representative flavonoids reported in *Catharanthus* species. (**b**) Two-dimensional and three-dimensional chemical structures (**155–161**) of representative flavonoids reported in *Catharanthus* species. (**c**) Two-dimensional and three-dimensional chemical structures of representative anthocyanidins and anthocyanins (**162–174**) and phenolic acids (**175–180**) reported in *Catharanthus* species

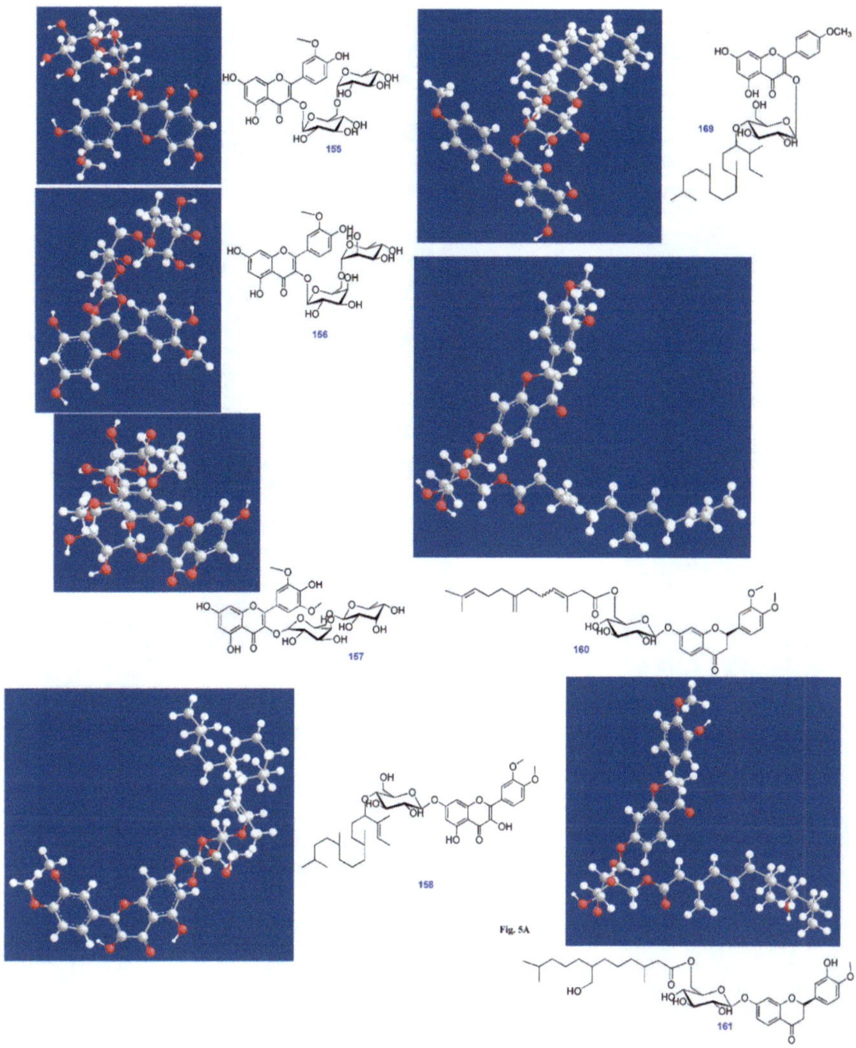

Fig. 9.3 (continued)

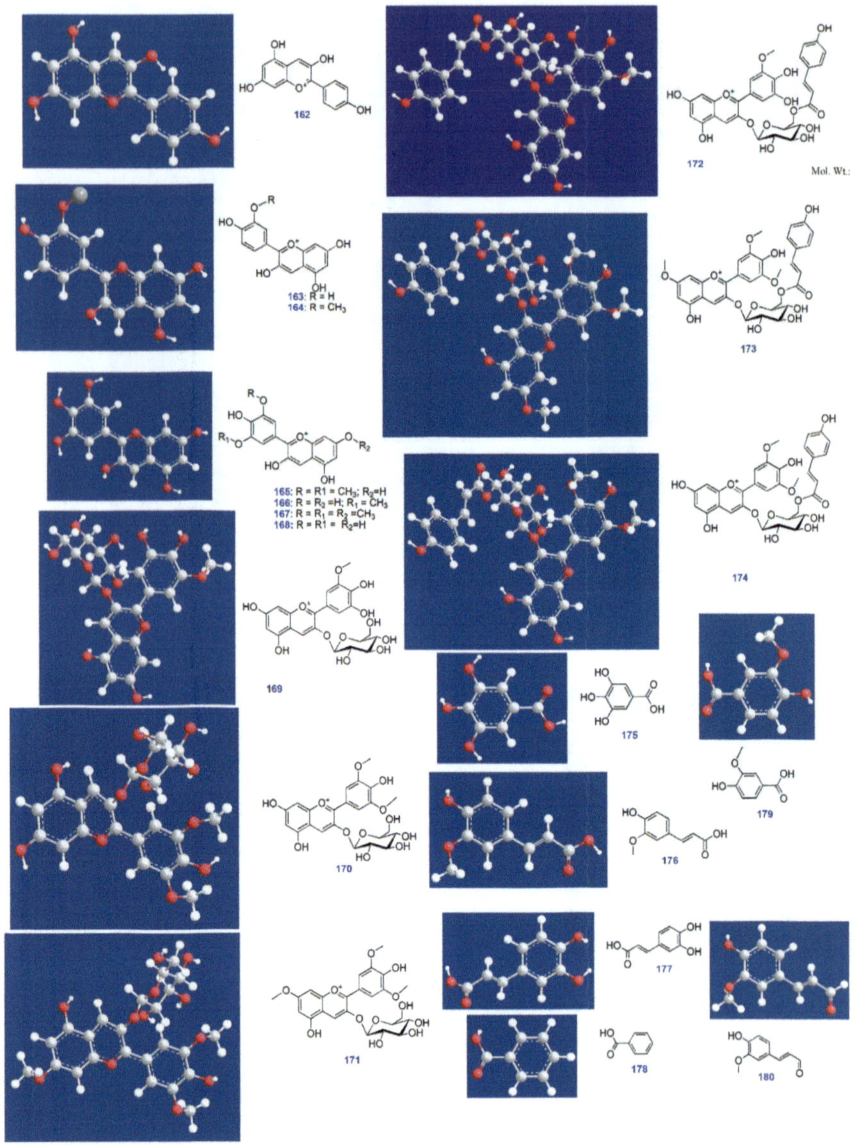

Fig. 9.3 (continued)

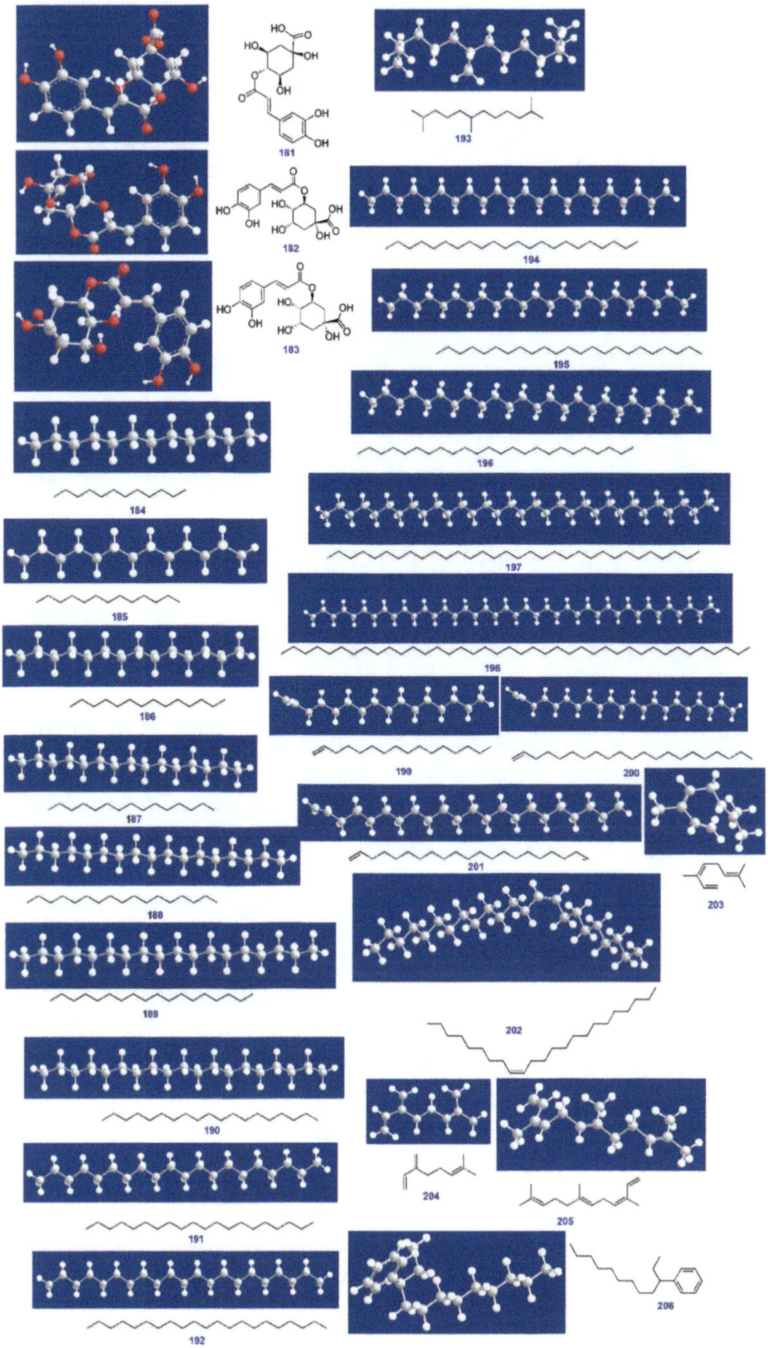

Fig. 9.4 (**a**) Two-dimensional and three-dimensional chemical structures of representative phenolic acids (**181–183**) and volatile compounds (**184–206**) reported in *Catharanthus* species. (**b**) Two-dimensional and three-dimensional chemical structures of representative volatile compounds (**207–243**) reported in *Catharanthus* species. (**c**) Two-dimensional and three-dimensional chemical

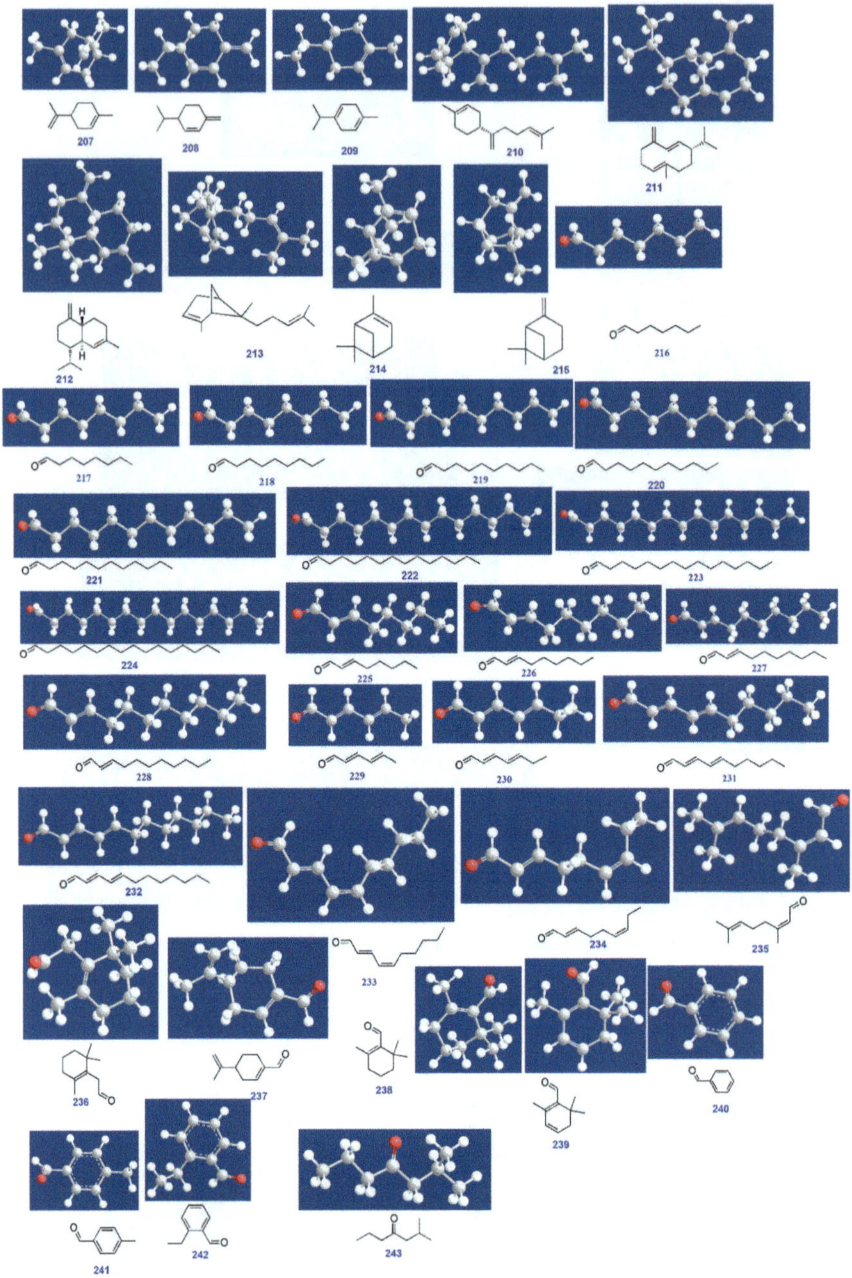

Fig. 9.4 (continued) structures of representative volatile compounds (**244–265**) reported in *Catharanthus* species. (**d**) Two-dimensional and three-dimensional chemical structures of representative volatile compounds (**267–295**) reported in *Catharanthus* species. (**e**) Two-dimensional and three-dimensional chemical structures of representative volatile compounds (**296–222**) reported in *Catharanthus* species. (**f**) Two-dimensional and three-dimensional chemical structures of representative volatile compounds (**297–339**) reported in *Catharanthus* species

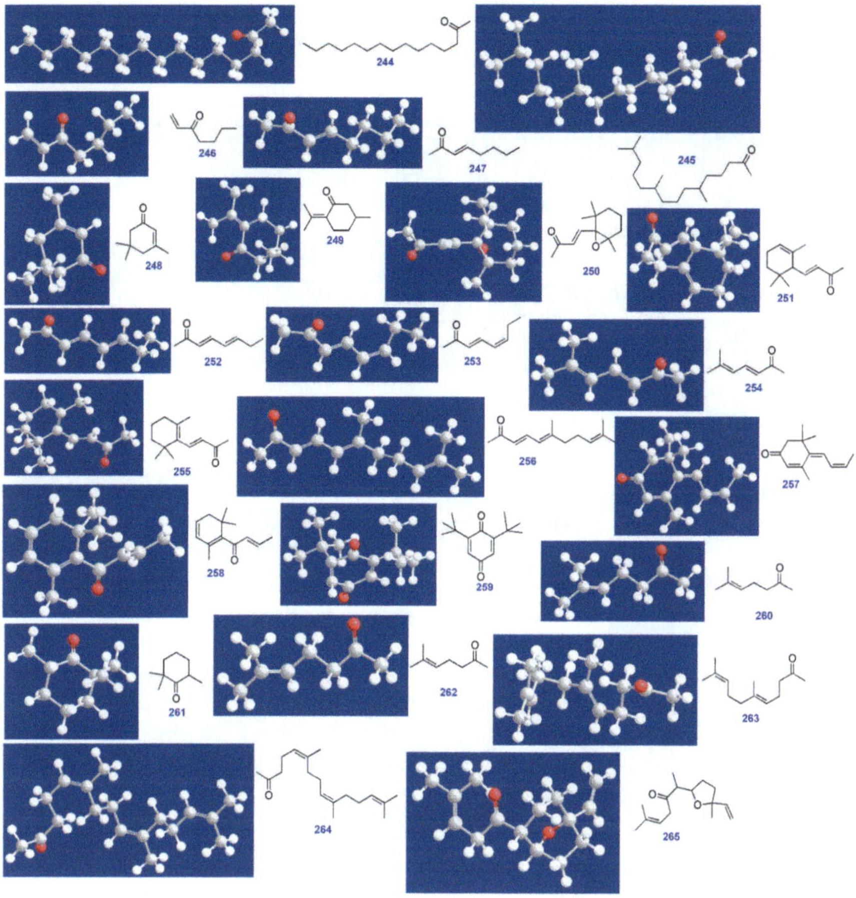

Fig. 9.4 (continued)

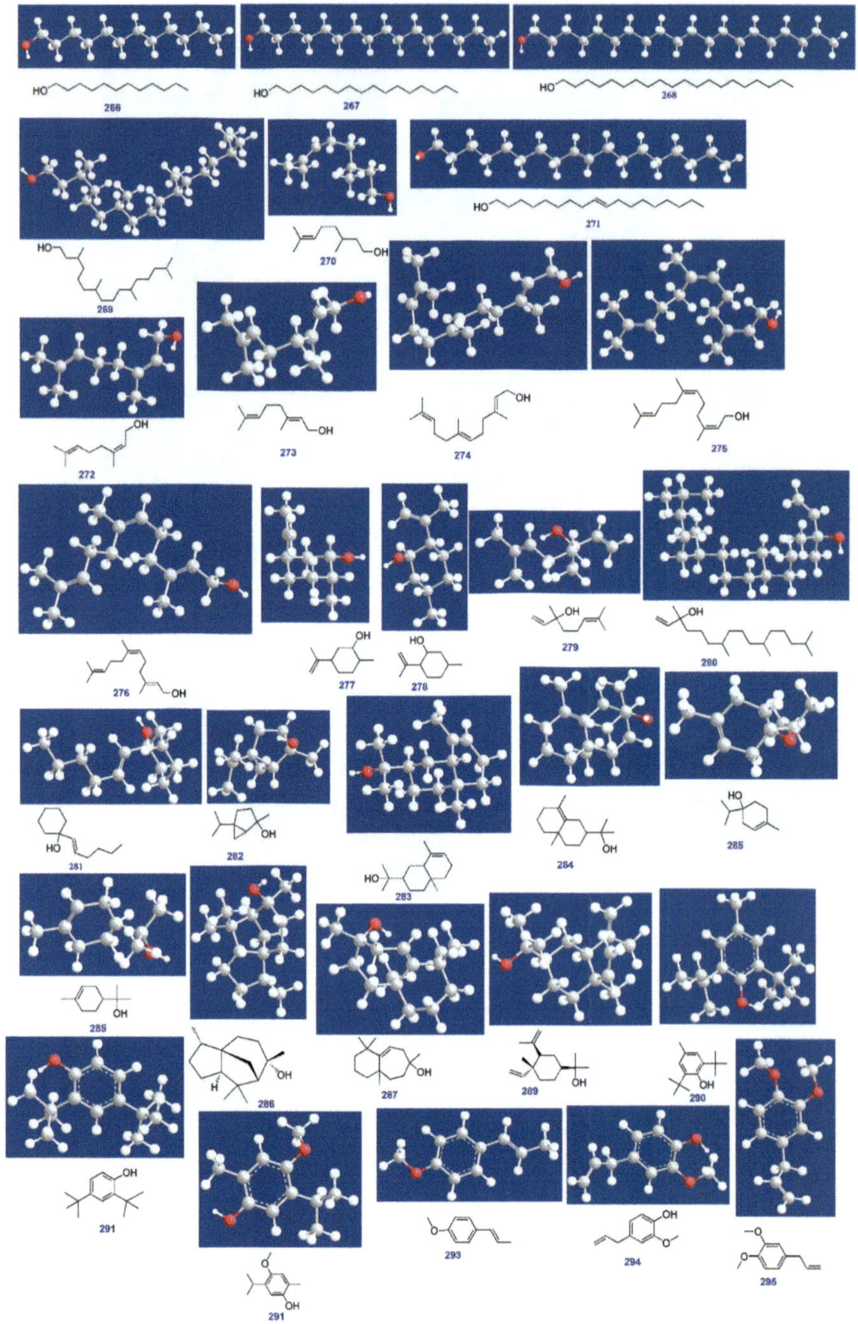

Fig. 9.4 (continued)

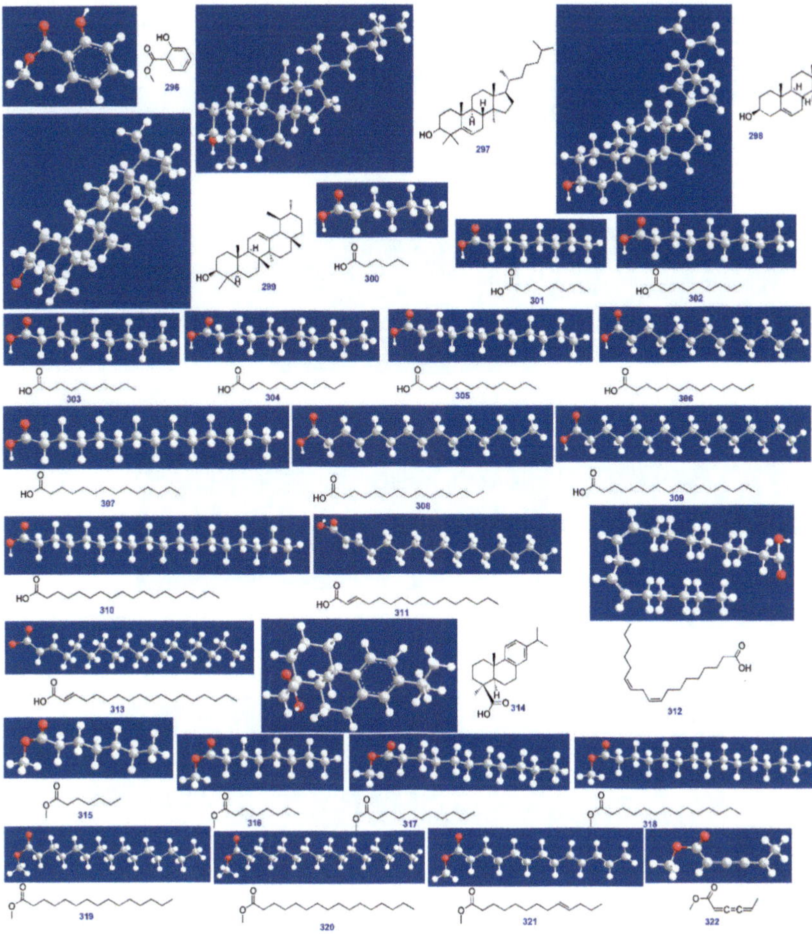

Fig. 9.4 (continued)

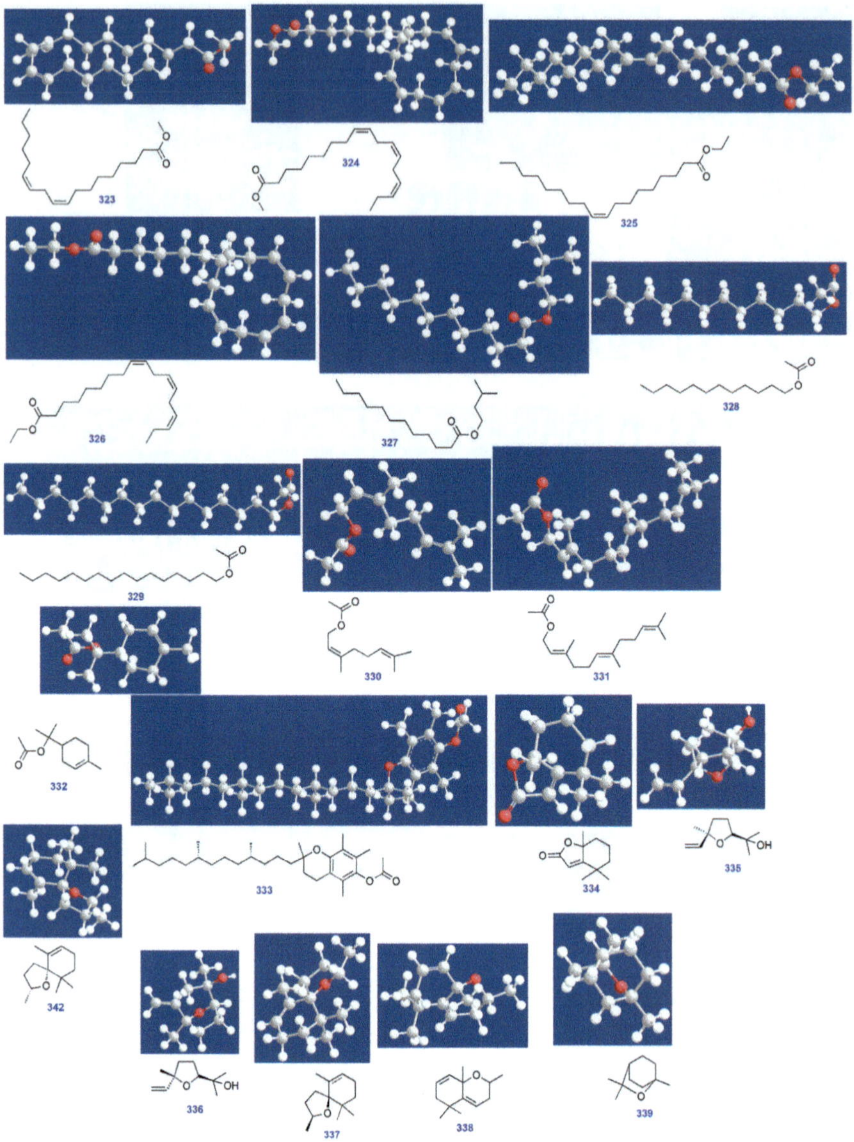

Fig. 9.4 (continued)

References

Bahia MS, Kaspi O, Touitou M, Binayev I, Dhail S, Spiegel J, Khazanov N, Yosipof A, Senderowitz H (2023) A comparison between 2D and 3D descriptors in QSAR modeling based on bio-active conformations. Mol Inform 42(4):2200186

Putz MV, Duda-Seiman C, Duda-Seiman D, Putz AM, Alexandrescu I, Mernea M, Avram S (2016) Chemical structure-biological activity models for pharmacophores' 3D-interactions. Int J Mol Sci 17(7):1087

Biosynthetic Pathways of Phytochemicals in *Catharanthus*: Function and Evolution

10

Abstract

The biosynthetic pathways of phytochemicals in *Catharanthus* involve complex, multistep processes responsible for producing diverse bioactive compounds, including TIAs, flavonoids, and phenolic acids. These pathways integrate primary and secondary metabolism, with key enzymes orchestrating the synthesis of compounds like vinblastine and vincristine. Evolutionary adaptations have shaped these pathways, enabling tissue-specific expression and metabolite diversity. Functional studies reveal their roles in plant defense, development, and ecological interactions. Understanding the evolution and regulation of these pathways provides insights into plant metabolic networks and supports metabolic engineering strategies for enhanced phytochemical production in medicinal and biotechnological applications.

Keywords

Biosynthetic pathways · Phytochemicals · Functions · *Catharanthus*

10.1 Introduction

The biosynthesis of specialized metabolites such as indole alkaloids, phenolic acids, flavonoids, and terpenes in plant families like Apocynaceae, Loganiaceae, and Rubiaceae is a highly intricate and multistep process. These biosynthetic pathways are tightly regulated and involve a network of enzymatic reactions responsible for structural diversity and pharmacological properties. To date, approximately 20 out of the estimated 50 enzymes required for these pathways have been identified and functionally characterized using biochemical, molecular, and genomic approaches. However, a significant number of key enzymes remain uncharacterized. Their discovery hinges on the isolation and cloning of the corresponding genes, followed by

S. Kumar, B. Singh, *Catharanthus in Drug Discovery*,
https://doi.org/10.1007/978-981-95-5048-7_10

functional validation (Liu et al. 2021; Williams and De Luca 2023; Le Pogam and Beniddir 2024).

Understanding the regulatory mechanisms governing the biosynthesis of therapeutically important terpenoid indole alkaloids (TIAs)—such as vinblastine, vincristine, ajmaline, catharanthine, tabersonine, and serpentine—is also essential. These regulatory processes operate at both cellular and molecular levels and include transcriptional control, metabolite channeling, and organ-specific expression patterns. Unraveling these regulatory networks is critical not only for understanding the physiological role of these compounds in plant defense and development but also for enhancing their production through metabolic engineering and biotechnological interventions. Advancements in genomics, transcriptomics, and proteomics, along with synthetic biology tools, offer promising avenues to manipulate these pathways. Future research must prioritize gene discovery, enzyme function elucidation, and pathway reconstitution to optimize TIA yields for pharmaceutical applications (Zhu et al. 2014; El-Sayed and Verpoorte 2007; Liu et al. 2007; Pandey et al. 2016).

10.2 Biosynthetic Pathways of Monoterpene Indole Alkaloids

More than 30 enzymes perform the biosynthetic process in multiple cellular compartments, such as chloroplasts, cytosol, endoplasmic reticulum, and vacuoles of plant cells (Fig. 10.1a–c). To initiate the indole branch of alkaloid biosynthesis, decarboxylases remove carbon dioxide from –COOH group of L-tryptophan to produce tryptamine. C–C and C–O bonds are synthesized in the presence of synthase enzymes, such as strictosidine synthase (STR). The lone pair of the nitrogen atom of the primary amine attacks the *C* atom of the aldehyde group to form an imine intermediate (Schiff base). The condensation reaction of tryptamine and secologanin produces strictosidine. Cyclization of the electrophilic iminium and the nucleophilic *C* atom of the indole ring forms the tetrahydro-*β*-carboline ring system via the Pictet–Spengler reaction. Proton transfer and rearrangement of the tetrahydro-*β*-carboline ring system produce the final product strictosidine. The corynanthean skeleton is formed after deglycosylation of strictosidine; it undergoes further rearrangements and 1,4-Michael-type addition (between the enamine and the activated alkene moiety on the side chain (often the C-16 = C-17 double bond)) creates a new ring, resulting in closure of the corynanthean skeleton. Therefore, biosynthesis of 19E-Geissoschizine from strictosidine is performed in the presence of strictosidine *β*-glucosidase (SGD) and geissoschizine synthase (GS). 19E-geissoschizine is oxidized by geissoschizine oxidase (GO) to form an unstable intermediate, which is subsequently reduced by two NADPH-dependent oxidoreductases (Redox1 and Redox2), leading to the formation of stable stemmadenine. Acetylation of the hydroxyl group of 19E-geissoschizine, facilitated by an acetyl group from acetyl-CoA, converts it into *O*-acetylstemmadenine by stemmadenine-O-acetyltransferase (SAT). A step for activating molecule is formed by

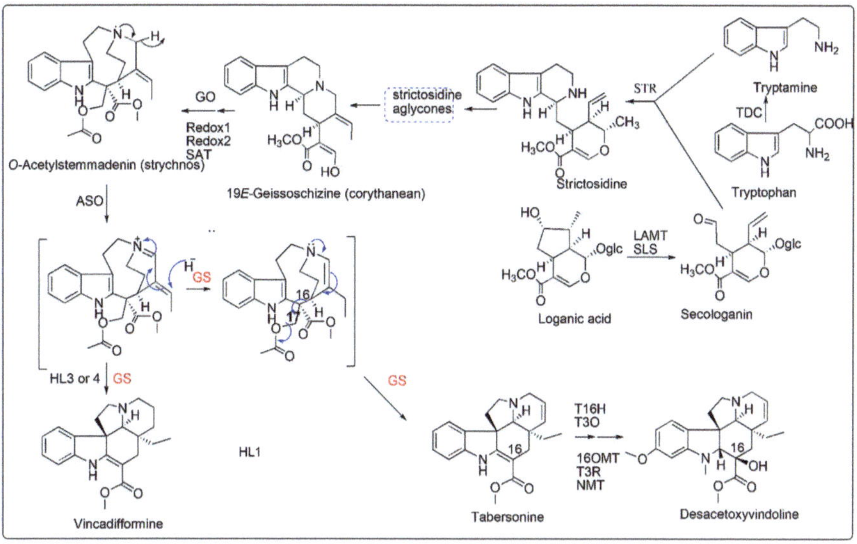

Fig. 10.1 (**a**) Biosynthetic pathway of monoterpene indole alkaloids. (**b**) Biosynthetic pathway of monoterpene indole alkaloids. (**c**) Biosynthetic pathway of monoterpene indole alkaloids

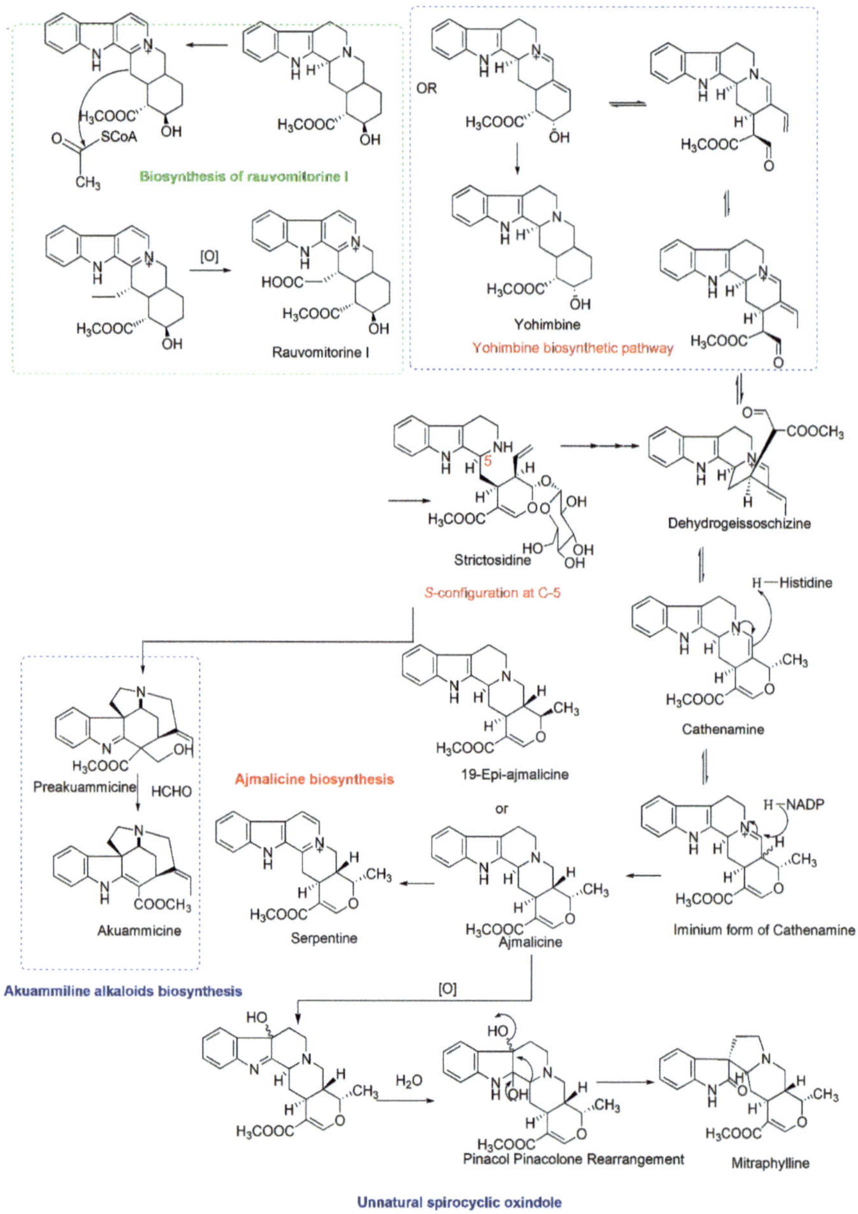

Fig. 10.1 (continued)

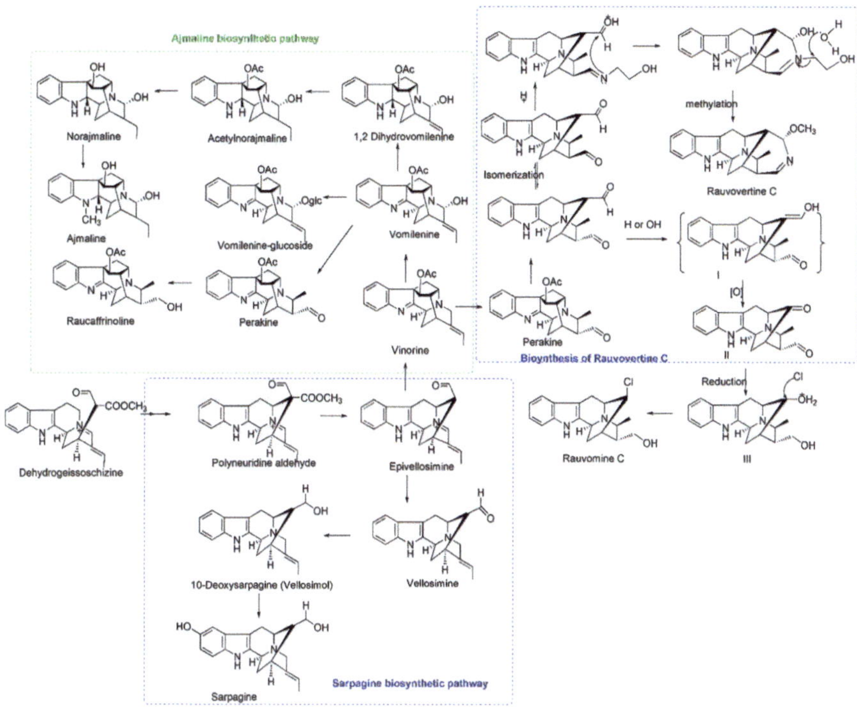

Fig. 10.1 (continued)

O-acetylstemmadenine oxidase (ASO) into a reactive iminium ion intermediate of *O*-acetylstemmadenine. The oxidized intermediate is reduced by geissoschizine synthase (GS), followed by cyclization by hydrolase 3 (HL3), producing the aspidosperma skeleton of (+)-vincadifformine. Tabersonine is synthesized through a process similar to that of (+)-vincadifformine, using the same enzymes as ASO, GS, and hydrolase 2 (HL2). This process resulted in the formation of the aspidosperma structure of tabersonine through a reactive iminium ion and cyclization. Hydroxylation at C-16, methylation of the –OH group, epoxidation at the C-2-C-3 double bond, ring opening and hydroxylation, reduction of the imine, and formation of desacetoxyvindoline are the steps involved in the synthesis of desacetoxyvindoline from tabersonine, facilitated by tabersonine 16-hydroxylase (T16H), 16-*O*-methyltransferase (16OMT), tabersonine 3-oxygenase (T3O), tabersonine 3-reductase (T3R), and N-methyltransferase (NMT), respectively. Rearrangement and intramolecular cyclization of geissoschizine in the presence of geissoschizine synthase (GS) facilitates formation of the corynanthe-type indole alkaloid tetrahydroalstonine. Epimerization at C-3 of tetrahydroalstonine, facilitated by tetrahydroalstonine epimerase, produces ajmalicine. Further oxidization of ajmalicine leads to spontaneous or enzyme-mediated lactamization to form serpentine Conversion of ajmalicine to yohimbine is a multistep process probably facilitated by enzymes (cytochrome P450s, oxidoreductases, and acyltransferases) involving hydroxylation, oxidation, and ring modifications. The most probable biosynthetic pathway of rauvomitorine I involves the synthesis of polyneuridine aldehyde, vomilenine, and rauvomitorine I in the presence of geissoschizine oxidase, polyneuridine aldehyde esterase, P450 monooxygenases, methyltransferases, cyclases, or rearrangement enzymes, respectively, via oxidation, hydroxylation, methylation, and rearrangement or cyclization (Memelink and Gantet 2007; Irmler et al. 2000).

Hydroxylation at the C-21 position of vinorine produced 21-hydroxyvinorine, facilated by vinorine hydroxylase. Rearrangement and oxidation of 21-hydroxyvinorine produce vomilenine. Subsequently, vomilenine 1,2-dihydrovomilenine, 17-O-acetylnorajmaline, norajmaline, and ajmaline are synthesized, facilitated by vomilenine reductase (VR), dihydrovomilenine reductase (DHVR), acetylajmalan esterase (AAE), and norajmaline methyltransferase (NAMT). Reduction, rearrangement, and methylation of vomilenine lead to the formation of the ajmalan skeleton in the presence of vomilenine reductase, dehydrogenase, and methyltransferase. Acetylenorajmaline is produced from dihydroajmaline via the reduction of an acetylenic triple bond and a double bond. Ajmaline can also be formed from dihydroajmaline and norajmaline, facilitated by rearrangement and S-adenosyl-L-methionine (SAM) (El-Sayed and Verpoorte 2007; Hedhili et al. 2007; Oudin et al. 2007; Zhu et al. 2015; Liu et al. 2021).

The sarpagan skeleton is synthesized via rearrangement and cyclization of the geissoschizine intermediate into the sarpagan alkaloid framework is facilitated by various reductases and isomerases. The akuammiline skeleton in the biosynthesis of preakuammicine from dehydrogeissoschizine undergoes C–C bond formation to form a fused indolo-quinolizidine skeleton. The steps in the biosynthesis of preakuammicine to akuammicine are not fully characterized. Mitraphylline (oxindole skeleton) is biosynthesized from a corynanthe intermediate in the presence of hydroxylases and oxidases via aromatic hydroxylation of the indole ring at C-2, followed by rearrangement to an oxindole moiety and further oxidation/cyclization to stabilize the pentacyclic oxindole skeleton (Salim and De Luca 2013).

10.3 Biosynthetic Pathways of Bis Terpeneindole Alkaloids

The dimerization of two MIAs, like catharanthine and vindoline, leads to the synthesis of vincristine and vinblastine via multiple steps involving deglycosylation, hydroxylation, acetylation, methylation, and ring rearrangement along with respective enzymes strictosidine synthase (STR), strictosidine β-glucosidase (SGD), desacetoxyvindoline 4-hydroxylase (D4H), deacetylvindoline 4-O-acetyltransferase (DAT), and anhydrovinblastine synthase (PRX1) (Fig. 10.2) (El-Sayed and Verpoorte 2007; Pan et al. 2016; Li et al. 2025).

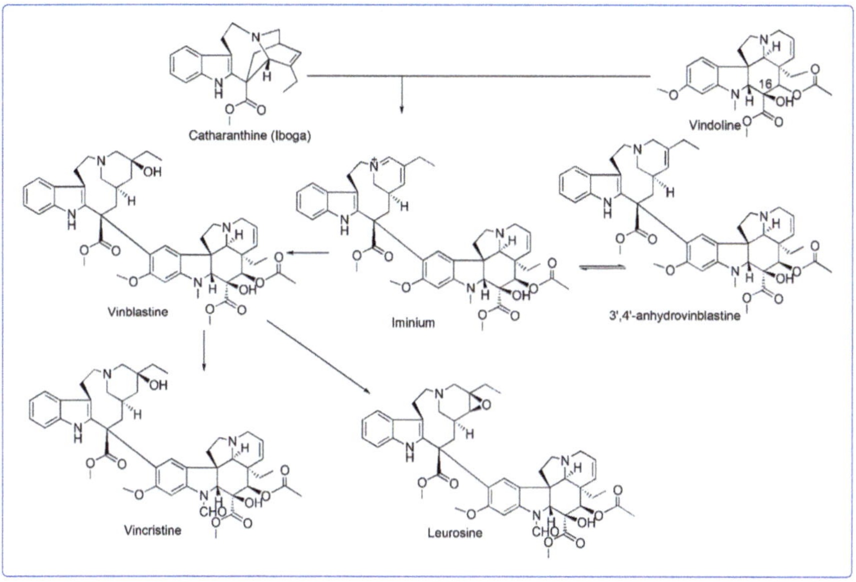

Fig. 10.2 Biosynthetic pathway of bisindole alkaloids

10.4 Biosynthetic Pathways of Phenolic Acids and Flavonoids

The biosynthesis of phenolic acids and flavonoids, flavones, isoflavones, anthocyanins, etc., arises from two pathways: the phenylpropanoid pathway and the polyketide pathway (Fig. 10.3a, b). Both pathways are interconnected but follow distinct metabolic routes. The C6–C3 and C15 (C6–C3–C6) backbones are the starting materials for the phenylpropanoid and polyketide pathways, respectively. Biosynthesis of phenolic acids and flavonoids occurs through deamination, hydroxylation, and activation in phenylpropanoid pathway, a process facilitated by oxygenases, lyases, and ligases (PAL, C4H, and 4CL). In the polyketide pathway, *p*-coumaroyl-CoA and malonyl-CoA are the starting molecules. The C15 flavonoid skeleton is synthesized via chain elongation and cyclization, facilitated Type III polyketide synthases (CHS and CHI). Methylation of flavonoids such as quercetin, luteolin, and chrysin into isorhamnetin, diosmetin, and 7-*O*-methylchrysin is facilitated by *O*-methyltransferases (OMTs) like flavonoid 3′-*O*-methyltransferase, flavonoid 4′-*O*-methyltransferase and flavonoid 7-*O*-methyltransferase using methyl donor S-adenosyl-L-methionine (SAM). Similarly, glycosylation of flavonoids (kaempferol, apigenin, and quercetin) proceeds through uridine diphosphate (UDP)-dependent glycosyltransferases (UGTs) such as flavonoid 3-*O*-glucosyltransferase, flavonoid 7-*O*-glucosyltransferase, and flavonoid 3-*O*-rhamnosyltransferase, producing respective glycosides (Winkel-Shirley 2002; Heller and Forkmann 2017; Patil and Murumkar 2024).

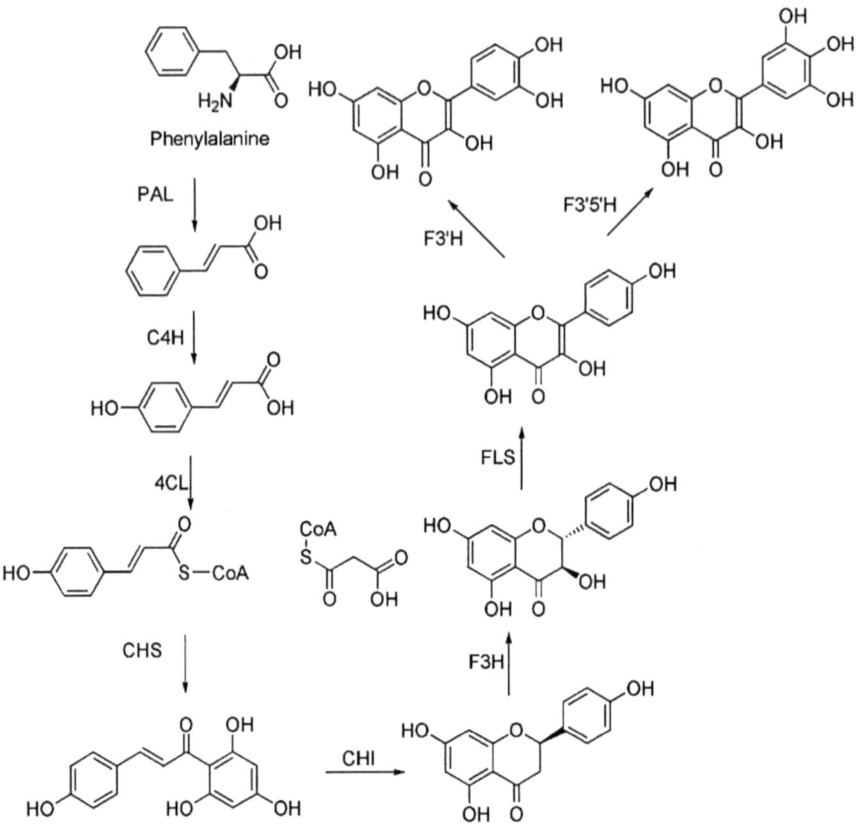

Fig. 10.3 (**a**) Biosynthetic pathways of flavonoids. (**b**) Biosynthetic pathways of flavonoids, anthocynanin, and cathachin

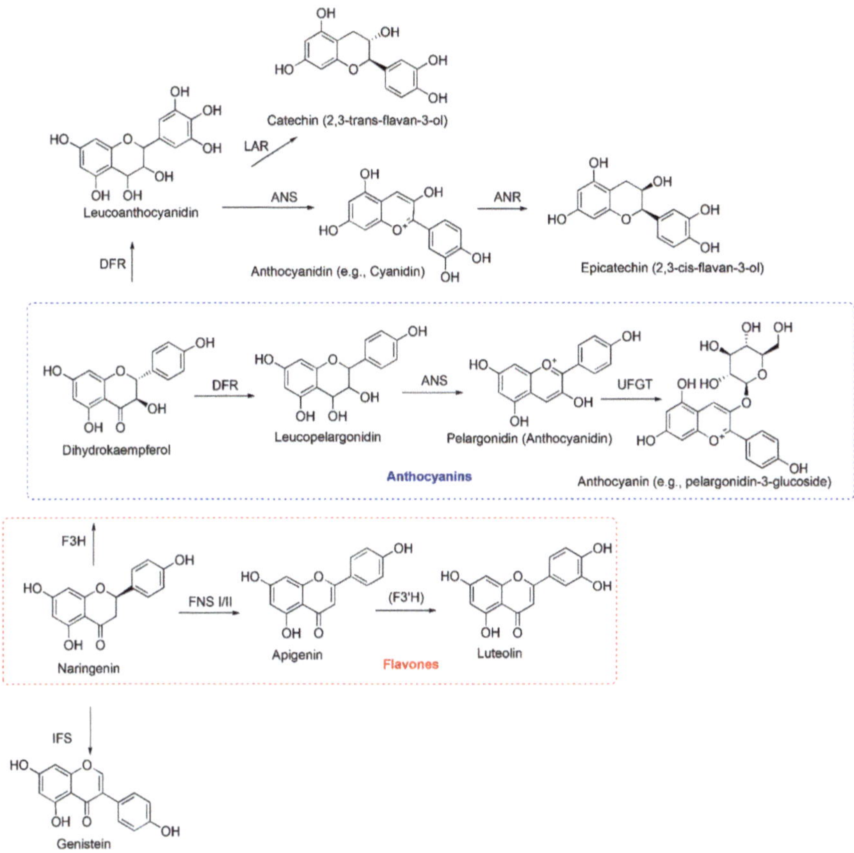

Fig. 10.3 (continued)

10.5 Biosynthetic Pathways of Pentacyclic Triterpenoids and Other Volatile Compounds

Terpenoids or isoprenoids are largest and most diverse class of secondary metabolites. All terpenoids are derived from isoprene units (C_5H_8). Biosynthetically, they are synthesized via mevalonate (MVA) and methylerythritol phosphate (MEP) pathways in cytosol and plastids, respectively. Sesquiterpenes, triterpenes, and steroids are synthesized through mevalonate (MVA) pathways. The methylerythritol phosphate (MEP) pathway produces monoterpenes, diterpenes, carotenoids, and chlorophyll side chains. Mevalonate and 1-deoxy-D-xylulose-5-P are the main intermediates of mevalonate (MVA) and methylerythritol phosphate (MEP) pathways, respectively (Figs. 10.4 and 10.5) (Tholl 2015; Huang et al. 2021; Dong and Qi 2025).

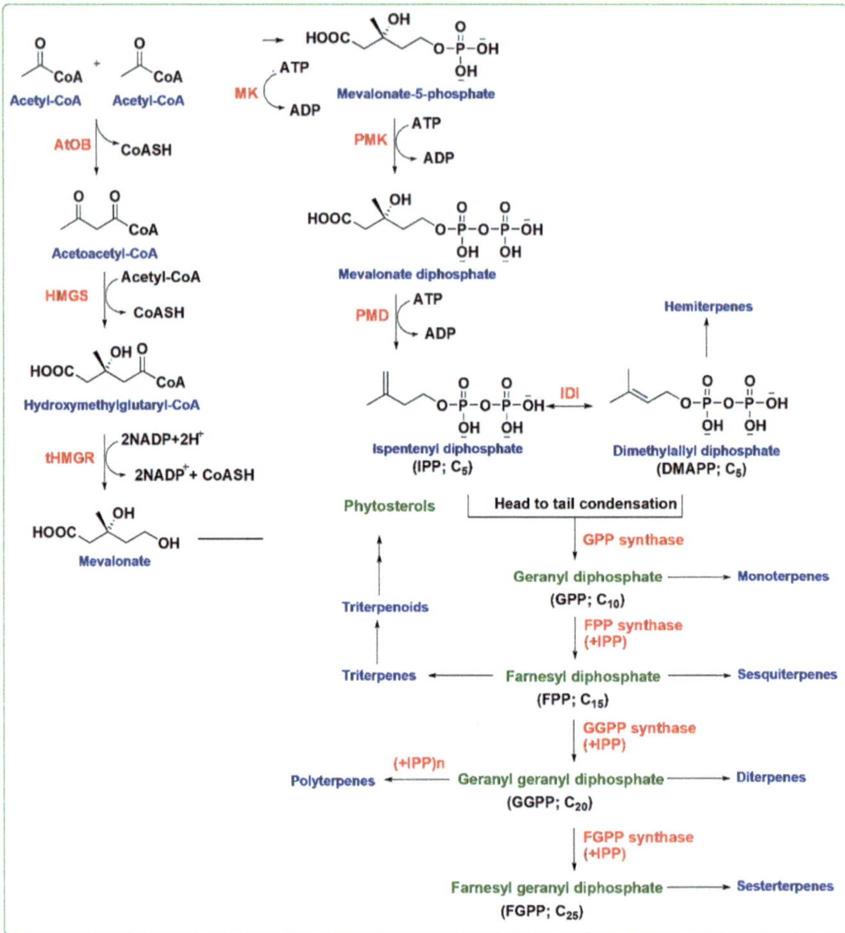

Fig. 10.4 Biosynthetic pathways of terpene class of compounds

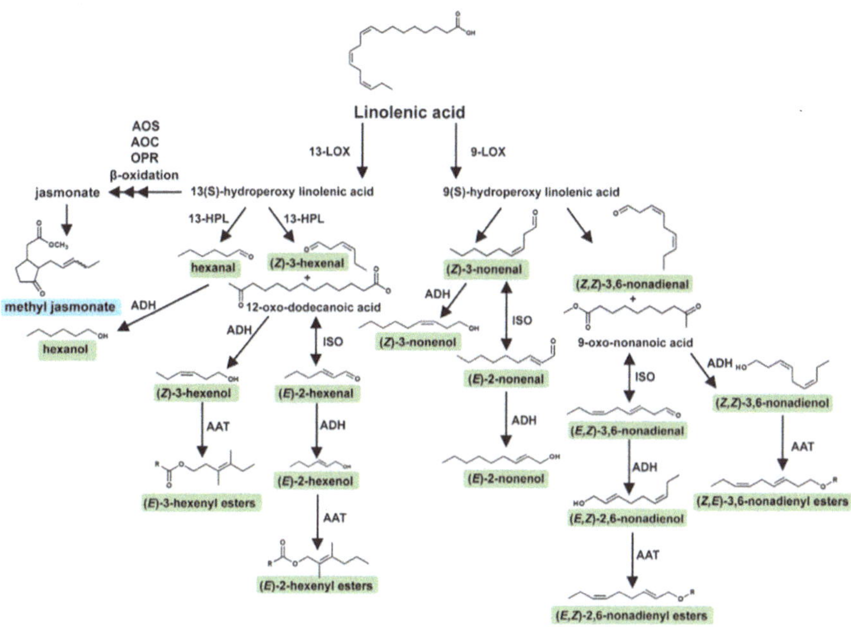

Fig. 10.5 Biosynthesis of fatty acid-derived volatile organic compounds (VOCs) and green leaf volatiles (GLVs). Abbreviations: AAT, alcohol acyltransferase; ADH, alcohol dehydrogenase; AOC, allene oxide cyclase; AOS, allene oxide synthase; 9-HPL, 9-hydroperoxide lyase; 13-HPL, 13- hydroperoxide lyase; ISO, isomerase; 9-LOX, 9-lipoxygenase; 13-LOX, 13-lipoxygenase; OPR, 12-oxophytodienoate reductase (Dudareva et al. 2013)

References

Dong H, Qi X (2025) Biosynthesis of triterpenoids in plants: pathways, regulation, and biological functions. Curr Opin Plant Biol 85:e102701

Dudareva N, Klempien A, Muhlemann JK, Kaplan I (2013) Biosynthesis, function and metabolic engineering of plant volatile organic compounds. New Phytol 198(1):16–32

El-Sayed M, Verpoorte R (2007) Catharanthus terpenoid indole alkaloids: biosynthesis and regulation. Phytochem Rev 6(2–3):277–305

Hedhili S, Courdavault V, Giglioli-Guivarc'h N, Gantet P (2007) Regulation of the terpene moiety biosynthesis of Catharanthus roseus terpene indole alkaloids. Phytochem Rev 6(2):341–351

Heller W, Forkmann G (2017) Biosynthesis of flavonoids. In: The flavonoids advances in research since 1986. Routledge, pp 499–535

Huang Y, Xie FJ, Cao X, Li MY (2021) Research progress in biosynthesis and regulation of plant terpenoids. Biotechnol Biotechnol Equip 35(1):1799–1808

Irmler S, Schröder G, St-Pierre B, Crouch NP, Hotze M, Schmidt J, Strack D, Matern U, Schröder J (2000) Indole alkaloid biosynthesis in Catharanthus roseus: new enzyme activities and identification of cytochrome P450 CYP72A1 as secologanin synthase. Plant J 24(6):797–804

Le Pogam P, Beniddir MA (2024) Structural diversity and chemical logic underlying the assembly of monoterpene indole alkaloids oligomers. Nat Prod Rep 41:1723–1765

Li C, Colinas M, Wood JC, Vaillancourt B, Hamilton JP, Jones SL, Caputi L, O'Connor SE, Buell CR (2025) Cell-type-aware regulatory landscapes governing monoterpene indole alkaloid biosynthesis in the medicinal plant Catharanthus roseus. New Phytol 245(1):347–362

Liu DH, Jin HB, Chen YH, Cui LJ, Ren WW, Gong YF, Tang KX (2007) Terpenoid indole alkaloids biosynthesis and metabolic engineering in Catharanthus roseus. J Integr Plant Biol 49(7):961–974

Liu Y, Patra B, Singh SK, Paul P, Zhou Y, Li Y, Wang Y, Pattanaik S, Yuan L (2021) Terpenoid indole alkaloid biosynthesis in Catharanthus roseus: effects and prospects of environmental factors in metabolic engineering. Biotechnol Lett 43(11):2085–2103

Memelink J, Gantet P (2007) Transcription factors involved in terpenoid indole alkaloid biosynthesis in Catharanthus roseus. Phytochem Rev 6(2):353–362

Oudin A, Courtois M, Rideau M, Clastre M (2007) The iridoid pathway in Catharanthus roseus alkaloid biosynthesis. Phytochem Rev 6(2):259–276

Pan Q, Mustafa NR, Tang K, Choi YH, Verpoorte R (2016) Monoterpenoid indole alkaloids biosynthesis and its regulation in Catharanthus roseus: a literature review from genes to metabolites. Phytochem Rev 15(2):221–250

Pandey SS, Singh S, Babu CV, Shanker K, Srivastava NK, Shukla AK, Kalra A (2016) Fungal endophytes of Catharanthus roseus enhance vindoline content by modulating structural and regulatory genes related to terpenoid indole alkaloid biosynthesis. Sci Rep 6:26583

Patil M, Murumkar C (2024) The classes and biosynthesis of flavonoids. In: Flavonoids as nutraceuticals. Apple Academic Press, Palm Bay, pp 233–251

Salim V, De Luca V (2013) Towards complete elucidation of monoterpene indole alkaloid biosynthesis pathway: Catharanthus roseus as a pioneer system. In: Advances in botanical research, vol 68. Academic Press, Elsevier, pp 1–37

Tholl D (2015) Biosynthesis and biological functions of terpenoids in plants. In: Biotechnology of isoprenoids. Springer, Cham, pp 63–106

Williams D, De Luca V (2023) Plant cytochrome P450s directing monoterpene indole alkaloid (MIA) and benzylisoquinoline alkaloid (BIA) biosynthesis. Phytochem Rev 22(2):309–338

Winkel-Shirley B (2002) Biosynthesis of flavonoids and effects of stress. Curr Opin Plant Biol 5(3):218–223

Zhu X, Zeng X, Sun C, Chen S (2014) Biosynthetic pathway of terpenoid indole alkaloids in Catharanthus roseus. Front Med 8(3):285–293

Zhu W, Yang B, Komatsu S, Lu X, Li X, Tian J (2015) Binary stress induces an increase in indole alkaloid biosynthesis in Catharanthus roseus. Front Plant Sci 6:582

Pharmacological Activities Associated with Catharanthus Species

11

Abstract

Catharanthus species are well recognized for diverse pharmacological activities due to the presence of bioactive phytochemicals and secondary metabolites, particularly terpenoid indole alkaloids like vincristine and vinblastine. These species possess potent anticancer, antimicrobial, antihypertensive, antidiabetic, antioxidant, and antiinflammatory properties. The compounds act through various mechanisms, including cell cycle arrest, enzyme inhibition, and free radical scavenging. Traditional medicinal systems have long utilized the plant *Catharanthus* for treating numerous ailments. Recent pharmacological studies support and expand these traditional uses, highlighting the therapeutic potential of different plant parts of member species of *Catharanthus*. Continued research into their bioactivity is crucial for developing novel drugs and validating their role in modern medicine. This communication deals with all pharmacological activities associated with *Catharanthus* contributing to drug discovery programmes.

Keywords

Anticancer · Antioxidant · Pharmacological activities · Catharanthus · Drug discovery

11.1 Introduction

The species of *Catharanthus* is widely recognized in ethnomedicine and modern pharmacology for its rich repertoire of bioactive compounds with diverse therapeutic potentials. Native to Madagascar and now cultivated worldwide, *Catharanthus* species have been extensively studied due to their broad spectrum of pharmacological activities and significant medicinal value. The plant produces around 130 different secondary metabolites in the form of characterized alkaloids, many of which are

S. Kumar, B. Singh, *Catharanthus in Drug Discovery*,
https://doi.org/10.1007/978-981-95-5048-7_11

pharmacologically active, making it the most chemically and biologically diverse plant of medicinal value. The various parts of *Catharanthus*—including the leaves, stems, whole plant parts, flowers, flower buds and roots—have been studied by various group of scientists and researcher for their therapeutic effects using different extracts and solvent fractions. These extracts have demonstrated different pharmacological properties such as antimicrobial, antifungal, antihypoglycemic, cytotoxic, hepatoprotective, hypolipidemic, and larvicidal activities. The nature and intensity of these activities often depend on the plant part used, the type of extract, the dosage administered, and the experimental model employed.

Notably, the anticancer properties of vincristine and vinblastine, two indole alkaloids derived from *C. roseus*, have gained global prominence in cancer chemotherapy. The pharmacological effects observed are largely due to secondary metabolites like terpenoid indole alkaloids, flavonoids, phenolic acids, and saponins. These compounds interact with specific molecular targets, exhibiting both therapeutic potential and selectivity. Table 11.1 summarizes key pharmacological activities associated with different plant parts, extract types, dosages, and corresponding outcomes.

11.2 Different Pharmacological Activities Associated with Catharanthus Species

11.2.1 Anthelmintic Activity

In an experiment, anthelmintic properties of *Catharanthus roseus* were evaluated through in-vitro methods using ethanol extracts of the whole plant. The investigation targeted the earthworm *Pheretima posthuma*, commonly employed in anthelmintic studies due to its physiological resemblance to parasitic helminths. The study tested various concentrations of the ethanol extract to determine its ability to cause paralysis and death in the worms. Results demonstrated a clear concentration-dependent response. At 200 mg/mL, the extract induced paralysis in 6.67 minutes, indicating a notable effect, while a concentration of 250 mg/mL proved most effective, causing worm death within 46.33 minutes. These outcomes suggest that *C. roseus* contains bioactive compounds with significant anthelmintic potential, particularly at higher concentrations. This research indicated the traditional medicinal use of the plant for treating helminthic infections and emphasizes its value as a potential source of novel anthelmintic agents. Additionally, the findings encourage further research to isolate and identify the specific active constituents responsible for the observed properties, as well as to explore their pharmacological profiles and safety through in vivo studies.

Table 11.1 Biological activities reported from *Catharanthus* species

Pharmacological activities	Compounds/plant part	Extract/class of compound	Dose(s)	Results	References
Antimicrobial	Crude extracts/ leaves, stem, root and flower	Methanolic and ethanolic extract, aqueous extract	100 μL	No activity was reported in the aqueous extracts.	Ramya (2008)
Antimicrobial	Hairy roots/ wild-type roots	Methanolic and ethanolic extract	100–300 μL	High antimicrobial activity	Hanafy et al. (2016)
Antibacterial	Leaf, stem, flower and root	Methanolic and ethanolic extract	100–300 μL	Antibacterial activity.	Govindasamy and Srinivasan (2012)
Antimicrobial	Leaves, stems and flowers	Methanolic and ethanolic extract, aqueous extract, chloroform extract	–	High antibacterial (ethanolic extract), high antifungal activity (metanolic extracts)	Chaman et al. (2013)
Antibacterial	Root	Hexane extract, chloroform extract, ethanolic extract, methanolic extract, aqueous extract	250–1000 μg	Chloroform extract was most active	Ethalsha and Retna (2014)
Antibacterial	Leaf, callus and root	Aqueous extract		Highest activity with stabilized nanoparticles	Malabadi et al. (2012)
Antimicrobial	Endophytic fungi	Fungal extracts		Activity due to detected antimicrobial compounds	Akpotu et al. (2017)
Antimicrobial	Leaves, shoot and root	Alkaloids		Multiphytopathoges	Vardhan et al. (2012)
Antibacterial	Stems	Granaticinic acid		Endophytic Streptomyces thermoviolaceus significant source	Roy and Banerjee (2015)
Antibiogram	Leaves, stems, roots and flowers	Methanol extract, acetone extract and ethyl acetate extract	250–1000 mg/L	Possesses the best antibiogram in ethyl acetate extracts	Sathiya et al. (2008)
Antibacterial	Seeds	–		Antibacterial activities	Kumar et al. (1997)

(continued)

Table 11.1 (continued)

Pharmacological activities	Compounds/plant part	Extract/class of compound	Dose(s)	Results	References
Antibacterial	Leaf, nodal callus and fruit	Methanolic extract		Strong microbial activity	Naz et al. (2015)
Antimicrobial	Leaf	Methanol		Exhibited a strong antimicrobial activity against several pathogens	Al-Shmgani et al. (2017)
Antimicrobial	Callus	Aqueous extract of callus		Antibacterial effect	Barkat et al. (2014)
Antimicrobial	Leaf	Ethanol		Strong antibacterial properties	
Antimicrobial, antifungal	Leaf	Aqueous extract, ethanol extract	25–100 μg/disc	Inhibit bacterial and fungal growth.	Balaabirami and Patharajan (2012)
Antibacterial	Leaf, stem, flower, root	95% ethanol		Exhibited broad-spectrum antibacterial properties	Raza et al. (2009)
Antimicrobial	Leaf	Acetone extract, ethanolic extract, chloroform extract	100 mg/mL	Inhibitory activity against most of the pathogenic microorganisms	Shanmugaraju and Bhakyaraj (2016)
Antimicrobial	Leaf	Methanol ethanol acetone		Antimicrobial activity	Patil and Ghosh (2010)
Antimicrobial	Actinomycetes/ leaf			Cr-12 and Cr-20 exhibited the highest activity against the test organism	Kafur and Basheer Khan (2011)
Antibacterial		Water and methanol		Best antimicrobial properties in methanol extract	Kabesh et al. (2015)
Antimicrobial	Leaf	Aq. extract	10 mL of 0.025 (M) $AgNO_3$	Good antimicrobial activity.	Kotakadi et al. (2013)
Antimicrobial	Stem	Methanol and its fractions n-butanol, *n*-hexane, aqueous		Highest antimicrobial properties (*n*-butanol fraction)	Pham et al. (2018)

Pharmacological activities	Compounds/plant part	Extract/class of compound	Dose(s)	Results	References
Antibacterial	Seed derived callus			Antibacterial properties	
Antihypoglycemic	Leaf	Methanol		Increases hypoglycemic level	Ohadoma and Michael (2011)
Antihyperglycemic	Leaf	Dichloromethane: methanol	500 mg/body weight	Exhibited anti hyperglycemic activity in alloxan induced diabetic rats	Jayanthi et al. (2010)
Antihyperglycemic	Leaf	Ethanol	150 and 300 mg/kg	Blood glucose lowering effect in allox an-induced hyperglycemic rats	Akhtar et al. (2007)
Antihyperglycemic	Leaf	Powder	55 mg/kg body wt	Hypolipidemic properties have a potential in this extract	Rasineni et al. (2010)
Antihyperglycemic	Leaf	Aq. extract	150 mg/kg bw	Antihyperglycemic effect in alloxan-induced diabetic rats	Manoharan et al. (2011)
Antiyperglycemic	Leaf	Petroleum ether, chloroform, ethyl acetate fractions	150 mg/kg	High antiyperglycemic activity (Chloroform fractions)	Islam et al. (2009b)
Antiyperglycemic	Leaf, twigs	Total alkaloids extract	100–150 mg/kg	Hyperglycemic properties in rats	Zhang et al. (2016)
Antidiabetic	Leaves	Aqueous	10 mg/100 g	Extracts were found to be effective as an anti- diabetic agent.	Muralidharan (2014)
Anti- diabetic	Leaves	Aqueous	3 and 500 mg/body wt	Significant antidiabetic effects	Prasad et al. (2009)
Antidiabetic	Leaves	Petroleum-ether, ethyl acetate, chloroform fractions	150 mg/kg	Ethyl acetate fractions best for diabetic	Islam et al. (2009a, b)

(continued)

Table 11.1 (continued)

Pharmacological activities	Compounds/plant part	Extract/class of compound	Dose(s)	Results	References
Antidiabetic	Leaves	Aqueous extract	1 g/kg and 500 mg/kg	Antidiabetic	Mostofa et al. (2007)
Antidiabetic	Leaf	Leaf juice or water decoction	0.5, 0.75, and 1.0 mL/kg body weight	Antidiabetic activity with the leaf juice	Nammi et al. (2003)
Antidiabetic	Leaf and flower	Ethanol extrac	12.5 mg/mL	Carbohydrate metabolism (α-amylase, α-glucosidase)	Malathi et al. (2010)
Antidiabetic	Whole plant	Methanolic	300 and 500 mg/kg	Antidiabetic in rats	Ahmed et al. (2010)
Antidiabetic	Leaf	Methanolic extract	200 and 400 mg/kg body weight	Significantly increases the wound contraction in mice	Singh et al. (2014)
Antidiabetic	Leaf	Powder	100 mg/kg body weight/day/60 days	Antidiabetic properties	Rasineni et al. (2010)
Antidiabetic	Leaves	Methanol	250 mg/kg body wt.	Reduce the blood glucose level	Aruljothi and Samipillai (2016)
Antidiabetic	Whole plant	Ethanolic	500 mg/kg body wt.	Antidiabetic activity (*C. pussillus*)	Navitha et al. (2012)
Antidiabetic	Leaves	Vindoline, vindolidine, vindolicine and vindolinine	25.0 μg/mL	High glucose uptake	Tiong et al. (2013)
Antidiabetic	Leaf	Ethanolic	100 and 200 mg/kg	Antidiabetic effect	Al-Shaqha et al. (2015)
Antidiabetic	Leaves and twigs	Dichloromethane:methanol extract (1:1)	500 mg/kg	Increased of glucose level	Singh et al. (2001)
Antidiabetic	Leaves	Aq. extract		Antidiabetic effect	Aadil et al. (2012)
Antidiarrheal	Leaf	Ethanolic	200 and 500 mg/kg	Significant antidiarrheal effect	Hassan et al. (2011)

Pharmacological activities	Compounds/plant part	Extract/class of compound	Dose(s)	Results	References
Antifungal	Leaf	Ethanol, acetone, and aqueous	25–50 μg/μL	Inhibition against microbes	Kumari and Gupta (2013)
Antifungal	Leaves	Methanol and its fractions	25–100 μg/disc	Used as prophylactic agent	Balaabirami and Patharajan (2012)
Antifungal	Leaves	Aqueous, ethanolic, chloroform	5–80 mg/mL	Highest antifungal activity in chloroform fraction	Roy and Chatterjee (2010)
Antioxidant	Root	Hexane, chloroform	100–1000 μg	High vitro antioxidant activity	Ethalsha and Retna (2014)
Antioxidant	Leaf callus cultures	Compound rf1	1.0 mg/mL	Compound rf1 possess strong antioxidant potential	Verma et al. (2012)
Antioxidant	Whole plant	Ethanol	3.9–125 μg/mL	No antioxidant property	Widowati et al. (2010)
Antioxidant	Shoots	N–hexane, chloroform	20–100%	Exhibit highest antioxidant property	Rasool et al. (2011)
Antioxidant	Leaves	Methanol, aqueous	50–250 μg/mL	Free radical scavenging and antioxidant property	Aadil et al. (2012)
Antioxidant	Root and shoot		0, 50, and 100 μM	Increased active constituents under stress	Rai et al. (2014)
Antioxidant	Leaf	Methanol	–	Synthesized AgNPs: strong in vitro antioxidant	
Antioxidative	Seeds	Crude enzyme extract	–	Salinity decreased GSH, enhanced ascorbate and peroxidase activities	Jaleel et al. (2007a, b, c)
Antioxidant	Cells	Enzymes extraction	0–50 T	Increased antioxidant activity in response to NaCl damage	Elkahoui et al. (2005)
Antioxidant	Roots	Aqueous extract	–	Antioxidant	Pereira et al. (2010)

(continued)

Table 11.1 (continued)

Pharmacological activities	Compounds/plant part	Extract/class of compound	Dose(s)	Results	References
Antioxidant	Whole parts	Aqueous extracts	300–1200	Strong antioxidant activity	Ferreres et al. (2008)
Antioxidant	Root, leaf, hairy root, callus	Ethanolic extract	200–400 μg/mL	Strong antioxidant activity	Mardani-Nejad et al. (2016)
Antioxidant	Roots	Extract of ethyl acetate of actinomycetes	500–1000 μg/mL	Antioxidant activity and Alpha glucosidase inhibition	Jasmine and Agastian (2013)
Antioxidant	Leaves	Vindoline, vindolidine, vindolicine and vindolinine	0–25.0 μg/mL	Strong antioxidant activity	Tiong et al. (2013)
Antioxidant	Whole parts	Ethyl acetate extract, aq. extract	–	Strong antioxidant activity	Pham et al. (2017)
Antioxidant	Stem	Methanol extract, n-butanol	–	Strong antioxidant properties	Pham et al. (2018)
Antioxidant	Leaves	Ethanolic extract	26,728 μg/mL	No antioxidant properties	Widowati et al. (2011)
Antioxidant	Cell suspensions	Crude aq. extract	50 μg/mL	Antioxidant properties	Verma et al. (2013)
Antioxidant	Leaves and twigs	Crude aq. extract	150 mg/kg	CTA antioxidant properties	Zhang et al. (2015)
Cytotoxic	Leaves	Methanol, *n*-hexane, chloroform, water	200 μg/mL	Cytotoxic properties	Siddiqui et al. (2010)
Cytotoxic	Leaves	Catharanthine	60 μg/mL	Cytotoxic properties	Siddiqui et al. (2010)
Cytotoxic		50 alkaloids	0.1–1 μg/mL	Dose-dependent cytotoxic effect on the cell lines	
Cytotoxic	Stem	Methanol, n-butanol (b); and aqueous fraction	5.2–21.0 μg/mL	Antitumor therapeutic agents.	Pham et al. (2018)
Cytotoxic		Vinblastine, vincristine		Biologically activities for tumour	Ruszkowska et al. (2003)

Pharmacological activities	Compounds/plant part	Extract/class of compound	Dose(s)	Results	References
Cytotoxic	*C. Roseus* suspension cultured-cells	Indole alkaloid-enriched extract	211 and 210 ng/mL	Antitumor properties	Fernández-Pérez et al. (2013)
Cytotoxic	Leaves	Ajmalicine, vindorosine, lochnerinine	–	Cytotoxic property (*C. pusillus*)	
Cytotoxic	Leaves	Aqueous	2.38–2.55 μg/mL	Differential inhibition of Jurkat cells; promoted normal immune cell proliferation	Ahmad et al. (2010)
Cytotoxic	Whole plants	Dimeric indole alkaloids	–	Cytotoxic properties for human breast cancer cell line	Wang et al. (2012a, b)
Hepatoprotective	Leaves	Ethanol	20 mg/kg	Hepatoprotective property	Ahmed and Rao (2013)
Hypoglycemic	Leaf	Methanol	250 mg/kg	Hypoglycemic effect of metformin	Ohadoma and Michael (2011)
Hypoglycemic	Leaves	Ethanol	150 mg/kg	Antidiabetic activity	Mohan et al. (2015)
Hypolipidemic	Leaves	Petroleum ether, chloroform	150 mg/kg	Hypoglycemic activity	Islam et al. (2009a, b)
Hypoglycemic	Leaf	Vindogentianine	200 μg/mL	α-Amylase and α-glucosidase inhibition	Tiong et al. (2015)
Hypoglycemic	Leaf	Aqueous extract	–	Seven products were developed using fresh leaves	Bisla et al. (2014)
Hypoglycemic	Leaf	Aqueous extract	250 mg/kg	Hypoglycemic property	Vega-Ávila et al. (2012)
Hypoglycemic	Flowers, tender stems	Aqueous extract	25 mg/mL	Significant glucose lowering compared to chlorpropamide	Iweala and Okeke (2005)

(continued)

Table 11.1 (continued)

Pharmacological activities	Compounds/plant part	Extract/class of compound	Dose(s)	Results	References
Hypoglycemic	Leaf	Aqueous extract	0.3, 1 mL/kg b.w.	Significant reduction over time in treated rats	Ahmed et al. (2007)
Hypoglycemic	Whole plant	Ethanolic	500 mg/kg	Significant hypoglycemic activity (*C. pusillus*)	Navitha et al. (2012)
Hypolipidemic	Leaf	Aqueous extract	150 mg/kg	No hypoglycemia effect observed	Muralidharan (2014)
Hypolipidemic	Leaf	Aqueous extract	0.5–1 mL/kg	Significant activity recorded for lowering lipid, also antiatherosclerotic activity	Patel et al. (2011)
Hypolipidemic	Leaf	Ethanol	150, 300 mg/kg	Lowered blood glucose	Akhtar et al. (2007)
Hypolipidemic	Leaves	Ethanolic extract	30 mg	Hypolipidemic effects	Ara et al. (2009)
Hypolipidemic	Leaf	Aqueous extract		Hipolipidemic activity in hypercholesterolemic rats	Suarsana et al. (2015)
Hypolipidemic	Leaf	Aqueous extract	100 mg/kg body	*C. roseus* showed potent hypolipidemic properties	Rasineni et al. (2010)
Larvicidal, pupicidal	Leaf	Aqueous extract, methanol	50 mg/mL	Control malarial vector	Subarani et al. (2013)
Larvicidal efficacy	Leaf	Petroleum ether	20 g/L	Larvicidal potentiality	Panneerselvam et al. (2013)
Larvicidal effect	Leaf	Aqueous, methanol crude, petroleum ether, methanol fraction, ethyl acetate fraction	1000 ppm	Exhibited larvicidal effects	Alaguchamy and Jayakumararaj (2015)
Larvicidal effect	Root	Silver nanoparticles	3.5–5 mg/L	Against dengue and filariasis larvae	Rajagopal et al. (2015)
Larvicidal effect	Leaf	Ethyl acetate extract, hexane and methanolic extract		Significant larvicidal effects	Quinquefasciatus (2017).
Larvicidal effect	Leaf	n-Hexane fraction of acetone extract	1 ppm	Produced 100% mortality in 1st 2nd instars, 94% in 3rd 4th instars	Kuppusamy et al. (2009)

11.2.2 Antiangiogenesis Activity

Anvar Sadique and Hashim (2019) reported significant antiangiogenic activity associated with the faction of ethyl acetate of a plant extract after 21-day incubation period. Their study revealed that this fraction strongly inhibited angiogenesis (formation of new blood vessels from pre-existing vasculature). This effect is particularly relevant in the context of pathological conditions such as cancer, where uncontrolled angiogenesis facilitates tumor growth and metastasis by ensuring a continuous supply of nutrients and oxygen to proliferating tumor cells. The researchers attributed the observed antiangiogenic activity primarily to the high concentration of flavonoids in the ethyl acetate fraction. Flavonoids are a class of polyphenolic compounds widely recognized for their broad spectrum of biological effects, including antioxidant, anti-inflammatory, and anticancer properties. In the context of this study, flavonoids likely interfered with critical steps in angiogenesis, such as endothelial cell proliferation, migration, and capillary tube formation. These steps are essential for neovascularization, and their disruption suggests that flavonoids can modulate key signaling pathways involved in vascular development. The study's findings underscore therapeutic potential of flavonoid-rich plant extracts in the development of natural antiangiogenic agents. In particular, the ethyl acetate fraction may act through mechanisms such as downregulation of VEGF (vascular endothelial growth factor) or inhibition of MMPs (matrix metalloproteinases), both crucial in angiogenesis. However, the study also emphasizes the need for further research to isolate and characterize the specific flavonoid compounds responsible for this activity. Future work involving in-vitro mechanistic studies and in-vivo models will be essential to better understand the pharmacodynamics and therapeutic applicability of these natural compounds in diseases where angiogenesis plays a pivotal role.

11.2.3 Antibacterial (Antiseptic) and Antimicrobial Activity

The crude extracts obtained from the various parts of *Catharanthus roseus* were evaluated for antimicrobial (bacterial) properties against a range of bacterial strains, including *Escherichia coli*, *Pseudomonas aeruginosa*, *Serratia marcescens*, *Salmonella typhi*, *Staphylococcus aureus*, *Streptococcus pyogenes*, *Bacillus cereus*, and *B. subtilis* using the disc diffusion assay. The inhibitory effects observed were influenced by several factors such as the part of plant used, extraction method, solvent type, and the microbial strain tested (Sibaoka 1969). Among the various plant parts, leaf extracts demonstrated the highest antibacterial efficacy. Notably, extracts obtained using organic solvents showed significantly greater antibacterial activity compared to those prepared with aqueous solvents, which were largely inactive against the tested strains. Ethanol-extracted samples exhibited the strongest inhibitory effect, with Gram-negative bacteria showing greater sensitivity than Gram +ve strains (Ramya 2008). Additionally, when Agrobacterium-infected leaves were used as explants, they produced transgenic hairy roots at a higher frequency than explants of stems. These hairy root cultures showed enhanced antimicrobial activity. Root development was assessed by culturing root slices in either half-strength or full-strength MS liquid medium, with fresh weight monitored over time in shake flask cultures. This study highlights the potential of both conventional and genetically modified plant tissues of *C. roseus* as sources of bioactive compounds with promising antibacterial properties.

The maximum biomass production and antimicrobial activity of *C. roseus* hairy roots were observed when cultured in full-strength MS liquid medium (Hanafy et al. 2016). The various solvent extracts, including ethanol and methanol, were tested against human pathogens such as *Bacillus subtilis*, *Escherichia coli*, *Staphylococcus aureus*, *Salmonella typhi*, and *Pseudomonas aeruginosa*. The ethanolic extract showed the highest zone of inhibition (21.15 ± 1.64 mm) against *S. typhi* and the lowest (6.24 ± 0.69 mm) against *S. aureus*. The methanolic extracts also showed notable activity, with the highest inhibition (15.61 ± 1.35 mm) against *S. typhi* and the lowest (5.20 ± 0.86 mm) against *E. coli* (Govindasamy and Srinivasan 2012). Further several studies evaluated ethanolic, methanolic, aqueous, and chloroform extracts of *C. roseus* leaves, stems, and flowers against a range of bacterial and fungal strains. The ethanolic extracts were effective against *C. albicans*, *P. aeruginosa*, and *A. niger*. The methanolic flower extracts exhibited the strongest antifungal activity against *C. albicans*, followed by leaf and stem extracts. The chloroform flower extracts displayed notable antibacterial activity, particularly against *P. aeruginosa*. The aqueous extracts of all plant parts were generally inactive against the tested microorganisms (Chaman et al. 2013). Additionally, the hexane, chloroform, ethyl acetate, methanolic, and aqueous root extracts were tested against *E. coli*, *P. vulgaris*, *K. pneumoniae*, *S. aureus*, *M. luteus*, and *B. subtilis*. Chloroform root extracts showed the most potent antibacterial activity with Gram −ve bacteria being more sensitive than Gram −ve strains (Ethalsha and Retna 2014).

The antimicrobial activity of *C. roseus* was also evaluated using extracts from its leaves, stems, roots and flowers prepared with different solvents—methanol, acetone, and ethyl acetate. Among these, the ethyl acetate extracts demonstrated the most significant antibacterial effects, followed by methanol and acetone extracts. The ethyl acetate's superior performance is attributed to its higher efficiency in extracting bioactive compounds from *C. roseus* (Sathiya et al. 2008). The root, stem, leaf, and flower extracts exhibited the highest antibacterial activity against *B. subtilis*, followed by *Klebsiella* species, while the weakest activity was observed against *Streptococcus* species. The *S. aureus* displayed moderate sensitivity to all tested extracts. In related studies, Kumar et al. (1997) screened the seeds from 40 plant species growing on cattle dung heaps for antibacterial activity against *B. subtilis*, *E. coli*, *Pseudomonas cichorii* and *Salmonella typhimurium*. Antibacterial activity was found in seeds of 36 species, with the highest inhibition against *P. cichorii* (97%), followed by *B. subtilis* (83%), *S. typhimurium* (69%) and *E. coli* (55%). Furthermore, *Vinca rosea* (synonymous with *C. roseus*) methanolic extracts from in vivo and in vitro leaves and calluses derived from leaf, nodal, and fruit explants exhibited notable antimicrobial activity (Naz et al. 2015). In vitro extracts were more effective than in vivo ones. The highest antibacterial activity (30.3 ± 0.58 mm) was observed against *B. licheniformis* using in vitro leaf callus extract at 2.0 mg/mL. The strongest antifungal activity (34.6 ± 0.57 mm, MIC 6.0 mg/mL) was recorded against *Aspergillus niger* using the same extract.

The ethanol leaf extract of *C. roseus* has demonstrated strong antimicrobial properties against several human pathogenic microorganisms and fungi, particularly *Staphylococcus aureus*, *Escherichia coli*, and *Candida albicans*, when evaluated using the disc diffusion method. According to study, the most significant inhibition was observed against *S. aureus*, with a 15 mm zone of inhibition at a concentration of 100 mg/mL, followed by *E. coli* (11 mm) and *C. albicans* (12 mm). This highlights the potential of ethanol-extracted leaf compounds in combating both bacterial and fungal pathogens. Further evidence of antimicrobial potential is provided by Balaabirami and Patharajan (2012), who reported broad-spectrum activity of *C. roseus* leaf extracts against a wide range of bacteria including *E. coli*, *Klebsiella oxytoca*, *K. pneumoniae*, *Proteus mirabilis*, *Pseudomonas aeruginosa*, *Salmonella typhimurium*, *S. paratyphi*, and *S. aureus*. Antifungal activity was also noted against *Aspergillus niger*, *A. fumigatus*, *C. albicans*, *Penicillium chrysogenum*, and *A. flavus*. These findings reinforce the antimicrobial potency of C. roseus, particularly the leaf extracts. However, the efficacy of different plant parts varied. Raza et al. (2009) reported that while leaf, stem, and flower extracts did not inhibit *S. aureus* or *P. aeruginosa*, root extracts displayed strong antibacterial activity. Notably, root extract showed broad-spectrum activity against *Salmonella typhi* and *Shigella boydii*, with inhibition zones of 24 and 22 mm, respectively. Additionally, flower extracts showed activity against *Corynebacterium diphtheriae*, whereas leaf extracts were ineffective against this pathogen. Comparative studies by Shanmugaraju and Bhakyaraj (2016) confirmed that ethanol extracts were more effective than acetone and chloroform extracts. Staphylococcus species were most susceptible, followed by E. coli, indicating solvent-dependent variations in antimicrobial effectiveness. Further studies by Patil and Ghosh (2010), as well as Kabesh et al. (2015), showed that ethanol, acetone, and methanol extracts of *C. roseus* possessed notable activity against *P. aeruginosa*, *S. typhimurium*, and *S. aureus*, with ethanol showing the highest efficacy. Similarly, the studies conducted by Pham et al. (2018) found that the n-butanol fraction of *C. roseus* effectively inhibited the growth of *E. coli* and *Staphylococcus lugdunensis*, indicating the potential of other solvent fractions beyond ethanol. A significant discovery was made by Kafur and Basheer Khan (2011), who isolated endophytic actinomycetes from surface-sterilized *C. roseus* leaves. Out of 38 isolates cultured on starch casein agar, 20 morphologically distinct strains were screened for antimicrobial activity. Approximately 65% showed positive activity, with isolates Cr-12 and Cr-20 demonstrating the strongest inhibition against both bacterial strains (*B. subtilis*, *S. aureus*, *P. aeruginosa*, *P. vulgaris*) and fungal strains (*C. albicans*, *Botrytis cinerea*, *Curvularia lunata*, *Fusarium oxysporum*, *F. solani, and Rhizoctonia solani*). These findings point to the rich microbial diversity harbored within *C. roseus*, particularly its endophytic communities, as promising sources for novel antimicrobial agents.

The silver nanoparticles (AgNPs) have emerged as highly valuable materials due to their broad-spectrum antimicrobial properties, rapid synthesis, and cost-effectiveness. In this context, the present study supports the development of an eco-friendly and sustainable approach for synthesizing AgNPs using plant extracts—a significant advancement in the application of nanotechnology. AgNPs were successfully synthesized from room-dried leaves of *C. roseus* through a green synthesis method. The nanoparticles were characterized using UV–visible spectroscopy, confirming surface plasmon resonance, while their morphology and elemental composition were further analyzed using scanning electron microscopy (SEM) coupled with energy-dispersive X-ray spectroscopy (EDX). X-ray diffraction (XRD) analysis confirmed the crystalline nature of the synthesized AgNPs. The particle size of the AgNPs was measured using the Hariba nanoparticle analyzer and found to be 27 ± 2 nm and 30 ± 2 nm for *V. rosea* and *C. roseus*, respectively. Stability of the nanoparticles was assessed using Zeta potential analysis, which revealed a high negative value of −63.1 mV, indicating strong repulsion between particles and excellent colloidal stability. These biosynthesized AgNPs exhibited potent antimicrobial activity against a range of pathogens (Kotakadi et al. 2013; Ohadoma and Michael 2011; Al-Shmgani et al. 2017; Barkat et al. 2014). In another study, Malabadi et al. (2012) synthesized AgNPs using aqueous leaf, callus, and root extracts of in vitro-derived *C. roseus* plants. The nanoparticles showed significant antibacterial activity against clinical strains such as *B. subtilis*, *S. aureus*, *E. coli*, *K. pneumoniae*, and the fungus *C. albicans*. Further insights into the antimicrobial potential of C. roseus were provided by Akpotu et al. (2017), who isolated endophytic fungi from *C. roseus* plants collected from the riverbanks of Amassoma, Southern Nigeria. The fungi were subjected to solid-state fermentation on rice media, and ethyl acetate extracts of their metabolites exhibited strong antibacterial and antifungal activities, with minimum inhibitory concentrations (MICs) ranging from 0.0625 to 1 mg/mL.

11.2.4 Antifertility Effect

The bioefficacy of ethanolic leaf extracts from three plant species—*C. roseus*, *O. sanctum*, and *L. camara*—was investigated for their impact on the fecundity and fertility of the red cotton bug (*D. koenigii*) (Kayesth et al. 2017). The study focused on oviposition behavior, egg numbers, and the fertility of females that emerged from nymphs treated with plant extracts. The results demonstrated a significant reduction in reproductive performance across all treatments, with treated females laying fewer egg batches and total eggs within a shortened lifespan. Among the tested extracts, the ethanolic leaf extract of *Catharanthus roseus* was most effective in suppressing oviposition. While mated females still laid eggs, indicating successful fertilization, the hatchability of those eggs was significantly reduced. Notably, the extract of O. sanctum showed the highest reduction in egg fertility, suggesting a potential disruption in embryonic development. Further supporting the antifertility potential of *C. roseus*, another studies carried out by Gupta (2009) found that its petroleum ether leaf extract inhibited estrogen-induced uterine weight gain when co-administered with estradiol-17β. In a similar study, Shaw et al. (2017)

investigated the long-term reproductive effects of this species leaf extract in male animals. Divided into control, low-dose, and high-dose groups, animals received daily oral administration for 30 days. Post-treatment analysis revealed a significant decline in epididymal sperm count and serum testosterone levels, confirming the extract's antifertility potential. These findings suggest that C. roseus contains bioactive compounds capable of altering reproductive physiology in both insects and mammals.

11.2.5 Antifungal Activity

The antifungal potential of *C. roseus* has been widely studied using various extraction solvents and fungal pathogens. A study carried out by Kumari and Gupta (2013) reported the antifungal activity of ethanol, acetone and aqueous leaf extracts of *C. roseus* against *C. albicans*, *A. fumigatus*, *A. niger*, and *F. moniliforme* using the paper disc diffusion method. Their results revealed that the efficacy of inhibition largely depended on the solvent used. The ethanolic and acetone extracts exhibited stronger antifungal activity compared to aqueous extracts. Notably, the ethanolic leaf extract showed excellent inhibition against *F. moniliforme*, with minimum inhibitory concentrations (MIC) ranging between 25 to 50 μg/μL. Similarly, in another study carried by Balaabirami and Patharajan (2012) demonstrated that leaf extracts of *C. roseus* effectively inhibited the growth of several fungal strains, including *A. niger*, *A. fumigatus*, *C. albicans*, *P. chrysogenum*, and *A. flavus*. Roy and Chatterjee (2010) assessed different solvent fractions of aqueous ethanolic leaf extracts and found the chloroform fraction exhibited the highest antifungal activity. Chaman et al. (2013) reported maximum antifungal activity against *C. albicans* using methanolic flower extracts, followed by leaf and stem extracts. However, aqueous extracts of stem, leaves, and flowers were found ineffective.

Several studies indicated the endophytic fungi isolated from *Catharnthus species* used to possess very strong antimicrobial properties. Akpotu et al. (2017) identified bioactive secondary metabolites from these fungi via solid-state fermentation and ethyl acetate extraction, which exhibited antibacterial and antifungal properties with MIC values ranging from 0.0625 to 1 mg/mL. In a study carried out by Kafur and Basheer Khan (2011) reported that the isolated 38 endophytic actinomycetes from *C. roseus* leaves, and 20 morphologically distinct strains species were screened, and found that 65% of these showed antimicrobial activity. Cr-12 and Cr-20 isolates were particularly effective against fungal pathogens like *C. albicans*, *F. oxysporum*, and *R. solani*, indicating high microbial diversity and biopotency within the plant's endophytes. Naz et al. (2015) evaluated methanolic extracts of in vivo and in vitro leaves, nodal, and fruit calluses. In vitro leaf callus extract exhibited superior antimicrobial activity, with a maximum inhibition zone of 30.3 mm against *B. licheniformis* and 34.6 mm against *A. niger*, with MIC values of 2.0 and 6.0 mg/mL, respectively. The study also confirmed the ethanol leaf extract's has strong antimicrobial properties against *S. aureus* (15 mm), *E. coli* (11 mm), and *C. albicans* (12 mm) at 100 mg/mL. These results reinforce the antifungal potential of *C. roseus* across different strains and extraction methods. Antifungal and broader antimicrobial activity of *C. roseus* is highly dependent on the extraction solvent, plant part used, and method of application. The presence of active secondary metabolites, including those from endophytes, positions *C. roseus* as a promising candidate for the development of natural antifungal agents.

11.2.6 Antihyperglycemic Activity

A study on the antidiabetic potential of *C. roseus* demonstrated that oral administration of a dichloromethane: methanol (1:1) leaf extract (500 mg/kg body weight) significantly improved metabolic parameters in alloxan-induced diabetic rats over 20 days. The treatment led to increase in the body weight and a significant reduction ($p < 0.05$) in blood glucose, urea, and cholesterol levels. Additionally, the treated diabetic rats showed elevated protein and glycogen levels ($p < 0.01$), which indicate improved carbohydrate metabolism (Jayanthi et al. 2010). The hepatic enzyme activity was also positively affected by the extract. There was a significant increase ($p < 0.01$) in hexokinase activity, an enzyme critical in glycolysis, while the key gluconeogenic enzymes—glucose-6-phosphatase and fructose-1,6-bisphosphatase—were significantly reduced ($p < 0.05$), suggesting improved insulin sensitivity and reduced glucose production in the liver. In a separate experiment, diabetes was induced in male Wistar rats using streptozotocin (STZ, 55 mg/kg bw). Animals were divided into control, control-treated, diabetic and diabetic-treated groups. Diabetic and control-treated groups received *C. roseus* leaf powder suspension (100 mg/kg/day) orally for 60 days. In diabetic rats, elevated plasma glucose and reduced insulin levels were observed. However, in the diabetic-treated group, plasma glucose levels significantly decreased after 15 days and neared normal levels by day 60, although the insulin levels remained lower than normal (Rasineni et al. 2010). The treatment also improved lipid profiles—total cholesterol, triglycerides, LDL and VLDL were normalized, along with the atherogenic index. Additionally, the hepatic and muscle glycogen depletion seen in the diabetic controls was prevented. The enzymatic alterations in glucose metabolism (glycogen phosphorylase, phosphofructokinase, pyruvate kinase, glucose-6-phosphate dehydrogenase) were significantly reversed. These studies, along with evidence of hypoglycemic effects from total alkaloids in *C. roseus* leaves and twigs, support its antidiabetic potential.

The antidiabetic and antioxidant properties of *C. roseus* have been extensively investigated, particularly in the models of chemically induced diabetes in rats. Oxidative stress as the major contributor to diabetic complications, was evaluated in heart, liver, and kidney tissues by measuring the levels of key antioxidant enzymes and oxidative markers, including superoxide dismutase (SOD), catalase (CAT), glutathione (GSH), and thiobarbituric acid reactive substances (TBARS). Oral administration of *C. roseus* leaf extract for 30 consecutive days in diabetic rats led to a significant ($p < 0.05$) decrease in fasting blood glucose levels and a corresponding increase in insulin secretion compared to untreated diabetic controls (Zhang et al.

2016). Additionally, the *C. roseus* treatment significantly reduced serum levels of hepatic enzymes—alanine aminotransferase (ALT), aspartate aminotransferase (AST), and alkaline phosphatase (ALP)—indicating a protective effect on liver function. Importantly, the administration of *C. roseus* also normalized oxidative stress markers, with significant reductions ($p < 0.05$) in SOD, CAT, and GSH levels, along with a decrease in TBARS in the heart, liver, and kidneys. These results suggested that *C. roseus* plays a key role in restoring antioxidant balance and mitigating oxidative damage in diabetic conditions through free radical scavenging, thus reducing the risk of diabetes-related complications. In a similar study, the diabetes mellitus was induced in overnight-fasted Wistar rats via a single intraperitoneal injection of alloxan monohydrate (150 mg/kg bw). The mechanism of antidiabetic action was evaluated by analyzing glucose, insulin, C-peptide, glycosylated hemoglobin (HbA1c), carbohydrate-metabolizing enzyme activities, and glycogen content in liver and muscle tissues. The ethanolic extracts of *C. roseus* leaves and *Coscinium fenestratum* stems exhibited potent antidiabetic activity, with efficacy comparable to the standard antidiabetic drug, glibenclamide (Manoharan et al. 2011). The experimentally (treated) diabetic rats showed improved insulin secretion, reduced HbA1c, enhanced glycogen storage, and normalization of enzymes involved in glycolysis and gluconeogenesis, such as hexokinase and glucose-6-phosphatase. Comparative studies further evaluated the antihyperglycemic and hypolipidemic effects of *C. roseus* and *Coccinia cordifolia* in alloxan-induced diabetic rats. Single intraperitoneal doses of 150 mg/kg and 300 mg/kg of the ethanol extracts were administered. On day 7, *C. roseus* (150 mg/kg) reduced fasting blood glucose levels by 37.1%, and by 48.5% on day 14 ($p < 0.01$), while C. cordifolia (300 mg/kg) showed even higher reductions of 47.4% and 59.7%, respectively. Lipid profiles also improved significantly: serum cholesterol and triglyceride levels were reduced by 43.3% and 39.4%, respectively, with *C. roseus* extract at 150 mg/kg by the 14th day. However, it was noted that the 300 mg/kg dose of *C. roseus* proved toxic, resulting in mortality within 2–5 days, highlighting the importance of dosage optimization.

In comparative studies on hypoglycemic activity, *Coccinia cordifolia* was found to be more effective than *C. roseus* in reducing blood glucose and serum triglyceride levels in alloxan-induced diabetic rats (AIDRs). However, *C. roseus* showed slightly better efficacy in lowering serum cholesterol over long-term treatment (Akhtar et al. 2007). Further investigations explored the hypoglycemic potential of different solvent fractions derived from the ethanolic extracts of *C. cordifolia* and *C. roseus*. The petroleum ether, chloroform, and ethyl acetate fractions (150 mg/kg bw) were administered intraperitoneally to both normal control and orally glucose-induced hyperglycemic rats. The blood glucose levels were measured at 0, 30, 60, 90, 150, and 270 minutes post-treatment using a glucometer via tail vein pricking. Among the tested fractions, the chloroform fraction of *C. cordifolia* exhibited the most pronounced hypoglycemic activity, showing a maximum blood glucose reduction of 21.94% at 60 minutes in glucose-induced hyperglycemic rats. In contrast, the petroleum ether fraction of *C. roseus* demonstrated a significant ($p < 0.05$) glucose-lowering effect, with a maximum reduction of 17.92% at 30 minutes. The metformin hydrochloride served as the standard reference drug in the experiment. Overall, the chloroform fraction of *C. cordifolia* appeared more potent among its respective solvent fractions, while the petroleum ether fraction was the most effective among *C. roseus* fractions. Phytochemical analysis revealed the presence of bioactive compounds such as saponins and flavonoids in the chloroform fraction of *C. cordifolia*, which are known for their hypoglycemic properties (Islam et al. 2009a, b). These findings suggest that specific solvent fractions may enhance the antihyperglycemic effects of medicinal plants through selective extraction of active phytoconstituents.

11.2.7 Antimutagenic Activity

Vindoline, an active alkaloid derived from *Catharanthus roseus*, demonstrated significant antimutagenic properties against a range of microbial strains such as *Pseudomonas aeruginosa*, *Saccharomyces cerevisiae*, *Candida auris*, *Escherichia coli*, *Staphylococcus aureus*, and *Bacillus subtilis* using dose 3 mg/kg.

11.2.8 Antioxidant Activity

Antioxidants act as radical scavengers that help protect the human body by inhibiting oxidative chain reactions caused by free radicals. At low concentrations, these compounds can significantly delay or prevent the oxidation of susceptible biological molecules. Their protective role is closely linked to a reduced risk of chronic conditions such as cardiovascular diseases, cancer, atherosclerosis, inflammatory bowel disease, and Alzheimer's disease. Due to their health benefits, extensive research has been conducted to identify new antioxidant compounds or radical scavengers. In this context, the chloroform root extract of *C. roseus* was evaluated for its antioxidant activity using the 2,2-diphenyl-1-picrylhydrazyl (DPPH) assay across concentrations ranging from 100 to 1000 μg/mL. The extract demonstrated strong antioxidant potential, which increased with concentration and surpassed that of the standard used. A significant activity was observed at concentrations of 180 and 200 μg/mL, with favourable IC_{50} values, indicating effective free radical inhibition (Ethalsha and Retna 2014). These findings suggest that *C. roseus* root extract could be a promising natural source of antioxidants for potential therapeutic applications.

Several studies have investigated the antioxidant potential of *C. roseus*) and revealing its capacity to scavenge free radicals and inhibit lipid peroxidation through various extracts and methods. A study conducted by Widowati et al. (2010) compared the antioxidant activity of *C. roseus* extracts with well-known standards such as quercetin and butylated hydroxyanisole (BHA) using the DPPH assay. The IC_{50} values for *C. roseus*, quercetin, and BHA were reported as 358.411, 19.200, and 94.178 μg/mL, respectively, indicating that while *C. roseus* exhibited antioxidant activity, it was less potent compared to the controls. Further exploration into *C. roseus* shoots using solvents of varying polarity—n-hexane, ethyl acetate, methanol, and chloroform—demonstrated significant free radical scavenging activity and lipid peroxidation inhibition in the linoleic acid system (Rasool et al. 2011). The extracts and fractions showed DPPH IC_{50} values ranging from 28.2 to 119 μg/mL, with inhibition of lipid peroxidation between 38.4% and 75.1%. Among these, the 100% methanolic extract consistently demonstrated the highest antioxidant capacity, and the variations observed were statistically significant ($p < 0.05$), underscoring the role of solvent polarity in extracting bioactive compounds. Another study by Aadil et al. (2012) confirmed the strong antioxidant potential of methanolic extracts through DPPH and H_2O_2 scavenging assays. These findings were in line with previous results, supporting methanol as an effective solvent for isolating antioxidant compounds from *C. roseus*. Pereira et al. (2010) focused on the antioxidant role of organic acids present in *C. roseus* root extracts. This study examined their effect on superoxide, DPPH, and nitric oxide radicals, revealing a concentration-dependent response. At concentrations up to 191 μg/mL, the mixture of organic acids reached its peak activity (29% inhibition), suggesting synergistic interactions among multiple compounds rather than the activity being attributed to any single major component. An additional comparative study evaluated lyophilized aqueous extracts of different *C. roseus* tissues—petals, seeds, leaves, and stems—for their antioxidant properties (IC_{50} values). The flower petals showed the highest antioxidant capacity ($IC_{50} = 197$ μg/mL), followed by seeds (265 μg/mL), leaves (447 μg/mL), and stems (476 μg/mL) in scavenging superoxide and nitric oxide radicals. In terms of superoxide scavenging, seeds were most effective ($IC_{50} = 74$ μg/mL), followed by leaves (90 μg/mL), stems (202 μg/mL), and petals (260 μg/mL). For nitric oxide scavenging, the order of effectiveness based on IC_{25} values was petals > seeds > leaves > stems. Overall, these studies highlight the diverse antioxidant capabilities of *C. roseus*, influenced by plant part, solvent used, and assay type. The methanolic extracts of shoots and petals, consistently demonstrated strong activity, while root-derived organic acid mixtures exhibited notable synergistic effects.

C. roseus has demonstrated antioxidant properties, particularly through its ability to scavenge a range of free radicals, including reactive oxygen species (ROS) such as the superoxide radical, and reactive nitrogen species (RNS) like nitric oxide. Several research studies have shown that this activity is not only concentration-dependent but also varies based on plant part and extraction method. In a study carried out by Ferreres et al. (2008) highlighted the strong scavenging capacity of *C. roseus* petals against DPPH, superoxide radicals, and nitric oxide. The petals displayed the most pronounced protective effect in a dose-dependent manner, suggesting their high antioxidant potential. Comprehensive assessments of *C. roseus* antioxidant activity have employed multiple assays including DPPH, reducing power, inhibition of lipid peroxidation, and malondialdehyde (MDA) production. In a comparative study by Mardani-Nejad et al. (2016), the ethanol extracts derived from hairy roots were found to be particularly rich in phenolics, flavonoids, and gallic acid—significantly more so ($p \leq 0.01$) than extracts obtained from natural roots, leaves and callus cultures. These biochemical compounds are well-known contributors to antioxidant activity.

The antioxidant activity of *C. roseus* is significantly influenced by extraction solvents, drying methods, and the plant's associated endophytes. A study carried out by Pham et al. (2017) investigated how different solvents and drying techniques affect the yield of bioactive compounds. The aqueous (water) was identified as the most effective solvent for extracting phenolics and flavonoids, while the methanol was optimal for extracting saponins and proanthocyanidins. Among drying methods, vacuum drying at 50 °C (VD_{50}) preserved high levels of phenolics and flavonoids in leaves and flowers. In contrast, infrared drying at 35 °C (IRD_{35}) was best suited for stems and roots, which are rich in saponins. Further analysis by Pham et al. (2018) revealed that the n-butanol fraction of *C. roseus* stem extract had significantly higher antioxidant activity than both aqueous fractions and crude extracts, particularly in DPPH, ABTS, and CUPRAC assays. This enhanced activity was attributed to the high levels of phenolics and saponins in the n-butanol fraction, suggesting a strong correlation between these compounds and antioxidant capacity. FRAP values did not follow the same trend. These findings aligned with previous studies showing saponins in *C. roseus* stems contributed significantly to DPPH radical scavenging. In addition to plant-derived compounds, endophytic actinomycetes (EEA) isolated from *C. roseus* roots also demonstrated promising antioxidant potential. At 1000 mg/mL, EEA showed 54.29% DPPH radical inhibition, with moderate activity (47.81%) at 500 mg/mL, in a dose-dependent manner. Hydroxyl radical scavenging was also high—71.15% at 1000 mg/mL and 61.5% at 500 mg/mL. These endophytes exhibited broad-spectrum antioxidant activity against hydroxyl radicals, hydrogen peroxide, nitric oxide, and DPPH, and also showed alpha-glucosidase inhibition, highlighting their potential in managing oxidative stress and hyperglycemia (Jasmine and Agastian 2013).

The oxygen radical absorbance capacity (ORAC), expressed as Trolox equivalent (TE), showed a strong linear correlation with Trolox dosage ($R^2 = 0.9668$). Among tested compounds, quercetin exhibited the highest TE value at 406.5 ± 2.1 μM. Among isolated alkaloids, vindolicine showed the strongest antioxidant activity with an ORAC value of 185.5 ± 4.9 μM, followed by vindolidine (92.0 ± 11.3 μM), compound IV (61.0 ± 19.8 μM), and vindoline (41.0 ± 12.7 μM). The vindolicine also demonstrated the most potent DPPH radical scavenging activity (IC_{50} = 52.9 ± 12.0 μM), surpassing vindoline (73.0 ± 14.7 μM), vindolidine (91.1 ± 20.9 μM), and vindolinine (165.5 ± 53.0 μM). Notably, vindoline, vindolidine, and vindolicine showed stronger DPPH scavenging than ascorbic acid. Vindolicine further reduced H_2O_2-induced oxidative damage in β-TC6 cells at concentrations of 12.5 and 25.0 μg/mL, indicating its protective potential (Tiong et al. 2013). In another study, the ethanol extracts of *Piper betle*, *Catharanthus roseus*, *Diospyros petandra*, and *Cinnamomum mangga* were evaluated for DPPH scavenging activity. The IC_{50} values were 5.49, 102.96, 4.74, and 277.79 μg/mL, respectively, with *P. betle* and *D. petandra* showing notably stronger antioxidant activity compared to *C. roseus* and *C. mangga* (Widowati et al. 2011). Additionally, *C. roseus* extracts from shake flask suspensions demonstrated higher DPPH radical scavenging activity (57.47%) across concentrations compared to bioreactor-grown cells. However, bioreactor-harvested crude alkaloid extracts supplemented with 5-methyl tryptophan (50 μg/mL) showed significantly higher total antioxidant capacity, reducing power, phenolic content, and ferric-reducing antioxidant power (increases of 65.21%, 17.75%, 97.0%, and 100%, respectively) than shake flask cultures. Interestingly, shake flask extracts still outperformed bioreactor extracts in DPPH scavenging (Verma et al. 2013).

The impact of chromium (Cr) on *C. roseus* was studied to understand its effects on both enzymatic and non-enzymatic antioxidant mechanisms, as well as on the production of important anticancer alkaloids vincristine and vinblastine. Two varieties of *C. roseus* were grown under pot culture conditions and exposed to different concentrations of Cr (0, 10, 50, 100 μM) to assess the plant's protective response against Cr-induced oxidative stress. The study revealed that Cr exposure adversely affected all measured parameters, including growth performance, which declined significantly with increasing Cr levels. Despite this, the contents of vincristine and vinblastine were enhanced under Cr stress, highlighting the plant's remarkable ability to maintain and even improve production of these active compounds despite toxic conditions. These findings suggest that *C. roseus* possesses robust antioxidant potential and adaptive mechanisms that allow it to tolerate heavy metal stress while enhancing valuable secondary metabolites (Rai et al. 2014). Salicylic acid (SA) has also been studied for its role in alleviating stress in plants, particularly under nickel (Ni) toxicity. In a separate experiment, thirty-day-old *C. roseus* plants were subjected to varying Ni concentrations (0, 50, 100, 150 mg/kg) with or without foliar applications of SA (10^{-5} M) under controlled net house conditions. Ni stress significantly impaired plant growth, reducing height, leaf area index, and both fresh and dry biomass of shoots and roots. Photosynthetic efficiency and activities of key enzymes like nitrate reductase and carbonic anhydrase also declined with increasing Ni concentration. In response to Ni-induced oxidative stress, the plants showed elevated activities of antioxidant enzymes such as superoxide dismutase, catalase, and peroxidase, alongside increased electrolyte leakage and proline accumulation, indicative of stress. Moreover, Ni toxicity caused a notable reduction in total alkaloid content. However, foliar application of SA mitigated many negative effects of Ni by promoting recovery in growth parameters and significantly boosting total alkaloid content, including a remarkable 22.2% increase in vincristine and a 50.0% increase in vinblastine levels in plants exposed to 150 mg/kg Ni (Idrees et al. 2013). In addition to these stress-related studies, the antioxidant activity of bioconjugated silver nanoparticles (AgNPs) synthesized using *C. roseus* leaf extracts was evaluated through the DPPH radical scavenging assay. The assay is based on the reduction of DPPH, a purple-colored stable radical, which turns yellow upon neutralization by antioxidants. The inhibition of DPPH radicals increased in a dose-dependent manner at concentrations of 50, 150, and 300 μg/mL of AgNPs, with the highest inhibition rate reaching 82% at 300 μg/mL. The disappearance of the purple color indicated the potent antioxidant activity of the synthesized nanoparticles, which is attributed to the functional groups derived from the bioactive compounds present in the *C. roseus* leaf extract that capped and stabilized the AgNPs (Al-Shmgani et al. 2015). Together, these studies demonstrate that *C. roseus* not only exhibits strong antioxidant defense mechanisms under heavy metal stress but also enhances valuable alkaloid production. Moreover, its leaf extracts can be effectively utilized in the green synthesis of antioxidant silver nanoparticles, broadening the potential applications of this versatile medicinal plant.

11.2.9 Antiplasmodial Activity

Silver nanoparticles (AgNPs) synthesized using the leaves of *C. roseus* were thoroughly characterized using several advanced techniques to confirm their formation, structure, and morphology. The ultraviolet-visible (UV-Vis) spectrophotometry was employed to detect the surface plasmon resonance peak, indicating the successful synthesis of AgNPs. The scanning electron microscopy (SEM) provided detailed images of the nanoparticles' surface morphology and size distribution. The energy dispersive X-ray analysis (EDX) was used to determine the elemental composition, confirming the presence of silver. Additionally, the X-ray diffraction (XRD) analysis revealed the crystalline nature and phase purity of the synthesized nanoparticles. These AgNPs demonstrated significant antiplasmodial activity against *Plasmodium falciparum*, the parasite responsible for the most severe form of malaria, highlighting the potential of *C. roseus*-mediated silver nanoparticles as effective antimalarial agents (Ponarulselvam et al. 2012).

11.2.10 Antidiabetic Activity

Diabetes mellitus is a metabolic disorder characterized by elevated blood glucose levels. Studies have shown that aqueous leaf extracts of *C. roseus* significantly reduce blood glucose, total cholesterol, LDL, VLDL, and triglycerides in diabetic rats without causing hypoglycemia or affecting body weight, demonstrating a high safety profile (Muralidharan 2014). Similarly, the oral administration of aqueous extracts of *C. roseus*, *Azadirachta indica*, and *Allium sativum* in rats for 14 days led to significant blood glucose reduction and increased body weight compared to the drug glimepiride (Mostofa et al. 2007). In streptozotocin-induced diabetic rats, aqueous extracts of *C. roseus*, *Murraya koenigii*, and *Psidium guajava* (500 mg/kg) administered for 15 days significantly improved body weight and lowered blood glucose levels, with effects comparable to glibenclamide (Prasad et al. 2009). These findings highlight the potential of *C. roseus* leaf extract as a safe and effective antidiabetic agent.

The various extracts of *C. roseus* have demonstrated significant antidiabetic and hypolipidemic properties across multiple studies. Islam et al. (2009a, b) reported that both petroleum ether and ethyl acetate fractions significantly reduced blood glucose in normoglycemic and streptozotocin-induced diabetic rats, with the ethyl acetate fraction showing the greatest reduction in triglycerides and superior activity compared to metformin (150 mg/kg). Similarly, leaf juice administered orally (0.5–1.0 mL/kg) caused a dose-dependent glucose reduction in both normal and alloxan-induced diabetic rabbits, comparable to glibenclamide (Nammi et al. 2003). In a study carried out by Malathi et al. (2010) showed that alcoholic extracts of

C. roseus leaves and flowers inhibit α-amylase and α-glucosidase enzymes in vitro, with the leaf extract exhibiting the highest inhibition at 10 mg/mL (IC_{50}), indicating a strong mechanism for blood glucose control. Ahmed et al. (2010) found that methanolic extracts of the whole plant, particularly at 500 mg/kg, significantly improved hyperglycemia, body weight, lipid profile, and pancreatic β-cell regeneration in diabetic rats, as supported by histopathological analysis. A study carried out by Singh et al. (2014) reported that the methanolic leaf extract (200 and 400 mg/kg) enhanced wound contraction and lowered blood glucose in diabetic mice, with optimal results at 200 mg/kg and increased collagen fiber formation ($p < 0.001$). Aruljothi and Samipillai (2016) confirmed that methanolic leaf extracts reduced blood glucose and restored serum protein levels in alloxan-induced diabetic rats. These findings collectively highlight *C. roseus* as a promising antidiabetic agent with additional benefits in lipid regulation, enzyme inhibition, and tissue healing.

Several studies have demonstrated the antidiabetic potential of *C. roseus* and related species through different mechanisms and formulations. Navitha et al. (2012) compared the ethanolic extract of *C. pusillus* with glibenclamide (5 mg/kg) in diabetic rats. A significant hypoglycemic effect ($p < 0.01, p < 0.001$) was observed at a dose of 500 mg/kg, highlighting the extract's effectiveness. Al-Shaqha et al. (2015) investigated the ethanolic extract of *C. roseus* (100 and 200 mg/kg) in streptozotocin-induced diabetic rats. The treatment enhanced the expression of glucose transporter genes (GLUT-2 and GLUT-4) in the liver and showed a positive correlation between intracellular calcium and insulin release in islets, suggesting a molecular basis for its antidiabetic action. Singh et al. (2001) found that a dichloromethane: methanol (1:1) extract of *C. roseus* leaves and twigs (500 mg/kg orally) reduced blood glucose by 48.6% and 57.6% after 7 and 15 days, respectively, in diabetic rats. Pre-treatment for 30 days offered full protection against streptozotocin (STZ) challenge. The extract also restored the activity of key hepatic enzymes and reduced oxidative stress markers, indicating improved glucose metabolism and antioxidant defense. Aadil et al. (2012) reported moderate α-amylase inhibition by *C. roseus* compared to other medicinal plants, with methanolic and aqueous extracts of four species tested. Benjamin et al. (1994) explored *C. roseus* suspension cultures and found that the CWS cell line extract significantly lowered blood glucose and produced ajmalicine, whereas other lines had minimal antidiabetic effect due to low alkaloid production. Rasineni et al. (2010) administered *C. roseus* leaf powder (100 mg/kg/day) to diabetic Wistar rats over 60 days. The treatment significantly reduced plasma glucose and improved insulin levels, lipid profile, and glycogen content, while also normalizing enzymes involved in glucose metabolism. Tiong et al. (2013) evaluated vindoline, vindolidine, vindolicine, and vindolinine for glucose uptake in pancreatic β-TC6 and C2C12 cells. Vindolicine showed the strongest activity, and all compounds inhibited protein tyrosine phosphatase-1B (PTP-1B), suggesting therapeutic relevance in type 2 diabetes. Collectively, these findings affirm *C. roseus* as a promising source of antidiabetic agents, acting through multiple mechanisms, including blood glucose regulation, insulin secretion, gene expression modulation, enzyme inhibition, and antioxidant activity.

11.2.11 Antidiarrheal Activity/Antidysenteric Activity

The antidiarrheal potential of *C. roseus* was evaluated using a castor oil-induced diarrhea model in rats. In this study, ethanolic extracts of *C. roseus* were administered orally at doses of 200 mg/kg and 500 mg/kg body weight to assess their inhibitory effects on diarrhea. Standard antidiarrheal drugs—loperamide and atropine sulphate—were used as positive controls for comparison in two separate experimental setups. The results demonstrated a clear dose-dependent antidiarrheal activity of the ethanolic extract. At the lower dose of 200 mg/kg, the extract produced a statistically significant reduction in the frequency and severity of diarrhea compared to the negative control group ($p < 0.05$). At the higher dose of 500 mg/kg, the extract showed an even more pronounced effect, with highly significant antidiarrheal activity ($p < 0.001$), indicating stronger inhibition of diarrhea symptoms. The observed effect is attributed to the phytochemical constituents present in *C. roseus*, such as alkaloids, flavonoids, and tannins, which are known to inhibit intestinal motility and fluid secretion—key mechanisms involved in diarrheal conditions. The ethanolic extract effectively delayed the onset of diarrhea, reduced the number of wet fecal outputs, and improved stool consistency in treated animals, with results comparable to those of standard drugs. These findings confirm that the ethanol extract of *C. roseus* possesses potent antidiarrheal properties, supporting its traditional use in the treatment of gastrointestinal disorders. The study by Hassan et al. (2011) provides scientific validation for the ethnomedicinal applications of *C. roseus* in managing diarrhea and suggests its potential as a natural therapeutic agent.

11.2.12 Cytotoxic Activity

The cytotoxic potential of *C. roseus* extracts and their alkaloidal fractions has been extensively studied across various cancer cell lines, highlighting the plant's promising anticancer properties. One study evaluated the methanolic extract of *C. roseus* and its fractions—n-hexane, chloroform, and methanol—using the MTT assay against the Human Colorectal Carcinoma Cell Line (HCT 116). Among these, the chloroform fraction exhibited the highest cytotoxic activity, whereas the water fraction showed only minor effects. Vindoline, an alkaloid isolated from the plant, demonstrated cytotoxicity at a concentration of 200 μg/mL, while catharanthine, another alkaloid, exhibited the most promising activity with an IC_{50} value observed at 60 μg/mL, suggesting its potent anticancer effect (Siddiqui et al. 2010). Further investigations on alkaloidal extracts revealed their dose-dependent cytotoxic effects on several cancer cell lines, including HT-29 (colon), Caco-2 (colon), T47D (breast), and NIH/3T3 (mouse fibroblast). Although all fractions affected cell proliferation, NIH/3T3 cells were less susceptible compared to the cancerous cell lines, indicating some selectivity in cytotoxicity (Khanavi et al. 2013). Another study reported that the n-butanol fraction of *C. roseus* displayed strong cytotoxicity across a wide range of cancer cell lines, such as ovarian (A2780), lung (H460), skin (A431), pancreas (MIA PaCa-2), prostate (Du145), colon (HT29), breast (MCF-7), neuroblastoma

(BE2-C), and glioblastoma (U87), with GI_{50} values ranging from 5.2 to 21.0 μg/mL, indicating potent anticancer activity at relatively low doses (Pham et al. 2018).

The antitumor properties of *C. roseus* are also linked to its indole alkaloid-enriched bioactive extracts. Suspension-cultured cells treated with methyl jasmonate (MJ) and cyclodextrins (CDs) produced extracts that inhibited growth of JURKAT E.6 human lymphocytic leukemia, THP-1 human monocytic leukemia, and BL 1395 non-tumor human B-cell lines. The extract's half-maximal inhibitory concentration (IC50) was approximately 210–211 ng/mL for the leukemia cell lines, showcasing strong cytotoxicity. This effect was attributed not to a single compound but to a synergistic action among four identified compounds within the extract (Fernández-Pérez et al. 2013). Historical reports also document cytotoxicity in *C. pusillus* alkaloid fractions, which showed activity against Eagle's 9 KB carcinoma of the nasopharynx in cell culture. The isolated alkaloid lochnerinine was particularly active against this carcinoma, illustrating the potential of alkaloids from species of *Catharanthus* in cancer treatment. More recent studies have assessed the effects of aqueous extracts of *C. roseus* on Jurkat cells (human T lymphocyte leukemia) and normal peripheral blood mononuclear cells (PBMCs). Using the MTS assay over 24, 48, and 72 hours, the extracts demonstrated a significant, dose- and time-dependent inhibition of Jurkat cell growth, with cell survival dropping to between 21.84% and 29% after 72 hours of treatment. At a concentration of 500 μg/mL, the extract showed minimal inhibitory effects after 24 hours but maximized inhibition at 72 hours. Interestingly, at 1000 μg/mL, the highest inhibitory effect occurred at 24 hours, though the effect lessened at 48 hours. The calculated IC50 values were 2.55 and 2.38 μg/mL at 48 and 72 hours, respectively. In contrast, the extract induced proliferation of normal PBMCs, particularly at 1000 μg/mL, increasing cell viability by over 165% at 24 hours and maintaining elevated proliferation at 48 and 72 hours. This dual effect highlights the extract's selective cytotoxicity against cancer cells while promoting growth in normal immune cells (Ahmad et al. 2010). Investigations into specific alkaloids have further clarified their cytotoxic profiles. Fourteen compounds including 14′,15′-didehydrocyclovinblastine, leurosine, 17-deacetoxycyclovinblastine, vinamidine, 17-deacetoxyvinamidine, catharine, cycloleurosine, and leurosidine, were tested against the MDA-MB-231 breast cancer cell line using the MTT assay. All compounds reduced cell viability after 72 hours, with varying IC50 values. Among them, one compound demonstrated the most potent inhibition, while others showed relatively weaker effects. Notably, compounds lacking the acetoxy group at the C-17 position displayed increased cytotoxicity, suggesting that structural modifications can enhance anticancer activity. This insight provides a valuable framework for future development of VLB-type alkaloid derivatives with improved therapeutic potential (Wang et al. 2012a, b). Therefore, *C. roseus* extracts and their alkaloidal fractions exhibit significant cytotoxic effects across a broad spectrum of cancer cell types. The activity is often dose-dependent and selective, with certain alkaloids showing enhanced potency due to specific structural features. The combination of multiple alkaloids often produces a synergistic antitumor effect, offering promising avenues for cancer therapy development. Additionally, the extracts' capacity to inhibit cancer cell proliferation while promoting normal immune cell growth underscores their therapeutic potential and safety profile. Ongoing research and structural modifications may further optimize these compounds for clinical application.

11.2.13 Hepatoprotective Activity

The liver is a vital organ responsible for metabolism, detoxification, and bile excretion. It plays a central role in filtering toxins and metabolizing xenobiotics, including therapeutic agents. Due to its constant exposure to harmful substances, the liver is vulnerable to damage, resulting in a range of liver diseases. These conditions remain a major global health concern, and currently available therapeutic options are often limited and insufficient. Therefore, there is a critical need to explore and develop more effective hepatoprotective agents, particularly from natural sources. Ahmed and Rao (2013) conducted a comparative study to evaluate the hepatoprotective efficacy of ethanolic leaf extracts of *Melia azedarach* L., *C. rosea*, and *Brassica oleracea* L. var. *capitata* against simvastatin-induced hepatotoxicity in rats. Hepatotoxic rats were treated with two doses (300 and 500 mg/kg, p.o.) of each extract. Biochemical analysis revealed that the extracts significantly improved altered liver function markers such as serum glutamate pyruvate transaminase (SGPT), serum glutamate oxaloacetate transaminase (SGOT), alkaline phosphatase (ALP), serum bilirubin, and total protein content. Histopathological examination further confirmed the hepatoprotective potential by showing improvements in liver tissue architecture compared to the hepatotoxic control group. Among the tested extracts, *M. azedarach* and *B. oleracea* demonstrated more pronounced hepatoprotective activity than *C. rosea*. The likely mechanism of action is attributed to their antioxidant and free radical scavenging properties, which help mitigate oxidative stress in liver tissues. This study supports the potential of plant-based treatments in managing liver disorders.

11.2.14 Hypoglycemic Activity

Diabetes is a chronic metabolic disorder characterized by persistently elevated blood glucose levels, often leading to complications such as cardiovascular disease and renal failure. Numerous studies have explored the hypoglycemic and hypolipidemic potential of *Catharanthus* species, particularly *C. roseus*, using various plant extracts and experimental models. In one study, the methanol leaf extract of *C. roseus* was evaluated in an alloxan-induced diabetic rat model. Six groups of five rats each were treated with either 250 mg/kg of *C. roseus* extract, 100 mg/kg metformin, 1 mg/kg glibenclamide, or combinations of extract with either metformin or glibenclamide. A control group received only distilled water. All treatments significantly lowered blood glucose levels compared to control ($p < 0.05$), with the metformin-extract combination producing the highest reduction (64.86%). The results indicated that *C. roseus* leaf extract enhances the hypoglycemic effect of standard antidiabetic drugs (Ohadoma and Michael 2011). Similarly, the ethanolic leaf extract of *C. roseus* was evaluated in diabetic rats induced with alloxan. The treated animals showed significant reductions in blood glucose, total cholesterol (TC), low-density lipoprotein (LDL), very low-density lipoprotein (VLDL), and triglycerides (TG), while high-density lipoprotein (HDL) levels were significantly increased. GC-MS analysis of the extract identified 11 bioactive compounds. These results were comparable to the effects of glibenclamide, suggesting that *C. roseus* has both hypoglycemic and hypolipidemic properties (Mohan et al. 2015).

Islam et al. (2009a, b) investigated the effects of three fractions—petroleum ether, ethyl acetate, and chloroform—of C. roseus leaf extracts in normal and streptozotocin-induced diabetic rats (SIDRs). Single intraperitoneal doses (150 mg/kg) significantly reduced fasting blood glucose in both normoglycemic and diabetic rats. All fractions decreased serum TC, but the ethyl acetate fraction had the most pronounced lipid-lowering effect, likely due to flavonoids and alkaloids that enhance glycogenesis and inhibit gluconeogenesis or intestinal glucose absorption. Vindogentianine, a compound isolated from *C. roseus*, also showed hypoglycemic activity in vitro by enhancing glucose uptake in β-TC6 and C2C12 cells and inhibiting protein tyrosine phosphatase 1B (PTP-1B). Although its inhibition of α-amylase and α-glucosidase was low, the enhancement of glucose uptake and PTP-1B inhibition suggests its potential for type 2 diabetes treatment (Tiong et al. 2015). Other alkaloids like vincristine and vinblastine may stimulate insulin production, contributing to the hypoglycemic effect (Bisla et al. 2014). Vega-Ávila et al. (2012) assessed the hypoglycemic activity of various organic (hexane, dichloromethane, and methanol) and aqueous extracts of C. roseus parts (flower, leaf, stem, and root) in healthy and alloxan-induced diabetic mice. Among them, the aqueous stem extract (250 mg/kg) and its alkaloid-free fraction (300 mg/kg) significantly reduced blood glucose by 52.9% and 51.2%, respectively—effects comparable to tolbutamide (58.1%). A similar study by Iweala and Okeke (2005) evaluated the effects of aqueous *C. roseus* extracts in alloxan-induced diabetic rats compared to chlorpropamide (Diabinese). Treated animals exhibited significant reductions ($p \leq 0.05$) in blood glucose, cholesterol, total protein, lipid peroxidation, and liver enzyme levels. The effect was more pronounced in *C. roseus* treated groups than in those receiving chlorpropamide. Ahmed et al. (2007) studied both crude leaf juice and aqueous extracts of *C. roseus* in streptozotocin-induced diabetic rats. Single oral doses of crude juice (0.5 and 1 mL/kg) and aqueous extract (0.30 and 0.45 g/kg) significantly reduced serum glucose levels at multiple time points—8, 12, and 24 hours—as well as on the 4th and 11th days of treatment. The effect was dose-dependent and comparable to that of glibenclamide (500 μg/kg), confirming the potent hypoglycemic activity of *C. roseus*. In addition, the ethanolic extract of *C. pusillus*, another species of *Catharanthus*, was tested in streptozotocin-induced diabetic rats using both fasting and oral glucose tolerance test models. At a dose of 500 mg/kg, the extract exhibited significant hypoglycemic effects ($p < 0.01$ and $p < 0.001$), comparable to glibenclamide (5 mg/kg), further supporting the antidiabetic potential of *Catharanthus* species (Navitha et al. 2012). Overall, these studies confirm the antidiabetic potential of *Catharanthus* extracts, primarily *C. roseus*, through multiple mechanisms—enhancing insulin secretion, improving lipid profiles, increasing glucose uptake, and modulating key metabolic enzymes. Both aqueous and ethanolic extracts, along with specific fractions like ethyl acetate, have shown promising results in animal models, often rivaling standard antidiabetic drugs in efficacy. These findings support the continued exploration of *Catharanthus* species for developing effective, plant-based therapies for managing diabetes and its associated metabolic complications.

11.2.15 Hypolipidemic Activity

Muralidharan (2014) evaluated the hypolipidemic and antidiabetic potential of the aqueous leaf extract of Catharanthus roseus in both alloxan-induced diabetic rats and guinea pigs. The study reported a high LD_{50} value for the extract, with no mortality observed even after administering a dose 30 times higher than normal over a 30-day period, indicating excellent safety. In diabetic rats, treatment with the leaf extract significantly ($p < 0.001$) reduced blood glucose levels, along with bringing total cholesterol (TC), low-density lipoprotein (LDL), very low-density lipoprotein (VLDL), and triglyceride (TG) levels close to normal. Importantly, no hypoglycemic episodes or significant changes in body weight were noted, supporting its potential antidiabetic use. In a separate experiment, adult guinea pigs were divided into seven groups to assess hypolipidemic effects. Groups received either normal or high-fat diets, with or without *C. roseus* leaf juice (0.5 or 1 mL/kg) or atorvastatin (3 mg/kg). After 6 weeks, serum lipid profiles showed notable improvements, particularly in groups treated with C. roseus.

Histological studies of the aorta, liver, and kidney were conducted in groups 1, 2, 6, and 7, along with blood cell counts before and after treatment with *C. roseus* juice. At 0.5 mL/kg, the juice prevented lipid elevation and fatty changes from a high-fat diet. At 1 mL/kg, it significantly ($p < 0.05$) reduced LDL-c and total cholesterol/HDL-c ratio. *C. roseus* leaf juice shows strong lipid-lowering and anti-atherosclerotic effects (Patel et al. 2011). Similarly different extracts of single doses (150 mg/kg, i.p.) of petroleum-ether, ethyl acetate, and chloroform of *C. roseus* in the fasting blood glucose (FBG) levels were tested in normal and SIDRs on 0, 1, 2, 3, 6, 10, 16, and 24th hour and serum triglyceride (TG) and serum total cholesterol (TC) levels were determined after 24th hour. In normoglycemic rats and in SIDRs, petroleum-ether and ethyl acetate fraction of *C. roseus* reduced blood glucose level significantly. The ethyl acetate fraction of *C. roseus* was the most effective than all fractions reduced serum total cholesterol. Similarly, the ethyl acetate fraction reduced triglyceride level at the highest. So ethyl acetate fraction was the best due to presence of flavonoids and alkaloids (Islam et al. 2009a, b).The hypolipidemic effects of ethanol-extract from the leaves of *C. cordifolia* and *C. roseus* were reported and compared in AIDRs. Single doses (150 and 300 mg/kg, i.p.) of both the *C. cordifolia* and *C. roseus* leaf extracts were given to diabetic rats. The FBG, serum TG, and serum TC levels were investigated in AIDRs on 0th, 7th and 14th days of treatment using single doses (150 and 300 mg/kg, i.p.) of both the *C. cordifolia* and

C. roseus leaf of ethanol-extract. Single dose treatment of *C. cordifolia* (300 mg/kg), *C. roseus* (150 mg/kg of body weight) significantly reduced blood glucose by 47.4% and 37.1% ($p < 0.01$) on 7th day and by 59.7% and 48.5% on 14th day. A significant reduction ($p < 0.01$) in serum total cholesterol of 31.7% and 43.3% and serum triglycerides of 45.5% and 39.4% was observed on the 14th day with a single dose of the extracts of *C. cordifolia* (300 mg/kg) and *C. roseus* (150 mg/kg of bw), respectively. These results of hypolipidemic activity were comparable with glibenclamide (0.05 mg/kg) and metformin HCl (100 mg/kg). *C. cordifolia* extract at the dose of 150 mg/kg did not show any significant effect whereas *C. roseus* extract at the dose of 300 mg/kg killed all the rats in 2–5 days. *C. cordifolia* was found to be better than *C. roseus* in reduced serum TC for reducing serum cholesterol on long-term treatment basis in AIDRs (Akhtar et al. 2007).

A series of experimental studies have demonstrated the significant hypolipidemic, hypoglycemic and hypotensive effects of *C. roseus* leaf extracts in animal models of metabolic disorders such as hypertension, diabetes, and hypercholesterolemia. In a crossover study conducted on adrenaline-induced hypertensive rats (AIHR), the ethanolic leaf extract of *C. roseus* was compared with the commercial antihypertensive drug Atenolol. Both were administered intraperitoneally (i.p.) for 1 week. The *C. roseus* extract was given at a dose of 30 mg per 155 ± 15 g of body weight, while Atenolol was administered at a dose of 0.1 mg per 155 ± 15 g, based on its pharmacokinetic properties. Parameters such as heart weight, body weight, blood glucose, serum cholesterol, and serum triglycerides were monitored throughout the treatment period. The results revealed that *C. roseus* produced significant hypotensive effects comparable to Atenolol, indicating its potential cardiovascular benefits (Ara et al. 2009). In another study, diabetes was induced in male Wistar rats via intraperitoneal injection of streptozotocin (STZ) at a dose of 55 mg/kg body weight. Diabetic rats were then treated with *C. roseus* leaf powder suspension (100 mg/kg body weight/day) orally for 60 days. This treatment led to a progressive reduction in plasma glucose and a partial restoration of plasma insulin levels by day 15, with glucose levels nearing normal by the end of the experiment. Furthermore, diabetic rats showed normalization of lipid profiles, including total cholesterol, LDL, VLDL, and triglycerides, along with a reduction in the atherogenic index. Additionally, the extract prevented depletion of hepatic and muscle glycogen and stabilized the activities of key glucose-metabolizing enzymes such as glycogen phosphorylase, hexokinase, phosphofructokinase, pyruvate kinase, and glucose-6-phosphate dehydrogenase (Rasineni et al. 2010). A separate study on hypercholesterolemic rats evaluated the effects of aqueous *C. roseus* leaf extract at varying concentrations. Fifteen male Sprague Dawley rats were divided into five groups. The hypercholesterolemic model was induced using a 1% (w/w) cholesterol-rich diet. Group K1 received a normal diet (control), K2 received a cholesterol-rich diet without treatment, and groups K3 to K5 received the high-cholesterol diet plus *C. roseus* extract at concentrations of 20%, 40%, and 80% (w/v), respectively, for 28 days. The 80% extract dose showed the most pronounced hypolipidemic effect, significantly lowering total cholesterol (79.33 ± 3.51 mg/dL), triglycerides (72.33 ± 6.65 mg/dL), LDL (9.87 ± 5.34 mg/dL), while increasing HDL

(55.00 ± 3.60 mg/dL). The cholesterol:HDL ratio was 1.4:1, and the atherogenic index dropped to 0.44, confirming the strong lipid-lowering effect of *C. roseus* at higher doses (Rasineni et al. 2010). Overall, these findings highlight the therapeutic potential of *C. roseus* as a natural agent for managing hypertension, diabetes, and hyperlipidemia through its multi-targeted biological effects.

11.2.16 Hypotensive Property

The leaf extract of *C. roseus* has been investigated for its hypotensive effects in adrenaline-induced hypertensive rats (AIHR) and compared to the standard antihypertensive drug, Atenolol. In this study, both the plant extract and Atenolol were administered intraperitoneally for 1 week. The *C. roseus* extract was given at a dose of 30 mg per 155 ± 15 g of body weight each morning throughout the treatment period. Several cardiovascular parameters were measured to assess efficacy, including heart weight, serum cholesterol, blood glucose, serum triglyceride levels, and body weight. The extract significantly altered all evaluated parameters in a manner indicative of cardiovascular improvement. The reductions in heart weight, blood pressure indicators and serum lipid levels were observed. These results suggest that *C. roseus* leaves possess notable antihypertensive and cardioprotective properties, likely due to the presence of bioactive alkaloids and flavonoids. This highlights its potential as a natural alternative for managing hypertension (Ara et al. 2009).

11.2.17 Larvicidal/Pupicidal Effect

Numerous studies have investigated the larvicidal and pupicidal properties of various *Catharanthus* species, particularly targeting mosquito vectors responsible for diseases such as malaria and filariasis. The extracts from *C. roseus* leaves—specifically aqueous, ethyl acetate, methanol, and petroleum ether—have demonstrated significant bioactivity against different developmental stages of *Anopheles stephensi* and *Culex quinquefasciatus*. Aqueous, ethyl acetate and methanol extracts of *C. roseus* were tested against fourth instar larvae and pupae of both mosquito species. Results showed varying levels of mortality after 24 and 48 hours of exposure, with no mortality in the untreated control groups. For *A. stephensi* larvae, LC_{50} values were 68.62 mg/mL (aqueous), 82.47 mg/mL (ethyl acetate), and 86.64 mg/mL (methanol). Against *C. quinquefasciatus* larvae, LC_{50} values were 85.21, 76.84, and 94.20 mg/mL for the respective extracts. The pupicidal activity also varied, with aqueous extract showing LC_{50} values of 118.08 and 146.20 mg/mL against *A. stephensi* and *C. quinquefasciatus*, respectively. The petroleum ether extracts of *C. roseus* exhibited stronger larvicidal activity against *A. stephensi*. LC_{50} values across the first to fourth instar larvae were 3.34, 4.48, 5.90, and 8.17 g/L, respectively. For comparison, *B. thuringiensis* yielded LC_{50} values of 1.72–2.42 g/L, while combined treatment with both agents improved efficacy (LC_{50} = 2.18–3.22 g/L). No mortality was observed in the controls. These results suggest the combination of

C. roseus extract and *B. thuringiensis* may offer enhanced control in vector management programs. In addition to mosquitoes, aqueous extracts of *C. roseus* were also tested against the agricultural pest *Helicoverpa armigera* (gram pod borer). At 1000 ppm, larvicidal activity increased progressively with the larval instar stage. Mortality rates ranged from 2.60% in first instar to 63.34% in sixth instar larvae after 24 hours, indicating moderate but stage-dependent effectiveness. Overall, the larvicidal and pupicidal effects of *C. roseus* extracts—especially in aqueous and petroleum ether forms—highlight its potential in integrated pest and vector control strategies (Subarani et al. 2013; Panneerselvam et al. 2013).

The Aqueous leaf extracts of *C. roseus* were found to have significant effects on *Helicoverpa armigera* larvae, with ED_{50} and LD_{50} values statistically significant at $p < 0.05$. This suggests potential for use as a botanical pesticide in pest management (Alaguchamy and Jayakumararaj 2015). Hexane, ethyl acetate, and methanol extracts of *C. roseus* and *Adenocalymma fruticosa* also showed dose-dependent larvicidal effects against *Culex quinquefasciatus*. The ethyl acetate extract of *A. fruticosa* showed the highest activity (LC_{50} = 253.08 ppm), followed by the hexane extract of *C. roseus* (LC_{50} = 645.33 ppm) (Quinquefasciatus 2017). Enzymatic analysis revealed increased β-carboxylesterase isozyme activity in larvae treated with these extracts, suggesting a biochemical mode of action. Additionally, treated larvae showed reduced total protein content, indicating physiological stress. Another potent bioactive compound, α-amyrin acetate (isolated from the hexane fraction of an acetone extract of *C. roseus*), demonstrated excellent larvicidal and pupicidal activity against *Anopheles stephensi*. At 1 ppm, it achieved 100% mortality in early instars and 94% in later instars. It also disrupted development by extending larval durations, reducing fecundity, and suppressing adult emergence—possibly by mimicking juvenile hormone activity. These findings suggest α-amyrin acetate could serve as an active ingredient in the development of novel, plant-based insecticides (Kuppusamy et al. 2009). In another innovative approach, silver nanoparticles (AgNPs) synthesized using aqueous root extracts of *C. roseus* exhibited strong larvicidal activity. Characterization using UV-Vis spectrophotometry, FTIR, XRD, and SEM confirmed the formation and stability of nanoparticles. These AgNPs achieved high mortality rates after 24 hours against *Aedes aegypti* (LC_{50} = 2.01 mg/L) and *Culex quinquefasciatus* (LC_{50} = 1.18 mg/L) (Rajagopal et al. 2015). The presence of aliphatic amines and alkanes was identified as crucial in nanoparticle formation. Together, these findings underscore the potential of *C. roseus* extracts and derivatives—including nanoparticles and isolated phytochemicals—as effective, natural alternatives to chemical insecticides for controlling mosquito vectors and agricultural pests

11.2.18 Memory Enhancement Activity

Vinpocetine derived from the species of *Catharanthus* has demonstrated a range of pharmacological actions that are theoretically beneficial in the context of Alzheimer's disease. These include neuroprotective, anti-inflammatory, and vasodilatory effects, which suggest potential therapeutic value. However, despite these promising mechanisms, clinical evidence supporting its effectiveness remains limited. Meta-analyses conducted in populations with broadly defined or poorly characterized dementia have concluded that current data are insufficient to justify the clinical use of vinpocetine for cognitive disorders, including Alzheimer's disease. While the results remain inconclusive, vinpocetine has shown a favourable safety profile in clinical trials involving patients with dementia and stroke. Doses of up to 60 mg/day have been well tolerated, with no significant adverse effects reported (Coleston and Hindmarch 1988). These findings highlight the need for more rigorous, targeted studies to better understand vinpocetine's efficacy and role in the management of neurodegenerative conditions.

11.2.19 Wound Healing Activity

A study conducted suggests that the ethanol extract derived from the flowers of *C. roseus* possesses significant wound healing properties, offering therapeutic benefits that surpass those observed in placebo-treated controls. Their findings suggest that the extract facilitates a more rapid recovery process in wounded tissues. This is supported by several physiological markers indicative of enhanced healing. For instance, studies report a notable improvement in tensile strength of the skin, which implies better structural integrity and resistance to further injury in the healing area. Additionally, there is an observed increase in hydroxyproline content—an amino acid integral to collagen synthesis—indicating elevated collagen deposition, which plays a crucial role in tissue regeneration and repair. Further reinforcing these outcomes, Senbagalakshmi et al. (2017) investigated the wound contraction process and confirmed that the extract contributes to faster wound closure. The presence of hypolipidemic active principles in the flower extract is also believed to support the healing process, potentially by modulating inflammation and promoting tissue repair mechanisms at the cellular level. Overall, the evidence highlights the promising role of *C. roseus* ethanol extract in enhancing wound healing through multiple mechanisms, including biochemical support, improved structural repair, and accelerated wound contraction.

References

Aadil KR, Barapatre A, Rathore N, Pottam S, Jha H (2012) Comparative study of in-vitro antioxidant and antidiabetic activity of plant extracts of Acacia arabica, Murraya koeingii, Catharanthus roseus and Rouwolfia serpentina. Int J Phytomed 4(4):543–551

Ahmad NH, Rahim RA, Mat I (2010) Catharanthus roseus aqueous extract is cytotoxic to Jurkat leukaemic T-cells but induces the proliferation of normal peripheral blood mononuclear cells. Trop Life Sci Res 21(2):101–113

Ahmed MF, Rao AS (2013) Comparative hepatoprotective activities of selected Indian medicinal plants. Glob J Med Res 13:15–21

Ahmed AU, Ferdous AH, Saha SK, Nahar S, Awal MA, Parvin F (2007) Hypoglycemic effect of Catharanthus roseus in normal and streptozotocin-induced diabetic rats. Mymensingh Med J 16(2):143–148

Ahmed MF, Kazim SM, Ghori SS, Mehjabeen SS, Ahmed SR, Ali SM, Ibrahim M (2010) Antidiabetic activity of Vinca rosea extracts in alloxan-induced diabetic rats. Int J Endocrinol 2010:841090

Akhtar MA, Rashid M, Wahed MII, Islam MR, Shaheen SM, Islam MA, Amran MS, Ahmed M (2007) Comparison of long-term antihyperglycemic and hypolipidemic effects between Coccinia cordifolia (Linn.) and Catharanthus roseus (Linn.) in alloxan-induced diabetic rats. Res J Med Med Sci 2(1):29–34

Akpotu MO, Eze PM, Abba CC, Umeokoli BO, Nwachukwu CU, Okoye FB, Esimone CO (2017) Antimicrobial activities of secondary metabolites of endophytic fungi isolated from Catharanthus roseus. J Health Sci 7(1):15–22

Alaguchamy N, Jayakumararaj R (2015) Larvicidal effect of Catharanthus roseus L. (G) Don. aqueous leaf extracts on the larvae of Helicoverpa armigera (Hübner). Int J Life Sci Educ Res 3(1):10–14

Al-Shaqha WM, Khan M, Salam N, Azzi A, Chaudhary AA (2015) Anti-diabetic potential of Catharanthus roseus Linn. and its effect on the glucose transport gene (GLUT-2 and GLUT-4) in streptozotocin induced diabetic wistar rats. BMC Complement Altern Med 15(1):379

Al-Shmgani HS, Mohammed WH, Sulaiman GM, Saadoon AH (2017) Biosynthesis of silver nanoparticles from Catharanthus roseus leaf extract and assessing their antioxidant, antimicrobial, and wound-healing activities. Artif Cells Nanomed Biotechnol 45(6):1234–1240

Anvar Sadique MP, Hashim KM (2019) Anti-angiogenesis studies of the aerial parts of Catharanthus roseus with special emphasis to ethyl acetate extraction. Int J Innov Sci Res Technol 4(6):376–378

Ara N, Rashid M, Amran S (2009) Comparison of hypotensive and hypolipidemic effects of Catharanthus roseus leaves extract with atenolol on adrenaline induced hypertensive rats. Pak J Pharm Sci 22(3):267–271

Aruljothi B, Samipillai SS (2016) Antidiabetic activity of Catharanthus roseus in alloxan induced diabetic rats. Int J Mod Res Rev 4(3):1121–1124

Balaabirami S, Patharajan S (2012) In vitro antimicrobial and antifungal activity of Catharanthus roseus leaves extract against important pathogenic organisms. Int J Pharm Pharm Sci 4(3):487–490

Barkat MA, Mujeeb M, Samim M, Verma S (2014) Biosynthesis of silver nanoparticles using callus extract of Catharanthus roseus var. alba and assessment of its antimicrobial activity. Br J Pharm Res 4(13):1591–1603

Benjamin BD, Kelkar SM, Pote MS, Kaklij GS, Sipahimalani AT, Heble MR (1994) Catharanthus roseus cell cultures: growth, alkaloid synthesis and antidiabetic activity. Phytother Res 8(3):185–186

Bisla G, Choudhary S, Chaudhary V (2014) Evaluation of the nutritive and organoleptic values of food products developed by incorporated Catharanthus roseus (Sadabahar) fresh leaves: explore their hypoglycemic potential. Sci World J 2014:304120

Chaman S, Sharma G, Anil S, Reshi K (2013) Study of antimicrobial properties of Catharanthus roseus by agar well diffusion method. Int Res J Pharm Appl Sci 5:65–68

Coleston DM, Hindmarch I (1988) Possible memory-enhancing properties of vinpocetine. Drug Dev Res 14(3–4):191–193

Elkahoui S, Hernández JA, Abdelly C, Ghrir R, Limam F (2005) Effects of salt on lipid peroxidation and antioxidant enzyme activities of Catharanthus roseus suspension cells. Plant Sci 168(3):607–613

Ethalsha P, Retna AM (2014) Evaluation of antioxidant potential and antibacterial activity of crude extracts of Catharanthus roseus. Int J Pharm Sci Res 5:3490–3495

Fernández-Pérez F, Almagro L, Pedreño MA, Gomez Ros LV (2013) Synergistic and cytotoxic action of indole alkaloids produced from elicited cell cultures of Catharanthus roseus. Pharm Biol 51(3):304–310

Ferreres F, Pereira DM, Valentão P, Andrade PB, Seabra RM, Sottomayor M (2008) New phenolic compounds and antioxidant potential of Catharanthus roseus. J Agric Food Chem 56(21):9967–9974

Govindasamy C, Srinivasan R (2012) In vitro antibacterial activity and phytochemical analysis of Catharanthus roseus (Linn.) G. Don. Asian Pac J Trop Biomed 2(1):155–158

Gupta P (2009) Antiestrogenic activity of petroleum ether extract of the leaves of Catharanthus roseus (Vinca rosea) in female albino mice. Asian J Exp Sci 23(1):313–316

Hanafy MS, Matter MA, Asker MS, Rady MR (2016) Production of indole alkaloids in hairy root cultures of Catharanthus roseus L. and their antimicrobial activity. S Afr J Bot 105:9–18

Hassan KA, Brenda AT, Patrick V, Patrick OE (2011) In vivo antidiarrheal activity of the ethanolic leaf extract of Catharanthus roseus Linn. (Apocyanaceae) in Wistar rats. Afr J Pharm Pharmacol 5(15):1797–1800

Idrees M, Naeem M, Aftab T, Khan MMA (2013) Salicylic acid restrains nickel toxicity, improves antioxidant defence system and enhances the production of anticancer alkaloids in Catharanthus roseus (L.). J Hazard Mater 252:367–374

Islam MA, Akhtar MA, Islam MR, Hossain MS, Alam MK, Wahed MII, Rahman BM, Anisuzzaman ASM, Shaheen SM, Ahmed M (2009a) Antidiabetic and hypolipidemic effects of different fractions of Catharanthus roseus (Linn.) on normal and streptozotocin-induced diabetic rats. J Sci Res 1(2):334–344

Islam MA, Akhtar MA, Khan MR, Hossain MS, Alam AH, Ibne-Wahed MI, Amran MS, Rahman BM, Ahmed M (2009b) Oral glucose tolerance test (OGTT) in normal control and glucose induced hyperglycemic rats with Coccinia cordifolia L. and Catharanthus roseus L. Pak J Pharm Sci 22(4):402–404

Iweala EE, Okeke CU (2005) Comparative study of the hypoglycemic and biochemical effects of Catharanthus roseus (Linn) G. Apocynaceae (Madagascar periwinkle) and chlorpropamide (Diabenese) on alloxan-induced diabetic rats. Biokemistri 17(2):149–156

Jaleel CA, Gopi R, Manivannan P, Panneerselvam R (2007a) Antioxidative potentials as a protective mechanism in Catharanthus roseus (L.) G. Don. plants under salinity stress. Turk J Bot 31(3):245–251

Jaleel CA, Gopi R, Manivannan P, Panneerselvam R (2007b) Responses of antioxidant defense system of Catharanthus roseus (L.) G. Don. to paclobutrazol treatment under salinity. Acta Physiol Plant 29(3):205–209

Jaleel CA, Manivannan P, Kishorekumar A, Sankar B, Gopi R, Somasundaram R, Panneerselvam R (2007c) Alterations in osmoregulation, antioxidant enzymes and indole alkaloid levels in Catharanthus roseus exposed to water deficit. Colloids Surf B: Biointerfaces 59(2):150–157

Jasmine DJ, Agastian P (2013) In vitro antioxidant activity and in vivo alpha glucosidase activity of endophytic actinomycetes isolated from Catharanthus roseus (L.) G. Don. J Pharm Res 6(6):674–678

Jayanthi M, Sowbala N, Rajalakshmi G, Kanagavalli U, Sivakumar V (2010) Study of antihyperglycemic effect of Catharanthus roseus in alloxan induced diabetic rats. Int J Pharm Pharm Sci 2(4):114–116

Kabesh K, Senthilkumar P, Ragunathan R, Kumar RR (2015) Phytochemical analysis of Catharanthus roseus plant extract and its antimicrobial activity. Int J Pure Appl Biosci 3(2):162–172

Kafur A, Basheer Khan A (2011) Isolation of endophytic actinomycetes from Catharanthus roseus (L.) G. Don leaves and their antimicrobial activity. Iran J Biotechnol 9(4):302–306

Kayesth S, Shazad M, Kumar S, Gupta KK (2017) Effect of ethanol extract of Catharanthus roseus, Ocimum sanctum and Lantana camara on fecundity and fertility of red cotton bug, Dysdercus koenigii Fabricius (Heteroptera: Pyrrhocoridae). J Entomol Zool Stud 5(4):348–353

Kotakadi VS, Rao YS, Gaddam SA, Prasad TNVKV, Reddy AV, Gopal DS (2013) Simple and rapid biosynthesis of stable silver nanoparticles using dried leaves of Catharanthus roseus Linn. G. Donn and its antimicrobial activity. Colloids Surf B: Biointerfaces 105:194–198

Kumar S, Bagchi GD, Darokar MP (1997) Antibacterial activity observed in the seeds of some coprophilous plants. Int J Pharmacogn 35(3):179–184

Kumari K, Gupta S (2013) Antifungal properties of leaf extract of Catharanthus roseus L. (G.) Don. Am J Phytomed Clin Ther 1(9):698–705

Kuppusamy C, Murugan K, Arul N, Yasodha P (2009) Larvicidal and insect growth regulator effect of α-amyrin acetate from Catharanthus roseus Linn against the malaria vector Anopheles stephensi Liston (Diptera: Culicidae). Entomol Res 39(1):78–83

Malabadi RB, Chalannavar RK, Meti NT, Mulgund GS, Nataraja K, Kumar SV (2012) Synthesis of antimicrobial silver nanoparticles by callus cultures and in vitro derived plants of Catharanthus roseus. Res Pharm 2(6):18–31

Malathi V, Devi SS, Revathi K (2010) Anti-diabetic activity by the in vitro alpha-amylase and alpha-glucosidase inhibitory activity of Catharanthus roseus. Bioscan 5(4):655–659

Manoharan S, Umadevi S, Jayanthi S, Baskaran N (2011) Antihyperglycemic effect of Coscinium fenestratum and Catharanthus roseus in alloxan-induced diabetic rats. Int J Nutr Pharmacol Neurol Dis 1(2):189–193

Mardani-Nejad S, Khavari-Nejad RA, Saadatmand S, Najafi F, Azar PA (2016) Potent antioxidant properties of rolB-transformed Catharanthus roseus (L.) G. Don. Iran J Pharm Res 15(2):537–550

Mohan SCT, Priyadarshini GS, Balamurugan V (2015) GC-MS analysis of phytochemicals and hypoglycemic effect of Catharanthus roseus in alloxan induced diabetic rats. Int J Pharm Sci Res 25:123–128

Mostofa M, Choudhury ME, Hossain MA, Islam MZ, Islam MS, Sumon MH (2007) Antidiabetic effects of Catharanthus roseus, Azadirachta indica, Allium sativum and glimepride in experimentally diabetic induced rat. Bangladesh J Vet Med 5(1&2):99–102

Muralidharan L (2014) Catharanthus roseus leaves as an anti-diabetic and hypolipidemic agent in alloxan-induced diabetic rats. Am J Phytomed Clin Ther 2(12):1393–1396

Nammi S, Boini MK, Lodagala SD, Behara RBS (2003) The juice of fresh leaves of Catharanthus roseus Linn. reduces blood glucose in normal and alloxan diabetic rabbits. BMC Complement Altern Med 3(1):1–4

Navitha A, Helen Sheeba DA, Ramesh C, Sartaj Banu M (2012) Hypoglycemic and anti-diabetic activity of ethanolic extract of Catharanthus pusillus (Murray) G. Don. IOSR J Pharm 2(4):17–21

Naz S, Haq R, Aslam F, Ilyas S (2015) Evaluation of antimicrobial activity of extracts of in vivo and in vitro grown Vinca rosea L. (Catharanthus roseus) against pathogens. Pak J Pharm Sci 28(3):849–853

Ohadoma SC, Michael HU (2011) Effects of co–administration of methanol leaf extract of Catharanthus roseus on the hypoglycemic activity of metformin and glibenclamide in rats. Asian Pac J Trop Med 4(6):475–477

Panneerselvam C, Murugan K, Kovendan K, Kumar PM, Ponarulselvam S, Amerasan D, Subramaniam J, Hwang JS (2013) Larvicidal efficacy of Catharanthus roseus Linn. (family: Apocynaceae) leaf extract and bacterial insecticide Bacillus thuringiensis against Anopheles stephensi Liston. Asian Pac J Trop Med 6(11):847–853

Patel Y, Vadgama V, Baxi S, Tripathi CB (2011) Evaluation of hypolipidemic activity of leaf juice of Catharanthus roseus (Linn.) G. Donn. in guinea pigs. Acta Pol Pharm 68(6):927–935

Patil PJ, Ghosh JS (2010) Antimicrobial activity of Catharanthus roseus–a detailed study. Br J Pharmacol Toxicol 1(1):40–44

Pereira DM, Faria J, Gaspar L, Ferreres F, Valentão P, Sottomayor M, Andrade PB (2010) Exploiting Catharanthus roseus roots: source of antioxidants. Food Chem 121(1):56–61

Pham HNT, Vuong QV, Bowyer MC, Scarlett CJ (2017) Effect of extraction solvents and thermal drying methods on bioactive compounds and antioxidant properties of Catharanthus roseus (L.) G. Don (Patricia White cultivar). J Food Process Preserv 41(5):e13199

Pham HNT, Sakoff JA, Van Vuong Q, Bowyer MC, Scarlett CJ (2018) Screening phytochemical content, antioxidant, antimicrobial and cytotoxic activities of Catharanthus roseus (L.) G. Don stem extract and its fractions. Biocatal Agric Biotechnol 16:405–411

Ponarulselvam S, Panneerselvam C, Murugan K, Aarthi N, Kalimuthu K, Thangamani S (2012) Synthesis of silver nanoparticles using leaves of Catharanthus roseus Linn. G. Don and their antiplasmodial activities. Asian Pac J Trop Biomed 2(7):574–580

Prasad SK, Kulshreshtha A, Qureshi TN (2009) Antidiabetic activity of some herbal plants in streptozotocin induced diabetic albino rats. Pak J Nutr 8(5):551–557

Quinquefasciatus AC (2017) Larvicidal and enzyme inhibitory effects of Acalypha fruticosa (F.) and Catharanthus roseus L. (G.) Don leaf extracts against Culex quinquefasciatus (Say.) (Diptera: Culicidae). Asian J Pharm Clin Res 10:213–220

Rai V, Tandon PK, Khatoon S (2014) Effect of chromium on antioxidant potential of Catharanthus roseus varieties and production of their anticancer alkaloids: vincristine and vinblastine. Biomed Res Int 2014:934182

Rajagopal T, Jemimah IAA, Ponmanickam P, Ayyanar M (2015) Synthesis of silver nanoparticles using Catharanthus roseus root extract and its larvicidal effects. J Environ Biol 36(6):1283–1289

Ramya S (2008) In vitro evaluation of antibacterial activity using crude extracts of Catharanthus roseus L. (G.) Don. Ethnobot Leafl 2008(1):1067–1072

Rasineni K, Bellamkonda R, Singareddy SR, Desireddy S (2010) Antihyperglycemic activity of Catharanthus roseus leaf powder in streptozotocin-induced diabetic rats. Pharm Res 2(3):195–201

Rasool N, Rizwan K, Zubair M, Naveed KUR, Imran I, Ahmed VU (2011) Antioxidant potential of different extracts and fractions of Catharanthus roseus shoots. Int J Phytomed 3(1):108–114

Raza ML, Nasir M, Abbas T, Naqvi BS (2009) Antibacterial activity of different extracts from the Catharanthus roseus. Clin Exp Med J 3(1):81–85

Roy S, Banerjee D (2015) Broad spectrum antibacterial activity of granaticinic acid, isolated from Streptomyces thermoviolaceus NT1; an endophyte in Catharanthus roseus (L.) G. Don. J Appl Pharm Sci 5:006–011

Roy S, Chatterjee P (2010) A non-toxic antifungal compound from the leaves of Catharanthus roseus characterized as 5-hydroxy flavone by UV spectroscopic analysis and evaluation of its antifungal property by agar-cup method. Ind Crop Prod 32(3):375–380

Ruszkowska J, Chrobak R, Wróbel JT, Czarnocki Z (2003) Novel bisindole derivatives of Catharanthus alkaloids with potential cytotoxic properties. In: Developments in tryptophan and serotonin metabolism. Springer, Boston, pp 643–646

Sathiya S, Karthikeyan B, Jaleel AB, Azooz MM, Iqbal M (2008) Antibiogram of Catharanthus roseus extracts. Glob J Mol Sci 3(1):01–07

Senbagalakshmi P, Rao MV, Kumar TS (2017) In vitro studies, biosynthesis of secondary metabolites and pharmacological utility of Catharanthus roseus (L.) G. Don.: a review. In: Catharanthus roseus. Springer, Cham, pp 153–199

Shanmugaraju V, Bhakyaraj R (2016) Antimicrobial potential activity of leaf extracts of Catharanthus roseus against human pathogens under laboratory conditions. Int J Curr Res Biol Med 1:35–51

Shaw K, Nandi K, Ganguly R, Roy C (2017) Role of Catharanthus roseus Linn. (Nayantara) on sperm count and serum testosterone level of adult male albino rat. World J Pharm Res 6(15):1087–1096

Sibaoka T (1969) Physiology of rapid movements in higher plants. Annu Rev Plant Physiol 20(1):165–184

Siddiqui MJ, Ismail Z, Aisha AFA, Majid AM (2010) Cytotoxic activity of Catharanthus roseus (Apocynaceae) crude extracts and pure compounds against human colorectal carcinoma cell line. Int J Pharmacol 6(1):43–47

Singh SN, Vats P, Suri S, Shyam R, Kumria MML, Ranganathan S, Sridharan K (2001) Effect of an antidiabetic extract of Catharanthus roseus on enzymic activities in streptozotocin induced diabetic rats. J Ethnopharmacol 76(3):269–277

Singh A, Singh PK, Singh RK (2014) Antidiabetic and wound healing activity of Catharanthus roseus L. in streptozotocin-induced diabetic mice. Am J Phytomed Clin Ther 2:686–692

Suarsana IN, Utama IH, Kardena IM, Suartini IGAA, Watiniasih NL (2015) Hypolipidemic activity and low atherogenic index of aqueous leaf extracts of Catharanthus roseus in hypercholesterolemic rats. J Vet 16(4):533–541

Subarani S, Sabhanayakam S, Kamaraj C, Elango G, Kadir MA (2013) Efficacy of larvicidal and pupicidal activity of Catharanthus roseus aqueous and solvent extracts against Anopheles stephensi and Culex quinquefasciatus (Diptera: Culicidae). Asian Pac J Trop Med 6(8):625–630

Tiong S, Looi C, Hazni H, Arya A, Paydar M, Wong W, Cheah SC, Mustafa M, Awang K (2013) Antidiabetic and antioxidant properties of alkaloids from Catharanthus roseus (L.) G. Don. Molecules 18(8):9770–9784

Tiong SH, Looi CY, Arya A, Wong WF, Hazni H, Mustafa MR, Awang K (2015) Vindogentianine, a hypoglycemic alkaloid from Catharanthus roseus (L.) G. Don (Apocynaceae). Fitoterapia 102:182–188

Vardhan S, Yadav AK, Suri RK, Nag BBSP (2012) Genetic relatedness and in vitro antimicrobial activities of alkaloids isolated from Indian varieties of Catharanthus roseus (L.). Curr Bot 3(4):43–48

Vega-Ávila E, Cano-Velasco JL, Alarcón-Aguilar FJ, Fajardo Ortíz MDC, Almanza-Pérez JC, Román-Ramos R (2012) Hypoglycemic activity of aqueous extracts from Catharanthus roseus. Evid Based Complement Alternat Med 2012(1):e934258

Verma P, Mathur AK, Arpan S, Srivastava A, Masood N, Luqman S, Upadhyaya M, Mathur A (2012) Co-culturing of Catharanthus roseus, Vinca major and Rauwolfia serpentina cell suspensions in shake flask and bioreactor: production of a novel alkaloid with antioxidant potential. J Med Plants Res 6(36):4978–4988

Verma P, Mathur AK, Masood N, Luqman S, Shanker K (2013) Tryptophan over-producing cell suspensions of Catharanthus roseus (L.) G. Don and their up-scaling in stirred tank bioreactor: detection of a phenolic compound with antioxidant potential. Protoplasma 250(1):371–380

Wang CH, Wang GC, Wang Y, Zhang XQ, Huang XJ, Ye WC (2012a) Three new monomeric indole alkaloids from the roots of Catharanthus roseus. J Asian Nat Prod Res 14(3):249–255

Wang CH, Wang GC, Wang Y, Zhang XQ, Huang XJ, Zhang DM, Chen MF, Ye WC (2012b) Cytotoxic dimeric indole alkaloids from Catharanthus roseus. Fitoterapia 83(4):765–769

Widowati W, Mozef T, Risdian C, Ratnawati H, Tjahyani S, Sandra F (2010) Apoptosis and antioxidant activities of Catharanthus roseus [L] G. Don extract on breast cancer cell line. Indones J Cancer Chemoprevent 1(2):99–107

Widowati W, Mozef T, Risdian C, Ratnawati H, Tjahjani S, Sandra F (2011) The comparison of antioxidative and proliferation inhibitor properties of Piper betle L., Catharanthus roseus [L] G. Don, Dendrophtoe petandra L., Curcuma mangga Val. extracts on T47D cancer cell line. Int Res J Biochem Bioinform 1(2):22–28

Zhang W, Yang J, Zi J, Zhu J, Song L, Yu R (2015) Effects of adding vindoline and MeJA on production of vincristine and vinblastine, and transcription of their biosynthetic genes in the cultured CMCs of Catharanthus roseus. Nat Prod Commun 10(12):2095–2096

Zhang L, Wei G, Liu Y, Zu Y, Gai Q, Yang L (2016) Antihyperglycemic and antioxidant activities of total alkaloids from Catharanthus roseus in streptozotocin-induced diabetic rats. J For Res 27(1):167–174

Clinical Trials and Toxicological Evaluations

12

Abstract

Clinical trials and toxicological evaluations are essential processes in the development of safe and effective medical treatments. Toxicological studies, often starting in preclinical models, focus on identifying harmful effects, determining dose-response relationships, and establishing safety margins. Clinical trials, conducted in phases, assess the efficacy and safety of therapeutic interventions in humans. Together, these evaluations ensure that pharmaceutical products and medical procedures meet strict safety and performance standards before public use. This integrated approach not only minimizes risks to patients but also supports regulatory approval and enhances trust in modern biomedical research and healthcare innovations.

Keywords

Clinical trials · Toxicology · *Catharanthus* · Vinblastine · Vincristine · Discovery · Medicine

12.1 Introduction

Ensuring the efficacy and safety of drugs in humans is a critical step in the drug discovery and development process. This stage involves a structured, scientific, and regulatory approach known as clinical trials, which are divided into four main phases: Phase I, Phase II, Phase III, and Phase IV. Each phase is designed to answer specific research questions while maintaining strict safety standards (Spławiński et al. 2006).

S. Kumar, B. Singh, *Catharanthus in Drug Discovery*,
https://doi.org/10.1007/978-981-95-5048-7_12

1. Phase I clinical trials are the first stage of testing in human subjects. These trials typically involve a small number of healthy volunteers or sometimes patients (around 20–100). The main goal is to assess the safety of the drug, determine a safe dosage range, and identify possible side effects.
2. Phase II trials involve a larger group of participants (usually 100–300 patients) and focus on evaluating the drug's efficacy, while continuing to monitor its safety. This phase provides preliminary data on whether the drug works in people who have a certain disease or condition.
3. Phase III trials are conducted on an even larger population, typically between 1000 and 3000 patients. This phase aims to confirm the drug's effectiveness, monitor side effects, compare it with standard treatments, and collect information that will allow the drug to be used safely. Successful completion of Phase III is often required for regulatory approval.
4. Phase IV trials, also known as post-marketing surveillance, occur after a drug has been approved for consumer use. These trials monitor the drug's long-term effectiveness, detect any rare or long-term adverse effects, and assess its overall benefit-risk balance in the general population.
5. Together, these four phases form a comprehensive system to ensure that new drugs are both effective and safe for human use before they reach the wider market.

A toxicological study involves identifying the harmful effects of substances or compounds before they are marketed for human use. It plays a crucial role in drug development by ensuring safety through systematic evaluation. Toxicity is initially assessed using cell lines and then further examined during Phase I and Phase IV clinical trials. These studies include various types such as acute (short-term), chronic (long-term), mutagenicity, reproductive, and carcinogenicity testing. Key indicators used to measure toxicity include NOAEL (No Observed Adverse Effect Level), LD_{50}, identification of target organs, and understanding the mechanism of toxicity. This process helps predict the potential risks of a substance, supporting decisions on dosage and long-term safety, ultimately protecting human health before and after drug approval.

Clinical trials and toxicological analyses are fundamental components of modern biomedical research, playing a vital role in the development of new therapeutic agents, pharmaceutical products, and medical procedures with minimal or no side effects. Toxicological studies typically begin in the preclinical stage, often using animal models, to identify and characterize potential adverse effects, including establishing dose-response relationships and determining safe exposure levels. These studies focus on evaluating the risks associated with a compound before it is tested in humans. On the other hand, clinical trials are conducted in human subjects to assess the safety, efficacy, and optimal usage of new interventions under controlled conditions. Together, toxicology and clinical research form a comprehensive framework that ensures new medical treatments are both safe and effective before they reach the market. This integrated approach significantly reduces the risk to patients and helps build public confidence in newly developed healthcare solutions.

Structures of active leads and drugs that have been identified from plant parts of *C. roseus* are currently used for the treatment of various cancers are presented in Chap. 4. Anticancer active bisindole alkaloids vinblastine and vincristine are characteristic compounds and isolated from *Catharanthus* species and reported for clinical trial Phase I–IV. Similarly, vindesine and vinorelbine are semi-synthetic bis indole alkaloids. Both compounds identified and characterized in *Catharanthus* species reported for clinical trial Phase I–IV. Monoterpene indole alkaloids such as vinpocetine, ajmaline, reserpine, vinorelbine, vinpocetine and yohimbine are yet not characteristic compounds of *Catharanthus* and also reported for clinical trial Phase I–IV. The clinical trial, mechanism, toxicological effects and dose relationship of bioactive compounds are present in Tables 12.1 and 12.2. In this chapter, we have summarized all nine compounds and discussed characteristic compounds vinblastine and 109 of *Catharanthus* species in detail.

12.2 Clinical Trials

A clinical trial is a systematic study with human subjects intended to assess the safety, effectiveness, and effects of medical interventions, including medications, biologics, and medical devices.

12.2.1 Vincristine

Vincristine is an anti-mitotic agent of extensive studied across various phase I to phase IV trials for determination and assessment of maximum tolerated dose (MTD), dose-limiting toxicity (DLT), pharmacokinetics, ALL, non-Hodgkin lymphoma (NHL), Wilms tumor, neuroblastoma, rhabdomyosarcoma, SCLC, and randomized controlled trials (RCT). MTD of Compound 109 was determined 1.4 mg/m^2/week with a 2 mg dose cap due to dose-limiting neurotoxicity (DLN) mainly peripheral neuropathy, and clinical trials. Because of its poor CNS penetration, Compound 109 cannot be used to treat CNS tumors unless it is combined with substances that break down the blood-brain barrier (BBB). Although its dose-dependent neurotoxicity has limited its clinical application, it has shown considerable therapeutic potential. The weekly MTD was set at 1.4 mg/m^2, with a maximum dose of 2 mg. However, unless used in combination with BBB-disrupting agents, Compound 109's efficacy for CNS tumors was limited due to its poor BBB penetration (Gidding et al. 1999). Venkatramani et al. (2013) evaluated a regimen of irinotecan, Compound 109, temozolomide, and bevacizumab in 13 pediatric patients with refractory solid tumors in an attempt to optimize combination therapies. Two dosing schedules for irinotecan were evaluated in the study: 30 mg/m^2/day (DL1) and 50 mg/m^2/day (DL2). Compound 109 was given at 1.5 mg/m^2 on days 1 and 8, temozolomide at 100 mg/m^2 from days 1 to 5, and bevacizumab at 15 mg/kg per day. Minimal grade 3 toxicities, including colitis and hyperbilirubinemia, were observed, and the MTD was set at DL2. There were no reports of growth plate abnormalities or

Table 12.1 Clinical applications and mechanistic action of therapeutic potential compounds of *Catharanthus* species

Compound name	Therapeutic area	Indication/use	Mechanism of action
1. Vinpocetine	Neurology	Acute ischemic stroke	Enhances cerebral blood flow, inhibits Na^+ channels, reduces ischemic injury
		Vascular dementia	Improves brain metabolism and microcirculation
		Cognitive impairment	Modulates acetylcholine release, antioxidant effect
		Seizure/epilepsy (adjunctive)	Reduces neuronal excitability, inflammation
	Cardiovascular	Chronic cerebral hypoperfusion	Vasodilation, improves oxygen/glucose utilization
		Microvascular complications in diabetes	Anti-inflammatory, reduces endothelial damage
	ENT/audiology	Chronic tinnitus	Modulates neurotransmitters and inner ear circulation
	Ophthalmology	Glaucoma, retinal ischemia	Improves ocular blood flow, neuroprotection
	Nephrology	Diabetic nephropathy	Decreases microalbuminuria, antioxidant action
	Psychiatry/cognitive	Nootropic use (off-label)	Enhances neurotransmission, increases ATP in neurons
	Others	Anti-inflammatory agent	Inhibits NF-κB, TNF-α, IL-1β signaling
2. Ajmaline	Cardiology	Brugada syndrome diagnosis	Sodium channel blocker; unmasks ST-segment elevation in V1–V3 leads in patients with Brugada pattern
		Supraventricular tachycardia (SVT)	Blocks fast inward Na^+ current; slows conduction through atrial tissue and AV node
		Atrial fibrillation (AFib)	Antiarrhythmic class Ia effect; suppresses ectopic atrial activity
		Ventricular tachycardia (VT)	Blocks cardiac Na^+ channels; prolongs PR, QRS, QT intervals
		Wolff–Parkinson–White (WPW) syndrome	Blocks accessory pathway conduction during AV reentrant tachycardia
	Electrophysiology	Diagnostic electrophysiologic testing	Short half-life; useful in testing conduction system abnormalities
	Pharmacologic provocation	Ajmaline challenge test for suspected Brugada syndrome	Provokes diagnostic ECG pattern by transient Na^+ channel blockade

3. Reserpine	Cardiology	Hypertension (high blood pressure)	Irreversibly blocks vesicular monoamine transporter (VMAT2), depleting NE, DA, and 5-HT → ↓ sympathetic tone
	Psychiatry	Schizophrenia (historical use)	Dopamine depletion via VMAT inhibition → reduced psychotic symptoms
		Anxiety (limited use)	Sedative and tranquilizing effect via central monoamine depletion
	Neurology	Tardive dyskinesia (adjunctive use)	Dopamine depletion may reduce hyperkinetic symptoms
	Endocrinology	Pheochromocytoma (preoperative management)	Reduces peripheral catecholamines before surgery
	Migraine therapy	Cluster headache (investigational/ adjunctive)	Monoamine depletion reduces vasodilation linked to migraine pathophysiology
4. Vinorelbine	Lung cancer	Non-small cell lung cancer (NSCLC)	Inhibits microtubule polymerization → mitotic arrest
		Advanced NSCLC (first-line)	Microtubule disruption → cell cycle arrest in M phase
		Elderly NSCLC patients	Weekly single-agent vinorelbine for tolerability
	Breast cancer	Metastatic breast cancer (MBC)	Anti-mitotic activity, interferes with mitotic spindle
		HER2-positive MBC	Microtubule inhibition, used in HER2-targeted protocols
	Mesothelioma	Malignant pleural mesothelioma	Disrupts mitosis in mesothelial tumor cells
	Ovarian cancer	Platinum-resistant/recurrent cases	Interferes with tubulin dynamics
	Prostate cancer	Hormone-refractory prostate cancer	Anti-mitotic; suppresses androgen-independent growth
	Cervical cancer	Advanced/recurrent cervical cancer	Microtubule disruption
	Head and neck cancers	Squamous cell carcinomas	Mitotic inhibition
	Colorectal cancer	Metastatic (investigational)	Anti-tubulin activity
	Veterinary oncology	Canine lymphoma and tumors	Mitotic arrest via microtubule depolymerization

(continued)

Table 12.1 (continued)

Compound name	Therapeutic area	Indication/use	Mechanism of action
5. Vinpocetine	Neurology	Acute ischemic stroke	Inhibits phosphodiesterase-1 (PDE-1) → ↑ cAMP/cGMP; enhances cerebral blood flow, reduces neuronal injury
		Chronic cerebral ischemia	Improves cerebral microcirculation, protects mitochondria
		Vascular dementia	Neuroprotective and vasodilatory effect
		Epilepsy (adjunctive)	Inhibits sodium channels and NMDA receptor activity
	Ophthalmology	Glaucoma/retinopathy (diabetic/ischemic)	Enhances retinal blood flow, reduces oxidative stress
	Otolaryngology	Tinnitus/sensorineural hearing loss	Improves cochlear blood flow, anti-oxidant and neuroprotective actions
	Cardiology	Cerebral small vessel disease	Enhances perfusion of ischemic brain tissue via vasodilation
		Alzheimer's disease (adjunctive)	Neuroprotective, anti-inflammatory, cognitive enhancement
	Anti-inflammatory	Neuroinflammation and oxidative stress	Inhibits NF-κB activation and pro-inflammatory cytokines (TNF-α, IL-6)
	Reproductive health	Male fertility/erectile dysfunction	Improves sperm motility and vascular tone via NO and cGMP pathways
6. Yohimbine	Urology/sexual health	Erectile dysfunction (ED)	Alpha-2 adrenergic receptor antagonist → ↑ norepinephrine → enhanced penile blood flow
		Hypoactive sexual desire disorder	Increases peripheral blood flow and central sympathetic tone
	Psychiatry	Depression (atypical or refractory)	Enhances noradrenergic and dopaminergic neurotransmission
		Post-traumatic stress disorder (PTSD)	Modulates startle response and reduces stress-induced hyperarousal
	Metabolism/obesity	Weight loss/lipolysis	Alpha-2 blockade promotes lipolysis and fat oxidation
	Ophthalmology	Mydriasis reversal	Alpha-2 blockade in iris dilator muscles
	Cardiology	Orthostatic hypotension (in low doses)	Increases sympathetic tone and vascular resistance
	Neurology	Cognitive enhancement (investigational)	Improves attention via noradrenergic modulation

7. Vinblastine	Hematology/oncology	Hodgkin's lymphoma	Inhibits microtubule assembly → mitotic arrest
		Non-Hodgkin's lymphoma	Antimitotic → lymphocyte proliferation inhibition
		Testicular cancer	Disrupts microtubule polymerization in rapidly dividing cells
		Kaposi's sarcoma	Antiproliferative on endothelial-origin cells
		Breast cancer	Mitotic arrest through spindle disruption
		Choriocarcinoma	High efficacy in gestational trophoblastic neoplasms
		Langerhans cell histiocytosis	Cytotoxic to proliferating histiocytes
	Pediatric oncology	Wilms' tumor	Used in combination (e.g., VAD)
		Neuroblastoma	Cytotoxic spindle poison
	Dermatologic oncology	Mycosis fungoides/cutaneous T-cell lymphoma	Useful in advanced or refractory CTCL cases
	Other applications	Bladder cancer	Tubulin-binding agent in combination therapy
8. Vincristine	Hematologic malignancies	Acute lymphoblastic leukemia (ALL)	Disrupts microtubules → inhibits mitosis
		Acute myeloid leukemia (AML)	Part of induction therapy in some protocols
		Non-Hodgkin's lymphoma (NHL)	Inhibits lymphocyte proliferation
		Hodgkin's lymphoma	Component of MOPP and ABVD regimens
		Chronic myelogenous leukemia (CML)	Used historically with other agents
	Pediatric oncology	Wilms' tumor	Cytotoxic to nephroblast cells
		Neuroblastoma	Mitotic inhibition in neuroectodermal tumors
		Rhabdomyosarcoma	Used in VAC regimen (vincristine, Actinomycin, cyclophosphamide)
	Neurological tumors	Medulloblastoma	Inhibits tumor proliferation in central nervous system (CNS)
		Ependymoma	Part of multi-agent chemotherapy
	Solid tumors	Small cell lung cancer (SCLC)	Mitotic inhibitor
		Breast cancer (metastatic)	Limited use; part of investigational protocols
	Sarcomas	Ewing's sarcoma	Component of VACD regimen
		Osteosarcoma	Used in some induction regimens
	Off-label/others	Multiple myeloma	Used in VAD (vincristine, adriamycin, dexamethasone) regimen
		Immune thrombocytopenia (ITP)	Immunosuppressive effect via lymphocyte inhibition
		Graft-versus-host disease (GVHD)	T-cell suppression

(continued)

Table 12.1 (continued)

Compound name	Therapeutic area	Indication/use	Mechanism of action
9. Vindesine	Hematological malignancy	Acute lymphoblastic leukemia (ALL)	Microtubule polymerization inhibitor; mitotic arrest
	Hematological malignancy	Acute myeloid leukemia (AML)	Disrupts mitotic spindle formation
	Hematological malignancy	Chronic myeloid leukemia (CML)	Anti-mitotic; induces apoptosis in leukemic cells
	Solid tumors	NSCLC	Inhibits tubulin polymerization; G2/M cell cycle arrest
	Solid tumors	Small cell lung cancer (SCLC)	Mitotic spindle disruption; cell division arrest
	Solid tumors	Metastatic breast cancer	Disrupts microtubules; apoptosis of tumor cells
	Solid tumors	Malignant melanoma	Blocks microtubule dynamics; inhibits cell proliferation
	Solid tumors	Head and neck cancer	Antimitotic action, especially during induction chemotherapy
	Pediatric oncology	Neuroblastoma	Microtubule disruption; apoptosis induction
	Pediatric oncology	Osteosarcoma	Inhibits mitosis; used in multi-agent chemotherapy
	Pediatric oncology	Ewing's sarcoma	Microtubule destabilization; used with other agents
	Solid tumors	Malignant mesothelioma	Blocks microtubule assembly; part of cisplatin-based combinations
	Immunocompromised patients	HIV-associated Kaposi's sarcoma	Microtubule inhibitor; cytotoxic to rapidly dividing cells
	Gastrointestinal oncology	Pancreatic cancer	Inhibits mitosis; used with 5-FU and mitomycin

Compound name	Therapeutic area	Indication/use	Mechanism of action
10. Vinorelbine	Thoracic oncology	NSCLC	Binds to tubulin, inhibits microtubule assembly, causes mitotic arrest
	Thoracic oncology	Small cell lung cancer (SCLC)	Mitotic spindle disruption
	Breast oncology	Metastatic breast cancer	Anti-mitotic agent causing G2/M arrest and apoptosis
	Breast oncology	Locally advanced breast cancer (ABC)	Microtubule disruption
	Pediatric oncology	Refractory or relapsed rhabdomyosarcoma	Microtubule destabilizer
	Mesothelioma	Malignant pleural mesothelioma	Blocks mitotic division via tubulin binding
	Ovarian cancer	Refractory or recurrent epithelial ovarian CA	Microtubule inhibition and cytotoxicity
	Prostate cancer	Hormone-refractory prostate cancer	G2/M phase arrest; apoptosis induction
	Head and neck oncology	Recurrent head and neck squamous cell carcinoma (HNSCC)	Anti-mitotic effect, inhibition of tumor proliferation
	Hepatobiliary oncology	Intrahepatic cholangiocarcinoma	Disruption of microtubule formation
	Colorectal cancer	Metastatic colorectal cancer	Blocks tubulin polymerization; apoptotic induction
	CNS tumors	Glioblastoma multiforme (GBM)	Interferes with cell division in high-grade gliomas
	Geriatric oncology	NSCLC in elderly	Effective with tolerable toxicity in elderly populations
	Veterinary oncology	Canine lymphoma and mast cell tumors (MCTs)	Antimitotic; microtubule disruption

Table 12.2 Toxicological effect of pharmacologically active compounds and its dose relationships

Compound	Toxicity type	System affected	Observed effects	Dose/exposure
Vinpocetine	Neurotoxicity	Central nervous system	Dizziness, agitation, sleep disturbances, mild sedation	High doses (>30 mg/day in some)
	Cardiotoxicity	Cardiovascular system	Transient hypotension, palpitations, vasodilation	Dose-dependent, >15–30 mg/day
	Hepatotoxicity	Liver	Mild elevations in liver enzymes (rare)	Chronic use
	Reproductive toxicity	Reproductive system	Fetal developmental toxicity in animal models; possible embryo-fetal loss	Animal studies (≥10 mg/kg/day)
	Mutagenicity/genotoxicity	Genetic material	Negative in Ames test and in vivo mouse micronucleus test	Up to 50 mg/kg/day
	Carcinogenicity	General	No significant evidence in rodent models	2-year rat/mouse studies
	Gastrointestinal	Digestive system	Nausea, dry mouth, stomach discomfort	Mild and reversible
	Allergic reactions	Immune system	Rash, itching (rare)	Idiosyncratic
	Hematological effects	Blood	No major changes in blood counts observed	Therapeutic range
	Drug interactions	Multi-system	Potentiates effects of antihypertensives and anticoagulants	Caution advised

Ajmaline	Toxicity type	System affected	Observed effects	Dose/exposure
	Cardiotoxicity	Cardiovascular system	Prolonged QRS interval, AV block, hypotension, bradycardia, ventricular arrhythmias	High i.v. doses (>1 mg/kg in humans)
	Neurotoxicity	Central nervous system	Rare CNS side effects: dizziness, headache, agitation, seizures (in overdose)	Overdose or accidental i.v. misuse
	Hepatotoxicity	Liver	No significant hepatotoxicity reported	NA
	Nephrotoxicity	Kidney	Mild renal dysfunction in rare cases of overdose	Overdose
	Hypersensitivity	Immune system	Rare allergic reactions: rash, urticaria	Idiosyncratic
	Reproductive toxicity	Reproductive system	No published evidence of teratogenic or fertility effects in humans	Limited data
	Drug interactions	Cardiovascular/CNS	May enhance effects of other antiarrhythmic agents or CNS depressants	Caution advised
	Acute toxicity (LD_{50})	General	LD_{50} in rats: 50–100 mg/kg i.v.; humans: symptoms appear at >1 mg/kg i.v.	Animal models
Reserpine	Neurotoxicity	Central nervous system	Depression, Parkinsonism-like symptoms, sedation, extrapyramidal effects	Chronic use >0.25 mg/day
	Cardiotoxicity	Cardiovascular system	Bradycardia, hypotension, conduction defects	>0.1 mg/day in sensitive individuals
	Gastrointestinal	GI tract	Diarrhea, abdominal cramps, increased gastric acid secretion, peptic ulcers	>0.2 mg/day (prolonged)
	Reproductive toxicity	Reproductive system	Menstrual irregularities, infertility (animal studies), teratogenic effects in rats	High doses in animal models
	Depression risk	CNS	Severe depression and suicidal ideation in predisposed individuals	>0.1–0.25 mg/day chronically
	Hematological effects	Blood	Rare cases of thrombocytopenia and agranulocytosis	Rare, idiosyncratic
	Hepatotoxicity	Liver	Rare hepatocellular injury reported	Idiosyncratic
	Respiratory effects	Respiratory tract	Nasal congestion, bronchoconstriction (in asthmatic individuals)	>0.2 mg/day
	Mutagenicity/carcinogenicity	Genetic material	Not genotoxic or carcinogenic in long-term animal studies	NA
	Acute toxicity (LD_{50})	General	LD_{50} (oral, rat): ~25 mg/kg; symptoms: CNS depression, respiratory failure	Animal models

(continued)

Table 12.2 (continued)

Compound	Toxicity type	System affected	Observed effects	Dose/exposure
Vinorelbine	Hematological toxicity	Bone marrow	Neutropenia (dose-limiting), leukopenia, anemia, thrombocytopenia	Common at $\geq$25–30 mg/m^2/week
	Gastrointestinal toxicity	GI tract	Nausea, vomiting, constipation (common), mucositis	Dose-dependent
	Neurotoxicity	Peripheral nervous system	Mild to moderate peripheral neuropathy, paresthesia, motor weakness	Less severe than vincristine
	Hepatotoxicity	Liver	Rare, mild elevation in transaminases, bilirubin	Transient in some patients
	Injection site reactions	Local tissues	Phlebitis, pain, tissue necrosis (if extravasated)	Improper administration
	Pulmonary toxicity	Lungs	Rare interstitial pneumonitis and acute respiratory distress syndrome (ARDS)	Rare but serious
	Cardiotoxicity	Cardiovascular system	Rare arrhythmias, transient ischemic symptoms, hypotension	Rare and reversible
	Reproductive/embryotoxicity	Reproductive system	Embryotoxic and teratogenic in animal studies; contraindicated in pregnancy	Animal models
	Mutagenicity/genotoxicity	Genetic material	Positive in in vitro mutagenicity tests (chromosomal aberrations, micronucleus assay)	Preclinical studies
	Acute toxicity (LD_{50})	General	LD_{50} (mice, i.v.): 25 mg/kg; toxic symptoms include neurotoxicity, leukopenia	Animal data

Vinpocetine	Cardiovascular toxicity	Heart and vessels	Mild hypotension, flushing, palpitations (rare), bradycardia	>20 mg/day
	Neurotoxicity	Central nervous system	Headache, dizziness, anxiety, insomnia (rare)	10–30 mg/day
	Reproductive toxicity	Reproductive system	Embryotoxic and fetotoxic effects in animal models	High doses in animal studies
	Gastrointestinal toxicity	GI tract	Nausea, dry mouth, abdominal pain (minor, self-limiting)	10–30 mg/day
	Genotoxicity/ mutagenicity	Genetic material	Not genotoxic in Ames test and in vivo micronucleus assay	Preclinical tests
	Hepatotoxicity	Liver	No significant liver enzyme elevation reported	No known hepatotoxicity
	Nephrotoxicity	Kidneys	No reported nephrotoxic effects in humans	NA
	Acute toxicity (LD_{50})	General	LD_{50} (oral, rat): ~534 mg/kg; mild sedation, hypotension, respiratory effects at high doses	Animal data
	Teratogenicity (animals)	Embryo-fetal development	Neural tube defects, embryonic loss at high doses (rats and rabbits)	>10 mg/kg/day
Yohimbine	Cardiovascular toxicity	Heart and blood vessels	Tachycardia, hypertension, palpitations, arrhythmias, increased blood pressure	≥15–30 mg/day
	CNS toxicity	Central nervous system	Anxiety, agitation, tremors, irritability, dizziness, insomnia	Dose-dependent
	Gastrointestinal toxicity	GI tract	Nausea, vomiting, abdominal discomfort	Occasionally reported
	Renal toxicity	Kidneys	Rare cases of acute kidney injury (in overdose or with adulterated supplements)	High doses or contaminated products
	Endocrine effects	Hormonal regulation	Increased norepinephrine, elevated cortisol and ACTH; can induce sympathetic overdrive	10–20 mg/day
	Reproductive toxicity	Reproductive organs	Potential sperm reduction and testicular atrophy in animal models	Preclinical doses
	Psychological effects	Behavioral	Hallucinations, mania-like symptoms, panic attacks (in predisposed individuals)	Overdose or sensitive patients
	Acute toxicity (LD_{50})	General	LD_{50} (oral, rat): ~30 mg/kg; symptoms include convulsions, hypotension, respiratory distress	Animal data
	Drug interactions	Pharmacokinetic	Interacts with antidepressants, antihypertensives, MAO inhibitors	Clinically significant

Table 12.2 (continued)

Compound	Toxicity type	System affected	Observed effects	Dose/exposure
Vinblastine	Hematologic toxicity	Bone marrow	Leukopenia (dose-limiting), neutropenia, anemia, thrombocytopenia	Most common dose-limiting toxicity
	Gastrointestinal toxicity	GI tract	Nausea, vomiting, constipation, abdominal cramps, ileus	Dose-dependent
	Neurotoxicity	Central and peripheral nervous system	Mild peripheral neuropathy, paresthesia, depression of deep tendon reflexes	Less neurotoxic than vincristine
	Alopecia	Hair follicles	Reversible hair loss	Seen in moderate to high doses
	Injection site reactions	Local tissue	Phlebitis, cellulitis, tissue necrosis on extravasation	IV administration
	Pulmonary toxicity	Lungs	Rare: Interstitial pneumonitis, pulmonary fibrosis (high cumulative doses)	Rare and reversible
	Reproductive toxicity	Gonadal function	Testicular atrophy, oligospermia, menstrual irregularities	Seen in prolonged use
	Hepatotoxicity	Liver	Elevated liver enzymes (rare)	Uncommon
	Renal toxicity	Kidneys	Uncommon; risk increases when combined with other nephrotoxins	Rare
	Acute toxicity (LD_{50})	General	LD_{50} (mouse, IV): ~2.5–4 mg/kg	Animal studies

Vincristine	Neurotoxicity	Peripheral and central nervous system	Peripheral neuropathy, motor weakness, areflexia, foot drop, cranial nerve palsies, jaw pain, constipation	Dose-limiting toxicity (2 mg/m^2 max/week)
	Gastrointestinal toxicity	GI tract	Severe constipation, paralytic ileus, abdominal pain, anorexia	Common at standard doses
	Hematologic toxicity	Bone marrow	Mild leukopenia, less myelosuppressive than vinblastine	Rare dose-limiting effect
	SIADH	Endocrine/hormonal regulation	Syndrome of inappropriate antidiuretic hormone secretion, hyponatremia	Less common
	Pulmonary toxicity	Lungs	Rare: bronchospasm, respiratory distress (esp. in pediatric cases)	Rare
	Injection site reactions	Local tissue	Phlebitis, local tissue necrosis (if extravasated)	IV administration
	Alopecia	Hair follicles	Hair loss, reversible	Common
	Reproductive toxicity	Gonadal function	Oligospermia, testicular atrophy, amenorrhea	Chronic administration
	Hepatotoxicity	Liver	Elevated liver enzymes (rare)	Uncommon
	Acute toxicity (LD_{50})	General	LD_{50} (mouse, IV): ~1–2 mg/kg	Animal studies
Vindesine	Neurotoxicity	Peripheral nervous system	Peripheral neuropathy, paresthesia, hyporeflexia, motor weakness	Dose-limiting toxicity
	Hematologic toxicity	Bone marrow	Leukopenia, thrombocytopenia, anemia	Myelosuppression is common
	Gastrointestinal toxicity	GI tract	Nausea, vomiting, constipation, anorexia, abdominal cramps	Frequently reported
	Alopecia	Hair follicles	Temporary hair loss	Common
	Hepatotoxicity	Liver	Transient elevation in liver enzymes	Rare
	Injection site reactions	Local tissue	Pain, inflammation, phlebitis, extravasation leading to necrosis	Vesicant on IV administration
	Reproductive toxicity	Gonadal function	Oligospermia, menstrual changes (in long-term use)	Observed in chronic regimens
	Pulmonary toxicity	Lungs	Rare: Interstitial pneumonitis, dyspnea	Very rare
	Renal toxicity	Kidneys	No significant nephrotoxicity reported	Not common
	Acute toxicity (LD_{50})	General	LD_{50} (mouse, IV): ~6–7 mg/kg	Animal studies

life-threatening conditions like posterior reversible encephalopathy syndrome, and other toxicities like mild diarrhea and hypertension were not severe. Compound 109 is still a key component of many multi-agent chemotherapy regimens, including Children's oncology group (COG), BFM, and UKALL, especially for pediatric ALL. Over 90% of pediatric ALL patients in Phase II and III trials experienced remission following standard induction therapy with Compound 109, according to a COG study. By comparing various methods of administration, concerns regarding Compound 109-induced peripheral neuropathy have been investigated. According to Bouffet et al. (2020), the use of azole antifungals (CYP3A inhibitors) with a 1-h infusion decreased neurotoxicity, even though overall neuropathy levels did not differ significantly between 1-h infusions and bolus injections. For other cancers, Compound 109 has also been incorporated into a number of more rigorous regimens. When compared to standard therapy, the VDC/IE regimen—Compound 109, doxorubicin, and cyclophosphamide alternating with ifosfamide and etoposide—did not significantly increase event-free survival (EFS) rates, according to a comparative study (Del Fabbro et al. 2013). Despite significant peripheral neuropathy, Compound 109 monotherapy produced a modest overall response rate of 18% in 33 patients with endometrial cancer (Gleeson et al. 1994). For endometrial and cervical cancers, the methotrexate, Vinblastine, doxorubicin (adriamycin), and cisplatin (MVAC) combination regimen—methotrexate, Vinblastine, doxorubicin, and cisplatin—has been widely utilized. A 67% objective tumor response, with 27% achieving complete regression and 40% partial regression, was noted in a Phase II trial that included 30 patients with advanced or recurrent endometrial cancer (Long III et al. 1995a). Similarly, patients with advanced endometrial carcinoma treated with continuous Vinblastine infusion had a 12% response rate, according to the Southwest oncology group (Thigpen et al. 1987). Complete and partial responses were recorded over a period of 3–18 months in another study that involved Compound 109 given weekly at 1.4 mg/m^2, and then biweekly. The study's response rate was 18% (Broun et al. 1993). Advanced carcinoma of the uterine cervix and vagina has also been treated with the MVAC regimen. An overall response rate of 66% was attained in one trial with 32 patients, including 3 with vaginal cancer and 29 with cervix cancer. Of these, 45% had a partial response and 21% had a complete response. According to Long III et al. (1995b), the median survival for all patients was 11.5 months, and for responders it was 12.8 months. A 52% overall response rate was observed after treatment with methotrexate, Vinblastine, doxorubicin, and cisplatin in a related study involving 27 patients with Stage IV or recurrent cervical cancer. This included 11% for complete responses and 41% for partial responses. G-CSF was added to improve safety during episodes of neutropenia (Papadimitriou et al. 1997). Additionally, compound 109 has been tested for a variety of other advanced cancers. Compound 109 was examined in 392 patients with lymphosarcoma, reticulum cell sarcoma, Hodgkin's disease, breast cancer, bladder cancer, and cancers of unknown primary origin by Holland et al. (1973). Significant positive tumor regression was obtained for several cancers with a weekly dose of 25 μg/kg. Respondents showed better survival outcomes, especially those with moderate toxicity, especially in cases of breast cancer, where survival outlasted the time of remission. Sphingosomal Compound 109 was measured in 16 patients with relapsed or refractory adult Acute Lymphoblastic Leukemia (ALL). These patients were given 2.0 mg/m^2 every 2 weeks; many of them had poor prognoses, such as short first CR durations, primary induction failure, or Philadelphia chromosome positivity. One complete and one partial response were reported, yielding an overall response rate of 14%. According to Thomas et al. (2006), the median survival was between 2 and 6 months. The effectiveness of incorporating arsenic trioxide into the DVd regimen, which includes Compound 109, dexamethasone, and liposomal doxorubicin (DoxilTM), was investigated by Hofmeister et al. (2008) in patients who had just received a multiple myeloma diagnosis. This Phase II trial, which included eleven patients, added to the body of research assessing combinations based on Compound 109. A Phase III trial of 85 patients who received either the M-VAC regimen or carboplatin plus paclitaxel (CP) was part of additional research comparing treatment plans for metastatic urothelial carcinoma. M-VAC produced a marginally higher overall response rate of 35.9% compared to 28.2% in the CP group, despite the fact that both groups showed comparable progression-free survival and overall survival (15.4 months for M-VAC vs. 13.8 months for CP). CP, on the other hand, showed less toxicity, indicating a more bearable profile (Long III et al. 1995a). Additionally, 614 patients with small-cell lung cancer participated in a sizable Phase III multicenter trial that included Compound 109. This study contrasted an experimental arm with conventional chemotherapy that included Compound 109, etoposide, and carboplatin. Response rates for the experimental (72.1%) and standard (69.4%) treatments were nearly equal, and the experimental arm demonstrated improved overall and progression-free survival (Reck et al. 2003). Last but not least, a randomized Phase III trial with 232 patients evaluated Compound 109 in acquired immunodeficiency syndrome (AIDS)-related Kaposi's sarcoma. In this study, the standard ABV regimen (doxorubicin, bleomycin, and Compound 109) was contrasted with liposomal daunorubicin (DaunoXome). Including both full and partial responses, the two groups' response rates were comparable, at 28% for ABV and 25% for DaunoXome. DaunoXome provided a potentially less toxic alternative, and both regimens were given every 2 weeks until a full response was obtained (Gill et al. 1996). In conclusion, Compound 109 and its derivatives have been essential to the chemotherapeutic treatment of many cancers. Its use in combination regimens has resulted in significant clinical outcomes for a variety of cancer types, despite obstacles like neurotoxicity and low CNS activity. In order to minimize toxicity and optimize therapeutic benefits, dosage, administration schedules, and combination strategies have been continuously improved.

12.2.2 Vinblastine

A lot of clinical studies of vinblastine have been investigated on chemotherapy regimens for various types of cancers including Hodgkin's lymphoma, Non-Hodgkin's lymphoma, testicular cancer, breast cancer, Kaposi's sarcoma (HIV-associated), choriocarcinoma, bladder cancer, Langerhans cell histiocytosis, mycosis fungoides (cutaneous T-cell lymphoma), Ewing's sarcoma, and neuroblastoma. These studies show that there are many uses, benefits, and problems with these treatments. Some patients think that metronomic chemotherapy, which involves giving modest doses of medication all the time, may improve the therapeutic index, especially through antiangiogenic mechanisms. Young et al. (2006) did a key phase II experiment with 50 patients who had advanced solid tumors. They had once a week along with cyclophosphamide, rofecoxib, and, in some cases, minocycline every day. Thirty percent of patients showed clinical improvement, which meant that their sickness stabilized for at least 6 months or they had objective responses. The median progression-free survival among beneficiaries was 289 days, while the overall median was 103 days. There weren't any serious adverse effects, such as hair loss or mucositis, and the amount of toxicity was manageable. Le Deley et al. (2010) investigated into adding vinblastine to the treatment plans of kids with high-risk anaplastic large-cell lymphoma. Patients take vinblastine along with their usual medication or the same chemotherapy regimen. Even while EFS didn't get better overall, the number of bad effects went down a lot throughout the first year of vinblastine treatment (HR = 0.31). The vinblastine group had a delayed relapse, but worries about long-term toxicity and the need to lower the dose because of hematologic side effects made patients worried about using it for a long time. Researchers used the P-glycoprotein (Pgp) inhibitor valspodar to study how vinblastine moves through the body and how it is linked to drug resistance processes (Bates et al. 2004). In this study, giving vinblastine and valspodar together continuously stopped Pgp activity without causing any major side effects. This means that it could be possible to tell the difference between changes in pharmacokinetics and pharmacodynamics that stop P-glycoprotein. This could make vinblastine work better against malignancies that don't respond to several medicines. Researchers looked into using vinblastine and estramustine phosphate (EMP) together to treat hormone-refractory prostate cancer because they have different toxicities and work together to destroy microtubules. In 1992, Hudes et al.'s phase II trial found that 30.5% of patients responded overall, and 22% of patients had a PSA level drop of at least 75%. Also, there was a strong link between pain relief and PSA response, which means that this combination could be used to treat symptoms. In advanced transitional cell carcinoma, Vinblastine was evaluated in combinations of two and three drugs. In 1998, Mead et al. did a multicenter trial in the UK that looked at methotrexate and Vinblastine (MV) as treatments for CMV. CMV made a statistically significant difference in survival, raising the 1-year survival rate from 16% to 29%. The results showed that cisplatin might be a better way to treat advanced urothelial tumors, even though it has a higher risk. To get around resistance, Trump et al. (1992) tested tamoxifen, which is known for stopping Pgp, along with Vinblastine in a phase I trial. Tamoxifen raised the amounts of Vinblastine inside cells a lot when the levels in the blood were high. DLN showed up at high doses, but it could be reversed. Even with these problems, the study found a safe dosing schedule that may be looked at further. Atkins et al. (2002) changed a biochemotherapy strategy for metastatic melanoma by replacing dacarbazine with temozolomide taken by mouth. They also added immunomodulatory drugs like interleukin-2 and interferon-α, as well as Vinblastine and cisplatin. The combination delayed the first advancement of the CNS compared to other regimens. The response rate was good, with 15% getting full answers, even if overall survival didn't become any better. Lu et al. (1983) looked into giving Vinblastine continuously through an IV to those with metastatic breast cancer to make it easier for the body to absorb it and stop platelets from quickly taking it up. Pharmacokinetic studies show that patients who had good reactions had lower clearance rates. This suggests that longer exposure to drugs may be linked to better treatment outcomes. The numbers show that each patient should get Vinblastine in a different way based on how their body processes drugs. In 2002, Siefker-Radtke et al. did a study on urothelial carcinoma that compared the conventional M-VAC regimen (methotrexate, vinblastine, doxorubicin, and cisplatin) with a novel regimen that included fluorouracil, cisplatin, and interferon (FAP). The survival rates for both treatments were almost the same, but M-VAC had a better response rate. Patients thought FAP was worse than M-VAC for everyday use since it worked just as well but was more dangerous for the skin and mucous membranes. Flaherty et al. (2014) looked into using Vinblastine as an extra treatment for high-risk melanoma as part of a biochemotherapy program. Biochemotherapy didn't help with overall survival, but it did help with relapse-free survival (RFS) compared to high-dose interferon (HDI). Biochemotherapy based on Vinblastine may delay the return of cancer, but it doesn't always lengthen survival because it has more hazards than HDI. Donaldson et al. (2002) looked at a therapy plan for kids with early-stage Hodgkin's disease that included Vinblastine (VAMP). Combining low-dose radiation with chemotherapy led to very high 5-year survival and event-free survival rates, above 90%. The method got around stronger radiation therapy and alkylating chemicals. This lowered the chance of long-term harm while keeping the option of treatment open. Pfister et al. (2024) talked about the 5-year findings of the cisplatin in muscle-invasive bladder cancer (VESPER) study, which looked at dose-dense MVAC (dd-MVAC) and gemcitabine-cisplatin (GC) in patients with muscle-invasive bladder cancer. In the neoadjuvant subgroup, dd-MVAC led to a longer time until death from bladder cancer, even though there was no statistically significant increase in overall survival during the perioperative phase. This means that adjuvant therapy needs to keep getting better and that EVO should back up what it says. Gill et al. (1996) looked at how well DaunoXome worked compared to standard ABV therapy in patients with advanced Kaposi's sarcoma and AIDS at the same time. ABV caused more hair loss and nerve damage, even though both groups had similar responder and survival rates. On the other hand, DaunoXome was better accepted generally, but it caused more neutropenia. The results showed that DaunoXome is a better and safer way to treat Kaposi's sarcoma. The DESMOPAZ trial, which was done by Toulmonde et al. (2018), looked at how well pazopanib and Vinblastine-methotrexate (MV) worked to treat desmoid tumors. Pazopanib had a much higher percentage of disease stability and tumor shrinkage than the MV arm. The results suggest that targeted medicines like pazopanib may work better and be easier to handle for this rare and difficult cancer, even though Vinblastine may have some benefits. Ogura et al. (2013) also looked at a modified ABVD regimen for Hodgkin lymphoma that didn't include dacarbazine. The fact that the progression-free survival was lower than with typical ABVD regimens, even though the complete response rate was higher with extra radiation therapy, shows how important dacarbazine is for getting long-lasting effects in Hodgkin lymphoma. The thorough review of this research shows that Vinblastine works to treat a wide spectrum of malignancies, including advanced bladder, breast, and prostate cancers, as well as lymphomas in children. The effectiveness of a medicine in a given setting depends on things including the person's pharmacokinetics, the timetable for taking the drug, resistance mechanisms, and combination therapies. The clinical data support the use of Vinblastine in different therapeutic regimens, especially when it is used with targeted delivery methods or other drugs. No matter how well it works, effective toxicity management and personalized doses are important to get the most advantages and the fewest problems. Further research into precision medicine techniques and resistance modulation may make Vinblastine more effective in treating different forms of cancer.

12.2.3 Vindesine

Vinblastine is used to synthesize from vindesine, a semi-synthetic vinca alkaloid. It has been found to have many different actions that fight cancer in both solid tumors and blood malignancies. It is a beneficial medicine in chemotherapy because it slows mitosis by messing with the dynamics of microtubules. We have learned a lot about its probable usefulness as a treatment, how well it is tolerated, and how dangerous it is for different types of malignancies through a number of Phase I and II trials. Early clinical trials indicated that vindesine worked against a number of malignancies, including ALL, Hodgkin's and non-Hodgkin's lymphomas, melanoma, testicular and esophageal carcinomas, bronchogenic carcinoma, and Wilms' tumor (Gralla et al. 1979). A weekly dose of 3–4 mg/m^2 IV was safe; however, leukopenia and peripheral neuropathy were the worst side effects that limited the amount. Patients with blood malignancies and germ cell tumors responded very well to daily doses of 1.0–1.3 mg/m^2 IV for 5–7 days, indicating considerable remissions. Vindesine had mild to moderate effects on the neurological system, although it didn't harm platelets and caused severe leukopenia in less than 5% of instances. More tests looked at the neurological and hematological damage profile in patients who had never undergone chemotherapy before. Weekly dosages of 0.5–3.0 mg/m^2 elicited neurological symptoms such as paresthesias, myalgias, and reflex diminishment, but there were no big alterations in nerve conduction or H-reflex responses. Leukopenia was one of the blood-related side effects, but the number of thrombocytes rose up, which is interesting (Obrist et al. 1979). In advanced hepatocellular carcinoma, however, vindesine did not cause any objective responses. It was nonetheless well tolerated and did not cause any major side effects (Falkson and Burger 1995). When given as a continuous IV infusion over 48 h instead of a bolus IV push, vindesine worked well in many types of leukemia and lymphosarcoma. When VDS was used to treat acute leukemias and lymphosarcomas, the response rates were 63% for ALL and 55% for chronic myeloid leukemia blastic crises. It's notable that

55% of patients who were resistant to Compound 109 responded to VDS. This suggests that there wasn't much cross-resistance (Bayssas et al. 1979). The toxicity was largely in the blood and neurological system, although it was still safe. A lot of research has gone into finding ways to make vindesine work better when used with other drugs. A treatment regimen that combined vindesine, cisplatin, and mitomycin showed partial response rates of 35% and longer response periods in patients with NSCLC; however, the improvement was only statistically significant. The toxicity profiles were the same for both groups, although the mitomycin group had greater leukopenia (Shinkai et al. 1991). For metastatic breast cancer, vindesine was also treated with mitoxantrone or epirubicin. The second combination worked better and caused speedier remissions, while the mitoxantrone-based treatment had fewer subjective adverse effects, like hair loss and nausea. This means that the optimum treatment relies on how severe the condition is and how many symptoms it produces (Hausmaninger et al. 1995). Researchers wanted to know if vindesine could overcome multidrug resistance when taken with lonidamine (LND), a metabolic modulator that makes tumor cell membranes more permeable. In a study involving multiple centers, adding LND to a treatment plan that included vindesine, cisplatin, and epirubicin for NSCLC patients resulted in much better outcomes, with higher response rates (43% compared to 24%), longer median time to progression (8 months versus 5 months), and improved overall survival (11 months compared to 7.6 months). The adverse effects weren't too bad and could generally be controlled with steroids; thus, this combination is good for patients who aren't doing well (Ianniello et al. 1996). Patients who have already had treatment for bronchogenic carcinoma have also seen some success with vindesine immunotherapy. Twenty-seven patients, predominantly with small-cell carcinoma, had partial responses. The neurologic side effects weren't too bad, and corticosteroids worked well to treat phlebitis (Vogelzang et al. 1982). There were no big differences in survival rates between vindesine alone and vindesine plus cisplatin and mitomycin in a randomized Phase III trial. This means that using vindesine by itself isn't as useful for NSCLC as using it with other drugs (Einhorn et al. 1986). A Phase II study that looked at weekly bolus and fractionated vindesine injections in individuals with blood cancers that didn't react to other treatments indicated that the bolus regimen had a greater response rate (42.8% vs. 16%). There were remissions in different kinds of lymphomas and leukemias. Myelosuppression remained the worst adverse effect, but neurotoxicity was rare and not very bad. This backs up the idea that vindesine can be used alongside other medications to treat hematologic problems that don't respond to previous treatments (Mandelli et al. 1982). Researchers looked at vindesine with epirubicin and epirubicin alone in individuals with ABC. Adding vindesine didn't help the responder or survival rates very significantly, but it was connected to greater neurotoxicity and less thrombocytopenia, which implies it, could be useful for some patients. It is a concern that cumulative anthracycline doses can make the heart more toxic, although vindesine did not accomplish that (Nielsen et al. 1990). Vindesine also helped with prostate cancer, but not very much. In a Phase II study, 19% of the patients got better, although only for a brief time. Neurotoxicity became a big problem, impacting more than half of the patients. Such effects often made it impossible to boost the dose and delayed treatment (Jones et al. 1983). A randomized study in advanced NSCLC also looked at vindesine alone or with other medications like mitomycin or ifosfamide. The three-drug combinations did improve objective response rates, but these changes weren't statistically significant, and neither regimen led to a complete remission. There were many things that could predict survival, such as performance level, stage of disease, weight loss, and serum albumin. This shows how crucial it is to pick the proper patients (Masutani et al. 1996). The most noteworthy difference between vindesine plus cisplatin and docetaxel plus cisplatin (DC) in untreated stage IV NSCLC was that DC had a higher response rate (37% vs. 21%), a longer median survival (11.3 months vs. 9.6 months), and a higher 2-year survival rate (24% vs. 12%).

Vindesine induced greater adverse effects relating to blood and lower quality-of-life scores. Docetaxel, on the other hand, had a superior balance between effectiveness and tolerance. This became the norm for first-line treatment for NSCLC (Kubota et al. 2004). In general, vindesine is still a useful chemotherapeutic medication that has been found to act against a number of solid tumors and blood malignancies. Adding it to combination medicines makes treatment work better for several cancers, such as breast cancer, lung cancer, and leukemia, but DLN and low effectiveness in some solid tumors usually make it less beneficial in the clinic. The studies we looked at here illustrate how vital it is to employ tailored regimens, classify patients by risk, and add additional medications like LND to get the most out of vindesine in the clinic. DiBella et al. (1984) designed a Phase II experiment in which 56 patients with advanced metastatic breast cancer, melanoma, and lymphoma were administered vindesine. Patients who didn't respond got the medicine once a week as a 3 mg/m^2 IV bolus and then as a 48-h infusion. In terms of clinical outcomes, it didn't help much. Only two breast cancer patients experienced partial responses, and four melanoma patients' illnesses became better without any objective responses. One patient with non-Hodgkin's lymphoma went into complete remission for 26 months, which was a major outlier. There were no deaths linked to the medications, and the adverse effects were easy to cure, with neutropenia being the most common. De Marinis et al. (1999) completed a groundbreaking randomized four-arm factorial study on 153 elderly individuals with advanced NSCLC who had never been treated before. The study looked at how vindesine and lonidamine worked on their own and together. Of the 104 patients who could be tested, just three gave partial responses, which was not good. The toxicity stayed low, although a lot of patients had testicular pain, tiredness, and muscle pain. There was no improvement in overall survival, which means that regimens based on vindesine may not be very helpful for older patients. This also shows that we need to do more cancer trials with older folks. Burkes et al. (1992) completed a small study in which vindesine was used as part of an MVP (mitomycin, vindesine, cisplatin) induction regimen for patients with resectable stage IIIA N2 NSCLC. Of the 39 patients who underwent two rounds of MVP chemotherapy and then surgery, 64% responded. Three of these answers were full. There were 18 patients in this group who had full surgical resections, and 40% of them lived for 3 years. The average length of time patients lived was 29.7 months, but the poison was very strong, and a few patients died from sepsis and bronchopleural fistula. The results nevertheless showed that neoadjuvant MVP therapy might help make surgery go better. In 1988, Roth and others undertook a randomized study with 39 patients who had esophageal cancer to find out more about vindesine. Patients who took vindesine, bleomycin, and cisplatin before and after surgery had the same rates of being able to have surgery and having problems as those who merely had surgery. Patients who responded to chemotherapy, on the other hand, survived a lot longer (median > 20 months). This shows that vindesine-based neoadjuvant therapy should assist doctors in identifying the best patients for treatment, especially those who had lost very little weight before treatment. Fukuoka et al. (1992) looked at a similar MVP regimen in 49 patients with advanced NSCLC who had not yet received treatment by cisplatin, mitomycin-C, and vindesine every 4 weeks. The overall response rate was 52%. The response rates for squamous cell, adenocarcinoma, and large-cell carcinoma were 52%, 45%, and 80%, respectively. Patients survived for an average of 10.6 months. A number of patients suffered significant leukopenia (85%) and thrombocytopenia (33%), but they were able to deal with it. There was only one death from sepsis. The results showed that MVP worked as a treatment strategy, but it needed to be compared to standard two-drug regimens. Kodani et al. (2002) undertook a randomized study with 132 NSCLC patients to evaluate the three-drug Vasoactive Intestinal Peptide (VIP) regimen (vindesine, ifosfamide, cisplatin) compared to the conventional VP regimen (vindesine, cisplatin). The objective response rates were roughly the same (49.3% for VIP and 44.6% for VP), but Cox multivariate analysis showed that the VIP group lived longer. VIP caused increased myelosuppression, which is a big deal. These results support VIP's survival advantage in advanced NSCLC, but they also suggest that it is more dangerous to the blood. Fukuoka et al. (1991) tested three different combinations of vindesine in 199 patients with NSCLC: VP, MVP, and a sequential EP/VM (etoposide/cisplatin alternating with vindesine/mitomycin). The MVP group had the highest response rate (43%); however, the overall survival rate was not particularly different across the groups. MVP produced a lot greater thrombocytopenia of grades III and IV. Multivariate studies found that the treatment plan, gender, and histology all had an effect on how well the patient responded and how long they lived. Better results were connected to squamous histology and higher hemoglobin levels.

Furuse et al. (1996b) undertook a randomized crossover trial with 204 NSCLC patients to assess the effects of vindesine with vinorelbine. Vinorelbine had a higher response rate (31.1% compared to 8.9%) and a longer response duration (18.5 weeks compared to 11.7 weeks) than vindesine. Vinorelbine caused a lot less damage to the nerves in the periphery. The VDS → VRB + cisplatin arm had a better second-line response than the other way around in patients who didn't respond to monotherapy. The study found that vinorelbine functioned better and was easier to use than other NSCLC therapies. Furuse et al. (1996b) studied 21 patients with NSCLC who could not be treated and gave them chemoradiation with vindesine and cisplatin at the same time. Patients received either conventional fractionated or accelerated hyperfractionated radiation together with chemotherapy. There were a lot of bad outcomes, such as hematologic toxicities and two deaths from radiation pneumonitis. DLT made it impossible to give more of the drug. The response and 1-year survival rates were good, but the regimen's toxicity demonstrated that radiation regimens need to be modified even more. Smith et al. (1995) examined the costs of vindesine/cisplatin, vinorelbine alone, and vinorelbine/cisplatin in NSCLC using data from clinical trials. Vinorelbine/cisplatin had the longest average survival time (49.6 weeks) and cost $17,700 for each year of life gained. Vindesine/cisplatin added 19 days to life at a cost of $1150, or $22,100/year. The study showed that vinorelbine/cisplatin worked better and cost less than combinations of vindesine. Rixe et al. (1995) studied the effects of combining vindesine, dacarbazine, and fotemustine in 43 patients with advanced malignant melanoma. The treatment worked for 32% of patients (14 out of 41 who could be tested), and the average time they lived was 10 months. The combination worked well, especially for visceral metastases, where it worked better than what is generally seen with dacarbazine or fotemustine alone. The toxicity was acceptable. Woods et al. (1990) undertook a randomized study to see how well vindesine/cisplatin chemotherapy worked compared to supportive care for NSCLC. Out of 188 patients that could be examined, 97 underwent chemotherapy and 91 did not. The response rate was 28%, and the median survival duration was 27 weeks. This is a small but not statistically significant change. Toxicity was very bad and common; thus, 13 of the 17 patients stopped treatment after one cycle. The authors emphasized that this combination shouldn't be utilized all the time and stressed how crucial it is to incorporate quality-of-life endpoints in future studies. All of these trials together show the whole clinical course of vindesine. Vindesine works well in some cases, but newer medications like vinorelbine have mainly replaced it since they function better and are less damaging to the neurological system. Still, combinations that include vindesine, notably for NSCLC and melanoma, have taught us a lot about how to improve treatment, time doses, and employ several medications at once. Mulder et al. (1982) were some of the first patients to look at vindesine as part of a combination treatment. They gave patients with advanced malignant melanoma vindesine and cisplatin. Of the 61 patients that answered, 21% did so, with 2 full responses and 11 partial responses, but it was hard to handle because of bad side effects like hair loss, nausea, renal damage, and low white blood cell count. The study showed that when provided the way they were, vindesine and cisplatin did not function together to destroy melanoma cells. In a randomized trial comparing gemcitabine monotherapy to the cisplatin-vindesine combination (PV regimen) in stage IIIB/IV NSCLC, Vansteenkiste et al. (2001) found that the gemcitabine arm had a better clinical benefit response (48.1% vs. 28.9%, $P = 0.03$) and fewer severe toxicities. Gemcitabine showed a lower risk of adverse effects in the blood and nervous system, which suggested a higher therapeutic index, even though the ORR and survival rates were the same. Aristu et al. (1997) investigated neoadjuvant chemotherapy with cisplatin, mitomycin, and vindesine in stage III NSCLC, followed by surgery and radiation therapy. In stage IIIA patients, the entire resection rate was high (85%), while the 5-year survival rate was 29%, but for stage IIIB patients, being able to have surgery didn't mean they would survive longer; only 7% of them lived for 5 years. The results demonstrated that these kinds of treatments don't function very well when the disease is further advanced, even while they do assist in keeping the disease under control in the area. Burkes et al. (2005) also completed a long-term follow-up of the Toronto phase II trial, which treated 65 patients with stage IIIA NSCLC with combinations of MVP or VLB/P. There were 35 complete resections, and the 10-year survival rate was 22%. The total response rate was 67.7%. Even though the pathology results were good, recurrence, especially in the brain, was common. This made many think about adding prophylactic cranial irradiation to future treatments. In 2002, Gebbia et al. examined the effectiveness of vinorelbine-cisplatin (VC) with MVP in advanced NSCLC. In this trial with 247 patients, the ORR was 39% for VC and 42% for MVP. The two groups did not have a statistically significant difference in survival. MVP, on the other hand, was connected to a lot more blood poisoning and hair loss. The study found that VC is just as effective but less hazardous. For patients with severely pre-treated metastatic breast cancer, vindesine and cisplatin combined worked 20% of the time (Paridaens et al. 1985). Most of the time, remissions happened in soft tissue metastases. The adverse effects on the gastrointestinal and neurological systems were widespread, although not too bad. The writers noted that more studies should be done on the disease's initial stages. Gralla et al. (1981) undertook another investigation on lung cancer in which they looked at cisplatin and vindesine at low and high doses. The high-dose group had a substantially longer median response time (12 months vs. 5.5 months, $P = 0.05$) and a much longer median survival time (21.7 months vs. 10 months, $P = 0.02$). However, they also had more neurotoxicity and azotemia. These results demonstrated that dosage strategies can help therapy work well. Iizuka et al. (1991) looked at 31 patients with esophageal cancer who were given vindesine and cisplatin and found that 16.1% of them reacted. The average life span of responders was 5.5 months. Hematologic toxicity was the most common kind. The study showed that this combination didn't work any better than monotherapies. Elliott et al. (1984) studied patients with NSCLC who couldn't be cured and were given either vindesine alone or a combination of vindesine and cisplatin. The combination made response rates (33% vs. 7%) and median survival (11 months vs. 4 months, $P = 0.008$) a lot better. Patients with a good performance status and squamous cell histology clearly benefited. These results showed that the combination worked better than vindesine on its own. Vokes et al. (1989) undertook a RCT with 27 patients who had regionally advanced NSCLC and gave them neoadjuvant vindesine, cisplatin, and etoposide. Out of the 23 patients that could be evaluated, 13 had partial responses. There were modest myelosuppression and neurotoxicity as side effects, and one person died from a hemorrhagic event. The procedure didn't prove statistically that it helped patients live longer, but it did support greater research into induction regimens. Ringborg et al. (1989) studied the use of dacarbazine and vindesine combined to treat malignant melanoma. The response rates were poor (25% for the combo and 18% for dacarbazine alone), and they weren't statistically significant. The results showed that vindesine doesn't help this type of tumor very significantly. Tellez-Bernal et al. (1990) investigated 33 patients with head and neck squamous cell carcinoma (SCC) a continuous infusion of vindesine and cisplatin. This caused a 52% overall response rate, with five patients going into full remission. It took an average of 6.4 months for a CR to end. The most prevalent adverse consequence was leukopenia. The study found that continuous infusion, which is a way to enhance schedules, can help boost activity. Sekine et al. (2002) studied providing 41 patients with stage III NSCLC chemotherapy (cisplatin and vindesine) and high-dose thoracic radiotherapy (TRT) on different days. There were partial responses in 88% of the patients, and 24% of the patients lived for 5 years. Even though the long-term survival rates were good, there were big difficulties with treatment delays and injury to the esophagus. Lastly, Jungnelius et al. (1998) completed a phase III study on dacarbazine-vindesine with or without cisplatin in patients who had extensive melanoma. The arm with cisplatin took longer to develop (4.2 months vs. 2.2 months, $P = 0.007$), but there was no difference in overall survival. The cisplatin arm had a lot more side effects, which makes many question how useful the triplet combination is in real life. These clinical studies suggest that vindesine is still a good chemotherapeutic medication for some solid tumors, such as NSCLC and metastatic breast cancer, especially when combined with cisplatin. The kind of tumor, the dose schedule, and the other treatments it is given with all have an effect on how effectively it works, but there are still issues with hematologic, gastrointestinal, and neurologic toxicity. Future research should focus on finding biomarkers that can predict how well a treatment will work, changing the way drugs are given, and adding tailored medicines to make treatment more accurate. Vindesine is a semisynthetic vinca alkaloid that comes from Vinblastine. It has been used a lot in chemotherapy for several types of cancer, like leukemia, NSCLC, and squamous-cell carcinoma of the cervix. It is dangerous because it breaks up microtubules, just like its parent compounds, but it is not as poisonous. Phase I to Phase III clinical trials have looked at vindesine with various medications and in varied doses throughout the course of decades. This chapter talks about critical clinical data on vindesine from several studies that utilized it alone and in combination with other drugs. Komp et al. (1980) did a major Phase I study on vindesine with 21 kids, most of whom had leukemia. Granulocytopenia was the DLT in 34 treatment courses that were given once or twice a week. Some patients felt pain in their jaw, neck, or bones for a short time after taking the medicine. Neurotoxicity was not very common and not very serious. Patients thought that a starting dose of 1.85 mg/m^2 twice a week would be fine. Ettinger et al. (1982) also undertook an early study on kids in which they fed 36 kids different amounts of vindesine. Because of significant neurotoxicity, the regimens for high doses were altered. The treatment was easier to deal with weekly infusions of 4 mg/m^2 for 4 h. There was some activity in leukemia, neuroblastoma, Hodgkin's disease, and Ewing's sarcoma. Vindesine has been a key element of treating NSCLC, frequently as part of a doublet or triplet treatment with platinum. Saito et al. (1993) looked at the effects of cisplatin, carboplatin, and vindesine on 40 patients with advanced NSCLC in 1993. Adding carboplatin didn't make the treatment work better; just 30% of patients responded partially, and the median survival time was 38 weeks. There weren't many hematologic toxicities, but there wasn't any more antitumor activity. Mori et al. (1992) also found that 30 NSCLC patients who got a 5-day continuous infusion of vindesine and cisplatin had a 50% partial response rate. The regimen was usually well-tolerated; however, there were a few times when therapy had to be stopped because of hematopoietic toxicity. About 30 weeks was the average time it took for patients to react. Studies have shown that a mixture of vindesine and gemcitabine that was supposed to treat NSCLC that couldn't be eliminated. There were a number of adverse effects, such as neutropenia and thrombocytopenia, but only 20% of patients responded overall. After learning that the median survival was just 31 weeks, the authors wondered if this combination should be researched further. Kiura et al. (2008) completed a better study that looked at 200 Locally Advanced Non-Small Cell Lung Cancer (LA-NSCLC) patients who were also having thoracic radiation therapy at the same time. They compared docetaxel-cisplatin (DP) to MVP. The DP group had a higher 2-year survival rate and fewer febrile neutropenia events, but the rates of non-hematological toxicities stayed the same. This study found that DP works better than MVP for LA-NSCLC. A number of studies have looked at vindesine in treatments that involve more than one approach. Van den Brande et al. (1998) investigated NSCLC patients a succession of VIP chemotherapy agents (vindesine, ifosfamide, and cisplatin) and then radiation that worked better because of cisplatin. The survival rates were still not good, with a median survival of 10.6 months, even if the toxicity was okay. Atagi et al. (2002) studied continuous concurrent chemotherapy with vindesine, mitomycin, and cisplatin together with thoracic radiation that didn't cease in Phase II clinical trials. The medicine worked well for 77% of patients, and the average survival duration was 19 months. However, it was hard to schedule because it was harmful to blood cells. Tanaka et al. (2005) tested adjuvant therapy on stage IIIA-N2 NSCLC that had been surgically removed. They compared vindesine-cisplatin followed by oral tegafur and uracil (UFT) to UFT alone. The combo arm had a higher rate of survival without recurrence, although overall survival didn't alter much. This suggests that the doublet with vindesine only had a small effect. Takigawa et al. (2000), on the other hand, studied vindesine and nedaplatin in NSCLC patients who had already had a relapse or were not responding to treatment. The overall response rate was low at 14.6%, although patients who had never undergone chemotherapy before did better. There were clear pharmacokinetic correlations between neutropenia and treatment outcome, which means that second-line use is limited. In the case of advanced cervical cancer, vindesine has also been utilized with other medications. Van Luijk et al. (2007) observed that 45% of patients with metastatic or recurrent SCCUC responded to the BEMP regimen, which included bleomycin, vindesine, mitomycin, and cisplatin. The treatment worked best on metastases in the lungs, lymph nodes, and skin, but more than one-fifth of the patients terminated treatment because the adverse effects were too bad. In a broader EORTC study, Vermorken et al. (2001) studied at BEMP and single-agent cisplatin side by side. BEMP had a higher response rate (42% vs. 25%), but this did not mean that patients lived longer overall. It was crucial to be vigilant since BEMP had greater toxicity rates. Cisplatin monotherapy remained the best choice for palliative care. Silva et al. (1996) introduced α-interferon to MVP chemotherapy for advanced NSCLC to make therapies based on vindesine stronger. The overall response rate was 51%, and the side effects were not too bad. The regimen showed promise even though not all of the responses were full. More research into interferon-based modulation is needed. Gridelli et al. (1996)

did a Phase III trial with 204 stage IV NSCLC patients and compared MVP (mitomycin, vindesine, cisplatin) to MEV (mitomycin, vindesine, etoposide). The response rates (28.7% for MVP vs. 21.4% for MEV) and survival rates were very similar for both groups, but MVP has more adverse effects, especially on the blood; therefore, MEV is a better choice.

Vindesine is a drug that works against a number of cancers, including NSCLC and leukemia, when used as part of multi-agent chemotherapy. It is used in a lot of different places, but not as much as newer treatments because it can be bad for blood cells and doesn't work as well. The most well-known combinations are still those with cisplatin or mitomycin. Some regimens, like MVP or VIP, have been around for a long period, but in current medicine, vindesine has generally been supplanted by newer treatments. More studies into its use, especially with novel pharmaceuticals or targeted therapies, could help make its role in cancer treatment clearer.

12.2.4 Vinorelbine

Vinorelbine is a semisynthetic compound which is identified and characterized in *C. roseus*. It is synthesize from vinblastine. Clinically, it is mainly used for treatment of NSCLC and metastatic breast cancer. Ovarian cancer, mesothelioma, prostate cancer, pediatric cancers, and veterinary oncology are also treated by vinorelbine or under investigation clinical trials. It works by preventing microtubule polymerization, which causes mitotic arrest and apoptotic cell death. Its potential as in immunotherapy or conjunction with other cytotoxic agents and radiation has been examined in several clinical trials. In a phase I clinical trial, when patients with chest cancers received single and combination of vinorelbine and cisplatin along with standard radiation therapy, the combination showed significant results. In addition to standard radiation (60 Gy in 30 fractions), patients received vinorelbine (20 mg/m^2/week) and cisplatin (100 mg/m^2 every 3 weeks). Higher doses resulted in neutropenia, while delayed esophagitis was the hallmark of DLT. Vinorelbine 15 mg/m^2 on days 1 and 8 and cisplatin 80 mg/m^2 every 3 weeks were the suggested phase II regimens (Masters et al. 1998). A phase I dose-escalation showed a novel "daily-times-three" vinorelbine schedule to investigate the pharmacokinetics and safety of vinorelbine in combination with ifosfamide and G-CSF. Vinorelbine (up to 35 mg/m^2) and ifosfamide (2.0 g/m^2) were administered to patients with stage IIIB/IV NSCLC for 3 days. Sepsis and severe neutropenia were identified as DLTs. Vinorelbine 30 mg/m^2 and ifosfamide 1.6 g/m^2 were the suggested dosages. According to Masters et al. (1997), the study found a 40% partial response rate, a median survival of 50 weeks, and a 48% 1-year survival rate. Vinorelbine clearance rose noticeably in this schedule in terms of pharmacokinetics. The gemcitabine-vinorelbine (GemVin) arm showed less toxicity in a phase III trial that compared cisplatin-based treatments with a GemVin combination in advanced NSCLC. Cisplatin regimens were linked to increased rates of vomiting, alopecia, and appetite loss, but overall quality of life (QoL) scores were comparable across arms. Although, there were few differences in survival (median survival: 32 weeks for GemVin versus 38 weeks for cisplatin-based), GemVin was thought to be better for patients who were more sensitive to toxicity (Gridelli et al. 2003). Vinorelbine's usefulness was further expanded in a phase I/II study that paired it with oxaliplatin in patients with NSCLC who had never received chemotherapy. Vinorelbine (beginning at 22 mg/m^2) was given on days 1 and 8 of a 21-day cycle, while oxaliplatin (130 mg/m^2) was given on day 1. The DLT was neutropenia, particularly at higher doses of vinorelbine (32 mg/m^2). The combination demonstrated a 37% response rate and was well tolerated on an outpatient basis, despite hematologic toxicity. Vinorelbine 26 mg/m^2 and oxaliplatin 130 mg/m^2 were the suggested dosages (Monnet et al. 2001). Researchers looked at how vinorelbine could be utilized to treat canine MCTs in dogs. Twenty-four dogs got one dosage of vinorelbine through an IV at a rate of 15 mg/m^2. Eighteen percent of dogs had stable disease, whereas 13% had partial or full responses. Neutropenia and gastrointestinal side effects were common, but they could be handled. The study indicated that further research should be done on combination regimens or repeated dosing schedules (Grant et al. 2008). A phase III trial compared vinorelbine to best supportive care (BSC) in older patients with NSCLC. Vinorelbine was given on the first and eighth days of a 21-day cycle. Vinorelbine made it much more likely that 161 individuals who could be assessed would live. The median survival duration was 28 weeks, which is longer than 21 weeks. The 1-year survival rate was 32%, which is higher than 14%. Patients who took vinorelbine said their pain, social, and cognitive function became improved, which means it was effective for this group of patients without being too hazardous (Gridelli 2001). On the other hand, a phase II trial of vinorelbine in metastatic colorectal cancer (*m*CRC) with the BRAF V600E mutation, which has been shown in preclinical research to be susceptible to mitotic spindle poisons, did not show any meaningful clinical efficacy. Twenty patients who had already had a lot of treatment had vinorelbine by IV (25 mg/m^2) or orally (60 mg/m^2) every 3 weeks on days 1 and 8. The median progression-free survival was barely 1 month, and only one patient proved that their condition was stable. The average length of life was 2.1 months. The fact that it didn't work revealed how crucial resistance and molecular heterogeneity are in BRAF-mutated *m*CRC (Cremolini et al. 2017). A phase II randomized study looked at how effectively vinorelbine worked to treat malignant pleural mesothelioma that had come back. Patients either had vinorelbine with active symptom control (ASC) or only ASC. Neutropenia was the most common grade III toxicity, and vinorelbine made progression-free survival go up a lot (4.2 months vs. 2.8 months). The data nonetheless show that vinorelbine is effective in treating mesothelioma, even if BRCA1 expression, which is expected to predict response, did not correspond with outcomes (Fennell et al. 2022). Authors ran a phase I trial to find out what the MTD and DLTs of vinorelbine were in cats. Vinorelbine was administered to nineteen cats with different forms of cancer once a week. The MTD was found to be 11.5 mg/m^2. Nephrotoxicity, vomiting, and neutropenia are some examples of acute DLTs. VRL is usually well tolerated; however, further research is needed to find out how well it works against tumors in cats (Pierro et al. 2013). In a randomized phase II trial, Jassem et al. (2001) compared oral and intravenous vinorelbine in patients with stage IIIB or IV NSCLC who had not received any prior treatment. Of the 115 patients who were enrolled, 76 were given oral vinorelbine (beginning at 60 mg/m²/week and increasing to 80 mg/m²), and 38 were given IV vinorelbine (30 mg/m²/week). Both the oral (14%) and IV (12%) arms had similar response rates, and overall survival (9.3 vs. 7.9 months) and progression-free survival (3.2 vs. 2.1 months) were comparable. The IV arm had a higher rate of neutropenia (62% of patients) than the oral arm (46%). In general, non-hematological toxicities, like gastrointestinal issues, were not severe. These results validated oral vinorelbine as a good substitute for the intravenous formulation, particularly for elderly or outpatient patients. According to pharmacokinetic studies, oral vinorelbine is extensively metabolized by the liver through CYP3A4, has a rapid absorption time (T_{max}: 1.5–3 h), and has an absolute bioavailability of roughly 40%. It has a strong 78% binding to platelets and a weak 13% binding to plasma proteins. Interestingly, its pharmacokinetics hold true for patients of all ages, including the elderly. The only active metabolite is deacetyl-vinorelbine, which is primarily eliminated through biliary excretion. Gebbia and Puozzo (2005) verified that the pharmacodynamic effects, intra-patient exposure reproducibility, and metabolism patterns of the oral and IV forms are similar, enabling conversion between forms using dose-equivalence ratios (e.g., 80 mg/m² oral ≈ 30 mg/m² IV). Total of 189 patients with stage IIIB/IV NSCLC were randomly assigned 2:1 to receive oral ($n = 120$) or intravenous ($n = 61$) vinorelbine in a different randomized study conducted by Hirsh et al. (2007). With a planned increase to 70 mg/m², oral administration started at 60 mg/m² per week. Weekly doses of 30 mg/m² of vinorelbine were given intravenously. About 45% of patients in both groups experienced disease stabilization, while the oral and IV arms had partial response rates of 4% and 13%, respectively. In patients with poor performance status (KPS 60–70), the median survival was longer in the IV group (40.9 vs. 26 weeks), indicating that oral vinorelbine may be less effective if dose escalation is not accomplished. Schilling et al. (1996) carried out a high-dose study in patients with solid tumors to further investigate vinorelbine pharmacokinetics. Every 21 days, increasing doses of vinorelbine (35, 40, and 45 mg/m²) were given as a bolus. Granulocytopenia and constipation were the dose-limiting toxicities, and 40 mg/m² was determined to be the MTD. There were no discernible antitumor responses. Pharmacokinetic data demonstrated Vinorelbine's high affinity for blood corpuscles, which was also consistent with earlier lower-dose investigations. Vinorelbine has demonstrated promising results in malignant pleural mesothelioma (MPM), where relapse following first-line chemotherapy is frequent and there is no established second-line therapy. Oral vinorelbine (beginning at 60 mg/m² weekly and increasing to 80 mg/m²) was compared to best BSC in the VIM phase II randomized trial (Fennell et al. 2021). The median progression-free survival (PFS) was considerably longer in the vinorelbine arm (4.2 vs. 2.8 months; HR = 0.59; $P = 0.0017$) among 154 patients who were randomized 2:1 to vinorelbine + BSC or BSC alone. Vinorelbine demonstrated a positive benefit in disease control and supported its off-label use in relapsed MPM, despite the fact that overall survival was not significantly different (9.3 vs. 9.1 months). In cases of metastatic breast cancer, vinorelbine has also shown single-agent activity. 54 women with metastatic or ABC who had never received chemotherapy participated in a phase II study by Garcia-Conde et al. (1994). A 50% overall response rate, comprising 2% complete and 48% partial responses, was obtained with weekly IV vinorelbine at 30 mg/m^2. The lymph nodes (64%) and lungs (66%) had the highest target-specific response rates. The median duration of response was 9 months, and the median survival was 15 months. 71% of patients experienced grade 3/4 granulocytopenia, indicating primarily hematologic toxicity. Vinorelbine's role in combination therapy for breast cancer was reinforced by the low incidence of gastrointestinal side effects. Vinorelbine and tariquidar, a strong P-glycoprotein (Pgp) inhibitor, were combined in a pharmacokinetic interaction study by Abraham et al. (2009) to evaluate drug resistance modulation. A total of 26 patients were enrolled. Tariquidar had no pharmacokinetic interaction with vinorelbine, but it changed the tumor uptake of radiolabeled sestamibi and prevented Pgp-mediated efflux in immune cells. Patients with breast and renal cancer showed only slight responses, and the combination was generally tolerable. Future vinorelbine regimens would benefit greatly from tariquidar's ability to overcome multidrug resistance. A semi-synthetic vinca alkaloid called vinorelbine has shown strong antitumor activity against a variety of cancers, most notably gynecologic tumors, lung cancer, and breast cancer. Its therapeutic promise, controllable toxicity, and, in certain situations, oral availability—which improves patient compliance and convenience—have all been confirmed by a number of clinical trials that evaluated its effectiveness either alone or in combination. Spielmann et al. (1994) conducted a seminal phase II study in which 97 women with advanced or metastatic breast cancer who had never received chemotherapy were tested with vinorelbine and doxorubicin. 66 (74%), including 21% of the 89 evaluable patients, obtained objective responses. Both visceral metastases (71%) and soft tissue/bone metastases (81%) had significant response rates. The median response duration was 12 months, and the median overall survival was 27.5 months. Cardiotoxicity was comparatively low (10%), whereas the primary toxicity was dose-limiting neutropenia. This combination's high response rate and longevity lend support to additional comparison studies with conventional regimens. 24 patients with small-cell lung cancer (SCLC) who had advanced following cisplatin-based treatment were included in a prospective phase II study conducted by Furuse et al. (1996a). Leukopenia (92%) and anemia (71%) were the main toxicities of weekly vinorelbine (25 mg/m^2), which produced a partial response in 12.5% of patients. In general, non-hematologic toxicities were not severe. These findings lend credence to vinorelbine's ongoing assessment, especially in patients who have never received treatment. In a phase II trial, Jahanzeb et al. (2002) investigated the use of vinorelbine in conjunction with trastuzumab for HER2-positive metastatic breast cancer. The response rate among 40 patients was 78% (11% complete, 68% partial). 72 weeks was the median time to progression. Patients with IHC 3+ tumors had a higher response rate (82%), which was correlated with HER2 status. The main side effect was grade 3/4 neutropenia, but there was no significant cardiac toxicity, suggesting a good therapeutic index and encouraging more research on this regimen. Vinorelbine monotherapy for advanced cervical squamous cell carcinoma was studied by Morris et al. (1998). The median treatment-related survival for the 35 patients was 11 months, and an

18% response rate (1 complete, 5 partial) was noted. Neurologic and gastrointestinal side effects were mild, but 61% of patients developed grade 3/4 leukopenia. These results point to vinorelbine's moderate activity in this context, which justifies further research on combination regimens. In a phase I/II trial, Luu et al. (2014) assessed the combination of vinorelbine and sorafenib in patients with metastatic breast cancer. At 4 months, 44% of the 54 patients who received treatment had a progression-free survival rate. The side effects, which included myelosuppression, fatigue, and nausea, were tolerable and in line with expectations. Cycle numbers were impacted by prior bevacizumab exposure. The combination was workable and could be strengthened by additional biomarker-guided investigation. Vinorelbine was assessed by Jassem et al. (1993) in patients with recurrent SCLC who had received chemotherapy in the past. A 16% response rate was noted among the 25 eligible patients. The main toxicity was non-cumulative leukopenia (80%). Two patients had their treatment stopped due to injection site reactions. Vinorelbine's moderate activity in relapsed SCLC is supported by the study. Johansen et al. (2006) carried out a phase I study of oral and intravenous vinorelbine in pediatric populations. In children without bone marrow involvement, weekly vinorelbine over 6 weeks showed that the MTD was 30 mg/m^2. In addition to mild nausea and hepatic toxicity, neutropenia was dose-limiting. Pharmacokinetic parameters showed greater clearance and distribution than in adults, and the oral formulation's bioavailability was 28.5%. Maass et al. (2014) evaluated the use of vinorelbine and imatinib together in metastatic breast cancer in a phase I/II study. Thirty-three patients who expressed either KIT or PDGFR were given daily imatinib and weekly vinorelbine. The median time to progression was 155 days, and the clinical benefit rate (CBR) was 50%. Although frequent, toxicities such as nausea, vomiting, infections, and neuropathy were usually treatable. There was no correlation between response and PDGFR/KIT expression. In patients who have already received treatment, the combination might be advantageous. Vinorelbine monotherapy was tested in 57 patients with metastatic breast cancer by Terenziani et al. (1996), who separated the patients based on their previous treatment status. Overall, 47% of patients responded, with 35% of those who had previously received treatment and 59% of those who had not. There was no cross-resistance between vinorelbine and previous CMF or anthracycline regimens. Flu-like symptoms and temporary neutropenia were the primary toxicities, and the 30 mg/m^2 regimen on days 1 and 8 every 3 weeks worked well and was well tolerated. In nonsquamous NSCLC, oral vinorelbine plus cisplatin was evaluated in the NAVoTrial 01 (Bennouna et al. 2014). Both regimens demonstrated similar efficacy in terms of overall survival (10.2–10.8 months), response rate (24%–31%), and progression-free survival (4.2–4.3 months) when compared to the reference arm of pemetrexed plus cisplatin. In the vinorelbine arm, neutropenia was more common (44% vs. 18%). The rates of febrile neutropenia (2%), however, were the same. The treatment plan included maintenance therapy. A workable substitute with tolerable toxicity was the oral vinorelbine/cisplatin doublet. A semi-synthetic vinca alkaloid called vinorelbine (VNB) has been thoroughly tested in solid tumors and shown to have a good safety record along with a wide range of antitumor effects. Breast cancer, NSCLC, ovarian, cervical, prostate, melanoma, and salivary gland tumors have all been successfully treated with it either alone or in conjunction with drugs like cisplatin and trastuzumab. Vinorelbine was studied as a single agent in patients with anthracycline-sensitive ABC in a multicenter, non-randomized Phase II study (Weber et al. 1995). Objective response rates (ORR) for the 107 women who were enrolled were 34% overall, with 35% occurring in first-line and 32% in second-line settings. Three second-line and nine first-line patients experienced complete responses (CR). Survival was 67 weeks with first-line therapy and 62 weeks with second-line therapy, with a median response duration of 34 weeks. The primary DLT that resulted in two treatment-related deaths was granulocytopenia. Vinorelbine's usefulness in managing ABC and its place in upcoming combination regimens were both validated by the study. Patients were randomized to receive vinorelbine alone or in conjunction with cisplatin in a Phase II randomized trial for recurrent salivary gland tumors (Airoldi et al. 2001). While 19% of the combination group experienced CR, 25% of the combination and 20% of the monotherapy arms showed partial responses (PR). Although there was some disease stabilization in both arms, cisplatin-vinorelbine produced a higher CR rate and longer survival, indicating superior efficacy. In the combination arm, intestinal toxicity was noticeably higher. In a two-stage Phase II trial, vinorelbine was also assessed in older patients (≥70 years) with advanced NSCLC (Gridelli et al. 1997). Vinorelbine 30 mg/m^2 was administered weekly to 43 patients. The main toxicity was hematologic, but it was controllable. Thirty-six percent of patients survived for more than a year, with a median survival of 36 weeks, and 23% of patients achieved PRs. Clinical symptoms like pain and cough significantly improved. A subsequent Phase III trial comparing vinorelbine plus best supportive care (BSC) versus BSC alone was supported by these results. Vinorelbine demonstrated moderate efficacy in ovarian cancer that was resistant to paclitaxel and platinum (Rothenberg et al. 2004). With an ORR of just 3%, the 6-month survival rate for the 79 patients was 65%. The therapy did not significantly improve physical function or quality of life, but it did not increase the burden of symptoms. When anthracyclines and taxanes were used to treat metastatic breast cancer, vinorelbine maintained its effectiveness (Seo et al. 2011). With a median response duration of 2.8 months and a median survival of 10.4 months, the ORR for 26 patients was 20.8%. Nearly 70% of patients developed grade 3/4 neutropenia, but peripheral neuropathy remained mild and curable. The results validated the effectiveness of vinorelbine in breast cancer that has undergone extensive treatment. Docetaxel–vinorelbine plus GM-CSF (DVS) demonstrated promise in treating metastatic melanoma (Eroglu et al. 2011). Of the 52 patients, 37% had stable disease and 15% had a PR. The median OS was 11.4 months, and the 1-year OS rate was 48.1%. Both febrile and manageable neutropenia were among the tolerable toxicities. DVS deserves more research in the treatment of melanoma. Vinorelbine provided a moderately beneficial effect in hormone-refractory prostate cancer (Oudard et al. 2001). In 47 patients, weekly dosing of 25 mg/m^2 produced a 17% PSA response rate and 32% clinical benefit (particularly a reduction in bone pain). 10.2 months was the median survival. Fifty-one percent of patients experienced grade 3/4 neutropenia, indicating dose-limiting hematologic toxicity. A Phase III study comparing vinorelbine as a single agent versus a combination with cisplatin further established its role in NSCLC (Depierre et al. 1994). The combination did not increase overall survival (median 32–33 weeks), despite producing a longer time to progression (20 vs. 10 weeks) and a higher ORR (43% vs. 16%). Cisplatin increased toxicity, particularly nephrotoxicity and neurotoxicity. However, vinorelbine by itself was still a good choice, particularly for patients who weren't a good fit for platinum-based regimens. Vinorelbine's efficacy as a first-line treatment for NSCLC was confirmed by a previous Phase II trial (Depierre et al. 1991). With median survival at 33 weeks and median response duration at 34 weeks, 29.4% of the 78 patients who had not received treatment before achieved PRs. Although the agent was generally well tolerated, 12.5% of cycles experienced grade 3–4 leukopenia. A semi-synthetic vinca alkaloid called vinorelbine (VNB) has shown significant clinical efficacy in treating a variety of cancers, such as solid tumors, breast, ovarian, prostate, and NSCLC. Its effectiveness as a monotherapy and in combination regimens has been assessed in numerous clinical trials, which have shown a well-tolerated toxicity profile and moderate-to-high response rates in certain patient populations. Fløtten et al. (2012) conducted a large randomized national trial to compare the effectiveness of vinorelbine plus gemcitabine (VG) and vinorelbine plus carboplatin (VC) in 444 patients with stage IIIB/IV NSCLC. The median overall survival was similar between the two regimens (VG: 6.3 months, VC: 7.0 months; $P = 0.802$), although VC had a higher incidence of grade IV neutropenia (19% vs. 7%, $P < 0.001$) and grade III/IV nausea/vomiting (12% vs. 4%, $P = 0.008$). While VG had a slightly better toxicity profile, both arms maintained comparable results in terms of HRQoL and radiotherapy use. Vinorelbine has also demonstrated encouraging outcomes in metastatic breast cancer. Thirty-four evaluable patients receiving weekly vinorelbine 25 mg/m^2 showed a 50% response rate (2 complete, 15 partial remissions) with a median response duration of 5.8 months in a Phase II study by Twelves et al. (1994). With one treatment-related death from neutropenic sepsis, neutropenia was the main toxicity. In spite of this, the treatment was generally well tolerated, with low rates of neuropathy and alopecia. A dose-finding study (Piatek et al. 2014) included 19 patients with various pretreated solid tumors. Temsirolimus was given once a week, and vinorelbine was given on days 1 and 15. Vinorelbine (20 mg/m^2) and temsirolimus (25 mg) were found to be the MTD. Eight patients had stable disease with a median disease control duration of 3.2 months, while two patients (prostate and NSCLC) showed partial responses. Anorexia, hypokalemia, and neutropenia were among the grade 3–4 toxicities. Further research was done on vinorelbine monotherapy in cases of platinum-resistant ovarian cancer. Bajetta et al. (1996) reported a 21% response rate in a subgroup of 24 out of 33 platinum-resistant patients who received vinorelbine at a dose of 25 mg/m^2/week. This subset included all responses, and 27% of them displayed a drop in CA-125 levels. Anemia and grade 3/4 neutropenia were among the toxicities; there was no treatment-related death. These results demonstrate vinorelbine's potential as a salvage treatment for ovarian cancer that has undergone extensive treatment. Morant et al. (2002) tested vinorelbine 30 mg/m^2 in 44 patients with advanced prostate cancer that doesn't depend on androgens, giving it on days 1 and 8 of a 21-day cycle. Three patients experienced partial remission, and the PSA response rate was 15.9%. With minimal non-hematologic side effects and controllable hematologic toxicities, the treatment was generally well tolerated. Clinical benefit (pain relief or analgesic reduction) was observed in 23% of patients with baseline pain. Briasoulis et al. (2013) investigated the metronomic oral administration of vinorelbine in patients with metastatic breast, prostate, or NSCLC. Briasoulis et al. (2013) administered doses of 30, 40, or 50 mg to 73 patients three times a week. Across the three arms, the time to treatment failure at 4 months varied from 18.2% to 33.3%, with 50 mg showing the highest efficacy. Four patients, the longest of whom lasted more than 100 weeks, displayed objective responses. In every dose group, the treatment was well tolerated. Biomarker analysis showed that patients with low levels of FGF2 and IL8 before treatment had better responses, while those with high levels of the TEK gene transcript were more likely to not respond. The best metronomic dose for upcoming research was suggested to be 50 mg three times per week due to its effectiveness, tolerability, and lack of drug accumulation.

12.2.5 Vinpocetine

The phase I to IV clinical trials undergone of vinpocetine should be found potential tolerability and therapeutic benefit mainly for acute ischemic stroke, vascular dementia/cognitive impairment, chronic ischemic retinopathy, sensorineural hearing loss, epilepsy (adjunctive treatment), glioblastoma (investigational), atherosclerosis/inflammation. A significant study was done to evaluate how safe and effective vinpocetine, a man-made version of apovincamine, is for treating thirty patients who had a recent acute ischemic stroke. All of these patients were diagnosed within 72 h of the stroke onset and had CT-confirmed diagnoses. Subsequently, they were randomly assigned to one of two cohorts: one cohort received low-molecular-weight dextran alone, while the other cohort received a combination of dextran and vinpocetine using the Barthel Index and the Rankin Scale. Poor outcomes were defined as death, a Barthel score of 70 or less, or a Rankin score of 3 or 5. At the 3-month follow-up, the relative risk (RR) of a poor outcome decreased by 60% based on the Rankin score and by 30% based on the modified Barthel Index. Furthermore, in the vinpocetine-treated group, the NIH Stroke Scale scores demonstrated a statistically significant but minor improvement ($P = 0.05$).

Vinpocetine may be a viable addition to larger, randomized, placebo-controlled stroke trials, as there were no reported adverse effects (Feigin et al. 2001). Vinpocetine has been utilized in clinical contexts for the treatment of cerebrovascular disorders, such as dementia, in addition to stroke. Recent research has underscored its anti-inflammatory and anti-remodeling properties, indicating a broader spectrum of potential therapeutic applications in cardiovascular and neurological disorders. These characteristics include reducing harmful changes in the heart, lowering responses to blood vessel damage, and decreasing the buildup of plaque in arteries. One of the first placebo-controlled investigations to demonstrate its efficacy involved eighty-four elderly patients with chronic cerebral dysfunction. Participants who were administered vinpocetine exhibited a substantial enhancement in their cognitive abilities on standardized assessments such as the MMSQ, SCAG, and CGI scales. This 90-day treatment regimen should not cause any significant adverse effects (Balestreri et al. 1987). The potential impact of vinpocetine on the immune system was reported when the subject it to significant clinical trials. The trial comprised 60 stroke patients who had experienced a blockage in the passage of blood to the frontal lobe of the brain. These patients received standard care alone or in addition to 30 mg of intravenous vinpocetine per day for 2 weeks. The treatment greatly reduced the activity of nuclear factor-kappa B, shown by higher levels of IκBα mRNA and less breakdown of IκBα. Nevertheless, there are no alternation of the count in lymphocytes in the body was reported. This change mainly estimated for due to reduction of body inflammation, along with an overdue progression of brain damage, and significant recovery of brain function 3 months after a stroke (Zhang et al. 2018). Researchers examined the efficacy of vinpocetine in treating acquired sensorineural hearing loss, a condition associated with sodium channel-mediated neurodegeneration, in an additional therapeutic context. Researchers administered 30 mg of vinpocetine daily to adult participants during a year-long study. Their hearing was assessed at 6 and 12 months using audiometry and logoaudiometry. The findings indicate that vinpocetine may be able to mitigate or even enhance the natural aging-related hearing loss, as evidenced by the improvement in hearing from 6 to 12 months. Despite the potential benefits of vinpocetine, it is not readily absorbed into the body when consumed orally, as approximately 75% of it is degraded prior to its efficacy. To resolve this issue, researchers implemented sugar esters as non-ionic surfactants in a novel proniosomal transdermal system. This formulation showed good lab results, such as effectively holding the drug, having the right vesicle sizes (0.63–2.52 μm), and being able to pass through the skin (91%). It also transformed into niosomes when the epidermis was hydrated. Clinical studies showed that it was absorbed much better (206%) than oral tablets and only caused slight skin irritation, suggesting it could be a good alternative to injections for delivering vinpocetine (El-Laithy et al. 2011). In addition, healthy volunteers and epileptic patients were considered to test the cognitive effects of vinpocetine. The 40 mg dose was reported as optimum dose to enhanced short-term memory by Sternberg test. Nevertheless, no discernible changes were observed in other thinking and emotion tests, such as the CFF and CRT. In a subsequent investigation, vinpocetine was evaluated in healthy adults and epileptic patients using both single and multiple dosing methodologies. Despite the exceptional tolerability, no discernible cognitive improvements were observed. Meador et al. (2021) showed that the amounts of vinpocetine and its breakdown product, apovincaminic acid, in human blood were lower than those found in rodents, based on pharmacokinetic analysis. This implies that to replicate preclinical efficacy in cognitive domains, higher dosages may be necessary. Finally, the investigation was conducted to determine the impact of vinpocetine on the rheology of the blood of patients with cerebrovascular disease. A small study was done to look at how blood thickness and related factors changed when patients took both vinpocetine and pyritinol together. Thirty normotensive individuals aged 50–65 participated in the study. The important improvements in low shear whole blood viscosity and other blood flow markers ($P < 0.05$) highlighted how vinpocetine might help improve blood flow to the brain, while high shear viscosity did not change (Alkuraishy et al. 2014).

References

Abraham J, Edgerly M, Wilson R, Chen C, Rutt A, Bakke S, Robey R, Dwyer A, Goldspiel B, Balis F, Van Tellingen O (2009) A phase I study of the P-glycoprotein antagonist tariquidar in combination with vinorelbine. Clin Cancer Res 15(10):3574–3582

Airoldi M, Pedani F, Succo G, Gabriele AM, Ragona R, Marchionatti S, Bumma C (2001) Phase II randomized trial comparing vinorelbine versus vinorelbine plus cisplatin in patients with recurrent salivary gland malignancies. Cancer 91(3):541–547

Alkuraishy HM, Al-Gareeb AI, Albuhadilly AK (2014) Vinpocetine and pyritinol: a new model for blood rheological modulation in cerebrovascular disorders—a randomized controlled clinical study. Biomed Res Int 2014(1):324307

Aristu J, Rebollo J, Martínez-Monge R, Aramendía JM, Viera JC, Azinovic I, Herreros J, Brugarolas A (1997) Cisplatin, mitomycin, and vindesine followed by intraoperative and post-operative radiotherapy for stage III non-small cell lung cancer: final results of a phase II study. Am J Clin Oncol 20(3):276–281

Atagi S, Kawahara M, Hosoe S, Ogawara M, Kawaguchi T, Okishio K, Naka N, Sunami T, Mitsuoka S, Akira M (2002) A phase II study of continuous concurrent thoracic radiotherapy in combination with mitomycin, vindesine and cisplatin in unresectable stage III non-small cell lung cancer. Lung Cancer 36(1):105–111

Atkins MB, Gollob JA, Sosman JA, McDermott DF, Tutin L, Sorokin P, Parker RA, Mier JW (2002) A phase II pilot trial of concurrent biochemotherapy with cisplatin, vinblastine, temozolomide, interleukin-2, and IFN-α2B in patients with metastatic melanoma. Clin Cancer Res 8(10):3075–3081

Bajetta E, Di Leo A, Biganzoli L, Mariani L, Cappuzzo F, Di Bartolomeo M, Zilembo N, Artale S, Magnani E, Celio L, Buzzoni R (1996) Phase II study of vinorelbine in patients with pretreated advanced ovarian cancer: activity in platinum-resistant disease. J Clin Oncol 14(9):2546–2551

Balestreri R, Fontana L, Astengo F (1987) A double-blind placebo controlled evaluation of the safety and efficacy of vinpocetine in the treatment of patients with chronic vascular senile cerebral dysfunction. J Am Geriatr Soc 35(5):425–430

Bates SE, Bakke S, Kang M, Robey RW, Zhai S, Thambi P, Chen CC, Patil S, Smith T, Steinberg SM, Merino M (2004) A phase I/II study of infusional vinblastine with the P-glycoprotein antagonist valspodar (PSC 833) in renal cell carcinoma. Clin Cancer Res 10(14):4724–4733

Bayssas M, Gouveia J, Ribaud P, Musset M, De Vassal F, Pico JL, De Luca L, Misset JL, Machover D, Belpomme D, Schwarzenberg L (1979) Phase-II trial with vindesine for regression induction in patients with leukemias and hematosarcomas. Cancer Chemother Pharmacol 2:247–255

Bennouna J, Havel L, Krzakowski M, Kollmeier J, Gervais R, Dansin E, Serke M, Favaretto A, Szczesna A, Cobo M, Ciuffreda L (2014) Oral vinorelbine plus cisplatin as first-line chemotherapy in nonsquamous non–small-cell lung cancer: final results of an international randomized phase II study (NAVotrial 01). Clin Lung Cancer 15(4):258–265

Bouffet E, Challinor J, Sullivan M, Biondi A, Rodriguez-Galindo C, Pritchard-Jones K (2020) Early advice on managing children with cancer during the COVID-19 pandemic and a call for sharing experiences. Pediatr Blood Cancer 67(7). https://doi.org/10.1002/pbc.28327

Briasoulis E, Aravantinos G, Kouvatseas G, Pappas P, Biziota E, Sainis I, Makatsoris T, Varthalitis I, Xanthakis I, Vassias A, Klouvas G (2013) Dose selection trial of metronomic oral vinorelbine monotherapy in patients with metastatic cancer: a Hellenic Cooperative Oncology Group clinical translational study. BMC Cancer 13:1–12

Broun GO, Blessing JA, Eddy GL, Adelson MD (1993) A phase II trial of vincristine in advanced or recurrent endometrial carcinoma: a Gynecologic Oncology Group Study. Am J Clin Oncol 16(1):18–21

Burkes RL, Ginsberg RJ, Shepherd FA, Blackstein ME, Goldberg ME, Waters PF, Patterson GA, Todd T, Pearson FG, Cooper JD (1992) Induction chemotherapy with mitomycin, vindesine, and cisplatin for stage III unresectable non-small-cell lung cancer: results of the Toronto Phase II Trial. J Clin Oncol 10(4):580–586

Burkes RL, Shepherd FA, Blackstein ME, Goldberg ME, Waters PF, Patterson GA, Todd T, Pearson FG, Jones D, Farooq S, McGlaughlin J (2005) Induction chemotherapy with mitomycin, vindesine, and cisplatin for stage IIIA (T1–3, N2) unresectable non-small-cell lung cancer: final results of the Toronto phase II trial. Lung Cancer 47(1):103–109

Cremolini C, Pietrantonio F, Tomasello G, Dadduzio V, Moretto R, Morano F, Schirripa M, Antoniotti C, Fucà G, Bergamo F, Rossini D (2017) Vinorelbine in BRAF V600E mutated metastatic colorectal cancer: a prospective multicentre phase II clinical study. ESMO Open 2(3):e000241

De Marinis F, Rinaldi M, Ardizzoni A, Bruzzi P, Pennucci MC, Portalone L, D'Aprile M, Ripanti P, Romano F, Belli M, Altavilla G (1999) The role of vindesine and lonidamine in the treatment of elderly patients with advanced non-small cell lung cancer: a phase III randomized FONICAP trial. Tumori J 85(3):177–182

Del Fabbro E, Dev R, Hui D, Palmer L, Bruera E (2013) Effects of melatonin on appetite and other symptoms in patients with advanced cancer and cachexia: a double-blind placebo-controlled trial. J Clin Oncol 31(10):1271–1276

Depierre A, Lemarie E, Dabouis G, Gamier G, Jacoulet P, Dalphin JC (1991) A phase II study of Navelbine (vinorelbine) in the treatment of non-small-cell lung cancer. Am J Clin Oncol 14(2):115–119

Depierre A, Chastang CL, Quoix E, Lebeau B, Blachon F, Paillot NAA, Lemarie E, Milleron B, Moro D, Clavier J, Herman D (1994) Vinorelbine versus vinorelbine plus cisplatin in advanced non-small cell lung cancer: a randomized trial. Ann Oncol 5(1):37–42

DiBella NJ, Berris R, Garfield D, Fink K, Speer J, Sakamoto A (1984) Vindesine in advanced breast cancer, lymphoma and melanoma: a Colorado Clinical Oncology Group Study. Investig New Drugs 2:323–328

Donaldson SS, Hudson MM, Lamborn KR, Link MP, Kun L, Billett AL, Marcus KC, Hurwitz CA, Young JA, Tarbell NJ, Weinstein HJ (2002) VAMP and low-dose, involved-field radiation for children and adolescents with favorable, early-stage Hodgkin's disease: results of a prospective clinical trial. J Clin Oncol 20(14):3081–3087

Einhorn LH, Loehrer PJ, Williams SD, Meyers S, Gabrys T, Nattan SR, Woodburn R, Drasga R, Songer J, Fisher W (1986) Random prospective study of vindesine versus vindesine plus high-dose cisplatin versus vindesine plus cisplatin plus mitomycin C in advanced non-small-cell lung cancer. J Clin Oncol 4(7):1037–1043

El-Laithy HM, Shoukry O, Mahran LG (2011) Novel sugar esters proniosomes for transdermal delivery of vinpocetine: preclinical and clinical studies. Eur J Pharm Biopharm 77(1):43–55

Elliott JA, Ahmedzai S, Hole D, Dorward AJ, Stevenson RD, Kaye SB, Banham SW, Stack BHR, Calman KC (1984) Vindesine and cisplatin combination chemotherapy compared with vindesine as a single agent in the management of non-small cell lung cancer: a randomized study. Eur J Cancer Clin Oncol 20(8):1025–1032

Eroglu Z, Kong KM, Jakowatz JG, Samlowski W, Fruehauf JP (2011) Phase II clinical trial evaluating docetaxel, vinorelbine and GM-CSF in stage IV melanoma. Cancer Chemother Pharmacol 68:1081–1087

Ettinger LJ, Brecher M, Coleman M, Smithson WA, Patterson R, Russell EC, Necheles T, Jones B, Ohnuma T (1982) Vindesine: a phase II study in childhood malignancies—a report for cancer and leukemia group B. Med Pediatr Oncol 10(1):35–43

Falkson G, Burger W (1995) A phase II trial of vindesine in hepatocellular cancer. Oncology 52(1):86–87

Feigin VL, Doronin BM, Popova TF, Gribatcheva EV, Tchervov DV (2001) Vinpocetine treatment in acute ischaemic stroke: a pilot single-blind randomized clinical trial. Eur J Neurol 8(1):81–85

Fennell DA, Casbard AC, Porter C, Rudd R, Lester JF, Nicolson M, Morgan B, Steele JP, Darlison L, Gardner GM, Nixon LS (2021) A randomized phase II trial of oral vinorelbine as second-line therapy for patients with malignant pleural mesothelioma. Mol Cancer Res 20(2):379

Fennell DA, Porter C, Lester J, Danson S, Taylor P, Sheaff M, Rudd RM, Gaba A, Busacca S, Nixon L, Gardner G (2022) Active symptom control with or without oral vinorelbine in patients with relapsed malignant pleural mesothelioma (VIM): a randomised, phase 2 trial. EClinicalMedicine 48:1–11

Flaherty LE, Othus M, Atkins MB, Tuthill RJ, Thompson JA, Vetto JT et al (2014) Southwest Oncology Group S0008: a phase III trial of high-dose interferon alfa-2b versus cisplatin, vinblastine, and dacarbazine, plus interleukin-2 and interferon in patients with high-risk melanoma. J Clin Oncol 32(33):3771–3778

Fløtten Ø, Grønberg BH, Bremnes R, Amundsen T, Sundstrøm S, Rolke H, Hornslien K, Wentzel-Larsen T, Aasebø U, von Plessen C (2012) Vinorelbine and gemcitabine vs vinorelbine and carboplatin as first-line treatment of advanced NSCLC. Br J Cancer 107(3):442–447

Fukuoka M, Masuda N, Furuse K, Negoro S, Takada M, Matsui K, Takifuji N, Kudoh S, Kawahara M, Ogawara M (1991) A randomized trial in inoperable non-small-cell lung cancer. J Clin Oncol 9(4):606–613

Fukuoka M, Negoro S, Masuda N, Kusunoki Y, Matsui K, Ryu S, Takifuji N, Kudoh S, Takada M (1992) Mitomycin C, vindesine, and cisplatin in advanced non-small-cell lung cancer. Am J Clin Oncol 15(1):18–22

Furuse K, Kubota K, Kawahara M, Takada M, Kimura I, Fujii M, Ohta M, Hasegawa K, Yoshida K, Nakajima S, Ogura T (1996a) Phase II study of vinorelbine in heavily pretreated small cell lung cancer. Oncology 53(2):169–172

Furuse K, Fukuoka M, Kuba M, Yamori S, Nakai Y, Negoro S, Katagami N, Takada Y, Kinuwaki E, Kawahara M, Kubota K (1996b) Randomized study of vinorelbine versus vindesine in untreated NSCLC. Ann Oncol 7(8):815–820

Garcia-Conde J, Lluch A, Martin M, Casado A, Gervasio H, De Oliveira C, De Pablo JL, Gorostiaga J, Giron GC, Cervantes A, Martinez A (1994) Phase II trial of weekly IV vinorelbine in first-line advanced breast cancer chemotherapy. Ann Oncol 5(9):854–857

Gebbia V, Puozzo C (2005) Oral versus intravenous vinorelbine: clinical safety profile. Expert Opin Drug Saf 4(5):915–928

Gebbia V, Galetta D, Riccardi F, Gridelli C, Durini E, Borsellino N, Gebbia N, Valdesi M, Caruso M, Valenza R, Pezzella G (2002) Vinorelbine plus cisplatin versus cisplatin plus vindesine and mitomycin C in stage IIIB–IV non-small cell lung carcinoma: a prospective randomized study. Lung Cancer 37(2):179–187

Gidding CE, Kellie SJ, Kamps WA, de Graaf SSN (1999) Vincristine revisited: a review of vincristine neurotoxicity in children. Pediatr Hematol Oncol 16(2):161–170

Gill PS, Wernz J, Scadden DT, Cohen P, Mukwaya GM, von Roenn JH, Jacobs M, Kempin S, Silverberg I, Gonzales G, Rarick MU (1996) Randomized phase III trial of liposomal daunorubicin versus doxorubicin, bleomycin, and vincristine in AIDS-related Kaposi's sarcoma. J Clin Oncol 14(8):2353–2364

Gleeson NC, Baile W, Roberts WS, Hoffman MS, Fiorica JV, Finan MA, Cavanagh D (1994) Pudendal thigh fasciocutaneous flaps for vaginal reconstruction in gynecologic oncology. Gynecol Oncol 54(3):269–274

Gralla RJ, Tan CTC, Young CW (1979) Vindesine: a review of phase-II trials. Cancer Chemother Pharmacol 2:271–274

Gralla RJ, Casper ES, Kelsen DP, Braun DW Jr, Dukeman ME, Martini N, Young CW, Golbey RB (1981) Cisplatin and vindesine combination chemotherapy for advanced carcinoma of the lung: a randomized trial investigating two dosage schedules. Ann Intern Med 95(4):414–420

Grant IA, Rodriguez CO, Kent MS, Sfilgoi G, Gordon I, Davis G, Lord L, London CA (2008) A phase II clinical trial of vinorelbine in dogs with cutaneous mast cell tumors. J Vet Intern Med 22(2):388–393

Gridelli C (2001) The ELVIS trial: a phase III study of single-agent vinorelbine as first-line treatment in elderly patients with advanced non-small cell lung cancer. Oncologist 6(S1):4–7

Gridelli C, Perron F, Palmeri S, D'Aprile M, Cognetti F, Rossi A, Gebbia V, Pepe R, Veltri E, Airoma G, Russo A (1996) Mitomycin C plus vindesine plus etoposide (MEV) versus mitomycin C plus vindesine plus cisplatin (MVP) in stage IV non-small-cell lung cancer: a phase III multicentre randomised trial. Ann Oncol 7(8):821–826

Gridelli C, Perrone F, Gallo C, De Marinis F, Ianniello G, Cigolari S, Cariello S, Di Costanzo F, D'Aprile M, Rossi A, Migliorino R (1997) Vinorelbine is well tolerated and active in elderly patients with advanced non-small cell lung cancer: a two-stage phase II study. Eur J Cancer 33(3):392–397

Gridelli C, Gallo C, Shepherd FA, Illiano A, Piantedosi F, Robbiati SF, Manzione L, Barbera S, Frontini L, Veltri E, Findlay B (2003) Gemcitabine plus vinorelbine vs. cisplatin plus vinorelbine or cisplatin plus gemcitabine for advanced non–small-cell lung cancer: a phase III trial. J Clin Oncol 21(16):3025–3034

Hausmaninger H, Lehnert M, Steger G, Sevelda P, Tschurtschenthaler G, Hehenwarter W, Fridrik M, Samonigg H, Schiller L, Manfreda D, Haidinger R (1995) Randomised phase II study of epirubicin-vindesine versus mitoxantrone-vindesine in metastatic breast cancer. Eur J Cancer 31(13–14):2169–2173

Hirsh V, Desjardins P, Needles BM, Rigas JR, Jahanzeb M, Nguyen L, Zembryki D, Leopold LH (2007) Oral versus intravenous administration of vinorelbine as a single agent for the first-line treatment of metastatic nonsmall cell lung carcinoma (NSCLC): a randomized phase II trial. Am J Clin Oncol 30(3):245–251

Hofmeister CC, Jansak B, Denlinger N, Kraut EH, Benson DM, Farag SS (2008) Phase II clinical trial of arsenic trioxide with liposomal doxorubicin, vincristine, and dexamethasone in newly diagnosed multiple myeloma. Leuk Res 32(8):1295–1300

Holland JF, Scharlau C, Gailani S, Krant MJ, Olson KB, Horton J, Shnider BI, Lynch JJ, Owens A, Carbone PP, Colsky J (1973) Vincristine treatment of advanced cancer: a cooperative study of 392 cases. Cancer Res 33(6):1258–1264

Hudes GR, Greenberg R, Krigel RL, Fox S, Scher R, Litwin S, Watts P, Speicher L, Tew K, Comis R (1992) Phase II study of estramustine and vinblastine, two microtubule inhibitors, in hormone-refractory prostate cancer. J Clin Oncol 10(11):1754–1761

Ianniello GP, Cataldis GD, Comella P, Scarpati MDV, Maiorino A, Brancaccio L, Cioffi R, Lombardi A, Carnicelli P, Tinessa V (1996) Cisplatin, epirubicin, and vindesine with or without lonidamine in the treatment of inoperable nonsmall cell lung carcinoma: a multicenter randomized clinical trial. Cancer 78(1):63–69

Iizuka T, Kakegawa T, Ida H, Ando N, Watanabe H, Takagi I (1991) Phase II evaluation of combined cisplatin and vindesine in advanced squamous cell carcinoma of the esophagus: Japanese Esophageal Oncology Group Trial. Jpn J Clin Oncol 21(3):176–179

Jahanzeb M, Mortimer JE, Yunus F, Irwin DH, Speyer J, Koletsky AJ, Klein P, Sabir T, Kronish L (2002) Phase II trial of weekly vinorelbine and trastuzumab as first-line therapy in patients with HER2+ metastatic breast cancer. Oncologist 7(5):410–417

Jassem J, Karnicka-Młodkowska H, Van Pottelsberghe CH, Van Glabbeke M, Noseda MA, Ardizzoni A, Gozzelino F, Planting A, van Zandwijk N (1993) Phase II study of vinorelbine (Navelbine) in previously treated small cell lung cancer patients. Eur J Cancer 29(12):1720–1722

Jassem J, Ramlau R, Karnicka-Młodkowska H, Krawczyk K, Krzakowski M, Zatloukal P, Lemarie E, Hartmann W, Novakova L, O'Brien M, Depierre A (2001) A multicenter randomized phase II study of oral vs. intravenous vinorelbine in advanced non-small-cell lung cancer patients. Ann Oncol 12(10):1375–1381

Johansen M, Kuttesch J, Bleyer WA, Krailo M, Ames M, Madden T (2006) Phase I evaluation of oral and intravenous vinorelbine in pediatric cancer patients: a report from the Children's Oncology Group. Clin Cancer Res 12(2):516–522

Jones WG, Fosså SD, Denis L, Coninx P, Glashan RW, Akdas A, De Pauw M, EORTC Genito-Urinary Cancer Cooperative Group (1983) An EORTC phase II study of vindesine in advanced prostate cancer. Eur J Cancer Clin Oncol 19(5):583–588

Jungnelius U, Ringborg U, Aamdal S, Mattsson J, Stierner U, Ingvar C, Malmström P, Andersson R, Karlsson M, Willman K, Wist E (1998) Dacarbazine–vindesine versus dacarbazine–vindesine–cisplatin in disseminated malignant melanoma. A randomised phase III trial. Eur J Cancer 34(9):1368–1374

Kiura K, Takigawa N, Segawa Y, Kamei H, Takemoto M, Tabata M, Ueoka H, Hiraki S, Matsuo K, Tanimoto M (2008) Randomized phase III trial of docetaxel and cisplatin combination chemotherapy versus mitomycin, vindesine, and cisplatin combination chemotherapy with concurrent thoracic radiation therapy for locally advanced non-small-cell lung cancer: OLCSG 0007. J Clin Oncol 26(15_suppl):7515–7515

Kodani T, Ueoka H, Kiura K, Tabata M, Takigawa N, Segawa Y, Moritaka T, Hiraki S, Harada M, Tanimoto M (2002) A phase III randomized trial comparing vindesine and cisplatin with or without ifosfamide in patients with advanced non-small-cell lung cancer: long-term follow-up results and analysis of prognostic factors. Lung Cancer 36(3):313–319

Komp D, Vats T, Shepherd DA, Vietti T, Hoogstraten B, Haas CD (1980) Phase I evaluation of vindesine in children: a Southwest Oncology Group pilot study. Med Pediatr Oncol 8(3):243–250

Kubota K, Watanabe K, Kunitoh H, Noda K, Ichinose Y, Katakami N, Sugiura T, Kawahara M, Yokoyama A, Yokota S, Yoneda S (2004) Phase III randomized trial of docetaxel plus cisplatin versus vindesine plus cisplatin in patients with stage IV non-small-cell lung cancer: the Japanese Taxotere Lung Cancer Study Group. J Clin Oncol 22(2):254–261

Le Deley MC, Rosolen A, Williams DM, Horibe K, Wrobel G, Attarbaschi A, Zsiros J, Uyttebroeck A, Marky IM, Lamant L, Woessmann W (2010) Vinblastine in children and adolescents with high-risk anaplastic large-cell lymphoma: results of the randomized ALCL99-vinblastine trial. J Clin Oncol 28(25):3987–3993

Long HJ III, Cross WG, Wieand HS, Webb MJ, Mailliard JA, Kugler JW, Tschetter LK, Kardinal CG, Ebbert LP, Rayson S (1995a) Phase II trial of methotrexate, vinblastine, doxorubicin, and cisplatin in advanced/recurrent carcinoma of the uterine cervix and vagina. Gynecol Oncol 57(2):235–239

Long HJ III, Cross WG, Wieand HS, Webb MJ, Mailliard JA, Kugler JW, Tschetter LK, Kardinal CG, Ebbert LP, Rayson S (1995b) Phase II trial of methotrexate, vinblastine, doxorubicin, and cisplatin in advanced/recurrent endometrial carcinoma. Gynecol Oncol 58(2):240–243

Lu K, Yap HY, Loo TL (1983) Clinical pharmacokinetics of vinblastine by continuous intravenous infusion. Cancer Res 43(3):1405–1408

Luu T, Frankel P, Chung C, Chow W, Mortimer J, Hurria A, Somlo G (2014) Phase I/II trial of vinorelbine and sorafenib in metastatic breast cancer. Clin Breast Cancer 14(2):94–100

Maass N, Schem C, Bauerschlag DO, Tiemann K, Schaefer FW, Hanson S, Muth M, Baier M, Weigel MT, Wenners AS, Alkatout I (2014) Final safety and efficacy analysis of a phase I/II trial with imatinib and vinorelbine for patients with metastatic breast cancer. Oncology 87(5):300–310

Mandelli F, Amadori S, Giona F, Antonietta M, Spiriti A, Pastore S, Meloni G, Paolucci G (1982) Vindesine in the treatment of refractory hematologic malignancies: a phase II study. Leuk Res 6(5):649–652

Masters GA, Hoffman PC, Hsieh AL, Drinkard LC, Mick R, Samuels BL, Guaspari A, Golomb HM, Vokes EE (1997) Phase I study of vinorelbine and ifosfamide in advanced non-small-cell lung cancer. J Clin Oncol 15(3):884–892

Masters GA, Haraf DJ, Hoffman PC, Drinkard LC, Krauss SA, Ferguson MK, Olak J, Samuels BL, Golomb HM, Vokes EE (1998) Phase I study of vinorelbine, cisplatin, and concomitant thoracic radiation in the treatment of advanced chest malignancies. J Clin Oncol 16(6):2157–2163

Masutani M, Akusawa H, Kadota A, Ohchi Y, Takahashi N, Tanigawa S, Koya Y, Horie T (1996) A phase III randomized trial of cisplatin plus vindesine versus cisplatin plus vindesine plus mitomycin C versus cisplatin plus vindesine plus ifosfamide for advanced non-small-cell lung cancer. Respirology 1(1):49–54

Mead GM, Russell M, Clark P, Harland SJ, Harper PG, Cowan R, Roberts JT, Uscinska BM, Griffiths GO, Parmar MKB (1998) A randomized trial comparing methotrexate and vinblastine (MV) with cisplatin, methotrexate and vinblastine (CMV) in advanced transitional cell carcinoma: results and a report on prognostic factors in a Medical Research Council study. Br J Cancer 78(8):1067–1075

Meador KJ, Leeman-Markowski B, Medina AE, Illamola SM, Seliger J, Novak G, Lin C, Ivanisevic M, Razavi B, Marino S, Boyd A (2021) Vinpocetine, cognition, and epilepsy. Epilepsy Behav 119:e107988

Monnet I, Soulié P, de Cremoux H, Saltiel-Voisin S, Bekradda M, Saltiel JC, Brain E, Dupont-André G, Cvitkovic E (2001) Phase I/II study of escalating doses of vinorelbine in combination with oxaliplatin in patients with advanced non–small-cell lung cancer. J Clin Oncol 19(2):458–463

Morant R, Schmitz SFH, Bernhard J, Thürlimann B, Borner M, Wernli M, Egli F, Forrer P, Streit A, Jacky E, Hanselmann S (2002) Vinorelbine in androgen-independent metastatic prostatic carcinoma—a phase II study. Eur J Cancer 38(12):1626–1632

Mori K, Saito Y, Tominaga K (1992) Phase II study of cisplatin continuous infusion plus vindesine in the treatment of non-small-cell lung cancer. Am J Clin Oncol 15(4):344–347

Morris M, Brader KR, Levenback C, Burke TW, Atkinson EN, Scott WR, Gershenson DM (1998) Phase II study of vinorelbine in advanced and recurrent squamous cell carcinoma of the cervix. J Clin Oncol 16(3):1094–1098

Mulder JH, Dodion P, Cavalli F, Czarnetzki B, Clavel M, Thomas D, Suciu S, Rozencweig M (1982) Cisplatin and vindesine combination chemotherapy in advanced malignant melanoma: an EORTC phase II study. Eur J Cancer Clin Oncol 18(12):1297–1301

Nielsen D, Dombernowsky P, Skovsgaard T, Jensen J, Andersen E, Engelholm SA, Hansen M (1990) Epirubicin or epirubicin and vindesine in advanced breast cancer: a phase III study. Ann Oncol 1(4):275–280

Obrist R, Paravicini U, Hartmann D, Nagel GA, Obrecht JP (1979) Vindesine: a clinical trial with special reference to neurological side effects. Cancer Chemother Pharmacol 2:233–237

Ogura M, Itoh K, Ishizawa K, Kobayashi Y, Tobinai K, Kinoshita T, Hirano M, Ueda R, Shibata T, Nakamura S, Tsukasaki K (2013) Phase II study of ABV (doxorubicin with increased dose, bleomycin and vinblastine) therapy in newly diagnosed advanced-stage Hodgkin lymphoma: Japan Clinical Oncology Group study (JCOG9705). Leuk Lymphoma 54(1):46–52

Oudard S, Caty A, Humblet Y, Beauduin M, Sue E, Piccart M, Rolland F, Fumoleau P, Bugat R, Houyau P, Monnier A (2001) Phase II study of vinorelbine in patients with androgen-independent prostate cancer. Ann Oncol 12(6):847–852

Papadimitriou CA, Dimopoulos MA, Giannakoulis N, Sarris K, Vassilakopoulos G, Akrivos T, Voulgaris Z, Vlahos G, Diakomanolis E, Michalas S (1997) A phase II trial of methotrexate, vinblastine, doxorubicin, and cisplatin in the treatment of metastatic carcinoma of the uterine cervix. Cancer 79(12):2391–2395

Paridaens R, Clarysse A, Rozencweig M, Rotmensz N, Heuson JC (1985) Cisplatin plus vindesine in advanced breast cancer: a phase II trial of the EORTC Breast Cancer Cooperative Group. Eur J Cancer Clin Oncol 21(5):595–599

Pfister C, Gravis G, Flechon A, Chevreau C, Mahammedi H, Laguerre B, Guillot A, Joly F, Soulie M, Allory Y, Harter V (2024) Perioperative dose-dense methotrexate, vinblastine, doxorubicin, and cisplatin in muscle-invasive bladder cancer (VESPER): survival endpoints at 5 years in an open-label, randomised, phase 3 study. Lancet Oncol 25(2):255–264

Piatek CI, Raja GL, Ji L, Gitlitz BJ, Dorff TB, Quinn DI, Hu J, El-Khoueiry AB, Pham HQ, Roman L, Garcia AA (2014) Phase I clinical trial of temsirolimus and vinorelbine in advanced solid tumors. Cancer Chemother Pharmacol 74:1227–1234

Pierro JA, Mallett CL, Saba CF (2013) Phase I clinical trial of vinorelbine in tumor-bearing cats. J Vet Intern Med 27(4):943–948

Reck M, von Pawel J, Macha HN, Kaukel E, Deppermann KM, Bonnet R, Ulm K, Hessler S, Gatzemeier U (2003) Randomized phase III trial of paclitaxel, etoposide, and carboplatin ver-

sus carboplatin, etoposide, and vincristine in patients with small-cell lung cancer. J Natl Cancer Inst 95(15):1118–1127
Ringborg U, Rudenstam CM, Hansson J, Hafström L, Stenstam B, Strander H (1989) Dacarbazine versus dacarbazine-vindesine in disseminated malignant melanoma: a randomized phase II study. Med Oncol Tumor Pharmacother 6:285–289
Rixe O, Borel C, Paraiso D, Benhammouda A, Petit T, Antoine E, Bizzari JP, Auclerc G, Soubrane C, Weil M, Giroux B (1995) Fotemustine, dacarbazine, vindesine combination chemotherapy in advanced malignant melanoma: a phase II study of 43 patients. Melanoma Res 5(6):419–424
Roth JA, Pass HI, Flanagan MM, Graeber GM, Rosenberg JC, Steinberg S (1988) Randomized clinical trial of preoperative and postoperative adjuvant chemotherapy with cisplatin, vindesine, and bleomycin for carcinoma of the esophagus. J Thorac Cardiovasc Surg 96(2):242–248
Rothenberg ML, Liu PY, Wilczynski S, Nahhas WA, Winakur GL, Jiang CS, Moinpour CM, Lyons B, Weiss GR, Essell JH, Smith HO (2004) Phase II trial of vinorelbine for relapsed ovarian cancer: a Southwest Oncology Group study. Gynecol Oncol 95(3):506–512
Saito H, Shimokata K, Saka H, Yamamoto M, Ogasawara T, Nomura F, Sakai S, Iwata M, Murate T, Miyachi T, Nakashima K (1993) Phase II study of carboplatin, cisplatin, and vindesine in advanced non-small-cell lung cancer. Cancer Chemother Pharmacol 33:154–156
Schilling T, Fiebig HH, Kerpel-Fronius S, Winterhalter B, Variol P, Tresca P, Heinrich B, Hanauske AR (1996) Clinical phase I and pharmacokinetic trial of vinorelbine administered as single intravenous bolus every 21 days in cancer patients. Investig New Drugs 14:371–378
Sekine I, Nishiwaki Y, Ogino T, Yokoyama A, Saito M, Mori K, Tsukiyama I, Tsuchiya S, Hayakawa K, Yoshimura K, Ishizuka N (2002) Phase II study of twice-daily high-dose thoracic radiotherapy alternating with cisplatin and vindesine for unresectable stage III non–small-cell lung cancer: Japan Clinical Oncology Group Study 9306. J Clin Oncol 20(3):797–803
Seo HY, Lee HJ, Woo OH, Park KH, Woo SU, Yang DS, Kim AR, Lee JB, Lee ES, Kim YH, Kim JS (2011) Phase II study of vinorelbine monotherapy in anthracycline and taxane pre-treated metastatic breast cancer. Investig New Drugs 29:360–365
Shinkai T, Eguchi K, Sasaki Y, Tamura T, Ohe Y, Kojima A, Oshita F, Saijo N (1991) A randomised clinical trial of vindesine plus cisplatin versus mitomycin plus vindesine and cisplatin in advanced non-small cell lung cancer. Eur J Cancer Clin Oncol 27(5):571–575
Siefker-Radtke AO, Millikan RE, Tu SM, Moore DF Jr, Smith TL, Williams D, Logothetis CJ (2002) Phase III trial of fluorouracil, interferon alfa-2b, and cisplatin versus methotrexate, vinblastine, doxorubicin, and cisplatin in metastatic or unresectable urothelial cancer. J Clin Oncol 20(5):1361–1367
Silva RR, Bascioni R, Rossini S, Zuccatosta L, Mattioli R, Pilone A, Delprete S, Battelli N, Gasparini S, Battelli T (1996) A phase II study of mitomycin C, vindesine and cisplatin combined with alpha interferon in advanced non-small cell lung cancer. Tumori J 82(1):68–71
Smith TJ, Hillner BE, Neighbors DM, McSorley PA, Le Chevalier T (1995) Economic evaluation of a randomized clinical trial comparing vinorelbine, vinorelbine plus cisplatin, and vindesine plus cisplatin for non-small-cell lung cancer. J Clin Oncol 13(9):2166–2173
Spielmann M, Dorval T, Turpin F, Antoine E, Jouve M, Maylevin F, Lacombe D, Rouesse J, Pouillart P, Tursz T (1994) Phase II trial of vinorelbine/doxorubicin as first-line therapy of advanced breast cancer. J Clin Oncol 12(9):1764–1770
Pławiński J, Kuźniar J, Filipiak K, Zieliński W (2006) Evaluation of drug toxicity in clinical trials. Sci Eng Ethics 12(1):139–145
Takigawa N, Segawa Y, Ueoka H, Kiura K, Tabata M, Shibayama T, Takata I, Miyamoto H, Eguchi K, Harada M (2000) Combination of nedaplatin and vindesine for treatment of relapsed or refractory non-small-cell lung cancer. Cancer Chemother Pharmacol 46:272–278
Tanaka F, Tsubota N, Namikawa S, Ohta M, Yamakido M, Hitomi S, Wada H (2005) A randomized phase III trial of adjuvant chemotherapy with cisplatin and vindesine followed by UFT for completely resected pathologic (p-) stage IIIA-N2 non-small cell lung cancer (NSCLC): West Japan Study Group for Lung Cancer Surgery (WJSG)-the 5th Study. J Clin Oncol 23(16_suppl):7262–7262

Tellez-Bernal E, Recondo G, Guillot T, Benhamed M, Domenge C, Izzo J, Cvitkovic E, Armand JP (1990) A phase II study of cisplatin and continuous infusion of vindesine in metastatic head and neck squamous cell cancer. Cancer 66(4):640–644

Terenziani M, Demicheli R, Brambilla C, Ferrari L, Moliterni A, Zambetti M, Caraceni A, Martini C, Bonadonna G (1996) Vinorelbine: an active, non cross-resistant drug in advanced breast cancer. Results from a phase II study. Breast Cancer Res Treat 39:285–291

Thigpen JT, Kronmal R, Vogel S, Hynes HE, Nahhas W, Belt RJ, Balcerzak SP, Epstein RB, Costanzi JJ (1987) A phase II trial of vinblastine in patients with advanced or recurrent endometrial carcinoma: a Southwest Oncology Group Study. Am J Clin Oncol 10(5):429–431

Thomas DA, Sarris AH, Cortes J, Faderl S, O'Brien S, Giles FJ, Garcia-Manero G, Rodriguez MA, Cabanillas F, Kantarjian H (2006) Phase II study of sphingosomal vincristine in patients with recurrent or refractory adult acute lymphocytic leukemia. Cancer 106(1):120–127

Toulmonde M, Ray-Coquard IL, Pulido M, Andre T, Isambert N, Chevreau C, Penel N, Bompas E, Thyss A, Bertucci F, Lebbe C (2018) DESMOPAZ pazopanib (PZ) versus IV methotrexate/vinblastine (MV) in adult patients with progressive desmoid tumors (DT): a randomized phase II study from the French Sarcoma Group. J Clin Oncol 36(15):11501–11501

Trump DL, Smith DC, Ellis PG, Rogers MP, Schold SC, Winer EP, Panella TJ, Jordan VC, Fine RL (1992) High-dose oral tamoxifen, a potential multidrug-resistance-reversal agent: phase I trial in combination with vinblastine. J Natl Cancer Inst 84(23):1811–1816

Twelves CJ, Dobbs NA, Curnow A, Coleman RE, Stewart AL, Tyrrell CJ, Canney P, Rubens RD (1994) A phase II, multicentre, UK study of vinorelbine in advanced breast cancer. Br J Cancer 70(5):990–993

Van den Brande P, De Ruysscher D, Vansteenkiste J, Spaas P, Specenier P, Demedts M (1998) Sequential treatment with vindesine–ifosfamide–platinum (VIP) chemotherapy followed by platinum sensitized radiotherapy in stage IIIB non-small cell lung cancer: a phase II trial. Lung Cancer 22(1):45–53

Van Luijk IF, Coens C, van der Burg MEL, Kobierska A, Namer M, Lhomme C, Zola P, Zanetta G, Vermorken JB (2007) Phase II study of bleomycin, vindesine, mitomycin C and cisplatin (BEMP) in recurrent or disseminated squamous cell carcinoma of the uterine cervix. Ann Oncol 18(2):275–281

Vansteenkiste JF, Vandebroek JE, Nackaerts KL, Weynants P, Valcke YJ, Verresen DA, Devogelaere RC, Marien SA, Humblet YP, Dams NL, Leuven Lung Cancer Group (2001) Clinical-benefit response in advanced non-small-cell lung cancer: a multicentre prospective randomised phase III study of single agent gemcitabine versus cisplatin–vindesine. Ann Oncol 12(9):1221–1230

Venkatramani R, Malogolowkin M, Davidson TB, May W, Sposto R, Mascarenhas L (2013) A phase I study of vincristine, irinotecan, temozolomide and bevacizumab (VITB) in pediatric patients with relapsed solid tumors. PLoS One 8(7):e68416

Vermorken JB, Zanetta G, De Oliveira CF, Van der Burg MEL, Lacave AJ, Teodorovic I, Boes GH, Colombo N (2001) Randomized phase III trial of bleomycin, vindesine, mitomycin-C, and cisplatin (BEMP) versus cisplatin (P) in disseminated squamous-cell carcinoma of the uterine cervix: an EORTC Gynecological Cancer Cooperative Group study. Ann Oncol 12(7):967–974

Vogelzang NJ, Peterson BA, Kennedy BJ, Vosika GJ, Conroy JA (1982) Vindesine in bronchogenic carcinoma: a phase II trial. Am J Clin Oncol 5(1):41–44

Vokes EE, Bitran JD, Hoffman PC, Ferguson MK, Weichselbaum RR, Golomb HM (1989) Neoadjuvant vindesine, etoposide, and cisplatin for locally advanced non-small cell lung cancer: final report of a phase 2 study. Chest 96(1):110–113

Weber BL, Vogel C, Jones S, Harvey H, Hutchins L, Bigley J, Hohneker J (1995) Intravenous vinorelbine as first-line and second-line therapy in advanced breast cancer. J Clin Oncol 13(11):2722–2730

Woods RL, Williams CJ, Levi J, Page J, Bell D, Byrne M, Kerestes ZL (1990) A randomised trial of cisplatin and vindesine versus supportive care only in advanced non-small cell lung cancer. Br J Cancer 61(4):608–611

Young SD, Whissell M, Noble JC, Cano PO, Lopez PG, Germond CJ (2006) Phase II clinical trial results involving treatment with low-dose daily oral cyclophosphamide, weekly vinblastine, and rofecoxib in patients with advanced solid tumors. Clin Cancer Res 12(10):3092–3098

Zhang QW, Lin LG, Ye WC (2018) Techniques for extraction and isolation of natural products: a comprehensive review. Chin Med 13(1):20

Success Story of Potential Anticancer Compounds: Vincristine and Vinblastine

13

Abstract

Vincristine and vinblastine are notable success stories in the development of anticancer compounds, derived from the Madagascar periwinkle plant (*C. roseus*). Discovered in the 1950s, these alkaloids demonstrated potent anticancer activity and in recent few year became essential components in chemotherapy regimens. Vincristine is primarily used to treat leukemia, lymphoma, and pediatric cancers, while vinblastine is effective against Hodgkin's lymphoma, breast cancer, and testicular cancer. Their mechanisms involve disrupting microtubule formation, halting cell division. Extensive toxicological evaluations and clinical trials confirmed their efficacy and safety. Today, they remain widely used, highlighting the value of natural products in cancer drug discovery development programmes.

Keywords

Vincristine · Vinblastine · Anticancer

13.1 Introduction

The discovery of plant-derived compound-based chemotherapy represents a significant milestone in the history of medical science. Vincristine and vinblastine play an important role in offering potent cytotoxicity for various cancer cells and hence are used for the treatment of various types of cancers. Both compounds, vincristine and vinblastine, were isolated from *C. roseus* by Robert Noble and Charles Beer in 1958 in Canada. Robert Noble and Charles Beer previously demonstrated and investigated vinblastine for its antineoplastic activity against Hodgkin's disease and testicular cancer in 1958 (Zhang et al. 2018).

Vincristine and vinblastine are both made of two parts called dimeric indole alkaloids, which are linked by a strong C–C bond and are commonly known as

S. Kumar, B. Singh, *Catharanthus in Drug Discovery*,
https://doi.org/10.1007/978-981-95-5048-7_13

vinca alkaloids. They are made from tryptamine and secologanin, and then the two parts, catharanthine and vindoline, combine to form the final structure of both compounds. They are classified as bis monoterpenoid indole alkaloids. The overall structure of both compounds has indole, dihydroindole, indoline, dihydroisoquinoline, and tetrahydropyran moieties. The whole structure of both compounds is the same except for the CH_3 and CHO groups on the nitrogen of the indole ring of the vindoline moiety of vinblastine and vincristine, respectively. This structural difference plays a significant role in the potential of pharmacological properties, especially toxicity and the therapeutic index. Both compounds have poor solubilities in water, so they are used as injections in the form of *sulfate salts* to improve solubility (Dhyani et al. 2022).

13.2 Distribution, Identification, Extraction and Isolation and Characterization

C. roseus (Madagascar periwinkle) derived alkaloids found predominantly in the leaves among the plant parts. Previously both compounds were detected by phytochemical screening and bioassay-guided studies. The powdered plant material is extracted with solvent extraction using methanol, ethanol, or chloroform-methanol mixtures. Acid-base extraction and fractionation techniques are used to isolate total alkaloid fractions. Flow chart of collection of plant material to characterization of vincristine and vinblastine both traditional and advanced by spectroscopic and spectrometric techniques and methods are presented in Figs. 13.1 and 13.2.

13.3 Chemical Structure and Properties

Vincristine and Vinblastine have molecular formula $C_{46}H_{56}N_4O_{10}$ and $C_{46}H_{58}N_4O_9$ with their corresponding molecular weight of 824.98 and 811.98 g/mol, respectively. 2D and 3D structures of both compounds are shown in Fig. 13.3.

The qualitative and quantitative analyses of vincristine and vinblastine have been reported from various extracts and fractions by HPTLC, HPLC, and HPLC-MS. In HPTLC methods, pre-coated HPTLC silica gel 60 F_{254} aluminum plates selected as the stationary phase with the common mobile phase chloroform: methanol (7:3 or 8:2, v/v) and toluene: ethyl acetate: methanol (6:3:1). There are three wavelengths 254 nm or 290 nm for vincristine and 254 nm or 305 nm for vinblastine for detection using CAMAG TLC Scanner. There are these alkaloids visualized by Dragendorff's reagent. The Rf value for the vincristine estimated around ~0.20–0.35 while vinblastine showed ranged from ~0.45–0.60 using calibration curve (; Barkat et al. 2015; Viju et al. 2023). The chromatographic separation of mixture vincristine and vinblastine carried out on C18 reverse-phase column with buffer of acetonitrile, methanol and water as mobile phase through 1.0–1.2 mL/min isocratic system. Gupta et al. (2005) have quantified both compounds. The chromatographic separation carried out on RP-18e monolithic column with the mobile phase

1. Collection of Catharanthus species (Mature Leaves)

↓

2. Shade Drying and Pulverization to Fine Powder

↓

3. Extraction with Ethanol or Methanol (70–90%)

↓

4. Filtration and Concentration of Extract (Rotary Evaporation/Open Evaporation)

↓

5. Acidification of crude Extract with HCl (pH ~2–3)

↓

6. Liquid–Liquid Partitioning:

a. Remove Non-Alkaloidal Impurities with Petroleum Ether/Hexane

↓

b. Alkalization of Aqueous Layer with NH_4OH to pH 9–10

↓

c. Extraction with Chloroform or Dichloromethane (3×)

↓

7. Drying of Organic Layer with Anhydrous Na_2SO_4

↓

8. Concentration to Obtain Total Alkaloid Mixture

↓

9. Fractionation via Column Chromatography:

Stationary Phase: Silica Gel

Mobile Phase: Chloroform:Methanol Gradient

↓

10. TLC-Based Screening of Fractions

↓

11. Pooling Fractions Showing Spots for Vincristine/Vinblastine (Rf Value Confirmation)

↓

12. Recrystallization (Methanol/Chloroform) for Crude Purification

↓

13. Final Purification:

Preparative TLC or Crystallization

↓

14. Characterization:

Melting Point, UV, IR, or Classic Color Tests (e.g., Dragendorff's Reagent)

Fig. 13.1 Flow chart of plant collection, drying, extraction, fraction, isolation and characterization of vincristine and vinblastine by traditional methods

1. Cultivation & Selection of High-Yielding Catharanthus roseus Varieties

↓

2. Harvesting of Young Mature Leaves (Optimal Alkaloid Content)

↓

3. Drying (Controlled Temperature) & Fine Pulverization

↓

4. Accelerated Solvent Extraction (ASE) or Ultrasound-Assisted Extraction (UAE)

• Solvent: Acidified Methanol or Ethanol (e.g., 70% MeOH + 0.1% HCl)

↓

5. Filtration and Rotary Evaporation under Reduced Pressure

↓

6. pH-Based Liquid-Liquid Partitioning:

a. Adjust to pH 2–3 → Extract with Hexane to remove lipids

↓

b. Adjust to pH 9–10 → Extract alkaloids with DCM/Chloroform

↓

7. Solvent Removal and Alkaloid-Rich Fraction Recovery

↓

8. Centrifugal Partition Chromatography (CPC) / High-Speed Countercurrent Chromatography (HSCCC)

↓

9. Fraction Analysis via Analytical HPLC or LC-MS

↓

10. Pooling of Vincristine & Vinblastine-Rich Fractions

↓

11. Preparative HPLC or Flash Chromatography for Final Purification

↓

12. Compound Identification and Purity Assessment:

• LC-MS/MS, NMR (1H, 13C), UV, IR, HRMS

↓

13. Freeze-Drying or Crystallization

↓

14. Pure Vincristine & Vinblastine Ready for Pharmacological Use

Fig. 13.2 Flow chart of plant collection, drying, extraction, fraction, isolation and characterization of vincristine and vinblastine by advanced methods

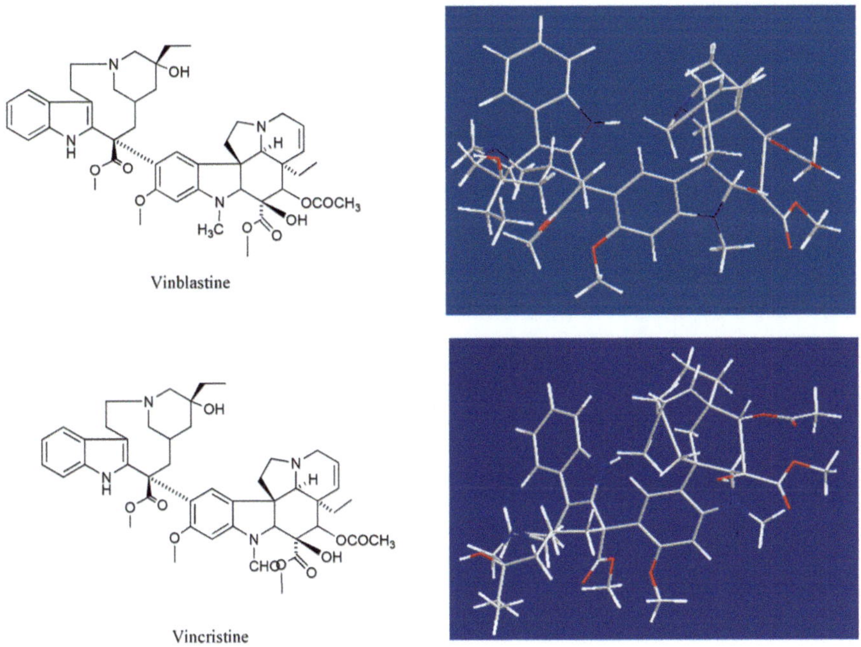

Fig. 13.3 Two (2D) and three (3D) structure of vincristine and vinblastine

Acetonitrile: 0.1 M phosphate buffer with 0.5% glacial acetic acid (21:79, pH 3.5) and maintained flow rate 1.2 mL/min. Both compounds are eluted at between ~3 and 5 min follow the linearity range 0.25–25 μg/mL. LOD/LOQ were determined at 8/25 μg/mL and 14/46 μg/mL vincristine and vinblastine, respectively. Precision is determined at 1.82% and 2.50% with the recovery range from ~97% to 98% (Gupta et al. 2005).

13.4 Mechanism of Action

The graphical representation of mechanism of action is presented in Fig. 13.4. Binding active sites of vincristine and vinblastine are bind to subunit of microtubules of β-tubulin. These bindings affect microtubule dynamics because of interaction is reversible but highly effective and prevent the polymerization of α- and β-tubulin into microtubules. Elongation and assembly of microtubules was inhibited due to prevent the polymerization is crucial. This inhibits many function including cell shape, intracellular transport, which is required mitotic spindle disorganization arresting the cell in metaphase of mitosis for accurate chromosome segregation during cell division. Finally, cell cycle arrested in metaphase (M phase) and apoptosis. Programmed cell death triggers due to intrinsic apoptotic pathways activation (Table 13.1).

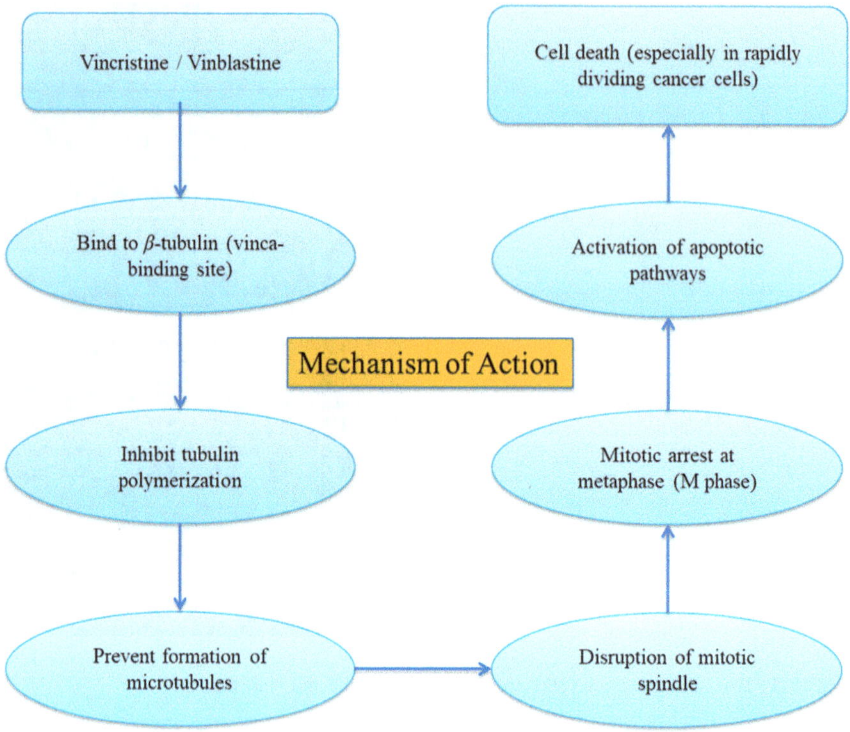

Fig. 13.4 Mechanism of action of vincristine and vinblastine

13.5 Pharmacological Claims, Clinical Trials and Approval

As discussed in the previous chapter, vincristine and vinblastine have emerged as potent anticancer agents with multiple pharmacological properties. Beyond their well-known cytotoxic activity, both compounds exhibit significant neurotoxic and myelosuppressive effects, which are dose-limiting in clinical use. Additionally, studies have reported their anti-angiogenic and immunomodulatory activities, contributing to their broader therapeutic potential (Rowinsky and Donehower 1991; Jordan and Wilson 2004; Staff et al. 2017). These pharmacological claims have been extensively supported by preclinical research and validated through various clinical trials. Their efficacy and safety profiles ultimately led to regulatory approval, reinforcing their value in modern oncology as vital components of combination chemotherapy regimens (Table 13.2).

Vincristine and vinblastine, two plant-derived anticancer agents, were approved by the U.S. FDA in July 1963 and July 1965, respectively. Vincristine was initially approved for the treatment of acute lymphoblastic leukemia (ALL), while vinblastine received approval for Hodgkin's lymphoma. Over time, their therapeutic applications have significantly expanded. Vincristine is now commonly used in the treatment of Hodgkin's and Non-Hodgkin's lymphoma, rhabdomyosarcoma,

Table 13.1 Standardization parameters for vincristine and vinblastine for quantitative analysis

Study/matrix		Column and mobile phase	Detection	Retention times (VCR/VBN)	Linearity and range (μg/mL)	LOD/LOQ (μg/mL)	Precision (%RSD)	Recovery/stability	References
C. roseus (purple, pink, white leaves)	Aqueous extract	Microsorb-MV C18 (150 × 4.6 mm, 5 μm); gradient mobile phase of methanol, 5 mM phosphate buffer (pH 6.0), and acetonitrile; Flow rate: 2.0 mL/min	PDA at 254 nm	VBN: 6.87 min (no data for VCR)	32–56 μg/mL	0.0230/0.0698	Intra-day: 1.11%–2.21% (area), 0.41%–1.12% (RT); Inter-day: 1.14%–2.37% (area), 1.59%–3.02% (RT)	Recovery: Purple 95%–103%; White 95%–99%; Pink 96%–101%; Stability not studied; simple, aqueous extraction	Rahim et al. (2018)
C. roseus leaves	extract	Merck chromolith performance RP-C18; acetonitrile: 0.1 M phosphate buffer with 0.5% glacial acetic acid (21: 79 v/v, pH 3.5); isocratic flow at 1.2 mL/min	UV at 254 nm	Vincristine/Vinblastine resolved; exact Rt not specified (~early, well-separated)	0.25–25 μg/mL	VCR: LOD 8/LOQ 25; VBN: LOD 14/LOQ 46	VCR: 1.82%; VBN: 2.50%	Recovery: Vindoline 98%; Catharanthine 98%; Vincristine 97%; Vinblastine 96%; peak purity confirmed via PDA detector	Rahim et al. (2018)

(continued)

Table 13.1 (continued)

Study/matrix		Column and mobile phase	Detection	Retention times (VCR/VBN)	Linearity and range (µg/mL)	LOD/LOQ (µg/mL)	Precision (%RSD)	Recovery/stability	References
Pharmaceutical preparations		C18 column; mobile phase: acetonitrile (32%): 50 mM potassium hydrogen phosphate buffer (68%), pH 3.2 ± 0.1; flow rate 1.5 mL/min	UV detection at $\lambda = 230$ nm		Vincristine: 1–5 µg/mL; Doxorubicin: 5–100 µg/mL	Vincristine: LOD 0.19/LOQ 0.63; Doxorubicin: LOD 0.51/LOQ 1.7			
Rat plasma		Dikma Dimonsil C_{18} (200 × 4.6 mm); 0.02 M sodium dihydrogen phosphate–methanol (36:64, v/v), pH 4.7; flow 1.0 mL/min	UV at 276 nm		0.05–5.0 µg/mL	Not explicitly stated; LOD approximate ≤0.05 µg/mL	Intra-day & inter-day accuracy: 92.6%–107.7%; RSD <9%	Recovery: ~66%–70% extraction; >83% for VCR; stable at −80 °C (1 month), room temp (8 h), freeze–thaw	Chen et al. (2013)

Plasma of Hodgkin's lymphoma patients		ODS Hypersil C_{18} (250 × 4.6 mm, 5 μm); mobile phase 30% (v/v) acetonitrile + 70% 0.05 M disodium hydrogen phosphate buffer with 0.5 mL TEA; pH adjusted to 3.7 with phosphoric acid; flow rate: 0.75 mL/min	UV at 230 nm	Vinblastine: ~6.19 min (other drugs: DOX 14.41 min; BLM 11.68 min; DTIC 9.54 min; IS nefopam 18.77 min) SciELO + 2Dialnet + 2Google Scholar + 2onesearch.slq.qld.gov.au + 7ResearchGate + 7SciELO + 7	Vinblastine: 0.075–50 μg/mL (others varied DOX: 0.05–50; BLM: 0.06–50; DTIC: 0.09–50 μg/mL)	Vinblastine: LOD 0.060/LOQ 0.075 μg/mL (others: DOX 0.020/0.050; BLM 0.045/0.060; DTIC 0.040/0.090)	Precision not reported per drug separately, overall method accuracy meets FDA guidelines (intra- and inter-day variance <3%)	% Recovery (mean): Vinblastine 99.08% (Adriamycin 99.24; Bleomycin 99.23; Dacarbazine 99.09); Analytes stable through validation procedures including freeze–thaw cycles	Zubair Malik et al. (2013)
Fusarium oxysporum culture filtrate		Preparative TLC followed by HPLC (likely C_{18} column; specifics not detailed in abstract)	UV-Vis spectroscopy at unspecified wavelength; identification also via ESI-MS/MS, UV-Vis, ^{1}H NMR	Vincristine: ~67 μg/L; Vinblastine: ~76 μg/L yield per liter (no Rt data reported)	Not specified	Not reported	Not reported	Purity confirmed by MS/MS and NMR; yields: VBN ≈ 76 μg/L, VCR ≈ 67 μg/L	Kumar et al. (2013)

(continued)

Table 13.1 (continued)

Study/matrix		Column and mobile phase	Detection	Retention times (VCR/VBN)	Linearity and range (μg/mL)	LOD/LOQ (μg/mL)	Precision (%RSD)	Recovery/stability	References
Dog plasma		Agela Venusil XBP C_{18} (2.1 × 30 mm, 3 μm); gradient elution; flow 0.4 mL/min	LC-MS/MS (ESI+), MRM transitions m/z 825.4 → 807.2 (VCR); 811.3 → 223.9 (IS, vinblastine sulfate)	Not specified visually; well-separated peaks	0.25–500 ng/mL ($r = 0.9943$)	LOQ: 0.25 ng/mL; LOD implies ≤0.25 ng/mL	Intra- and inter-run RSDs <15%		Li (2015)
C. roseus leaf		C18 column; methanol: 15 nmol/L ammonium acetate + 0.02% formic acid (65: 35 v/v)	HPLC-ESI-MS/MS, MRM mode	Exact retention times not disclosed; all five compounds were baseline-separated	Validated for trace levels (low ng/mL)	LOD/LOQ not stated explicitly in abstract	1.2%–11.5%	Recoveries: 79.9%–91.5%; stable 2 h at RT, 12 h at 4 °C, 2 weeks at −20 °C	Zhang et al. (2014)

C. roseus (leaf, stem, root)	Ethanolic extracts	ACQUITY UPLC BEH C_{18} (1.7 µm, 2.1 × 50 mm); gradient methanol with 15 nmol/L ammonium acetate + 0.02% formic acid	UHPLC-ESI-QqQ_{LIT}-MS/MS, in MRM mode	Baseline-separated; exact times not disclosed	0.5–1000 for all alkaloids ($R^2 \approx 1.00$)	0.039–0.583 (LOD); 0.118–1.767 (LOQ)	Intra-day: 0.23%–2.71%; Inter-day: 0.40%–2.90%	Recoveries: 99.63%–104.30% (±RSD ≤ 3.03%); stability RSD: 0.69%–3.45%	Kumar et al. (2018a, b)
Urine		RP-C18 column; gradient of 0.5% formic acid/methanol	LC-ESI-MS/MS (positive ion MRM scan mode)	Not specified numerically	LOQ: 0.5; LOD: ~0.25 (for both VCR and VBN)	64.2%–86.9% (six urine samples)	Vincristine: up to 63; Vinblastine: up to 111		

Table 13.2 Comparative potential and applications of vincristine and vinblastine

Pharmacological claim	Vincristine	Vinblastine
Anticancer cytotoxicity	Broad	Broad
Neurotoxicity	High	Low
Myelosuppression	Mild	High
Anti-angiogenic	Minimal	Strong
Immunomodulatory	Emerging	Moderate
Combination chemotherapy use	CHOP, MOPP	ABVD
Veterinary application	TVT	MCTs

Wilms' tumor, and neuroblastoma. Vinblastine, on the other hand, is used for Hodgkin's and Non-Hodgkin's lymphoma, testicular cancer, Kaposi's sarcoma, and is also applied off-label for breast cancer and choriocarcinoma. Detailed information about marketed formulations and brand names of both vincristine and vinblastine is provided in Table 13.3, highlighting their continued relevance in oncology.

13.6 Challenges in Production

The production of vincristine and vinblastine faces significant challenges, primarily due to their extremely low natural abundance in *C. roseus*. The extraction yield of vincristine is approximately 0.0003%, and vinblastine around 0.01%, requiring large volumes of plant biomass to obtain even small quantities of the active compounds. This creates major scalability and cost-related issues. Their complex biosynthetic pathway involves the dimerization of two precursor molecules, vindoline and catharanthine, through more than 30 enzymatic steps, further complicating large-scale production. Because of their trace content in the plant, advanced and sophisticated chromatographic techniques such as HPLC and counter-current chromatography are necessary to isolate vindoline and catharanthine with sufficient purity. These multi-step isolation processes are time-consuming and expensive. Additional challenges include the reliance on seasonal plant cultivation, batch-to-batch variability, Good Manufacturing Practice (GMP) requirements, regulatory hurdles, safe handling of toxic substances, and worker safety concerns. Furthermore, the lack of fully developed biotechnological or synthetic production alternatives has hindered the consistent and cost-effective supply of these vital anticancer compounds.

13.7 Side Effects and Management

Although vincristine and vinblastine share similar chemical structures, they exhibit distinct toxicity profiles. Vincristine is primarily associated with neurotoxicity, particularly peripheral neuropathy, while vinblastine is more commonly linked to myelosuppression, including bone marrow suppression and leukopenia. These differing side effects are summarized in Table 13.4, and a comparative overview is

Table 13.3 Global overview of marketed brand names, manufacturers, and formulations of vincristine and vinblastine

Drug	Brand name	Manufacturer	Dosage form	Strength(s)	Common markets/regions
Vincristine	Oncovin®	Eli Lilly	Injectable solution (IV)	1 mg/mL	USA, Europe, India, Worldwide
	Vincristine Sulfate	Hospira (Pfizer), Sandoz	Injectable solution (IV)	1 mg/mL	USA, Canada, India, Global
	Vincristine Sulfate Injection, USP	Fresenius Kabi, Baxter	IV injection	1 mg/10 mL, 1 mg/mL	USA, Europe
	Cytocristin®	Cipla (India)	IV injection	1 mg, 2 mg	India
	Vincasar PFS®	Teva Pharmaceuticals	Pre-filled syringe (IV)	1 mg/mL	USA, Europe
	Celcristine®	Celon Labs	IV Injection	1 mg, 2 mg	India
	Vincrist®	Intas Pharmaceuticals	IV injection	1 mg	India, Middle East
	Vincreta®	Getwell Oncology	IV injection	1 mg/1 mL, 1 mg/10 mL	India, South-East Asia
Vinblastine	Velban®	Eli Lilly	Injectable solution (IV)	10 mg/10 mL	USA, Europe, Asia
	Vinblastine Sulfate	Hospira, Teva, Sandoz	IV injection	10 mg/10 mL	USA, Canada, Europe
	Vinblastin®	Cipla	IV injection	10 mg	India
	Blaztere®	Getwell Oncology	IV injection	10 mg	India, SE Asia
	Vinsar®	Celon Labs	IV injection	10 mg	India
	Vinblast®	Intas Pharmaceuticals	IV injection	10 mg	India, Africa
	Cytoblastin®	Emcure Pharmaceuticals	IV injection	10 mg	India

Table 13.4 Side effects and management during treatment by vincristine and vinblastine

Side effect	Description
Vincristine	
Neurotoxicity	Most significant; includes peripheral neuropathy (tingling, numbness, loss of reflexes), cranial nerve palsies (e.g., ptosis, hoarseness), autonomic neuropathy (constipation, urinary retention)
Constipation	Due to autonomic nervous system involvement; can be severe, leading to paralytic ileus
Jaw pain/muscle cramps	Cramping or painful spasms in jaw, throat, or limbs
SIADH	Syndrome of inappropriate antidiuretic hormone secretion—causes hyponatremia
Mild myelosuppression	Less common than with vinblastine
Alopecia	Reversible hair loss in many patients
Injection site reaction	Local irritation, extravasation injury if drug leaks outside the vein
Vinblastine	
Myelosuppression	Dose-limiting toxicity. Includes leukopenia, neutropenia, and thrombocytopenia
Nausea and vomiting	Moderate emetogenic potential
Alopecia	Common and reversible
Neurotoxicity	Less frequent and less severe than vincristine
Stomatitis	Inflammation and soreness in the mouth
Injection site reaction	Risk of phlebitis and extravasation injury
Fatigue and malaise	Common in high-dose or long-term regimens

provided in Table 13.5. Management strategies for these adverse effects are essential to minimize patient discomfort and maintain treatment efficacy. For vincristine-induced peripheral neuropathy, supportive therapies such as vitamins B1, B6, and B12, gabapentin, and duloxetine are commonly used to reduce nerve-related symptoms. Constipation, another frequent side effect, is managed using stool softeners like docusate, laxatives such as senna or lactulose, and ensuring adequate hydration. In some cases, vincristine may also cause Syndrome of Inappropriate Antidiuretic Hormone Secretion (SIADH), which requires careful electrolyte monitoring and management. Vinblastine's myelosuppressive effects necessitate close monitoring of blood cell counts and may require dose adjustments, growth factor support, or temporary discontinuation of therapy. Proper management improves patient outcomes and enhances treatment tolerability.

Table 13.5 Comparative side effects of vincristine and vinblastine

Toxicity type	Vincristine	Vinblastine
Neurotoxicity	High (dose-limiting)	Mild to moderate
Myelosuppression	Mild	Severe (dose-limiting)
Constipation	Very common	Less common
Nausea/vomiting	Mild to moderate	Moderate
Alopecia	Yes	Yes
SIADH	Reported	Rare
Injection site injury	Yes (vesicant)	Yes (vesicant)

References

Barkat MA, Mohd Mujeeb MM, Harshita H, Rahman MA (2015) Analysis of vincristine in leaf and callus extract by validated HPTLC technique and its stress degradation studies. Br J Pharm Res 7(5):308–318

Chen Q, Zhang W, Zhang Y, Chen J, Chen Z (2013) Identification and quantification of active alkaloids in Catharanthus roseus by liquid chromatography–ion trap mass spectrometry. Food Chem 139(1–4):845–852

Dhyani P, Quispe C, Sharma E, Bahukhandi A, Sati P, Attri DC, Szopa A, Sharifi-Rad J, Docea AO, Mardare I, Calina D (2022) Anticancer potential of alkaloids: a key emphasis to colchicine, vinblastine, vincristine, vindesine, vinorelbine and vincamine. Cancer Cell Int 22(1):206

Gupta MM, Singh DV, Tripathi AK, Pandey R, Verma RK, Singh S, Shasany AK, Khanuja SPS (2005) Simultaneous determination of vincristine, vinblastine, catharanthine, and vindoline in leaves of Catharanthus roseus by high-performance liquid chromatography. J Chromatogr Sci 43(9):450–453

Jordan MA, Wilson L (2004) Microtubules as a target for anticancer drugs. Nat Rev Cancer 4(4):253–265

Kumar A, Patil D, Rajamohanan PR, Ahmad A (2013) Isolation, purification and characterization of vinblastine and vincristine from endophytic fungus Fusarium oxysporum isolated from Catharanthus roseus. PLoS One 8(9):e71805

Kumar S, Bajpai V, Singh A, Kumar B (2018a) Identification, characterization and distribution of terpene indole alkaloids in ethanolic extracts of Catharanthus roseus using high-performance liquid chromatography/electrospray ionization quadrupole time-of-flight tandem mass spectrometry and the study of their geographical variation. Rapid Commun Mass Spectrom 32(4):319–332

Kumar S, Singh A, Kumar B, Singh B, Bahadur L, Lal M (2018b) Simultaneous quantitative determination of bioactive terpene indole alkaloids in ethanolic extracts of Catharanthus roseus (L.) G. Don by ultra high performance liquid chromatography–tandem mass spectrometry. J Pharm Biomed Anal 151:32–41

Li BS (2015) Quantification of vincristine sulfate in dog plasma by HPLC-MS/MS and pharmacokinetics of vincristine sulfate thermosensitive liposomes in dogs. Chin Pharm J 24:705–708

Rahim RA, Ahmad NH, Al Azzam KM, Mat I (2018) Determination and quantification of the vinblastine content in purple, red, and white Catharanthus roseus leaves using RP-HPLC method. Adv Pharm Bull 8(1):157–161

Rowinsky EK, Donehower RC (1991) The clinical pharmacology and use of antimicrotubule agents in cancer chemotherapeutics. Pharmacol Ther 52(1):35–84

Staff NP, Grisold A, Grisold W, Windebank AJ (2017) Chemotherapy-induced peripheral neuropathy: a current review. Ann Neurol 81(6):772–781

Viju C, Agrawal SA, Jadhav AP (2023) Development and validation of a novel HPTLC method for quantitative estimation of vincamine from Catharanthus roseus Linn. Leaves extract. Indian Drugs 60(10):48

Zhang L, Gai Q, Yang L, Ma Y, Liu Y (2014) Simultaneous quantitative determination of five alkaloids in Catharanthus roseus by HPLC-ESI-MS/MS. Chin J Nat Med 12(10):786–793

Zhang F, Yan C, Wei C, Yao Y, Ma X, Gong Z, Liu S, Zang D, Chen J, Shi FD, Hao J (2018) Vinpocetine inhibits NF-κB-dependent inflammation in acute ischemic stroke patients. Transl Stroke Res 9:174–184

Zubair Malik M, Ahmad M, Muahammad S (2013) Rapid and simultaneous determination of adriamycin, bleomycin, vinblastine and dacarbazine in plasma of Hodgkin's lymphoma patients by a reversed phase HPLC method. J Chil Chem Soc 58(2):1674–1677

Future Research and Recommendations 14

Abstract

Future research on *Catharanthus* plants-derived drugs should focus on enhancing the key alkaloids (vincristine and vinblastine) yield and purity, which is crucial in cancer treatment. Advanced biotechnological approaches, including metabolic engineering, CRISPR gene editing, and synthetic biology, can optimize alkaloid biosynthesis in plant or microbial systems. Exploring plant's lesser-known compounds may reveal new pharmacological properties. Clinical trials should evaluate the efficacy and safety of novel formulations and delivery systems, such as nanoparticles or liposomes, to improve bioavailability and reduce side effects. Continued investment in research will help in fully realize the therapeutic potential of *Catharanthus* in modern medicine.

Keywords

Cathranthus · Future prospectives · Alkaloids · Yield and purity · Drug discovery · Molecules

For over half a century, *Catharanthus* alkaloids such as vincristine and vinblastine have been widely used and officially approved in chemotherapy for various cancers. *C. roseus* alone has yielded more than 130 alkaloids, many with potential therapeutic value. However, other species within the *Catharanthus* genus remain largely unexplored, presenting untapped opportunities for novel drug discovery. Since these valuable bisindole alkaloids are present only in trace amounts, it is crucial to optimize methods for extraction, isolation, and biosynthesis. Additionally, developing reliable qualitative and quantitative analytical techniques and investigating the pharmacological activities of both crude extracts and isolated compounds is essential to

S. Kumar, B. Singh, *Catharanthus in Drug Discovery*,
https://doi.org/10.1007/978-981-95-5048-7_14

enhance yields and therapeutic potential (Mustafa and Verpoorte 2007; Kumar et al. 2022).

There are traditional methods such as acid-base extraction method, Stas–Otto method, Dragendorff precipitation method, soxhlet extraction; maceration, and percolation have been used for isolation of phytochemical from *Catharanthus* species. These methods have drawback like low selectivity, time-consuming, high solvent utilization, thermal degradation, low efficiency/yield, lack of standardization, potential for artifact formation. These traditional extract methods can take hours to days and non-target compounds co-extraction with impurity with large volumes of toxic or flammable organic solvent. Prolong heating of may degrade heat-sensitive phytochemicals like alkaloids, flavonoids, and essential oils. Low efficiency and yield due to poor solvent penetration, improper pH control. These methods increase cost, health hazards, and environmental concerns. There is a need to replace or supplement traditional methods with more efficient and selective modern techniques like UAE, MAE, SFE, PLE, and EAE. There is a need to use green solvents such as water and methanol and solvent-free techniques (supercritical CO_2). The hazards solvent such as chloroform, benzene, and methanol are replaced with ethanol, water, or ionic liquids (Zhang et al. 2018a, b, c, d, e, f; Mondal et al. 2024; Shrivastav et al. 2025).

There is collection of time drying, grinding, and storage; freeze-drying (lyophilization) plant is a required proper method. All parameters required for optimization of traditional extraction and isolation methods are not available; therefore there is a need to optimize pH, temperature, solvent polarity, and extraction time with the use of response surface methodology (RSM) or design of experiments (DoE) (Jones and Case 1990; Keith 2000).

The survey of traditional uses of *Catharanthus* species mainly reported for *C. roseus* and researchers selected very limit area of study. There is an urge required for comprehensive ethnobotanical surveys and proper documentation particularly diversity of *Catharanthus* species across cultures in tribal and rural communities. Claimed traditional medicinal uses provided by tribal and rural communities and also researchers and rigorous are required validating by pharmacological and toxicological studies along with standardization of traditional formulations for consistency, reproducibility, safety, and efficacy. Combination of efficient traditional and advanced methods with advanced green techniques offer improves overall yield, selectivity, and cost-effectiveness. Miniaturize, automated, and high-throughput methods are required to reduce solvent usage, sample size, and time especially in early-stage drug discovery screening. The trace or low content of bis indole alkaloids has been synthesized by *Catharanthus* species. Therefore, new and sensitive methods should be developed to isolate minor compounds such as bisindole alkaloids or rare glycosides that are typically present in trace amounts. The solvent system plays an important role in extraction, isolation, and separation methods. Therefore, smart solvents and designer eluents such as ionic liquids, deep eutectic solvents should offer higher selectivity and sustainability over conventional solvents with cutting-edge Advanced Chromatographic Technologies (prep-HPLC, flash chromatography, centrifugal partition chromatography (CPC), and orthogonal

2D chromatography) (Shushan 2010; van den Ouweland and Kema 2012; Thomas et al. 2022; Yost 2022).

The WHO promoted the establishment of globally accepted validation standards and standardization methods for herbal formulations to assure precision, recovery, repeatability, reproducibility, and regulatory compliance (Indrayanto 2024).

Biosynthetic Pathways of MIAs and bisindole alkaloids are partially known. Many of the steps remain partially characterized. Future studies should be aimed to fully characterize biosynthetic pathways with help of advance chromatographic methods. Many steps such as functional characterization of key enzymes, genomic and transcriptomic profiling, metabolic flux analysis and regulation studies, synthetic biology and metabolic engineering, and integration of multi-omics approaches of biosynthesis remain partially touched or untouched.

Development of HTS methods for rapid analysis of MIAs and other classes of compounds should be establish rapid, sensitive, and cost-effective high-throughput methods to screen large plant material of *Catharanthus* and identify new or trace MIAs. Advanced integrate cutting-edge hyphenated Analytical techniques such as HPLC-MS/MS, UHPLC-QTOF-MS, and MS^n should be required for structural characterization and quantitation of trace and new compounds.

Most of the studies have been reported based on in vitro assays without any proper mechanistic insights. Therefore future studies should focus on rigorous validation using in vivo studies of diseases along with translational potential elucidating detailed mechanisms of action TIAs, bisindole alkaloids and phenolics, at the molecular and cellular levels using omics and bioinformatics tools. The dose optimization and toxicity studies are very crucial in drug discovery, so systematic dose-response and long-term toxicity studies should be conducted to define therapeutic windows and safety profiles of active constituents. Novel drug delivery systems must be developed to optimize solubility, bioavailability, and targeted administration of active components. There is very limited data on the pharmacokinetic and bioavailability profiles *Catharanthus* compounds. Mandatory nanoformulations or improved drug delivery technologies should be accompanied with ADME studies in both animals and humans. Phytoconstituents exert a synergistic impact and influence the efficacy of drugs. Combinatorial research should be conducted and validated through in vitro, in vivo, and clinical studies for chemotherapeutic drugs that can mitigate toxicity and enhance efficacy. The different plant extracts and compound also have potential and significant neuroprotection, antidiabetic, anti-inflammatory, antimicrobial, and cardiovascular indications. Such pharmacological activities should be explored with advanced in vitro, in vivo, and clinical studies. Enhanced preclinical models are essential for converting laboratory discoveries into therapeutic applications. Studies that include pharmacodynamics and toxicity will facilitate IND (Investigational New Drug) submissions. The traditional applications of *Catharanthus* in ethnomedicine necessitate scientific validation and incorporation into pharmacological research.

It is essential to examine structural characteristics like as polarity, steric bulk, π-systems, and hydrogen bonding potential in both 2D and 3D structures that affect chromatographic biological activities. It ought to develop customized stationary phases for specific compound classes (MIAs and bis indole alkaloids, flavonoids) according to their three-dimensional molecular conformations. It is necessary to develop MI models utilizing 2D/3D molecular descriptors (e.g., LogP, PSA, molecular volume) to forecast retention periods in various chromatographic systems (RP-HPLC, HILIC (Hydrophilic Interaction Liquid Chromatography), Supercritical Fluid Chromatography (SFC); perform comprehensive analyses on the influence of chiral centers and three-dimensional topology on enantioseparation in chiral chromatography. Investigate binding interactions between three-dimensional conformers and chiral selectors to improve resolution; examine the separation of dynamic molecular forms, including tautomers and atropisomers, by temperature-controlled or pH-modulated chromatography; employ 2D NMR and molecular dynamics simulations to connect isomeric populations with chromatographic elution characteristics; apply 2D-LC (e.g., NP-HPLC × RP-HPLC or HILIC × SFC) to resolve complicated natural product extracts where compound separation depends on both 2D and 3D characteristics; examine the impact of structural and physicochemical characteristics on separation efficacy utilizing green solvents, ionic liquids, or supercritical CO_2 in environmentally sustainable chromatography systems; establish extensive SRR databases that correlate structural characteristics (e.g., rotatable bonds, TPSA, ring systems) with chromatographic retention periods under standardized settings; employ computational modeling and molecular docking (utilizing stationary phase analogs) to forecast and enhance compound-stationary phase interactions; urge researchers to consistently disclose SMILES/InChI keys, three-dimensional coordinates, and essential physicochemical descriptors in conjunction with chromatographic results in their publications; employ QSRR in conjunction with 3D modeling to enhance peak annotation in LC–MS-based metabolomics of plant extracts; classify and elaborate on chiral columns and selectors according to their compatibility with the common stereochemical characteristics of natural substances, such as indole alkaloids, lignans, and sugars; create intuitive software applications that employ 2D and 3D models to recommend ideal mobile phases, columns, and detecting techniques; and integrate modules addressing the influence of molecular geometry and physicochemical parameters on chromatography into training curricula for phytochemists and analytical scientists.

Prospective research on clinical trials and toxicological evaluations should consider the integration of multi-omics, including genomics, proteomics, metabolomics, and transcriptomics, in toxicological studies. AI and MI should be incorporated into clinical and toxicological predictions, including in vitro and in silico methods. Natural products and polyherbal formulations derived from traditional medicine systems, such as Ayurveda, TCM, and Unani, require rigorous, placebo-controlled, multicentric clinical trials to substantiate their therapeutic claims. Expansion of post-marketing surveillance and pharmacovigilance studies for traditional products is necessary to obtain long-term safety data through real-world evidence and digital health records. Cross-national alignment of regulations and protocols for the approval of phytopharmaceuticals is necessary, especially regarding clinical trial conduct, safety thresholds, and ethical review processes. Research must prioritize the identification of clinical, genomic, or metabolic biomarkers that can function as surrogate endpoints for efficacy and early indicators of toxicity in herbal interventions.

Systematic research is necessary to document and evaluate the clinical significance of herb–drug interactions, especially in populations using conventional medications in conjunction with herbal supplements. Establish mandatory and systematically organized protocols for acute, sub-chronic, and chronic toxicity assessments prior to advancing to clinical phases, thereby ensuring the safety of compounds. It is advisable to utilize chemically characterized, standardized extracts to enhance reproducibility, ensure regulatory compliance, and promote patient safety in clinical trials. Advocate for the implementation of randomized, double-blind, placebo-controlled, and multicenter study designs that utilize sufficient sample sizes to produce statistically significant results. Promote the registration of trials in established databases (e.g., CTRI, ClinicalTrials.gov) and the public sharing of negative or inconclusive findings to mitigate duplication and bias. Promote collaboration among phytochemists, pharmacologists, clinicians, toxicologists, and regulatory experts to ensure the comprehensive development of safe and effective natural therapeutics. Train human resources in good clinical practice (GCP), good laboratory practice (GLP), and regulatory toxicology, emphasizing natural products. Employ electronic data capture systems, mobile applications, and wearable devices in clinical trials to improve data accuracy, patient adherence, and real-time monitoring of adverse events. It is advisable to enhance financial backing for long-term toxicology studies and multicenter clinical trials focused on promising natural compounds, especially those aimed at chronic diseases.

References

Indrayanto G (2024) Regulation and standardization of herbal drugs: current status, limitation, challenge's and future prospective. Profiles Drug Subst Excip Related Method 49:153–199

Jones JB Jr, Case VW (1990) Sampling, handling, and analyzing plant tissue samples. Soil Test Plant Anal 3:389–427

Keith BDA (2000) Sampling designs, field techniques and analytical methods for systematic plant population surveys. Ecol Manage Restor 1(2):125–139

Kumar S, Singh B, Singh R (2022) Catharanthus roseus (L.) G. Don: a review of its ethnobotany, phytochemistry, ethnopharmacology and toxicities. J Ethnopharmacol 284:e114647

Mondal S, Das M, Debnath S, Sarkar BK, Babu G (2024) An overview of extraction, isolation and characterization techniques of phytocompounds from medicinal plants. Nat Prod Res:1–23. https://doi.org/10.1080/14786419.2024.2426059

Mustafa NR, Verpoorte R (2007) Phenolic compounds in Catharanthus roseus. Phytochem Rev 6(2–3):243–258

Shrivastav G, Prava Jyoti T, Chandel S, Singh R (2025) Eco-friendly extraction: innovations, principles, and comparison with traditional methods. Sep Purif Rev 54(3):241–257

Shushan B (2010) A review of clinical diagnostic applications of liquid chromatography–tandem mass spectrometry. Mass Spectrom Rev 29(6):930–944

Thomas SN, French D, Jannetto PJ, Rappold BA, Clarke WA (2022) Liquid chromatography–tandem mass spectrometry for clinical diagnostics. Nat Rev Methods Primers 2(1):96

van den Ouweland JM, Kema IP (2012) The role of liquid chromatography–tandem mass spectrometry in the clinical laboratory. J Chromatogr B 883:18–32

Yost RA (2022) Why tandem mass spectrometry for trace analysis: concepts of tandem analytical techniques. Rapid Commun Mass Spectrom 36(13):e9310

Zhang A, Sun H, Wang X (2018a) Mass spectrometry-driven drug discovery for development of herbal medicine. Mass Spectrom Rev 37(3):307–320

Zhang QW, Lin LG, Ye WC (2018b) Techniques for extraction and isolation of natural products: a comprehensive review. Chin Med 13(1):20

Zhang D, Li X, Hu Y, Jiang H, Wu Y, Ding Y, Yu K, He H, Xu J, Sun L, Qian F (2018c) Tabersonine attenuates lipopolysaccharide-induced acute lung injury via suppressing TRAF6 ubiquitination. Biochem Pharmacol 154:183–192

Zhang F, Yan C, Wei C, Yao Y, Ma X, Gong Z, Liu S, Zang D, Chen J, Shi FD, Hao J (2018d) Vinpocetine inhibits NF-κB-dependent inflammation in acute ischemic stroke patients. Transl Stroke Res 9:174–184

Zhang Y, Zhao Y, Guo J, Cui H, Liu S (2018e) Anticancer activity of safranal against colon carcinoma is due to induction of apoptosis and G2/M cell cycle arrest mediated by suppression of mTOR/PI3K/Akt pathway. J BU ON 23(3):574–578

Zhang YW, Kong XY, Wang JH, Du GH (2018f) Vinblastine and vincristine. In: Natural small molecule drugs from plants. Springer, Singapore, pp 551–557

Glossary

Accuracy (Trueness) The closeness of agreement between the measured value and the true value (expressed as % recovery). It's how close a measured value is to the actual or true value.

Accurate Mass The mass of an ion measured with high precision, used for determining elemental composition.

Achiral The geometric symmetry of being indistinguishable from its own mirror image; lacking chirality.

Activation energy The minimum energy which must be available to a chemical system with potential reactants in order to result in a particular chemical reaction.

Additive effect The combined effect of two or more drugs is equal to the sum of their individual effects.

Adduct Ion An ion formed by the addition of a small molecule or ion (e.g., $[M + Na]^+$) to the analyte.

Adsorption Process where compounds adhere to the surface of the stationary phase.

Affinity The measure of a drug's ability to bind to a specific receptor.

Affinity Chromatography A separation method based on specific interactions between analyte and a binding agent (e.g., antigen-antibody).

Agonist A drug that binds to and activates a specific receptor, thereby producing a biological response similar to that of the endogenous ligand.

Alcohol Any organic compound consisting of at least one hydroxyl group attached to a saturated carbon atom. Alcohols have the general formula R–OH.

Alkaloid Any of a loosely defined class of organic compounds found in the tissues of many species of plants. Alkaloid molecules have one or more alkaline-reacting nitrogen atoms in their carbon structures. Many alkaloids are commercially important as drugs or poisons, e.g., caffeine, morphine, quinine, and strychnine. Alkaloids: Nitrogen-containing secondary metabolites with diverse pharmacological effects (e.g., morphine, vincristine).

Allelopathy The secretion by a plant of biochemicals which influence the growth and reproduction of nearby plants.

Allopatric Having geographically separate, non-overlapping ranges of distribution

S. Kumar, B. Singh, *Catharanthus in Drug Discovery*,
https://doi.org/10.1007/978-981-95-5048-7

Allosteric modulator A molecule that binds to a site on a receptor or enzyme separate from the active site, thereby altering its activity or affinity for the ligand or substrate.

Analyte The compound or substance of interest being analyzed or separated.

Animal disease models Naturally occurring or experimentally induced animal diseases with pathological processes sufficiently similar to those of human diseases. They are used as study models for human diseases.

Annual A plant that completes its life cycle (i.e. germinates, reproduces, and dies) within a single year or growing season

Antagonist A drug that binds to a receptor but does not activate it, blocking the effects of agonists.

Anthocyanins Water-soluble pigments responsible for red, purple, and blue colors in plants; part of the flavonoid group.

Antidote A substance that counteracts the toxic effects of a drug or poison.

Antioxidants Compounds that inhibit oxidation and neutralize free radicals (e.g., ascorbic acid, polyphenols).

Asymmetry Factor A measure of peak tailing or fronting in chromatographic peaks.

Attrition High attrition or failure rates in pharmaceutical R&D continue to be a challenge and ways to reduce these are increasingly required.

Band broadening Widening of the chromatographic peak due to dispersion of analyte molecules, reducing resolution.

Base peak The most intense peak in a mass spectrum, assigned a relative intensity of 100%.

Base peak chromatogram (BPC) A chromatogram that displays the most intense ion at each scan over time.

Baseline The flat signal level on a chromatogram when no analyte is eluting.

Biennial A plant which completes its life cycle (i.e. germinates, reproduces, and dies) within 2 years or growing seasons. Biennial plants usually form a basal rosette of leaves in the first year and then flower and fruit in the second year.

Bioavailability The fraction of an administered drug that reaches the systemic circulation and is available to exert its therapeutic effect is influenced by factors such as absorption, metabolism, and distribution.

Bond Any persistent attraction between atoms, ions, or molecules that enables the formation of chemical compounds.

Breakthrough therapies Breakthrough Therapy designation is a process designed to expedite the development and review of drugs that are intended to treat a serious condition and preliminary clinical evidence indicates that the drug may demonstrate substantial improvement over available therapy on a clinically significant endpoint(s).

Buffer A solution used in the mobile phase to maintain constant pH.

Calibration The process of adjusting the mass spectrometer to ensure accurate m/z measurements.

Capacity Factor (k') Also known as retention factor; describes the retention of a compound relative to the void volume.

Carotenoids Pigments (like beta-carotene, lycopene) that serve as antioxidants and precursors to vitamin A.

Carrier Gas In gas chromatography (GC), an inert gas (e.g., helium or nitrogen) that carries analytes through the column.

Carryover The presence of analyte in a blank run after a high concentration sample, indicating contamination.

Catechins A type of flavonoid with strong antioxidant activity, found in green tea and cocoa.

Chlorophyll Any of a variety of different chemical pigments in chloroplasts that are essential for photosynthesis.

Chromatogram A graphical output showing detector response (*y*-axis) vs. time or volume (*x*-axis).

Chromatograph The instrument used for chromatography (e.g., HPLC, GC).

Clearance The rate at which a drug is eliminated from the body typically expressed as the volume of blood or plasma that is cleared of the drug per unit of time.

Clinical trial A research study is conducted in humans to evaluate the safety, efficacy, and appropriate dosage of a new drug or treatment.

Column The tube packed with stationary phase through which the mobile phase and analyte travel.

Column back-pressure The pressure developed by the system due to flow resistance in the column.

Column conditioning Stabilizing the column with mobile phase prior to analysis to ensure reproducible results.

Column efficiency (N) A measure of the sharpness of a chromatographic peak, represented in terms of theoretical plates.

Coumarins Aromatic lactones with potential anticoagulant, anti-inflammatory, and antimicrobial activities.

Dead time (t_0) Time taken by an unretained compound to pass through the system.

Detector A device used to identify and measure the eluting analytes (e.g., UV, PDA, RI, MS); in mass spectrometry, it converts ion impacts into a measurable signal.

Dosage form The physical form in which a drug is administered, such as tablets, capsules, solutions, or suspensions.

Dose-response relationship The relationship between the dose of a drug administered and the magnitude of the resulting biological effect.

Drug A chemical substance that interacts with the body to produce a physiological effect, either for therapeutic purposes, diagnostic purposes, or experimental research. The World Health Organization (WHO) defines a drug as "any substance or product that is used or intended to be used to modify or explore physiological systems or pathological states for the benefit of the recipient“. This definition encompasses a wide range of substances, including prescription medications, over-the-counter drugs, traditional medicines, and other substances that have a physiological effect when ingested or otherwise introduced into the body.

It's a chemical substance that modifies physiological or pathological conditions in the body for therapeutic, diagnostic, or experimental purposes.

Drug addiction A chronic, relapsing disorder characterized by compulsive drug-seeking behavior and drug use, despite harmful consequences.

Drug allergy An immune-mediated adverse reaction to a drug may manifest as skin rashes, respiratory symptoms, or even life-threatening anaphylaxis.

Drug antagonism The interaction between two drugs that results in a reduction of the overall effect, which may occur when one drug inhibits the action of another drug or competes for the same receptor or target.

Drug delivery system A technology or formulation designed to enhance the targeted delivery of a drug to specific tissues or cells within the body, potentially improving efficacy and reducing systemic side effects.

Drug dependence A state in which an individual requires the presence of a drug to function normally, often accompanied by withdrawal symptoms upon cessation of drug use.

Drug development The process of bringing a new drug to market, including preclinical and clinical testing, regulatory approval, and post-marketing surveillance.

Drug discovery The process of identifying and developing new pharmaceutical compounds with potential therapeutic applications typically involves target identification, lead optimization, and preclinical testing.

Drug discovery pipeline The process of drug development has evolved into an extremely complex procedure. The average drug takes 12 years and $270 million from initial discovery to public usage. A pipeline is best defined as what the company potentially has to offer for a product.

Drug formulation The process of combining a drug with various excipients to create a dosage form suitable for administration, taking into account factors such as stability, solubility, and bioavailability.

Drug interaction A situation in which one drug affects the activity of another drug when both are administered together, potentially altering the therapeutic or adverse effects of either drug.

Drug metabolism The biochemical modification of a drug by enzymes in the body, typically in the liver, leads to the formation of metabolites, which may be more or less active than the parent drug.

Drug monitoring The process of measuring drug concentrations in the body, typically in blood or plasma, to ensure that they remain within the therapeutic window and to assess compliance with prescribed regimens.

Drug resistance A phenomenon in which a drug becomes less effective over time, often due to changes in the target organism or the development of compensatory mechanisms.

Drug schedule A classification system is used by regulatory authorities to categorize drugs based on their potential for abuse, addiction, or medical use, with varying degrees of restriction on their availability and prescription.

Drug screening The process of testing a large number of compounds for potential therapeutic activity, often using high-throughput techniques in the early stages of drug discovery.

Drug selection Traditionally the movement of compounds along the pipeline has been a fairly linear process. Because ways of speeding up the process can be enormously economically rewarding, greater attention is being paid to moving compounds along faster, trying to insure that compounds which will eventually fail, fail earlier, and looking at ways of revising the process to perform some evaluations in parallel rather than sequentially.

Drug solubility The ability of a drug to dissolve in a particular solvent, such as water or lipid, which can affect its absorption and bioavailability.

Drug tolerance A decrease in the effectiveness of a drug over time, resulting from the body's adaptation to the presence of the drug.

Drug-drug interaction The modification of the effect of one drug by the concurrent administration of another drug.

Drug-food interaction The alteration of a drug's effect by the presence of food in the gastrointestinal tract, which may enhance or inhibit the absorption or metabolism of the drug.

Drug-like Describes compounds that have physicochemical and structural properties consistent with known drugs (approximately 10,000 known).

Druggable Genome The subset of genes encoding proteins capable of binding drug-like molecules (estimated ~3000 genes).

Efficacy The ability of a drug to produce a maximal biological response after binding to its receptor.

Efficiency (*N*) A measure of how well a chromatography column separates analyte peaks, expressed in theoretical plates.

Electron Ionization (EI) A hard ionization technique in mass spectrometry where high-energy electrons bombard molecules, causing them to ionize and fragment.

Electrospray Ionization (ESI) A soft ionization technique in which a high voltage is applied to a liquid to create an aerosol, forming charged droplets for MS analysis. (Note: MALDI, not ESI, uses a laser to ionize samples in a matrix — this part was likely misplaced.)

Eluent The solvent or mobile phase used to carry analytes through a chromatographic column.

Elution The process by which analytes are transported through the column and eluted by the mobile phase.

Ethical drugs The old term ethical drugs signified drugs advertised only to doctors. The expression refers to the original 1847 code of ethics of the AMA, which deemed advertising directly to the public to be unethical. Over time, the term came to mean legal drugs.

Excipient An inactive ingredient in a pharmaceutical product that serves as a vehicle or medium for delivering the active drug, providing stability, enhancing absorption, or improving taste.

Extracted Ion Chromatogram (EIC/XIC) A chromatogram showing only the signal intensity of a specific ion (*m/z*) extracted from full-scan MS data.

Flavonoids Plant-derived polyphenolic compounds with antioxidant, anti-inflammatory, and anticancer properties (e.g., quercetin, kaempferol).

Flow Rate The speed at which the mobile phase passes through the chromatographic column, usually measured in mL/min.

Fourier Transform Ion Cyclotron Resonance (FT-ICR) A high-resolution mass spectrometry technique that uses magnetic fields and ion cyclotron resonance for precise *m/z* detection.

Fragmentation The process in MS where ions break into smaller pieces, providing structural information about the analyte.

Fronting A chromatographic peak distortion where the leading edge rises too steeply, often due to column overloading or active site interactions.

Gas chromatography (GC) A separation technique used to analyze volatile compounds based on differences in retention time and interaction with a gaseous mobile phase.

Generic drug A pharmaceutical drug that has the same chemical substance as the drug that was originally developed, patented and innovated. Generic drugs are allowed for sale after the expiry of the patent of the original drugs. Because the active chemical substance is the same, the medical profile of generics is believed to be equivalent in performance.

Gradient elution A chromatographic technique in which the mobile phase composition is varied during the run to improve resolution and peak separation.

Half-life The time it takes for the concentration of a drug in the bloodstream to decrease by half.

HETP (Height Equivalent to a Theoretical Plate) Column length required for one theoretical plate; used to assess efficiency.

High-Performance Liquid Chromatography (HPLC) An analytical technique used to separate, identify, and quantify components in a mixture using a liquid mobile phase under high pressure.

Injection repeatability Assessment of peak area consistency across multiple injections of the same sample.

Injection volume The volume of sample introduced into the HPLC system.

Intermediate Precision (inter-day) Variations within the same lab (different analysts, days, or instruments).

Internal standard A compound added in known quantity to calibrate or normalize MS data.

Inverse agonist A drug that binds to the same receptor as an agonist but produces an effect opposite to that of the agonist, stabilizing the receptor in an inactive state.

Investigational New Drug (IND) A drug that has not yet been approved for marketing but is being studied in clinical trials to evaluate its safety and efficacy.

Ion source The part of the mass spectrometer where ions are generated.

Ion suppression A decrease in ion signal due to competition among analytes during ionization.

Ion trap A type of mass analyzer that captures ions using electric fields and sequentially ejects them based on *m/z*.

Ionization The process by which neutral molecules are converted into ions. Common ionization techniques include ESI, MALDI, and EI.

Isocratic elution Chromatography method where the mobile phase composition remains constant during the run.

Isotopic peak A peak in the mass spectrum due to the presence of isotopes of elements in a molecule.

K **(Partition Coefficient)** Ratio of analyte concentrations in stationary and mobile phases.

Lifestyle drugs Drugs treating non-life-threatening conditions such as erectile dysfunction, baldness, and some aging therapies.

Ligand A molecule that binds to a specific target, such as a receptor or enzyme, to produce a biological response. Drugs can be considered ligands.

Limit of detection (LOD) The lowest concentration or amount of analyte that can be reliably detected but not necessarily quantified (commonly defined by S/N ~ 3:1).

Limit of quantification (LOQ) The lowest amount of analyte that can be quantitatively determined with acceptable precision and accuracy (commonly defined by S/N ~ 10:1).

Linear velocity Speed at which the mobile phase travels through the chromatographic column.

Linearity Ability of an analytical method to produce results that are directly proportional to analyte concentration over a specific range.

Loading dose A higher initial dose of a drug is administered to rapidly achieve a therapeutic concentration in the body, followed by maintenance doses to maintain the desired effect.

Maintenance dose A dose given at regular intervals to maintain a therapeutic drug level after the loading dose.

Mass Spectral Library-Based Methods Techniques that use reference mass spectra databases to assist in identifying compounds within complex mixtures.

Mass Spectrometry (MS) An analytical technique used to measure the mass-to-charge ratio (*m/z*) of ions, providing molecular weight and structural information.

Mass-to-Charge Ratio (*m/z*) The ratio of an ion's mass to its charge; the fundamental measurement in mass spectrometry.

Matrix All other components of a sample apart from the analyte(s) of interest.

Matrix Effect The influence of other substances in a sample that may enhance or suppress the analytical response of the analyte.

Medicinal Plants These are plants that provide health-promoting characteristics and temporary relief or curative properties

Metabolism The process by which the body breaks down and converts drugs into more water-soluble forms for excretion.

Method development The process of designing and optimizing chromatographic conditions to achieve accurate and efficient analyte separation.

Microdosing A technique for studying the behaviour of drugs in humans through the administration of doses so low ("sub-therapeutic") they are unlikely to produce whole-body effects, but high enough to allow the cellular response to be studied. This is called a "Phase 0study" and is usually conducted before clinical Phase I to predict whether a drug is viable for the next phase of testing. Human microdosing aims to reduce the resources spent on non-viable drugs and the amount of testing done on animals.

Mobile Phase The solvent or mixture that flows through the column, transporting analytes (e.g., liquids in HPLC, gases in GC).

Molecular Ion ($M^{+\bullet}$) An ion formed by the removal of an electron from a molecule without fragmentation; indicates the compound's molecular weight.

Monoclonal antibody A drug derived from a single type of immune cell is designed to bind to a specific target, such as a protein or receptor, with high specificity and affinity.

Neutral loss The loss of a small neutral fragment (e.g., H_2O, NH_3) from a precursor ion during fragmentation in mass spectrometry.

Orbitrap A high-resolution mass analyzer that traps ions using an electrostatic field and measures their oscillation frequency.

Organic modifier An organic solvent (e.g., methanol, acetonitrile) added to the mobile phase to improve chromatographic separation.

Orphan drugs Drugs intended for treatment, diagnosis, or prevention of rare diseases affecting fewer than 200,000 people in the U.S., or where recovery of development costs is unlikely.

Parenteral administration The delivery of a drug into the body via a route that bypasses the gastrointestinal tract, such as intravenous, intramuscular, or subcutaneous injection.

Partial agonist A drug that binds to and activates a receptor but produces a submaximal response compared to a full agonist.

Partition chromatography Separation based on different solubilities of analytes in mobile and stationary phases.

Peak area The integrated area under a chromatographic peak, proportional to analyte concentration. Used for quantification.

Peak fronting Opposite of tailing; where the peak ascends too sharply due to overloading or poor column packing.

Peak height The vertical distance from the baseline to the top of the chromatographic peak.

Peak intensity The height or area of a peak in the mass spectrum, correlating with ion abundance.

Peak symmetry (Tailing Factor, *T*) Describes the shape of a chromatographic peak. Ideal peaks are symmetrical ($T \approx 1.0$).

Peak tailing A form of peak distortion where the tail end of a peak extends further than expected; usually due to column or analyte interactions.

pH of mobile phase Important for controlling the ionization state of analytes and column performance.

Pharmacodynamics The study of the biochemical and physiological effects of drugs on the body and their mechanisms of action.

Pharmacogenetics The study of genetic variations that influence an individual's response to drugs; aims to optimise drug therapy based on a person's unique genetic makeup.

Pharmacokinetics The study of the absorption, distribution, metabolism, and excretion of drugs within the body.

Placebo An inactive substance that resembles the drug being tested is used as a control in clinical trials to account for the psychological effects of treatment and assess the true efficacy of the drug.

Polypharmacy The use of multiple medications by a patient, which increases the potential for drug interactions and adverse effects.

Potency A measure of the amount of a drug required to produce a given effect; the more potent a drug, the lower the dose required to achieve the desired effect.

Precision The closeness of individual measurements under specified conditions.

Preclinical development The area of preclinical development is critical to ensuring the safety of all drugs entering clinical trials and ultimately the marketplace. Molecular approaches to assess drug candidate toxicity and human metabolism will help to better predict safety and efficacy early in the development process. Developing more specific assays for genetic toxicology will be a key contribution to this effort. The use of modeling could shed light on the relationship between PK/PD and toxicity, and help to translate results to human in vivo response. Improvements and innovations in preclinical in vitro assays and animal models will lead to greater compound yield and lower costs.

Preclinical studies Studies that test a drug on animals and other nonhuman test systems. They must comply with FDA's good laboratory practices. Data about a drug's activities and effects in animals help establish boundaries for safe use of the drug in subsequent human testing (clinical studies). Also, because animals have a much shorter lifespan than humans, valuable information can be gained about a drug's possible toxic effects over an animal's life cycle and on offspring.

Preclinical testing Testing of compounds on cell lines and live animals for effectiveness, toxicity, and pharmacological properties.

Precursor Ion An ion selected for fragmentation in MS/MS.

Preparative chromatography Chromatography designed for the isolation of large quantities of pure compounds.

Prescription drug A medication requiring healthcare provider authorization due to potential risks or need for monitoring.

Product ion Ions produced from the fragmentation of the precursor ion in MS/MS.

Prototypes First in class compounds usually directed at a specific molecular target.

Quadrupole A mass analyzer that uses oscillating electric fields to filter ions based on their m/z.

Range The interval between the upper and lower concentration of analyte for which the method is shown to be suitable.

Receptor A protein on or within a cell that binds to specific molecules, such as hormones or neurotransmitters, and mediates their effects within the body.

Recovery The process of evaluating the accuracy of the method by measuring the amount of analyte recovered from a spiked sample.

Repeatability (intra-day) Precision under the same operating conditions over a short interval of time.

Reproducibility (inter-lab) Precision under conditions where measurements are made in different laboratories.

Resolution (Rs) A measure of how well two peaks are separated in chromatography; Rs $\geq$1.5 is generally considered baseline resolution. Also used in mass spectrometry to describe the ability to distinguish between closely spaced *m/z* values.

Retention factor (k') Indicates how long a compound is retained on the column relative to the unretained peak.

Retention indices (RIs) These are calculated based on the distance between two consecutive alkanes in the chromatogram, which helps in identifying compounds.

Retention time (tR) The time taken for an analyte to elute from the column and reach the detector.

Robustness The method's capacity to remain unaffected by small but deliberate variations in method parameters (e.g., pH, flow rate).

Route of administration The method by which a drug is introduced into the body, such as oral, intravenous, or topical.

Ruggedness Reproducibility of the test under variable conditions such as different analysts, instruments, or laboratories.

Sample Matrix The components in the sample other than the analyte(s) of interest.

Scalable Capable of being expanded for high-throughput applications; suitable for larger scale processes.

Selected Ion Monitoring (SIM) A technique where only specific *m/z* values are monitored, increasing sensitivity.

Selectivity (α) Ratio of retention factors of two analytes; indicates how easily they can be separated.

Side effect An unintended and often undesirable effect of a drug, which may occur in addition to the intended therapeutic effect.

Signal-to-Noise Ratio (S/N) A measure of signal strength relative to background noise; used in LOD and LOQ determination.

SimMR Method A method that evaluates mass spectrum similarity and retention index distance for compound identification in GC-MS.

Simultaneous evaluation Methods that improve identification accuracy by evaluating multiple parameters at once.

Small molecules Small molecules, often with molecular weights of 500 or below, are extremely important for researchers to explore function at the molecular, cellular, and in vivo level. Such molecules have also proven valuable for treating diseases, and most medicines marketed today are from this class. A key challenge is to identify small molecules effective at modulating a given biological process or disease state. Size of small molecules is generally under 1000 Daltons, but many estimates seem to range between 300 and 700 Daltons.

Solid Phase Extraction (SPE) A sample preparation method using solid phase materials to isolate analytes.

Solvent Strength The ability of the mobile phase to elute analytes from the stationary phase.

Specificity The ability of the method to measure the analyte response in the presence of potential interferences.

Stability The ability of the analyte in the sample or standard solution to remain unchanged under certain conditions over time.

Stationary Phase The material packed inside the HPLC column that interacts with analytes to aid separation (e.g., C18 silica).

Steady-state concentration The point at which the rate of drug administration equals the rate of drug elimination, resulting in a constant drug concentration in the body.

Synergism The interaction between two or more drugs results in a greater effect than the sum of their individual effects.

System Precision Repeatability of results using the same instrument, method, and analyst, usually expressed as %RSD.

System Suitability Test (SST) Tests performed to ensure that the analytical system is working correctly before analyzing samples.

Tailing Factor (*T*) A metric to assess peak symmetry; ideal value is around 1.

Tandem Mass Spectrometry (MS/MS) A technique involving multiple rounds of mass spectrometry with fragmentation steps in between to provide structural information.

Terpenes/Terpenoids Largest class of natural products derived from isoprene units; includes monoterpenes (10C), sesquiterpenes (15C), diterpenes (20C), etc. (e.g., menthol, artemisinin, taxol).

Theoretical Plates (*N*) Hypothetical zones or steps in which solute equilibrates between mobile and stationary phases; used to describe column efficiency.

Therapeutic index The ratio of the toxic dose to the effective dose of a drug, is used to assess the safety and efficacy of a drug.

Thin-Layer Chromatography (TLC) A simple and quick method for compound separation on a flat plate coated with adsorbent; commonly used to analyze mixtures like *Catharanthus* alkaloids.

Time-of-Flight (TOF) A mass analyzer that separates ions based on the time it takes for them to travel a fixed distance.

Total Ion Chromatogram (TIC) A plot of the total ion current detected over time during a chromatographic separation.

UV Detector A common detector in HPLC that identifies compounds absorbing ultraviolet light.

Validation A documented process to demonstrate that an analytical method is suitable for its intended purpose.

Void Volume (Vm) Volume of mobile phase within the column; also called dead volume.

Volatile Oils (Essential Oils) Mixtures of volatile aromatic compounds derived from plants, used in aromatherapy and pharmaceuticals (e.g., limonene, eugenol, thymol).

Index

S. Kumar, B. Singh, *Catharanthus in Drug Discovery*,
https://doi.org/10.1007/978-981-95-5048-7

MIX
Papier aus verantwortungsvollen Quellen
Paper from responsible sources
FSC® C105338

If you have any concerns about our products,
you can contact us on
ProductSafety@springernature.com

In case Publisher is established outside the EU,
the EU authorized representative is:
Springer Nature Customer Service Center GmbH
Europaplatz 3, 69115 Heidelberg, Germany

Printed by Libri Plureos GmbH
in Hamburg, Germany